Atlas of Mediterranean Environments in Europe

The Desertification Context

Edited by
Paola Mairota,
John B. Thornes
and
Nichola Geeson
King's College, University of London, UK

JOHN WILEY & SONS
Chichester · New York · Weinheim · Brisbane · Singapore · Toronto

Baffins Lane, Chichester,
West Sussex PO19 1UD, England

National 01243 779777
International (+44) 1243 779777
e-mail (for orders and customer service enquiries): cs-books@wiley.co.uk
Visit our Home Page on http://www.wiley.co.uk
or http://www.wiley.com

Other Wiley Editorial Offices

John Wiley & Sons, Inc., 605 Third Avenue,
New York, NY 10158–0012, USA

WILEY-VCH GmbH, Pappelallee 3,
D–69469 Weinheim, Germany

Jacaranda Wiley Ltd, 33 Park Road, Milton,
Queensland 4064, Australia

John Wiley & Sons (Asia) Pte Ltd, 2 Clementi Loop #02–01,
Jin Xing Distripark, Singapore 129809

John Wiley & Sons (Canada) Ltd, 22 Worcester Road,
Rexdale, Ontario M9W 1L1, Canada

Library of Congress Cataloging-in-Publication Data

Atlas of Mediterranean environments in Europe: the desertification context/edited by Paola Mairota, John B. Thornes and Nichola Geeson.
p. cm.
Includes bibliographical references and index.
ISBN 0-471-96092-6
1. Desertification—Europe—Maps. 2. Desertification—Mediterranean Region—Maps.
3. Soil degradation—Europe—Maps. 4. Environmental degradation—Europe—Maps. I. Mairota, Paola.
II. Thornes, John B. III. Geeson, Nichola.
G1797.21.C813A7 1996 <G&M>
551.4'15'094022—dc20 95-48761
CIP
MAP

British Library Cataloguing in Publication Data

A catalogue record for this book is available from the British Library

ISBN 0-471-96092-6

Typeset in 11/13pt Times by Cambridge University Press
Printed in Great Britain by Cambridge University Press

Atlas of Mediterranean Environments in Europe

PREFACE

The Mediterranean countries of Europe have been undergoing rapid change and development over the last thirty years and especially the last ten. These changes have been in all aspects of life and livelihood, and for many of these countries have included joining the European Union, which has had far-reaching social and economic impacts. Accompanying these changes, the results of centuries of distinctive man–environment relationships in a climatically marginal region, have been swept aside and the landscape itself is undergoing a rapid transformation. Large areas are being abandoned, temporarily or permanently by rural populations, urban growth is rampant, agriculture has been intensified and water scarcity, fed by the severe droughts of the 1980s and 1990s, has become a major environmental and resource concern. At the same time the threat of global climate change, and the indications of deeper and more sustained future droughts, has brought the issues into sharp focus.

This Atlas is designed to bring to a wide audience the background, the contemporary understanding and the future prospects of solution of the environmental crisis which is now facing the Mediterranean regions of Europe. The 'desertification' we see here is not the fearful, deeply embedded affliction of the drylands of the economically disadvantaged developing nations in Africa and the Indian subcontinent. Rather it is the culmination of centuries of environmental change under the difficult social, political, economic and physical conditions of Mediterranean Europe. Hardship here, especially in rural areas, began to ease forty years ago. Although slow in coming, the current relatively high standards of living have brought with them a fresh crop of environmental problems to be heaped on past land degradation. It is to investigate the roots of the present crisis that the European Commission developed and funded a programme of research in Mediterranean Desertification under the Environment Programme of Directorate General XII (DGXII).

Mediterranean countries in general, and the MEDALUS researchers in particular, owe a great deal to Roberto Fantechi, recently retired from DGXII, who early recognised and persistently championed the cause of Mediterranean European desertification. It is to him that this Atlas is dedicated.

Paola Mairota
John B. Thornes
Nichola Geeson
King's College, University of London
December 1996

CONTENTS

EDITORS' ACKNOWLEDGEMENTS

The Atlas is being published under the auspices of the European Communities thanks to Panagiotis Balabanis and Denis Peter, Research Officers of the Commission of the European Communities DGXII, and competent scientists in their own right, who have supported the MEDALUS (Mediterranean Desertification and Land Use) Project from its inception and helped it through some difficult periods. The MEDALUS Project, details of which are contained in Section 1.3, is entering its seventh year as the Atlas goes to press and has involved a large number scientists, mainly from European Mediterranean countries, and would be much weaker without continuing advice and support from Panagiotis Balabanis and Denis Peter.

We would also like to thank all the authors for their enthusiastic acceptance of and support to the initiative which has encouraged us to undertake the production of this Atlas in a very short time, as well as for their efforts to comply with very tight schedules.

The contributions in this Atlas have been written mainly by participants in the MEDALUS Project. Those of us from parts of Europe beyond the Mediterranean, who have had the privilege of working with our Mediterranean colleagues in their own countries, in some cases for decades, have learned much from them and owe much to them for their kindness and hospitality.

In particular we must thank Jean Palutikof, Clare Goodess, Jean Poesen, Bas van Wesemael, Rosemary Thornes, Fivos Papadimitriou and Corinna Hawkes for their help with reading and improving sections of the text. To Fivos Papadimitriou, Paola Mairota is also particularly indebted for his constant encouragement, suggestions and assistance, and to him she dedicates her first editorial effort.

Thanks also go to Jane Brandt and the MEDALUS Office for advice and for providing the list of MEDALUS publications; to Sandie Clarke for patiently typing some of the contributions; to Roma Beaumont and Gordon Reynell for their cartographic advice; to Peter Howard for photographic assistance; to Joy Obando and Sophia Burke for their assistance in the editing of the glossary; to Julia Clasby for her help in the copy-editing of the reference list; and to Helen Bailey at John Wiley & Sons for her help and guidance during the preparation of the material for publication.

We are particularly grateful to Julie Shannon for her help in compiling the plant species glossary and to John Diamantopoulos, Roberto Lazaro, Alberto Solé, Giovanna Pisano and Jorge Mourao for their assistance with this task. Julie Shannon has also assisted in the organisation of the subject index and the treatment of the material for the preliminary page design.

Last, but not least, we must thank Alison and Tim Aspden for taking our diagram and map drafts, computer diskettes and even scrappy graphs and turning them into the high quality drawings that appear in the Atlas.

Paola Mairota
John B. Thornes
Nichola Geeson
King's College, University of London
December 1996

AUTHORS' ACKNOWLEDGEMENTS

Hans-Jürgen Bolle would like to thank all the participants in the EFEDA Programme: J.-C. Andr, J. L. Arrue, H. K. Barth, P. Bessemoulin, A. Brasa, H. A. R. de Bruin, G. Dugdale, G. Ehret, E. T. Engman, D. L. Evans, R. Fantechi, F. Fiedler, A. van de Griend, A. C. Imeson, A. Jochum, P. Kabat, Chr. Kiemle, T. Kratzsch, J.-P. Lagouarde, I. Langer, R. Llamas, E. Lopez-Baeza, J. Meliá Miralles, L. S. Muniosguren, F. Nerry, J. Noilhan, H. R. Oliver, R. Roth, J. Sanchez Diaz, M. de Santa Olalla, W. J. Shuttleworth, G. Simmet, H. Sögaard, H. Stricker, J. B. Thornes, M. Vauclin and D. Wickland.

Jacques Lavabre would like to thank all the participants in the DM2E Programme: C. Obled, C. Martin, B. Blavoux, G. Lechapt, K. Beven, J. P. Feuvrier, F. Gallart, V. Castillo, J. Bellot and J. Bathurst.

Clare Goodess and Jean Palutikof would like to thank the Climate Impacts LINK Project (Department of the Environment Contract PECD7/12/96) on behalf of the UK Meteorological Office for supplying the data from the UKTR model.

Maria Luisa Gentileschi would like to thank Lida Viganoni and Monica Jorio for data processing.

Ian Woodward and Susannah Diamond would like to thank Peter Mitchell and John Sheehy for their assistance.

Giuseppe Pulina, Massimo d'Angelo, Salvatore Madrau, Marcello Niedda and Claudio Porqueddu would like to thank Giuseppe Enne, Fiorenzo Mancini, Paolo Baldaccini, Salvatore Caredda, Bruno Dettori, Giovanni Masala, Giovanna Pisano and Giovanna Dessí.

NOTES ON THE CONTRIBUTORS

Editors

Paola Mairota. Forester. Her research interests are in the ecology of Mediterranean landscapes and application of landscape ecology to the management of semi-natural vegetation in Mediterranean environments. Department of Geography, King's College, University of London, Strand, London WC2R 2LS, UK.

John B. Thornes. Professor of Physical Geography, Co-ordinator of the MEDALUS Project. Main research interests are in the interaction of erosion and vegetation cover. Department of Geography, King's College, University of London, Strand, London WC2R 2LS, UK.

Nichola Geeson. Research interests in soil science and geochemistry. MEDALUS Project Manager (with Jane Brandt). Department of Geography, King's College, University of London, Strand, London WC2R 2LS, UK.

Authors

Robert Abrahart. Research officer. His research interests include geographical information systems, environmental modelling, and neurohydrology. School of Geography, University of Leeds, Leeds LS2 9JT, UK.

Claudio Aliventi. Expert in statistics of territorial data and hydrological models. CSIM - Regione Veneto, Via Marconi 55, I-35037 Teolo (PD), Italy.

Angelo Aru. Professor of Geopedology at the University of Cagliari. He has 35 years' experience in soil surveying, teaching and consultancy, including work in Venezuela, Somalia and China. Department of Earth Sciences, Cagliari University, Via Trentino 51, I-09127 Cagliari, Italy.

Paolo Baldaccini. Professor of Soil Science, Department of Territorial Engineering, Sassari University, Via De Nicola, 07100 Sassari, Italy.

Giovanni Barrocu. Professor of Engineering Geology at the University of Cagliari. Currently researching into hydrogeology of fractured and porous media, groundwater conservation, and salt water intrusion in coastal aquifers. Department of Territorial Engineering, Cagliari University, Via Trentino 51, I-09127 Cagliari, Italy.

Francesco Basso. Professor of Herbaceous Cultivation at the University of Basilicata. Research interests in agronomic behaviour of various crop cultivation species, innovative land management and conservation techniques, and sustainable crops reducing energy inputs in marginal areas. Dipartimento di Produzione Vegetale, University of Basilicata, Via Nazario Sauro 85, I-85100 Potenza, Italy.

James C. Bathurst. Principal research officer with 20 years' experience in physically based hydrological and erosion modelling, mountain river hydraulics and sediment transport, including experience in Europe, USA, New Zealand, Chile, Indonesia and Yemen. Water Resource System Research Unit, Department of Civil Engineering, University of Newcastle upon Tyne, Newcastle upon Tyne NE1 7RU, UK.

Fabrizio Battistini. System designer. Specialising in numerical analysis and statistics of territorial data. Alenia - Aeritalia & Selenia, Via Vitorchiano 81, I-00189 Roma, Italy.

Adam Becker. Expert in numerical analysis and statistics of territorial data. Finsiel SpA, Via Giulio Vincenzo Bona 65, I-00156 Roma, Italy.

Matthias Böer. Physical geographer, specialising in the use of terrain analysis and remote sensing for land degradation hazard assessment purposes. CSIC Estación Experimental de Zonas Aridas General Segura, 1, 04001 Almería, Spain.

Carolina Boix. Physical geographer working in the ERMES and EHABUA (Erosion Hazard on Mediterranean Burnt Areas) projects. Landscape and Environmental Research Group, University of Amsterdam, Nieuwe, Prinsengracht 130, 1018 VZ Amsterdam, The Netherlands.

Hans-Jürgen Bolle. Co-ordinator of the EFEDA project. Retired professor of meteorology (topics: remote sensing, radiation and climate) from the Freie Universität Berlin. EFEDA Project Office, Institut für Meteorologie, Freie Universität Berlin, Carl-Heinrich-Becker-Weg 6-10, D-81247 Berlin, Germany.

Alberto Bonini Baraldi. Agronomist, specialising in agronomy and climatology. Consultant on land management and no-point water pollution. CSIM - Regione Veneto, Via Marconi 55, I-35037 Teolo (PD), Italy.

Lorenzo Bottai. Forester, specialising in the implementation of remote sense data into geographic information systems for vegetation and environmental analysis. Nuova Telespazio SpA, Via Tiburtina 965, I-00156 Roma, Italy.

Ettore Bove. Professor of Agricultural Policy and Economics at the University of Basilicata. His scientific interests are the socio-economic issues of marginal lands and rural areas. Dipartimento Tecnico-Economico per la Gestione del Territorio Agricolo-Forestale, University of Basilicata, Via Nazario Sauro 85, I-85100 Potenza, Italy.

Francesco Bovo. Electronical engineer, specialising in geographical information systems and graphical software. CSIM - Regione Veneto, Via Marconi 55, I-35037 Teolo (PD), Italy.

Kristin Bunte. Research associate at the KU Leuven, MEDALUS I project, specialising in soil erosion by overland flow in Mediterranean environments. Laboratory for Experimental Geomorphology, Katholieke Universiteit Leuven, Redingenstraat 16bis, 3000 Leuven, Belgium.

Sophia Burke. PhD student in geography. Her research interests are in impacts of land use and climate changes on hydrology and vegetation in semi-arid regions. Department of Geography, King's College London, Strand, London WC2R 2LS, UK.

Francísco Cabezas-Calvo-Rubio. Civil engineer, expert in watershed management. Confederación Hidrográfica del Segura, Plaza Fontes, 30004 Murcia, Spain.

Lucas H. Cammeraat. Research associate working on the role of soil water and soil properties related to land degradation and desertification. Landscape and Environmental Research Group, University of Amsterdam, Nieuwe, Prinsengracht 130, 1018 VZ Amsterdam, The Netherlands.

Aldolpho Calvo-Cases. Geomorphologist. Department of Geography. Universitat de València, Dr Moliner 50, 46100-Burjassot, València, Spain.

Maria Assunta Campanile. Electronic engineer, specialising in numerical analysis and territorial models. CSIM - Regione Veneto, Via Marconi 55, I-35037 Teolo (PD), Italy.

Antonio Capraro. Head of Environmental Systems of Alenia. Expert in image processing and photo interpretation. Alenia - Aeritalia & Selenia SpA, Via Vitorchiano 89, I-00189 Roma, Italy.

Leandro Chiarantini. Physicist. His current research activities concern remote sensing of soil by microwave and infrared sensors for applications in hydrological modelling. Fondazione Meteorologia Applicata (Applied Meteorology Foundation). c/o SMA Spa, Via del Ferrone n.5, 50124 Firenze, Italy.

Bruno Cittarelli. System designer specialising in metereological and agrometeorological models. Alenia - Aeritalia & Selenia SpA, Via Vitorchiano 89, I-00189 Roma, Italy.

Simon C. Clark. Lecturer in plant ecology (retired). Research interests in the ecology of extreme environments, in particular metalliferous and drought liable habitats. Department of Pure and Applied Biology, University of Leeds, Leeds LS2 9JT, UK.

Jean Jacques Collin. Hydrogeologist in the public survey, specialist of groundwater management; in the MEDALUS Project he was the co-ordinator of the Ground Water theme. Département Utilisation et Protection de l'Environnement Service Géologique, National Centre Thématique Eau, Montpellier, France.

Claudio Conese. MEDALUS II - PJ 4, Abruzzo Group Co-ordinator. He is responsible for the Remote Sensing and Land Information System Group of IATA-CNR. He is also Director of Centre for Computer Science Application in Agriculture (CeSIA), Florence. Main interests in remote sensing. CeSIA, Piazzale delle Cascine 18, I-50144 Firenze, Italy.

Michele Conte. Physicist. Former chief of the climatological branch of the National Meteorological Service of Italy. Senior scientist of the National Research Council CeSIA. Accademia dei Georgofili Loggia degli Uffizi Corti, I-50122 Firenze, Italy.

Catherine M. Cook. Post-doctoral researcher in plant physiology, has interests in photosynthesis and the effect of environmental stress on plants. Department of Botany, School of Biology, Aristotle University, 54006 Thessaloniki, Greece.

Peter Coppo. His current activity concerns microwave emission and scattering from rough surfaces. Fondazione Meteorologia Applicata (Applied Meteorology Foundation). c/o SMA Spa, Via del Ferrone n.5, I-50124 Firenze, Italy.

João Corte-Real. Associate Professor of Meteorology, University of Lisbon, Campo Grande, Ed. C1, Piso 4, 1700 Lisbon, Portugal.

Pedro Cortesão Casimiro. Lecturer in the human impact on the natural environment, risk mapping. Researching into land use, remote sensing for environmental degradation assessment. Departamento de Geografia e Planeamento Regional, Faculdade de Ciências socials and humanas, Universidade Nova de Lisboa, Portugal.

Massimo Crespi. Director of the CSIM. Expert consultant on hydrology of the Italian Ministry for Civil Protection. He is manager of environmental monitoring projects in Italy and Latin America. CSIM - Regione Veneto, Via Marconi 55, I-35037 Teolo (PD), Italy.

Miguel Cueto. Post-doctoral researcher. Research interests in the chorology and conservation of plant species of semi-arid environments and vegetation evolution with remote sensing. Estación Experimental de Zonas Áridas (CSIC) General Segura, 104001 Almería, Spain.

Anastasia G. Dalaka. PhD student, studying the role of grasses in evergreen sclerophyll formations. She has interests in plant photosynthesis and experience in Automated Weather Station operation. Department of Botany, School of Biology, Aristotle University, 54006 Thessaloniki, Greece.

Nicholas Danalatos. Post-doctoral researcher. Research in land evaluation and protection of land resources. Agricultural University of Athens. Laboratory of Soils and Agricultural Chemistry, Iera Odos 75, Botanicos 118 55, Athens, Greece.

Massimo d'Angelo. Forester, specialising in remote sensing. Researcher at Consorzio Ricerche Sardegna. Nucleo Ricerca Desertificazione, Università degli Studi di Sassari, Facoltà di Agraria Istituto di Zootecnica, Via E. De Nicola, I-07100 Sassari, Italy.

Tiziana De Filippis. Main research interests concern integration of multi-source data for modelling soil erosion processes. She is on the staff of several National and EC projects on environmental studies. CeSIA, Piazzale delle Cascine, 18, I-50144 Firenze, Italy.

Gabriel Del Barrio. Ecologist, specialising in the quantification of spatial properties of semi-arid landscapes and of their relation to land degradation.

CSIC Estación Experimental de Zonas Aridas General Segura, 1, 04001Almería, Spain.

Mario Del Prete. Professor of Geology and Engineering. He has conducted research on landslides and erosion in clay and flysch soils in Italy, Scotland, Spain and Argentina. Dipartimento Tecnico-Economico per la Gestione del Territorio Agricolo-Forestale, University of Basilicata, Via Nazario Sauro 85, I-85100 Potenza, Italy.

John Diamantopoulos. Lecturer in ecology, has interests in vegetation science, landscape ecology and remote sensing. Department of Botany, School of Biology, Aristotle University, 54006 Thessaloniki, Greece.

Susannah Diamond. Vegetation modeller, has assisted in the theoretical development and computer coding of vegetation and evapotranspiration models for MEDALUS, and explored the effect of climate manipulation experiments on vegetation. Department of Animal and Plant Sciences, University of Sheffield, Sheffield S10 2TN, UK.

Francisco Domingo. Post-doctoral researcher, Consejo Superior de Investigaciones Científicas. Estación Experimental de Zonas Áridas (CSIC) General Segura, 104001 Almería, Spain.

Massimo Fagnano. Graduate technician at the University of Naples. Presently researching crop productivity in Mediterranean environments, agriculture and pollution relationships, ozone effects on vegetation, agriculture and greenhouse gas reduction. Dipartimento di Scienze Agronomiche e Genetica Vegetale, Università di Napoli, via Università 100, I-80055 Portici (Napoli), Italy.

Francesca Fantola. Currently working on requalification of urban areas throughout the analysis and evaluation of the physical environment. Dipartimento Scienze della Terra, Università di Cagliari, Via Trentino 51, 09127 Cagliari, Italy.

Silvio Gagliani. His main activity is providing technical assistance for management of remote sensing campaigns, sensors and instrumentation, data acquisition and analysis. Fondazione Meteorologia Applicata (Applied Meteorology Foundation), c/o SMA SpA, Via del Ferrone n.5, I-50124 Firenze, Italy.

Javier García-Haro. Specialist in spectral mixture modelling. Departament de Termodinàmica, Unidad de Teledección, Facultat de Física, Universitat de València, Dr Moliner 50, 46100-Burjassot, València, Spain.

M. Luisa Gentileschi. Professor of economic and political geography. Dipartimento di Richerche Economiche e Sociali, Università di Cagliari, Viale Fra Ignazio da Laconi 78, 019123 Cagliari, Italy.

Maria Amparo Gilabert. Her current research interests include reflectance modelling, vegetation indices and field radiometry. Remote Sensing Unit, University of Valencia, Dr. Moliner 50, 46100-Burjassot, Valencia, Spain.

Christina E. Giourga. Agronomist. University of the Aegean, Karantoni 17, 81 100 Mytilini, Lesvos, Greece.

Aurora Girolamo. Mathematician. Expert in meteorological and agrometeorological models. Finsiel SpA, Via Giulio Vincenzo Bona 65, I-00156 Roma, Italy.

Clare M. Goodess. Senior Research Associate with over 13 years' research experience. She has considerable experience of the study of climate change in the Mediterranean region and the construction of regional scenarios of future change. She is a joint author, with Jean P. Palutikof, of *The Nature and Causes of Climate Change: Assessing the Long Term Future* (Belhaven Press, 1992).

Arthur T. Grove. Emeritus Fellow of Downing College, Cambridge, and Senior Research Associate Department of Geography, Downing Place, Cambridge CB2 3EN. Geographer. He has published on the palaeoclimatology and land use-history of Africa and the Mediterranean, and has reported on desertification in both regions.

Leonardo Gutiérrez. Research Associate (Consejo Superior de Investigaciones Científicas). Botanist. He has research experience in structure and biomass production of plant communities in semi-arid environments and its relations with the land use. Estación Experimental de ZonasÁridas (CSIC) General Segura, 104001 Almería, Spain.

Andrew Harrison. Lecturer in remote sensing. Department of Geography, University of Bristol, University Road, Bristol BS8 1SS, U.K.

Corinna Hawkes. PhD student in geography. Her research interests are in parameterisation of hydrological models and vegetation dynamics in semi-arid environments. Department of Geography, King's College London, Strand, London WC2R 2LS, UK.

Tómás Huszár. PhD student. Geographical Research Institute, Hungarian Academy of Sciences, 1062 Budapest, VI, Andrássy út 62, Hungary.

George D. Iatrou. He has interests in soil biology, experimental soil ecology and soil analysis. Department of Botany, School of Biology, Aristotle University, I-54006 Thessaloniki, Greece.

Anton C. Imeson. Heads a research group in Amsterdam working on soil degradation, erosion and environmental change mainly in Mediterranean Europe. Co-ordinator of the ERMES and MEDALUS 1 projects. Landscape and Environmental Research Group, University of Amsterdam, Nieuwe, Prinsengracht 130, 1018 VZ Amsterdam, The Netherlands.

Ádám Kertész. Head of the Department for Physical Geography, Geographical Reseach Institute, Hungarian Academy of Sciences, Professor at Eötvös Lóránd University, specialising in soil erosion and in GIS applications. Geographical Research Institute, Hungarian Academy of Sciences, 1062 Budapest, VI, Andrássy út 62, Hungary.

Chris Kilsby. Senior research associate working on physically based and large-scale hydrological modelling, with particular interests in climate change and meteorology. Water Resource System Research Unit, Department of Civil Engineering, University of Newcastle upon Tyne, Newcastle upon Tyne NE1 7RU, UK.

Mike J. Kirkby. Professor of Physical Geography and Managing Editor of *Earth Surface Processes and Landforms*. His research interests include hillslope hydrology; soil erosion; landform modelling and evolution. School of Geography, University of Leeds, Leeds LS2 9TJ, UK.

Costantinos S. Kosmas. Assistant Professor at the Agricultural University of Athens. Research experience in soil survey and soil genesis, land resources and the environment mainly in agricultural ecosystems, soil erosion on land degradation and desertification. Agricultural University of Athens. Laboratory of Soils and Agricultural Chemistry, Iera Odos 75, Botanicos 118 55, Athens, Greece.

Eugenia J. Koutsidou. Biologist. University of the Aegean, Karantoni 17, 81 100 Mytilini, Lesvos, Greece.

Maria Rita Lai. Research experience in geomorphology and processing environmental-related thematic maps, soil surveying and soil suitability studies. ERSAT, Via Caprera 8, I-09123 Cagliari, Italy.

Tom Lanaras. Reader in Plant Physiology, has interests in photosynthesis and small bioactive plant molecules. Department of Botany, School of Biology, Aristotle University, 54006 Thessaloniki, Greece.

Giuseppe Landi. Research studies on nitrogen loss and erosion related to soil tillage techniques in hilly areas of southern Italy, and eco-physiology of forage in relation to management especially in relation to management for pasture rehabilitation. Dipartimento Tecnico-Economico per la Gestione del Territorio Agricolo-Forestale, University of Basilicata, Via Nazario Sauro 85, I-85100 Potenza, Italy.

Jacques Lavabre. Research Engineer in Cemagref, Groupement d'Aix-en-Provence. Leader of research group in hydrology. Co-ordinator of the DM2E project. Professeur à l'Ecole Nationale du Génie Rural des Eaux et des Forêts. CEMAGREF - Le Tholonet BP31 13612 Aix-en-Provence, Cedex 1, France.

Hanoch Lavee. Professor of geomorphology. Laboratory of Geomorphology and Soils, Department of Geography, Bar-Ilan University, Ramat-Gan 52900, Israel.

Colin Legg. Lecturer in plant ecology. Research interests include post-fire successional processes, bryophyte ecology, rare plant autecology and vegetation modelling. Institute of Ecology and Resource Management (IERM), The University of Edinburgh, King's Buildings, Mayfield Road, Edinburgh EH9 3JU, UK.

Lila Leontidou. Senior lecturer in geography. Consultant and co-ordinator of planning and development studies, expertise in social change and urban development. Department of Geography, King's College London, University of London, The Strand, London WC2R 2LS, UK.

Susanna Lessi. Agronomist, specialising in geographical information systems and remote sensing with application to agriculture. She participates in the working group of the EC Project Corine Land Cover. CSIM - Regione Veneto, via Marconi 55, I-35037 Teolo (PD), Italy.

Dénes Lóczy. Senior research fellow, Geographical Research Institute, Hungarian Academy of Sciences, specialist in land capability analysis and in landscape ecology. Geographical Research Institute, Hungarian Academy of Sciences, 1062 Budapest, VI, Andrássy út 62, Hungary.

Francísco López-Bermúdez. Professor of Physical Geography, expert in arid and semi-arid environments. University of Murcia, Department of Physical Geography, 30001-Murcia, Spain.

Guido Luzi. Main research interests are in microwave emission and scattering from soil and vegetation by multiband microwave radiometric and SAR systems. Fondazione Meteorologia Applicata, c/o SMA Spa, Via del Ferrone n.5, I-50124 Firenze, Italy.

Malcolm McMahon. Research officer. His research interests include new developments in computing, software engineering, and numerical modelling. School of Geography, University of Leeds, Leeds LS2 9TJ, UK.

Salvatore Madrau. Pedologist. Researcher at the University of Sassari. Nucleo Ricerca Desertificazione, Università degli Studi di Sassari, Facoltà di Agraria Istituto di Zootecnica, Via E. De Nicola, I-07100 Sassari, Italy.

Theodore A. Mardiris. He has work experience in ecosystems management, environmental impact assessment, vegetation studies, soil analysis, reforestation impacts, and fire and grazing impacts. Department of Botany, School of Biology, Aristotle University, 54006 Thessaloniki, Greece.

Nikos S. Margaris. Professor of ecosystems management. University of the Aegean, Karantoni 17, 81 100 Mytilini, Lesvos, Greece.

Luigi Mariani. Director of Meteorological Office of ERSAL (Ente Regionale di Sviluppo Agricolo della Lombardia). He is an expert in meteorology, climatology and data processing. ERSAL, Palazzo Canova, Milanodue, I-20090 Milano, Italy.

José Martínez-Fernandez. Physical geographer, specialising in Mediterranean soils. University of Murcia, Department of Physical Geography, 30001 Murcia, Spain.

Julia Martínez-Fernandez. Ecologist, specialising in Mediterranean vegetation. University of Murcia, Department of Physical Geography, 30001 Murcia, Spain.

Fabio Maselli. Researcher at the Institute of Agrometeorology and Environmental Analysis for Agriculture of the National Council of Research of Italy, specialising in remote sensing data processing. CeSIA, Piazzale delle Cascine, 18, I-50144 Firenze, Italy.

István Mattyasovszky. Assistant professor at the Eötvös Lóránd University, specialising in meteorology and in climate change. Geographical Research Institute, Hungarian Academy of Sciences, 1062 Budapest, VI, Andrássy út 62, Hungary.

Stefano Mazzoleni. Associate professor in plant ecology at the University of Molise. Research interests include Mediterranean vegetation dynamics and land-use history. Dipartimento di Scienze Animali, Vegetali e dell'Ambiente (DiSAVA), Università del Molise, Via Cavour 50, I-86100 Campobasso, Italy.

Joaquín Meliá Miralles. Professor of physics and leader of the Remote Sensing Unit, University of Valencia, Dr. Moliner, 50, 46100-Burjassot, Valencia, Spain.

Valeria Menichini. Computer scientist, specialising in climatology in the ERSAL Climatological Section. ERSAL, Palazzo Canova, Milanodue, I-20090 Milano, Italy.

János Mika. Senior research fellow, Institute for Atmospheric Physics. Specialist in meteorology and climate change research. Geographical Research Institute, Hungarian Academy of Sciences, 1062 Budapest, VI, Andrássy út 62, Hungary.

Gian Paolo Minardi. His main research work in MEDALUS concerns the climatological dynamic and the application of neural networks for the weather types classifications. ERSAL Institute of Meteorology, Palazzo Canova, Milanodue, I-20090 Milano, Italy.

Katalin Molnár. Research Fellow specialising in climatology and in landscape ecology. Geographical Research Institute, Hungarian Academy of Sciences, 1062 Budapest, VI, Andrássy út 62, Hungary.

Roy P. C. Morgan. Professor of Soil Erosion Control, Department of Rural Land Use, Silsoe College, Cranfield University, Silsoe, Bedford MK45 4DT, UK.

Christophe Mouvet. Research on the transfer of organic micropollutants from the soil to the groundwater involving experimental design, process identification and analytical skills. Dept. de Géochiminie, BRGM, 36600, Pessac, France.

Marcello Niedda. Expert in hydraulics and hydrogeological, climatic and erosive processes. Nucleo Ricerca Desertificazione, Università degli Studi di Sassari, Facoltà di Agraria Istituto di Zootecnica, Via E. De Nicola, I-07100 Sassari, Italy.

Joy Obando. PhD student in geography. Her research interests are in the modelling of the relationship between vegetation and erosion on abandoned land under fluctuating rainfall regimes. Department of Geography, King's College London, Strand, London WC2R 2LS, UK.

Jean P. Palutikof. Reader and assistant director (Research) in the Climatic Research Unit. Research interests in climatic change impacts, particularly related to the greenhouse effect, and the application of climatic data to economic and planning issues. Climatic Research Unit, University of East Anglia, Norwich, UK.

John D. Pantis. Assistant Professor of Ecology, with interests in degradation and desertification processes in Mediterranean type ecosystems and biomonitoring of urban environments. Department of Botany, School of Biology, Aristotle University, 54006 Thessaloniki, Greece.

Fivos Papadimitriou. Geologist, meteorologist, DPhil student in geography. His research interests and publications are in mathematical modelling of land use change and land degradation. He has consulting experience in land use planning. School of Geography, University of Oxford, Mansfield Road, Oxford OX1 3TB, UK.

Effi Papatheodorou. Ecologist specialising in plant nutrients. Department of Ecology, School of Biology, Aristotelian University, 540 06 Thessaloniki, Greece.

Ljiljana Petkov. Agrometeorologist specialising in the development of new methodologies for the classification of satellite remote sensing images. Consultant for ETSAF. ETSAF, Via Verdi, 16, I-50122 Firenze, Italy.

Stergios A. Pirintsos. Research fellow. He has interests in applied ecology and lichen ecology. Department of Botany, School of Biology, Aristotle University, 54006 Thessaloniki, Greece.

Michele Pisante. Researcher. He is interested in the use of remote sensing techniques for agricultural purposes, plant physiology and crop productivity in southern Italian environments. Dipartimento di Produzione Vegetale, University of Basilicata, Via Nazario Sauro 85, I-85100 Potenza, Italy.

Jean Poesen. Professor and research leader of the Belgian National Fund for Scientific Research, specialising in soil erosion in relation to environmental change. Laboratory for Experimental Geomorphology, Katholieke Universiteit Leuven, Redingenstraat 16 bis, 3000 Leuven, Belgium.

Claudio Porqueddu. Agronomist, specialising in pasture and fodder crops. Nucleo Ricerca Desertificazione, Università degli Studi di Sassari, Facoltà di Agraria Istituto di Zootecnica, Via E. De Nicola, I-07100 Sassari, Italy.

Luigi Postiglione. Professor of General Agronomy and Field Crops. Director of the Department of Agronomic Sciences and Plant Genetics. His research activities are in agronomy and applied ecology, soil protection, and biological and agronomical problems of selected crops. Dipartimento di Scienze Agronomiche e Genetica Vegetale, Università di Napoli, Via Università 100, I-80055 Portici (Napoli), Italy.

Hein Prinsen. PhD student, working on the dynamics of soil–water–vegetation interactions after environmental disturbance of Mediterranean ecosystems. Landscape and Environmental Research Group, University of Amsterdam, Nieuwe Prinsengracht 130, 1018 VZ Amsterdam, The Netherlands.

Rita Puddu. Freelance geologist, has collaborated for the past three years with the MEDALUS project in Sardinia and has had various work experience in soil-related areas. Department of Earth Sciences, Cagliari University, Via Trentino 51, I-09127 Cagliari, Italy.

Juan Puigdefábregas. Research scientist. Consejo Superior de Investigaciones Científicas. He has extensive research experience in the ecology and vegetation–erosion interactions of Mediterranean semi-arid environments. Estación Experimental de Zonas Áridas (CSIC) General Segura, 104001 Almería, Spain.

Giuseppe Pulina. Associate professor. Expert in animal production systems. Nucleo Ricerca Desertificazione, Università degli Studi di Sassari, Facoltà di Agraria Istituto di Zootecnica, Via E. De Nicola, I-07100 Sassari, Italy.

Gloria Pungetti. Research fellow involved in landscape planning, especially in relation to the conservation of the ecological and cultural heritage in the Mediterranean region. European Center for Nature Conservation, Warandelaan 2, PO Box 1352, 5004 BJ Tilburg, The Netherlands.

John N. Quinton. Lecturer in soil and water management, has interests in soil degradation, non-point source pollution and modelling and has worked in Israel, Spain and Swaziland. Department of Water Management, Silsoe College, Cranfield University, Silsoe, Bedford MK45 4DT, UK.

Oliver Rackham. Botanist and ecological historian. He has published on ancient woodland in the British Isles and on the landscape history of England and Greece. Fellow of Corpus Christi College, Cambridge, and Senior Research Associate. Department of Geography, Downing Place, Cambridge CB2 3EN, UK.

Alessandra Rodolfi. Her main research work concerns the integration of remote sensing data in the Geographic Information Systems for environmental analysis. CeSIA, Piazzale delle Cascine 18, I-50144 Firenze, Italy.

Leopoldo Rojo-Serrano. Director of LUCDEME Project. Expert in watershed management. ICONA, Gran Via de San Francísco 4, 28005 Madrid, Spain.

Fabio Roli. Electronic engineer. His primary research activity is in the area of image processing and classification, especially artificial neural networks. DIBE University of Genoa, Via Opera Pia, 11A, I-16145 Genoa, Italy.

Maria Asunción Romero-Díaz. Physical geographer, expert in soil erosion processes. University of Murcia, Department of Physical Geography, 30001 Murcia, Spain.

Maria José Roxo. Lecturer in geomorphology, natural hazards and slope dynamics. Researching into environmental issues related to human activity. Departamento de Geografia e Planeamento Regional, Faculdade de Ciências socials and humanas, Universidade Nova de Lisboa, Portugal.

Celestino Ruggiero. Botanist specialising in plant root systems. Department of Agronomy and Plant Genetics, Università di Napoli, Via Università 100, 80055 Portici (Napoli), Italy.

Gerardo Sánchez. Research associate in ecology. He has research experience in the ecology of plant perennial species and its relations with soil and erosion. Estación Experimental de Zonas Áridas (CSIC) General Segura, 104001 Almería, Spain.

Sebastiano Bruno Serpico. He is currently the manager of the research team on Signal Processing and Understanding. His main research activities are in the area of image processing and classification. DIBE University of Genoa, Via Opera Pia, 11A, I-16145 Genoa, Italy.

Stefanos P. Sgardelis. Assistant Professor of Ecology. Research interests in demography, population dynamics and modelling of soil fauna and grasses in Mediterranean type ecosystems. Department of Botany, School of Biology, Aristotle University, 54006 Thessaloniki, Greece.

Jim Shao. He is a post-doctoral research fellow in the Department of Geography, Kings College London, Strand, London WC2R 2LS, UK.

Albert Solé. Research scientist. Consejo Superior de Investigaciones Científicas. Research in soil micromorphology and soil physics applied to geomorphology and hydrology to monitor landscape evolution and soil degradation. Estación Experimental de Zonas Áridas (CSIC) General Segura, 104001 Almería, Spain.

Roberto Sorani. Physicist. Chief of Research and Development of the CeSIA. Accademia dei Georgofili Loggia degli Uffizi Corti, I-50122 Florence, Italy.

George P. Stamou. Professor of Ecology. Research experience in the structure and dynamics of degraded Mediterranean type ecosystems, soil biology, population dynamics and ecological modelling. Department of Botany, School of Biology, Aristotle University, 54006 Thessaloniki, Greece.

László Szalai. Research fellow at the Geographical Research Institute, Hungarian Academy of Sciences, specialist in GIS and remote sensing applications. Geographical Research Institute, Hungarian Academy of Sciences, 1062 Budapest, VI, Andrássy út 62, Hungary.

József Szalay. Research fellow, Water Resources Research Center. Specialist in ground water research. Geographical Research Institute, Hungarian Academy of Sciences, 1062 Budapest, VI, Andrássy út 62, Hungary.

Daniele Tomasi. Geologist. Dipartimento di Scienze della Terra, Università di Cagliari, Via Trentino 51, I-09127 Cagliari, Italy.

Andrea Vacca. Research experience in soil surveying and soil suitability studies. ERSAT, Via Caprera 8, 09123 Cagliari, Italy.

Gianni Vernazza. Professor of Telecommunications, University of Genoa, Italy. His research interests are in the area of signal processing and recognition. DIBE University of Genoa, Via Opera Pia, 11A, I-16145 Genoa, Italy.

Patrizio Vignaroli. ETSAF Consultant for socio-economic studies for the costal zone of the Abruzzo Region. ETSAF, Via Verdi 16, I-50122 Firenze, Italy.

Bas Van Wesemael. Research associate at the KU Leuven, specialising in soil formation and degradation in Mediterranean environments. Laboratory for Experimental Geomorphology, Katholieke Universiteit Leuven, Redingenstraat 16bis, 3000 Leuven, Belgium.

Sue N. M. White. EC research fellow working on hydrological modelling in mountain regions, with interests in soil erosion, sediment yield and sustainable land use. Her experience in sediment transport and catchment hydrology includes work in the UK, Africa, SE Asia, India and Nepal. Istituto Pirenaico de Ecologia, CSIC, Zaragoza, Spain.

Ian Woodward. Professor of Ecology. Research in predicting the influence of global environmental change on the distribution and functioning of vegetation at the global scale. Department of Animal and Plant Sciences, University of Sheffield, Sheffield, UK.

Nicholas Yassoglou. Professor of Soil Science. Expert in fields of soil resources and environment, agricultural development, land planning, and major development projects in Greece. Agricultural University of Athens Laboratory of Soils and Agricultural Chemistry, Iera Odos 75, Botanicos 118 55, Athens, Greece.

Meiad Thanoon Younis. Currently involved in geological mapping and digital image processing. Remote Sensing Unit, University of València, S/N Dr. Moliner, 46100 Burjassot, Valencia, Spain.

1

INTRODUCTION

1.1 Mediterranean Desertification

J. B. Thornes

Desertification

Desertification has become a major environmental issue in scientific, political and even popular circles, to be set alongside blue whales, global warming and acid rain. The term captures a sense of moving deserts, drying lakes and especially starving people and a sense of impending or actual crisis. It is above all thought of as a problem of the Sahel, in Africa north of the Equator, and one that is far removed from the daily livelihood and concerns of the states of Europe. In discussing the 'myth' of desertification, Thomas and Middleton (1994) draw attention to three potential misunderstandings about desertification: first, that statements about its extent and activity are at best inaccurate and at worst based on nothing better than guesswork; second, that drylands are necessarily fragile ecosystems and highly susceptible to degradation and desertification, arising from the confusion between loss of vegetation cover on the one hand and soil degradation on the other; and third, that desertification is a primary, if not the main cause of human suffering and misery in drylands. Although Thomas and Middleton are mainly addressing the issue of the role of international agencies (and especially the United Nations) in resolving the problem, these misunderstandings often obscure the real physical, social and economic issues and therefore the efforts to mitigate them.

This confusion has led to a plethora of attempts to uniquely define desertification in a way that will apply to all cases. This is probably impossible, in part because some definitions deal with a condition, some with a cause and some with a set of processes. Moreover, as Perez-Trejo (1992) notes, the phenomenon is relative to particular geographical and physical conditions, the processes are context specific and climate sensitive, and the probability and/or onset of desertification is a function of local biotic and abiotic exchanges at the regional level, and human activity at the local level. Notwithstanding these difficulties and without accepting it as unique, the definition used by UNEP (1994) as 'land degradation in arid, semi-arid and dry subhumid areas resulting from various factors, including climatic variations and human activities' provides a workable definition for the evaluation of the phenomena in the European Mediterranean. 'Land' means the terrestrial bio-productive system that comprises soil, vegetation, other biota and the ecological and hydrological processes that operate within the system. 'Land degradation' means reduction and loss, in arid, semi-arid and dry subhumid areas, of the biological or economic productivity and complexity of rain-fed cropland, irrigated cropland, or range, pasture, forest and woodlands resulting from land uses or from a process or combination of processes, including processes arising from human activities and habitation patterns, such as soil erosion, deterioration of the physical, chemical and biological or economic properties of the soil and long-term loss of vegetation. With this as our benchmark we can move ahead to ask whether or not there exists a problem in Mediterranean Europe, and if so, what can be done about it.

Figure 1.1.1 A conventional view of desertification. The desert sand dunes march over agricultural land in Argentina.

The European Mediterranean Basin

The lands surrounding the Mediterranean Sea have long been recognised as having a distinctive character that arises both from the physiographic conditions and the history of development. In particular climatologists recognise a 'Mediterranean climate' with winter rainfalls, mild winters and high summer temperatures. The rainfalls are characteristically intense, especially in the drier season and in the drier parts of the region. Storms coming from the west are dominantly cyclonic in character and occur mainly in winter, whereas storms originating in the Mediterranean Basin itself are largely convective in origin and create heavy rainfalls when they meet the land. The annual rainfall varies from about 1000 mm in the more northerly areas, to 250 mm in the southern drylands. South-east Spain is one of the driest regions, with rainfall (250–300 mm) coming in the winter months and with 5–7 dry months in a year. Mediterranean drylands (those with less than about 600 mm) also have a large inter-annual variability, so that for a long record period the rainfall may vary from 150 to 600 mm. Runs of wet and dry years are also a characteristic climatological feature of these regions, with the early to mid 1980s being the last

Figure 1.1.2 A traditional view of Mediterranean desertification. Centuries of grazing of marginal land in marginal climates has left soil productivity in a very poor condition.

very dry period. Generally the rainfall has decreased since the end of the last century though the eastern and western parts of the Mediterranean tend to behave somewhat differently (Conte and Guifredda, 1991).

These climatic characteristics also vary greatly over short distances in the Mediterranean because of the basin and range character of the landscape and because of the impact of the sea in coastal areas. Onshore and offshore breezes play an important role in microclimates near the coast.

The mountains play a key role in providing a great diversity of land types, and generally they reflect the underlying geological structure. The mountains of the Mediterranean are very recent in geological time and in many areas are still tectonically active, as reflected by earthquakes and subsidence. The great river basins are normally filled with the erosional products of the mountain-building and are therefore even younger and many are unconsolidated. During the Tertiary large areas of the land surrounding the Mediterranean Sea were also inundated and as a consequence the basins often have salty or fresh water limestones.

There is virtually no natural vegetation left in the Mediterranean since humans have been cutting, grazing and burning for at least the last 10 000 years. None the less, there is an important and distinctive set of vegetation types that are ubiquitous today. Essentially, apart from cultivation, three dominant vegetation types occur: woodlands, typically of trees such as oak and chestnut; high scrub, usually about 2 m in height, with bushes and occasional trees; and sparse scrub, comprising lands that have been cultivated and abandoned, and including aromatic plants such as thyme and rosemary. The process of use, regeneration and destruction has occurred many times throughout the Mediterranean region, so that even the most natural looking forests may have regenerated since the Middle Ages.

As a result of both the natural conditions and human impact, land degradation is not a new issue in the Mediterranean. It is clear that strong erosion was already extensive, at least in some parts of Spain, 4000 years ago. Some of the lands eroded were later re-colonised by forest, cleared and then eroded again. However, it is clear that throughout history, as today, variability over small distances is the main feature of the Mediterranean environment. This should become evident from the descriptions of the field sites and target areas in the second half of the Atlas.

Desertification in Mediterranean Europe

The history of land degradation in the Mediterranean has been the subject of speculation, debate and research for centuries. It is only in recent years however that a serious effort has been made to identify and understand the phenomenon in the context of desertification and its implications. This effort dates from the Symposium of the Commission of the European Communities on *Desertification in Europe* held in Mytelene, Greece, in 1984 (Fantechi and Margaris, 1986). In the opening remarks to this conference, Fantechi noted the aptness of the title of the paper by Professor H. G. Mensching, 'Desertification in Europe?', where the question mark could take on all shades of meaning from candid astonishment to perplexity and beyond. Mensching himself concluded that in Europe, desertification is only widespread in the semi-arid regions of Mediterranean countries. These include the southern and eastern parts of the Iberian Peninsula, Mediterranean France, Italian Mezzogiorno and almost the whole of Greece, including the islands. Generally the most critical areas were identified as having less than 600 mm of rainfall per year, distributed in a few months, with a long dry hot summer.

In the 10 years since then, the issue of Mediterranean desertification has received much greater prominence at both an international and a European level. Individual countries, notably Spain with its Campaign Against Desertification in Mediterranean Areas, had initiated efforts to deal with the problem identified in the UNEP Conference on Desertification held in 1977. The UNEP *Atlas of Desertification* (UNEP, 1992) identified most of the Mediterranean as having very high, high or medium levels of soil degradation severity and the Fourth Annexe to the International Convention on Desertification (UNEP, 1994) identified Portugal, Spain and Greece as countries with a marked problem of desertification, which, while not needing international financial support, require national action programmes. At the same time, the European Parliament has called for substantial effort by the European Union to mitigate the economic and social problems arising from desertification in the southern European states. The reasons for this prominence are not difficult to see. In the last 30 years there has been an enormous shift in economic priorities and the corresponding regime of natural resource

Figure 1.1.3 Contemporary desertification. Badlands are in close juxtaposition to olive plantations.

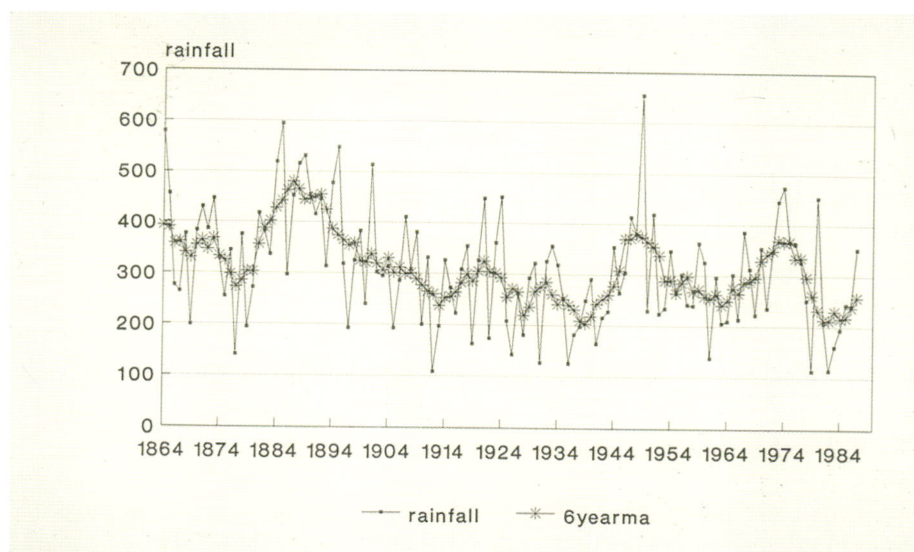

Figure 1.1.4 Contemporary desertification. Large fluctuations of annual rainfall amounts and persistent trends over long runs of years provide problems for overstretched water and biotic resources.

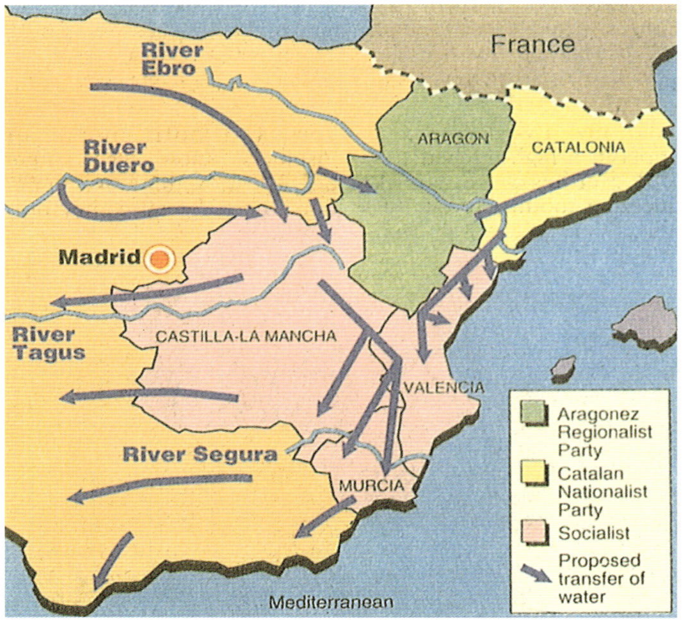

Figure 1.1.5 Contemporary desertification.

(a) Water transfers are a serious economic, social and political problem in Spain today. This map, reproduced from the *Sunday Observer,* shows actual and proposed water transfers. The most contentious is the transfer from Castilla-La Mancha to the south-east via the Tajo-Segura canal.

(b) Drip irrigation means that water use is no longer confined to flat areas.

(c) 'Let us close the door to the desert'.

use. Improved standards of living, the rise of tourism, the intensification of agriculture, the shift to irrigated agricultural production for export and the continued drift to an urban society, accompanied by rural depopulation, have all led to an emphasis on land-use conflicts ultimately linked to the problem of water (Grenon and Batisse, 1989). At the same time the growth of the conservation/environment movements in Mediterranean countries has sharpened the debate over land management by individuals and the state and the ever-deepening crisis of groundwater withdrawals has led to inter-regional conflicts exacerbated by the droughts of the early and mid 1980s and the renewed droughts in Spain during the 1990s.

Water as the root of the problem

Given these social and economic pressures, it is hardly surprising that the issues over desertification in these advanced and complex societies have become ever more confused. Added to this, however, is the nature of the interactions between hydrological processes (in all its manifestations), surface cover, soil degradation and human activity. Not only are variations and changes in climate crucial in determining water supply and the propensity to flooding, but so too are the storage capacity of soil, the vegetation cover and the capacity of the underlying rocks to hold water and yield it for use. Many aspects of this problem reach back into antiquity: by Plato's time erosion had already been identified as a major cause of flooding and the removal of forest and intensive grazing had been identified as the main culprits. That the picture is undoubtedly more complex than this has been demonstrated many times since. As long as populations remained small and demand remained at subsistence levels, problems of land degradation were not perceived to be critical. Now that water is the mainstay of tourism and agriculture, and hence a key component of Mediterranean regional and even national economies, the issue of desertification has become critical. It is a remarkable testimony to the perceptiveness of Roberto Fantechi that he foresaw the seriousness of this problem some 10 years ahead of the rest of the European Union and initiated a research programme that provides the basis for addressing it.

About the Atlas

Our intention in this Atlas is to give a sense of the problem of land degradation in Mediterranean Europe and the research being undertaken in the European Communities and the MEDALUS Project to understand the processes and determine what efforts can and should be made toward its mitigation. The Atlas is intended to emphasise the visual and graphic aspects of the problem and to reach an audience who have a concern and responsibility for the issues but who are not experts in the field. To this end we have provided a glossary of technical terms, a guide to the plant species names used in the different languages, and a fairly comprehensive list of the references used.

Following the Introduction, Section 1 reviews other work on the Mediterranean desertification problem funded by the European Commission. The second Section describes the main natural processes that are occurring in the region today and why and how they are modelled in the MEDALUS Project. The third Section then turns to the socio-economic processes and changes, an understanding of which is crucial for an appreciation of desertification. The fourth Section reports the nature of the field sites and the research undertaken there, and the fifth Section reviews the problems of desertification in three target areas in southern Spain and Italy. The final Section summarises the results of integrated research, and prospects for the future.

1.2 The European Commission Programmes and Projects on Desertification

J. B. Thornes

In the context of its Framework programmes, the European Commission (EC), by means of the Specific Programmes EPOCH (1989–1992), Environment (1991–1994) and the most recent Environment and Climate (1994–1998), recognised as a growing concern the problem of land degradation and desertification. The objective of the first programme was to achieve a better understanding of the processes and problems of desertification in southern Europe in relation to natural hazards. In subsequent phases, as desertification was recognised as an important environmental issue, it was specifically identified in the research programmes. Emphasis was given to a comprehensive, integrated and multidisciplinary approach to the problem. In the Environment programme (1991–1994), 29 projects, some larger than others, were funded by the EC for a total of 21 million ECU. A major effort of cross-border co-operation between European universities, research centres and Small and Medium Enterprises (SMEs) has hence been initiated and will continue, and will surely contribute to improve significantly our knowledge on this issue.

The sister project to MEDALUS, EFEDA (European Field Experiment in Desertification Threatened Areas) had as its primary objective the understanding of the role of near-surface conditions on the atmospheric boundary layer (Bolle *et al.*, 1993). The intention here was to ask whether or not, in Mediterranean type conditions, the land use at the surface has a significant control on the climate above the ground. Previous experiments had shown that extensive areas of uniform land use, such as the pine forests of the Bordeaux area, the prairie grasslands of Kansas or the Amazon rain forest, can have an important effects on the upward movement of energy and water vapour. These fluxes are of sufficient magnitude to perturb the atmosphere so that changes of ground-level land use could be important for climatic change. The Mediterranean land-use is a much more varied and complex mosaic, and it seems unlikely that, perhaps with the exception of very extensive irrigation, land use changes are likely to have much impact on the atmosphere. Moreover, the topography of the Mediterranean environment is extremely varied, so that the boundary layer is much more complex than in other areas, say, for example the Sahara, where it has been argued that desertification processes are re-inforcing, as the drying out and dying of the vegetation cover leads to lower soil moisture storage capacity. Further discussion of EFEDA appears in Section 1.4.

Another example is the ARCHAEOMEDES project, where the objective is to look at the changes in agriculture through time and attempt to evaluate the varying human response to environmental change. Thus, for example, in the Vera Basin, in south-east Spain, Gilman and Thornes (1985) showed that by Bronze Age times (approximately 4000 years ago) extensive gullying had already occurred in the area but also that local communities were adapting to irrigation practices and that social differences were emerging. These social differences were reflected in the objects found in the Bronze Age burials and the ARCHAEOMEDES project has demonstrated the more or less continuous and complex character of human response to environmental stress in the area in the subsequent 4000 years.

In addition to the research programmes directly involved with desertification, many other programmes of DGXII contribute to our knowledge of the processes and conditions that contribute to the problem. These include research into water resources, climate change and forest fires. Moreover, other Directorates General such as DGVI (Agriculture), DGXI (Environment) and DGXVI (Regional Development) have vested interests in the desertification problem in southern Europe, and in the future will be called upon to develop the techniques required to mitigate its effect. So far the thrust of the Commission's activities has been in research in the European Union, though the European Parliament and the southern European member states are now under great political pressure to fund the mitigation effort. The MEDALUS Project is directed at providing the knowledge base from which this could be achieved.

Table 1.2.1 Projects in the Environment Programme (1991–1994). Topic IV.3 Desertification in the Mediterranean Area.

1	Contract number	EV5V–CT91–0021	1.500.000*
	Coordinator	S.E. Van Der Leeuw – Cambridge	
	Title	Understanding natural and anthropogenic causes of desertification and land degradation in the Mediterranean Basin	
2	Contract number	EV5V-CT91-0023	400.000*
	Coordinator	A. Imeson – Amsterdam	
	Title	Modelling and exploring the impact of climate change on ecosystem degradation, hydrology and land use along a transect across Mediterranean	
3	Contract number	EV5V–CT91–0025	500.000*
	Coordinator	G. Nakos – Athens	
	Title	A GIS support system for the prevention of desertification resulting from forest fire	
4	Contract number	EV5V–CT91–0027	300.000*
	Coordinator	M. Follieri – Rome	
	Title	Palynological study on desertification in south-western Europe: Timing, natural trends and human impact	
5	Contract number	EV5V–CT91–0029	598.000*
	Coordinator	A. Rosema – Delft	
	Title	Assessment of remote sensing techniques for monitoring the extent and progression of desertification in the Mediterranean area (ASMODE)	
6	Contract number	EV5V–CT91–0031	200.00*
	Coordinator	A. Strid – Copenhagen	
	Title	Desertification in the White Mountains of Crete. A botanical study with special reference to the effects of grazing	
7	Contract number	EV5V–CT91–0033	999.000
	Coordinator	J.P. Goutorbe – Toulouse	
	Title	HAPEX-Sahel	
8	Contract number	EV5V–CT91–0035	249.800*
	Coordinator	B. Lacaze – Montpellier	
	Title	An integrated approach to Mediterranean land degradation mapping and monitoring by remote sensing (DEMON)	
9	Contract number	EV5V–CT91–0037	350.000*
	Coordinator	R. Julia – Barcelona	
	Title	Origin and evolution of desertification in the Mediterranean environment in Spain	

Table 1.2.1 *Continued/...*

10	Contract number	EV5V–CT91–0039	1.000.000*
	Coordinator	J. Lavabre – Aix-en-Provence	
	Title	Assisting the management of degraded Mediterranean areas by runoff and erosion modelling on representative catchments.	
11	Contract number	EV5V–CT91–0041	400.000*
	Coordinator	C. De Oliveira Alves Coelho – Aveiro	
	Title	Land management practice and erosion limitation in contrasting wildfire and gullied locations in the Iberian Peninsula	
12	Contract number	EV5V–CT91–0043	250.000*
	Coordinator	M. Sala – Barcelona	
	Title	Desertification risk assessment and land use planning in a Mediterranean coastal area	
13	Contract number	EV5V–CT91–0045	700.000*
	Coordinator	M. Dubost – Corte	
	Title	MEDIMONT – A multidisciplinary research programme on the role of and the place of the mountains in the desertification of Mediterranean regions	
14	Contract number	EV5V–CT92–0128	2.450.000*
	Coordinator	A. Imeson – Amsterdam	
	Title	MEDALUS II: Project I. Basic field programme	
15	Contract number	EV5V–CT92–0164	2.199.000*
	Coordinator	M. Kirkby – Leeds	
	Title	MEDALUS II: Project II. Modelling and regionalisation	
16	Contract number	EV5V–CT92–0165	1.400.000*
	Coordinator	A. Aru – Cagliari	
	Title	MEDALUS II: Project II. Managing desertification	
17	Contract number	EV5V–CT92–0166	3.006.700*
	Coordinator	J.B. Thornes – London	
	Title	MEDALUS II. Project IV. Research and policy interfacing in selected regions	
18	Contract number	EV5V–CT93–0269	179.130*
	Coordinator	J. Noihan – Toulouse	
	Title	EFEDA II – Surface processes and atmospheric modelling for the Echival-Efeda Phase II programme	
19	Contract number	EV5V–CT92–0270	970.000*
	Coordinator	F. Fielder – Karlsruhe	
	Title	EFEDA II – Aerology and measurement from aircraft	
20	Contract number	EV5V–CT93–0271	59.531*
	Coordinator	H. De Bruin	
	Title	EFEDA II – Atmospheric boundary layer, surface energy budget and soil-vegetation-atmosphere transfer	
21	Contract number	EV5V–CT93–0272	800.000*
	Coordinator	F. Martin De Santa Olalla	
	Title	EFEDA II – Vegetation, soil physics, inventory and impacts	
22	Contract number	EV5V–CT93–0282	739.000*
	Coordinator	J. Bromley – Wallingford	
	Title	EFEDA II – Hydrological response to land use change and over exploitation of water resources in a semi-arid area of Spain	
23	Contract number	EV5V–CT93–0284	1.040.000*
	Coordinator	H.-J. Bolle – Berlin	
	Title	EFEDA II – Assessment of desertification from space	
24	Contract number	EV5V–CT94–0466	366.000*
	Coordinator	I. Vardavas – Heraklion	
	Title	An integrated monitoring and modelling study of desertification and climatic change impacts in the Massara Valley of Crete	
25	Contract number	EV5V–CT94–0485	265.000*
	Coordinator	R. Jordana Butticaz – Pamplona	
	Title	Biodiversity of the ecotone Pine forest-Mediterranean shrubland as a tool against erosion in Mediterranean areas	
26	Contract number	EV5V–CT94–0486	200.000*
	Coordinator	S.E. Van Der Leeuw – Cambridge	
	Title	Environmental perception and policy making – cultural and natural heritage and the preservation of degradation sensitive environments in southern Europe	
27	Contract number	EV5V–CT94–0487	180.000*
	Coordinator	V.C. Lull Santiago – Bellaterra	
	Title	Palaeoclimatic reconstruction and the dynamics of human settlement and land use in the area of the middle Aguas (Almeria), of the south-cast Iberian peninsula	
28	Contract number	EV5V–CT94–0488	306.000*
	Coordinator	P. Jeffries – Canterbury	
	Title	Restoration of environmental diversity by effective ecosymbiont monitoring (REDEEM)	
29	Contract number	EV5V–CT94–0489	400.000*
	Coordinator	S. Mazzoleni – Campobasso	
	Title	Modelling vegetation dynamics and degradation in Mediterranean ecosystems (MEDMOD)	

* EC contribution (ECU)

Source:

Commission of the European Communities – Directorate General XII. Science. Research and Development, 1994. Environment Research Programme 1991–1994. Catalogue of Contracts Vol. 1

1.3 The MEDALUS Project

J. B. Thornes

MEDALUS is the acronym for Mediterranean Desertification and Land Use. The project, as described in this Atlas, has been undertaken in two phases, Phase I during 1991–1992 and Phase II during 1993–1995. A third phase, MEDALUS III, began in January 1996, running for three years. The first phase involved 18 groups from universities and research institutes, and the second involved 44 groups, including some from industry and commerce. 30 groups are working on the third phase. For each period a work plan was developed, the research undertaken and a series of plenary sessions convened to review the results.

MEDALUS I overview

In the first phase the main task was to improve our understanding of the climatic, hydrological, geomorphological and ecological processes of desertification mainly under semi-natural conditions on land that had largely been withdrawn from agriculture. In order to do this the project adopted five approaches (Brandt and Thornes, 1996):

1. Climatological studies of present, past and future trends in rainfall and temperature to obtain some idea of the range of fluctuations of contemporary climates within the Mediterranean region and the likely range of future climatic conditions under various model scenarios. Investigations of climatological series of anomalies in all countries have been used together with predictive models to estimate immediate future trends. In addition, scenarios from global circulation models have been downscaled to provide regional scenarios at the subgrid scale using regression techniques. The climatological studies have also provided syntheses of extreme event data (rainfall and temperature) for the field sites.
2. Field observations of climatic, ecological, erosional and hydrological processes at a set of ten field sites throughout the Mediterranean:

- *Almocreva Vale Formoso* in Alentejo, Portugal
- *El Ardal*, Murcia, and *Rambla Honda*, Almeria, Spain
- *Var* and *Roussillon,* Pyrenees Oriental, France
- *Santa Lucia,* Sardinia, Italy
- *Spata*, Attica, and *Hortiatis* and *Petralona*, Makedonia, Greece

Each field site, with automatic weather-recording facilities to collect a set of basic climatic, soil and ecological parameters, had a special topic of research relating to desertification. The special site studies include the ecophysiology of Mediterranean plant species, the structure of Mediterranean soils at different length scales, the impact of storm intensity, vegetation cover types and soil characteristics, fire, grazing and different agricultural treatments on runoff and erosion and the impact of stony soils on soil erosion, soil water budgets and field moisture capacity. Some agricultural work was also done on rainfall exclusion and stony soils. The sites are described in Section 4 of the Atlas and the data collected are archived on a CD-ROM (see the box in this section). The results of these research studies and the data

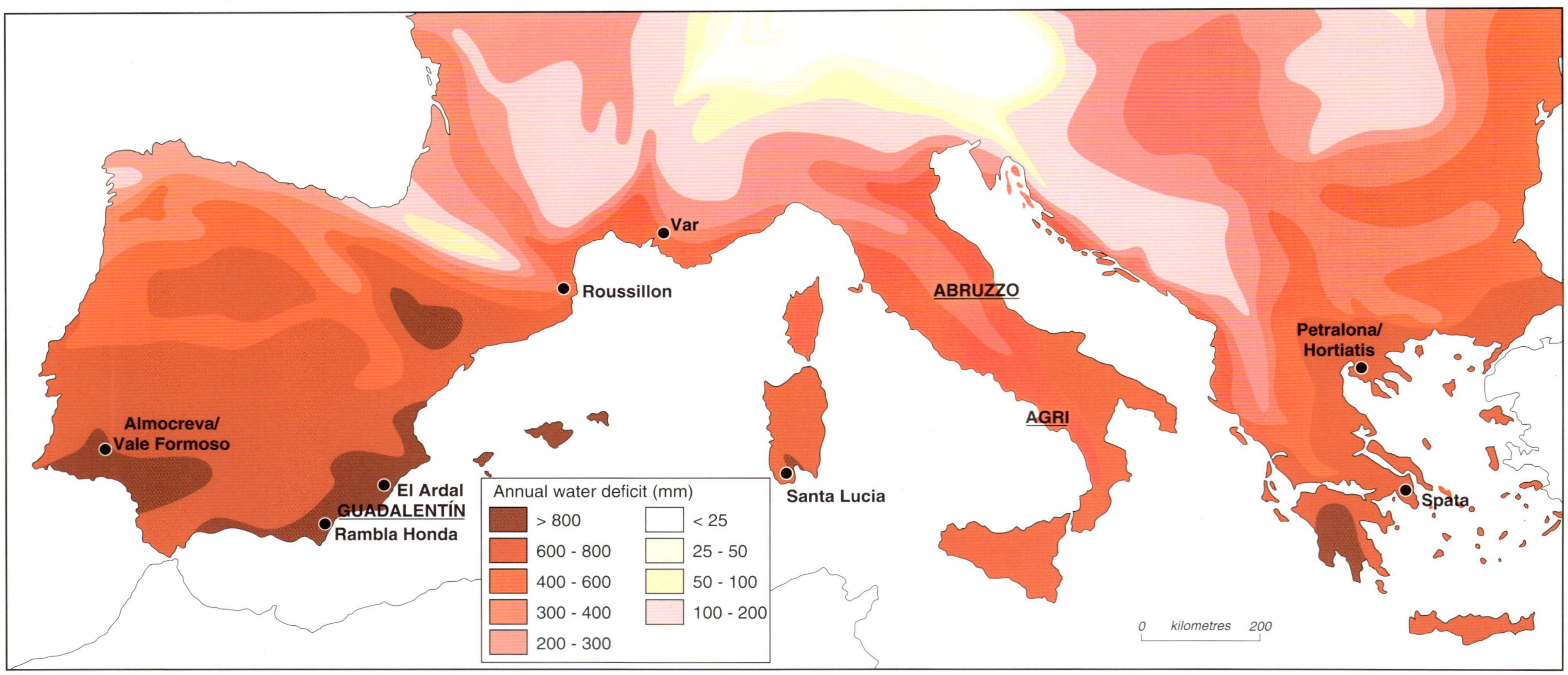

Figure 1.3.1 This map is based on Thran and Broekhuizen (1965), *Agro-climatic Atlas of Europe*, with permission. It shows the areas of water deficit and the locations of the MEDALUS sites and target areas (underlined).

provided by the core programme are the key products of the first phase of the programme.

3. Modelling investigations through the development and implementation of the MEDALUS hillslope and models of climatic change scenarios. The MEDALUS hillslope model (see Section 2.8) provides a mechanism for examining the role of changes in climate, vegetation and/or soils on runoff and sediment yield at the scale of a field, hillslope or erosion plot. It is based on inputs of hourly climatic variables, especially rainfall. The vegetation cover (if any) determines the interception, evapotranspiration and throughfall, after which infiltration and overland flow is partitioned by soil properties. Soil moisture determines available moisture for plant growth. Plant growth is determined for functional types (i.e. trees, bushes, grasses and bare ground). Overland flow determines erosion rates over an irregular surface, and the capacity for sediment transport of different size fractions determines deposition. This model has been tested in a variety of situations, using inputs from the field sites and validation against non-MEDALUS sites, and looking at the impact of climate change on vegetation cover, runoff and erosion.
4. Evaluation of the impact of various scenarios of climatic, economic and social change on population dynamics using complex systems models. These models allow for unexpected changes in the structural relationships underlying desertification processes and have been parametrised for the case of Crete.
5. Development of remote sensing techniques, calibrated against the field sites, for the analysis of vegetation and soil conditions where vegetation is sparse and the underlying lithology makes up an important part of the signal emitted back from the ground to the satellites. These remote sensing studies provided a wider context into which to set the field studies and the possibility of extending the models beyond the field sites themselves.

MEDALUS II

Although the time available for MEDALUS I was very short (two years), it became clear that (1) more effort at integration of the different components was needed; (2) a larger scale was required to make the techniques usable over a larger area; and (3) more effort was needed to incorporate economic and social effects on and responses to desertification. At the same time the need to sustain field sites providing a unique data set on land degradation was recognised.

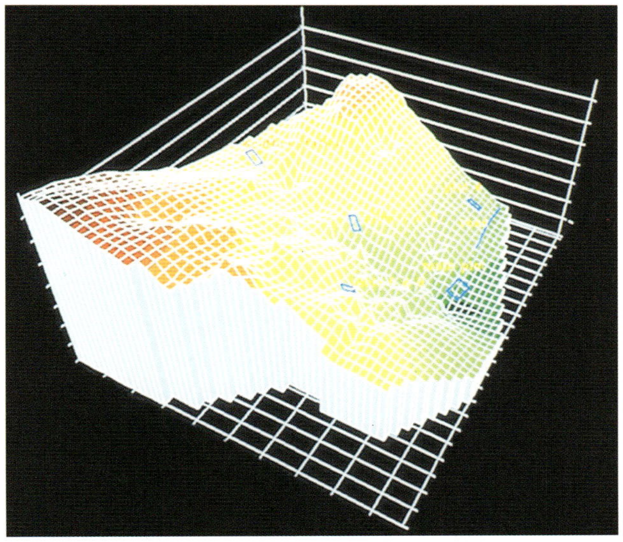

Figure 1.3.2 One of the MEDALUS field sites at El Ardal, Murcia, Spain (described in detail in Section 4.3).

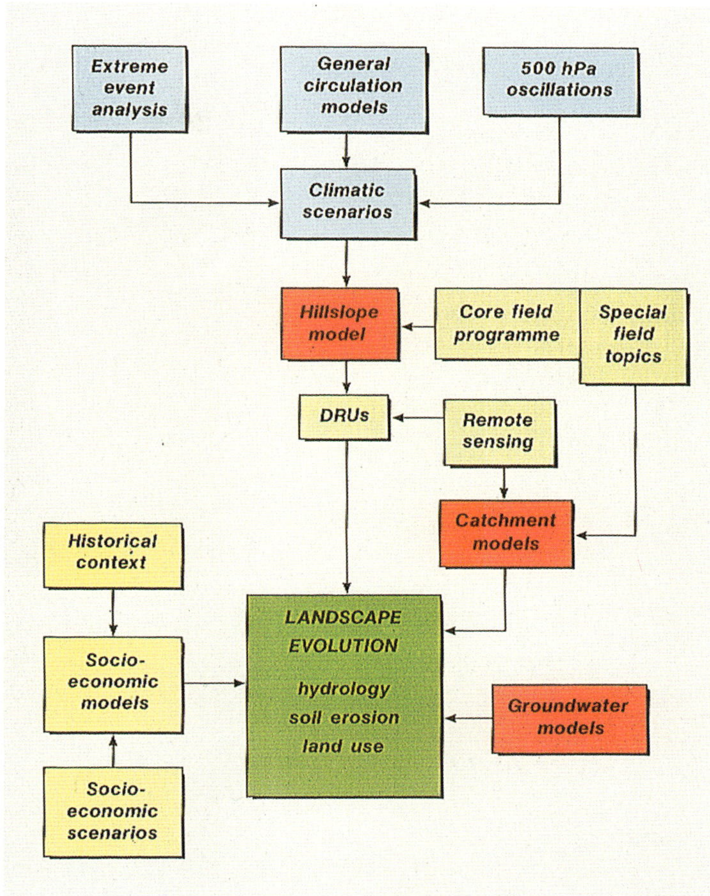

Figure 1.3.3 A flow chart showing the overall approach to the first stage of the MEDALUS programme.

The main field sites were therefore retained, but in addition all other research was focused on three 'target areas': the Guadalentín Basin, Spain (Section 5.1), the Agri Valley, Italy (Section 5.2), and the Abruzzo, Italy (Section 5.3). The first is an extreme case of desertification with critical water, land degradation and flooding problems; the second is a typical summer-dry Mediterranean area with intensive erosion, land use and water use conflicts, coupled with regional underdevelopment; and the third is a potentially desertification-threatened area in eastern Italy, chosen to provide a marginal case. All three areas are less-favoured regions of Europe. By concentrating research effort in these areas and co-operating with local authorities responsible for regional development and environmental management, a higher level of integration of the desertification studies has been achieved. Moreover, it has enabled the development of a better understanding of the regional variability, as opposed to the local (plot and field) scale, transnational (European economic) scale or global (climatic) scale approaches to desertification. This avoids the preoccupation with a single response to either changing socio-economic or climatic conditions which tends to occur in policy development at the European scale (Thornes, 1995).

In the modelling work the principles embodied in the MEDALUS hillslope model were upscaled and a new basin-scale model was developed called MEDRUSH. This incorporates the idea of a catchment made up of characteristic

Figure 1.3.4 A photograph of Bassu field site in the Astimini-Fiume Santo Basin in Sardinia (described in detail in Section 4.5).

'strips' on which land use can change, and down which water and sediment flow. By modelling characteristic strips and integrating their effects across the catchment and routeing the water and sediment from the catchment, the impact of climate and land-use changes on quite large catchments (up to 5000 km^2) can be evaluated. Upscaling and comparative analysis of the target areas have also been achieved by undertaking socio-economic analysis of the three target areas, on the basis of land-use changes. Finally, the shift to a regional scale has laid much greater emphasis on techniques applicable at that level. Critical among these has been the extensive use of remote sensing, the development of regional weather type analysis as a basis for improving the climatic change scenarios, and the use of multivariate classification for sampling and analysing regional change characteristics.

As the second phase of the MEDALUS project comes to an end there is a great deal of analysis still to be undertaken. Nevertheless, some important new insights into the desertification problem are emerging. Many of these are dealt with in the rest of this Atlas; others are dealt with in the project reports, working papers, conference proceedings and journal articles listed at the end of the Atlas.

...and MEDALUS III

The third and final phase of the MEDALUS programme will apply the models and research in selected areas and at the national scale, extending the techniques developed in the study of the target areas. The number of field sites has been reduced to three and the target areas to two. In addition, two new areas of activity have been introduced, partly to fill obvious gaps, partly to maintain an innovative approach to the research programme. The first of these is to develop Mediterranean-wide regional desertification indicators, the second is to investigate the implication of desertification for ephemeral gullies and channels.

The MEDALUS archive on CD-ROM

L. H. Cammeraat and H. Prinsen

In the MEDALUS Project data have been collected at the eight core project field sites to establish a database that could be used for data-sharing and as an input for the modelling efforts being carried out within the project. The data were gathered in a harmonised way following standardised procedures at standardised field sites (Cammeraat, 1992, 1996). The data ranged from single measurements of static properties of the fields sites to dynamic event data and covered vegetation and physical properties of the field sites. A database was set up to enable the dissemination of data between the various field groups and to make the data available to the modelling groups within the project.

Initially, during the establishment of the database, data were disseminated by diskettes on request, or via e-mail/Internet. As the database grew at a steady rate, the need to develop user-friendly access to all the data was recognised. The database contains many types of data, including (thematic) maps, aerial photographs, bibliographies, reports, climatological data, photographs and, of course, the field measurements.

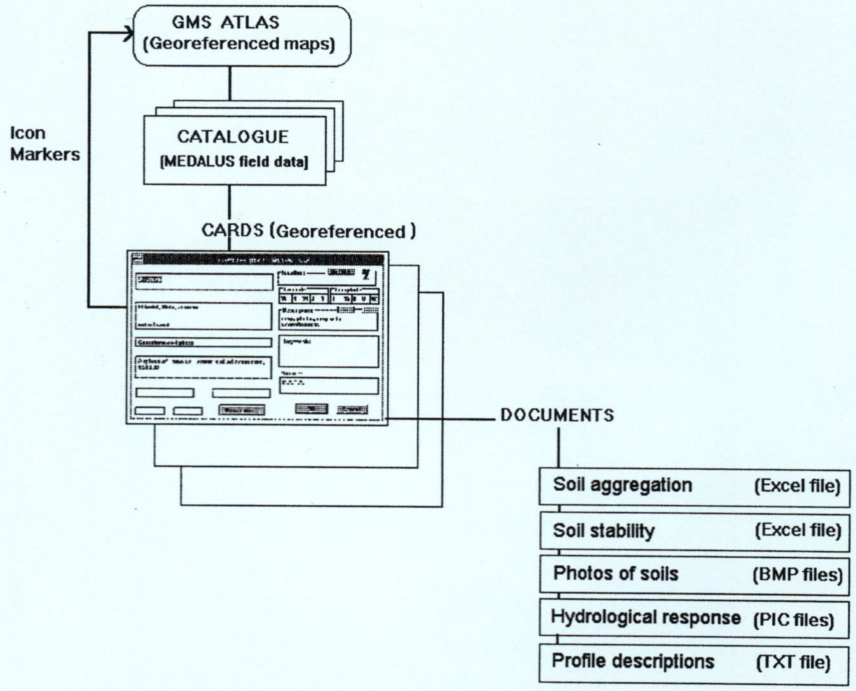

Figure 1.3.5 Structure of the geomanagement system. After opening the desired catalogue, a map is attached, and georeferenced cards are selected, giving access to the data files.

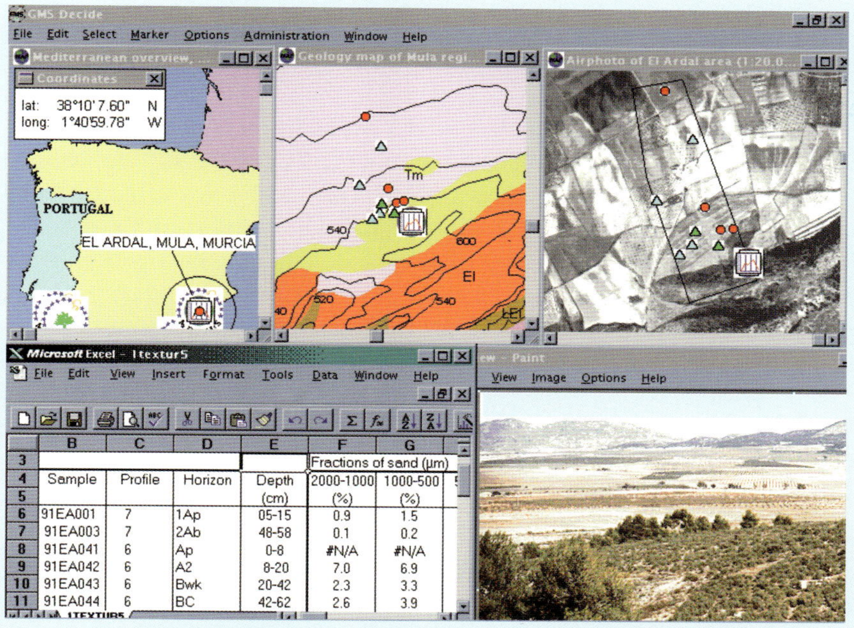

Figure 1.3.6 Typical screen during a GMS-session. From the map of the Mediterranean, the user zooms into the core field site of El Ardal, Murcia, Spain. All kinds of information can be retrieved, analysed and combined (in this case, an aerial photograph, a geological sketchmap, an overview of the area and a spreadsheet containing soil texture data).

Instead of a classical relational database, an easier-to-use geomanagement system (GMS) was developed and produced by Da Vinci SA (GMS-DECIDE, 1994). It takes only a day or so to learn how to use the program, provided that one is familiar with Windows™ and its general applications.

This system can be considered as a shell running under Windows, which uses the georeference (geographical coordinates) of the data as their first index. The data themselves are accessed by other applications (word processor, spreadsheet, GIS, and bitmap viewer) running under Windows via GMS. Data, no matter what type, are indicated on maps by icons with attached index lists. The use of Windows and the design of the database make it possible to undertake direct data analysis and interpretation on the screen, as all data are readily accessible.

By selecting areas on the maps available, data within the geographical areas can be accessed. Queries can be carried out between and within zones on predefined keywords, and between different areas or maps on similar properties. The same areas can be looked at on different maps or aerial photographs from the same regions. This allows the projection, for instance, of a selected land unit from an aerial photograph onto the soil map. For this land unit, the soil texture, soil physical and soil chemical data can be retrieved for the associated soil unit, giving information on the soil water properties and nutrient availability for the selected land unit.

The system also allows the user to employ other applications to process data. One of these applications is the MEDALUS hillslope simulation model (Kirkby *et al.*, 1996). Once the model is set up for the specific field site, it can use contemporary on-the-spot measured data, which are stored in the georeferenced database.

The data and the GMS system have been distributed on CD-ROM (Cammeraat and Prinsen, 1995) to the MEDALUS participants. It also contains the MEDALUS hillslope model, climatic data and meteosat images of the Mediterranean for 1993.

1.4 Examples from Other European Commission Projects

THE EFEDA PROJECT

H. J. Bolle

The state of the continental surface with its vegetation is determined by both climatic boundary conditions and human activity. The effect of the climate system is due to precipitation, solar radiation, and wind. The surface responds to these agents by the transfer of heat and matter into the atmosphere. The division of the absorbed solar energy into other forms of energy leaving the surface depends on the characteristics of the surface and the structure of the atmospheric boundary layer. Human activity affects the surface properties and thus the exchange of momentum and the splitting of the solar energy into sensible and latent heat fluxes and, to a much smaller degree, into photosynthetic energy in the form of biomass production. Changing the climatic and human-induced boundary conditions therefore affects the surface conditions and hence the coupling between the land surfaces and the atmosphere. This causes an ongoing circle, a feedback from the modified conditions at the land surface to the atmospheric climate system that, in turn, responds with a readjustment of the climatic boundary condition. To understand the processes that lead to land-surface changes it is necessary to keep in mind the whole system of interactions. The water cycle is very important in this context. A regional distortion of the water budget may lead to aridification and desertification.

The transfers between the surface and the atmosphere can be studied by so-called SVAT (Soil–Vegetation–Atmosphere–Transfer) models. These models require a detailed knowledge of the soil and vegetation properties as well as the state of the overlying atmospheric boundary layer. These properties, especially the vegetation cover, may change rapidly in space. Quantities measured with substantial effort at selected locations may be representative of only a few tens of thousands of square metres and SVAT models fed with these data simulate the fluxes with great accuracy. Climate models operate on sizes of the order of a hundred thousand square kilometres and consequently land-surface properties, that may be highly variable within one gridbox, have to be aggregated over large areas. In the Mediterranean area not only do the characteristics of the ecosystems vary quickly over space, but topography and land–sea distribution are also extremely complicated, at scales that cannot be resolved in climate models. For the computation of future scenarios it is therefore crucial to know the accuracy of these parametrisations that represent the land-surface processes at the larger scale and to what degree scenarios developed on the basis of these parametrisations may reflect the real conditions. To test the accuracy of the process descriptions it is necessary to conduct measurements at the scale of the model grid areas.

EFEDA was established by the EC in the framework of the European international project on Climatic and Hydrological Interactions between Vegetation, Atmosphere and Land-surfaces (ECHIVAL) as the ECHIVAL Field Experiments in Desertification-Threatened Areas to study the interaction between surface and atmosphere at a scale that comes close to the resolution of global models. A first pilot study was established in a flat and homogeneous area, the plain of Castilla-La Mancha in Spain to avoid complications due to topography. More than 25 European research institutions as well as five groups from the USA participated in the field experiment (Bolle *et al.*, 1993). More heterogeneous landscapes are currently under investigation.

During June 1991 intensive field studies of all quantities needed to test SVAT models were conducted in three representative areas forming an equilateral triangle: the partly irrigated agricultural area near Barrax, the dry farming vine culture around Tomelloso, and the slightly hilly dry farming and *matorral* area near Rada de Haro.

Below the surface of this area are aquifers, with a layered structure, that are fed by the precipitation falling over surrounding mountains. The groundwater

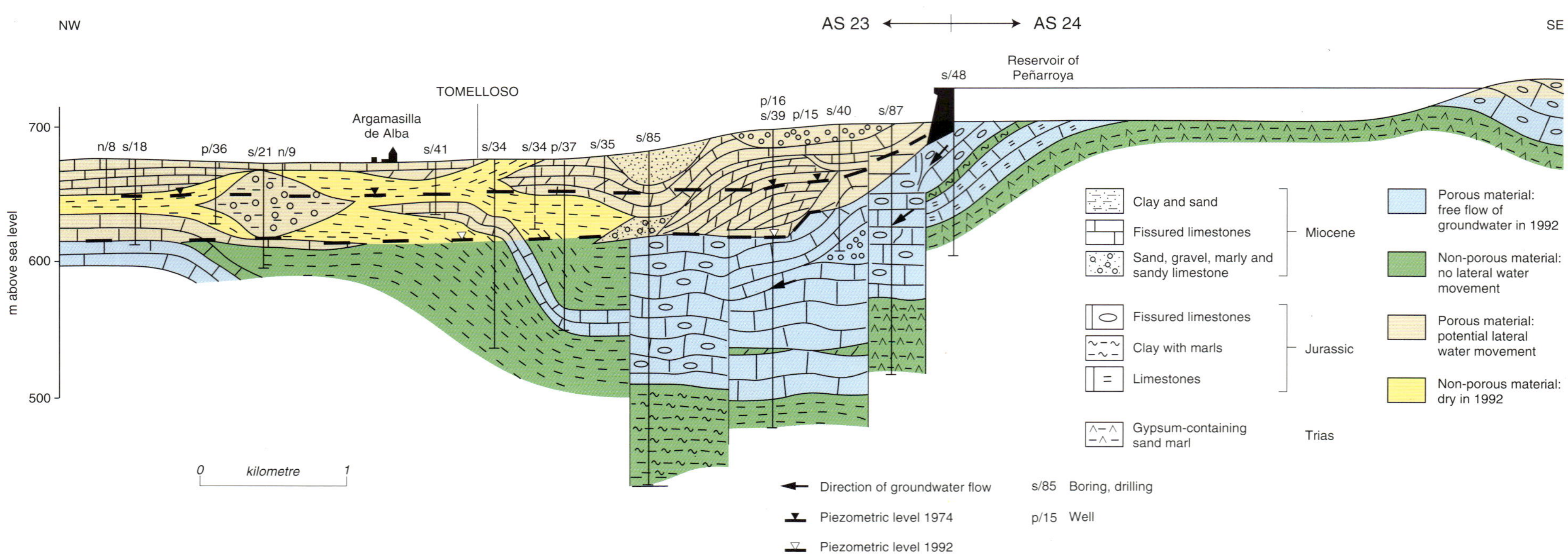

Figure 1.4.1 Soil structure and groundwater distribution in the Tomelloso area after Llamas *et al.* (1993) with permission.

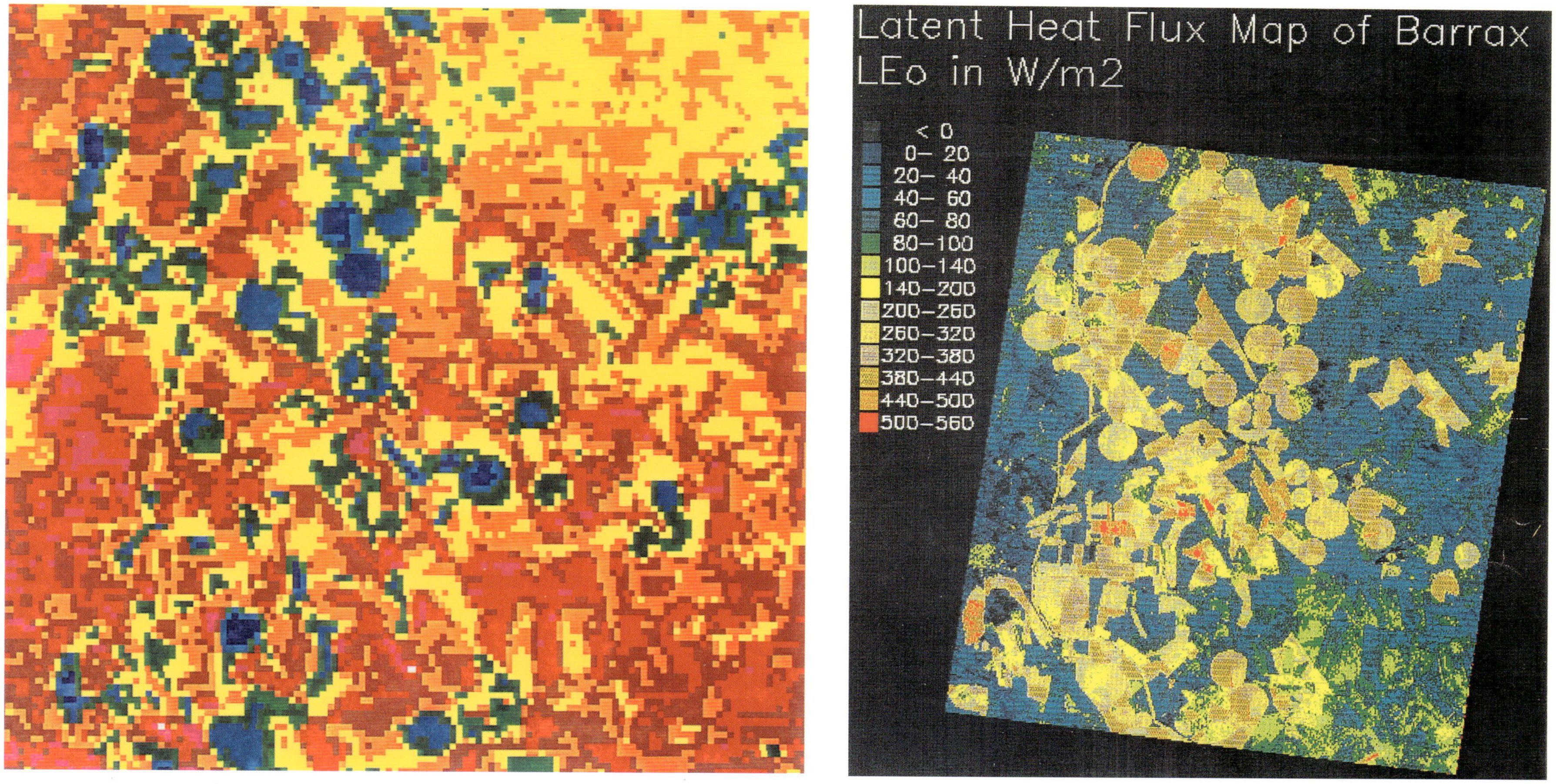

temperature (°C)	colour	temperature (°C)	colour
25 – 27	dark blue	40 – 42	orange
28 – 30	light blue	43 – 45	dark red
31 – 33	dark green	46 – 48	red
34 – 36	light green	49 – 51	magenta
37 – 39	yellow	52 – 54	white

Figure 1.4.2 Effective surface temperatures and the distribution of latent heat fluxes in the Barrax area as inferred from remote sensing data. The left side shows the temperature distribution inferred from LANDSAT-TM channel 6 data (Bolle, 1997). The surface of the irrigated fields is close to the air temperature. Over the mostly bare ground the gradient between surface temperature and air temperature is large, giving rise to high sensible heat fluxes. The latent heat flux, as shown on the right of the figure (after Bastiaanssen and Roebeling, 1993) is very small over these dry areas but at a maximum over the irrigated fields.

level in these aquifers is currently sinking by approximately 1 m per year due to an over-exploitation of the water resources. The assessment of the ground-water situation and the study of soil types and the water transfer through the unsaturated zone was therefore a major research effort within EFEDA.

The instrumentation was concentrated in the centre of each of the three selected test sites with additional single stations at distances up to a few kilometres away to assess the inhomogeneities of the area. The 'super-sites' are each about 70 km from each other. Movable equipment, satellite and aircraft data were used to bridge the gaps between them. Detailed maps of the land use have been generated using both remote sensing and ground observations.

The evaluation of remote sensing data, 'anchored' at super-sites, resulted in maps of land-surface characteristics and flux-distributions. The term 'anchored' means that corroborative measurements at the ground have been used to relate the signals measured in space to the surface properties. If such maps of physical quantities are produced periodically, they may indicate the vulnerable zones where aridification progresses, vegetation retreats and the hydrological processes change. Though maps of physical quantities have not yet been produced for the whole of the Mediterranean region, the feasibility of producing them has been demonstrated. 'Anchor stations' are needed around the Mediterranean to 'calibrate' the satellite data; that is, to adjust them to the various ecosystems. Vegetation index maps that provide a relative measure of the state of vegetation are precursors for the mapping of desertification in the Mediterranean. The first vegetation index map of a resolution of only little more than one kilometre for the whole area has been compiled.

The vulnerability of the areas in the Mediterranean in relation to mesoscale topographical features and the general atmospheric circulation system is a basic question. Over hot, dry land surfaces, areas of low pressure (heat lows) develop because the sensible heat fluxes from the surface cause the overlying air to warm up and to rise. To compensate for this loss, air is pulled in from the surroundings near the surface. The boundary layer extends high into the atmosphere and is often filled with dust raised from the dry surface by the winds accompanying the low level air transport. The state of the atmosphere over the plain of Castilla La Mancha was studied in detail by a dense network of radiosonde stations that were sent up frequently. With these data, mesoscale

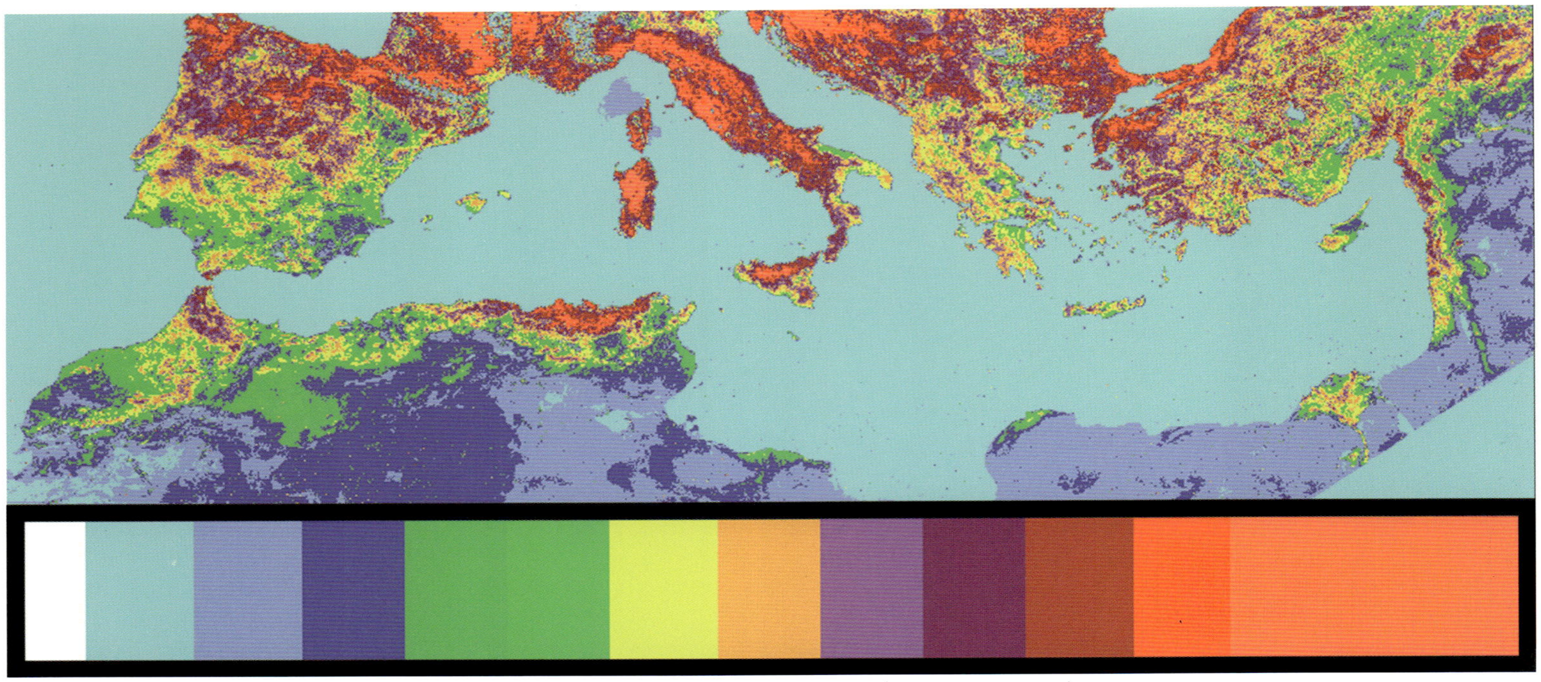

Figure 1.4.3 The normalised difference vegetation index map of the whole Mediterranean region composed of several NOAA overpasses during May 1994 at full AVHRR resolution (1.4 km^2). Key to normalised difference vegetation index (NDVI) classes: blues, 0–0.2; greens, 0.2–0.3; yellow/brown, 0.3–0.4; mauve/violet, 0.4–0.5; reds, 0.5–0.7.

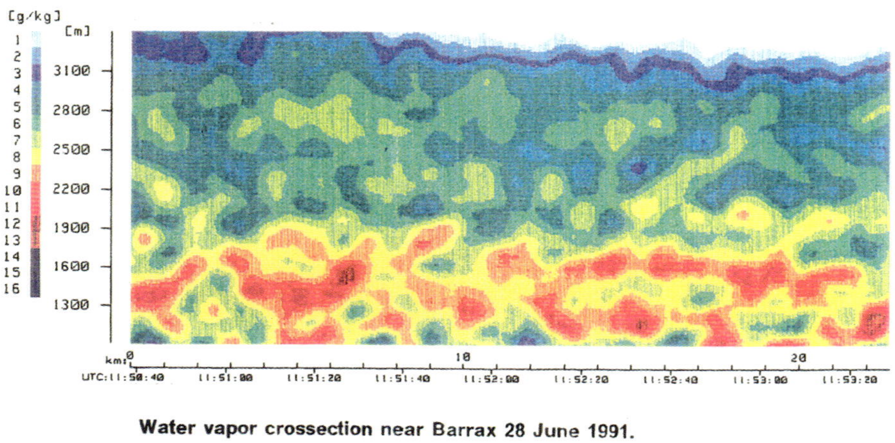

Figure 1.4.4 Height and structure of the atmospheric boundary layer as measured by an airborne laser scatterometer over the EFEDA test area (after Ehret *et al*., 1993). The figure shows the water vapour mixing ratio. The high reaching atmospheric boundary layer is clearly separated from the free troposphere.

models were operated to study the circulation system over the area. One interesting observation was that Mediterranean air is channelled through gaps of the coastal mountain chain, sometimes into the centre of the experimental area about 200 km from the coast. An example is provided by the situation experienced on 29 June 1991. These model simulations demonstrate the importance of the topography for the regional climate. In many areas of the Mediterranean dry 'islands' exist in areas controlled by topography, though wet air is transported by the sea breezes. Over sparsely vegetated hot surfaces this air is warmed up so much that even the uplifting often does not lead to condensation. This occurs primarily over mountains, where rainfall is gathered and percolates into the aquifers. From these aquifers the vegetation in the dry areas can obtain water, provided that the groundwater level can be maintained close to the root level. Changes of the large-scale wind systems are regionally very important because of the interaction with mountains to produce rainfall. Small changes in the wind direction may locally change precipitation and the water budget. Therefore the next research programme needs to investigate land-surface processes in hilly terrain more closely.

MODMED: MODELLING VEGETATION DYNAMICS AND DEGRADATION IN MEDITERRANEAN ECOSYSTEMS

S. Mazzoleni and C. Legg

Introduction

Mediterranean ecosystems around the world have received considerable scientific attention in recent decades because of their high biological diversity and risk of degradation and desertification.

Nowadays, in the Mediterranean region proper, i.e. the land surrounding the Mediterranean Basin, new problems are faced in the management of natural resources because of the recent changes in the ancient human-induced land-use patterns. In fact, while increasing demographic pressure has been leading to desertification and biodiversity losses in many coastal areas of the Mediterranean, in other marginal and inland areas, recent agriculture abandonment and decreasing grazing pressure are now opening up new spaces for re-colonisation by natural plant populations.

The key to maintaining and managing ecosystems and, particularly, the prevention and reversal of desertification, is an understanding of the processes which influence vegetation structure, as it is plant cover that protects the soil from erosion and most determines ecosystem functions and dynamics. Moreover, the vegetation responses to environmental changes (including both climatic changes and land-use management) ultimately depend on the auto-

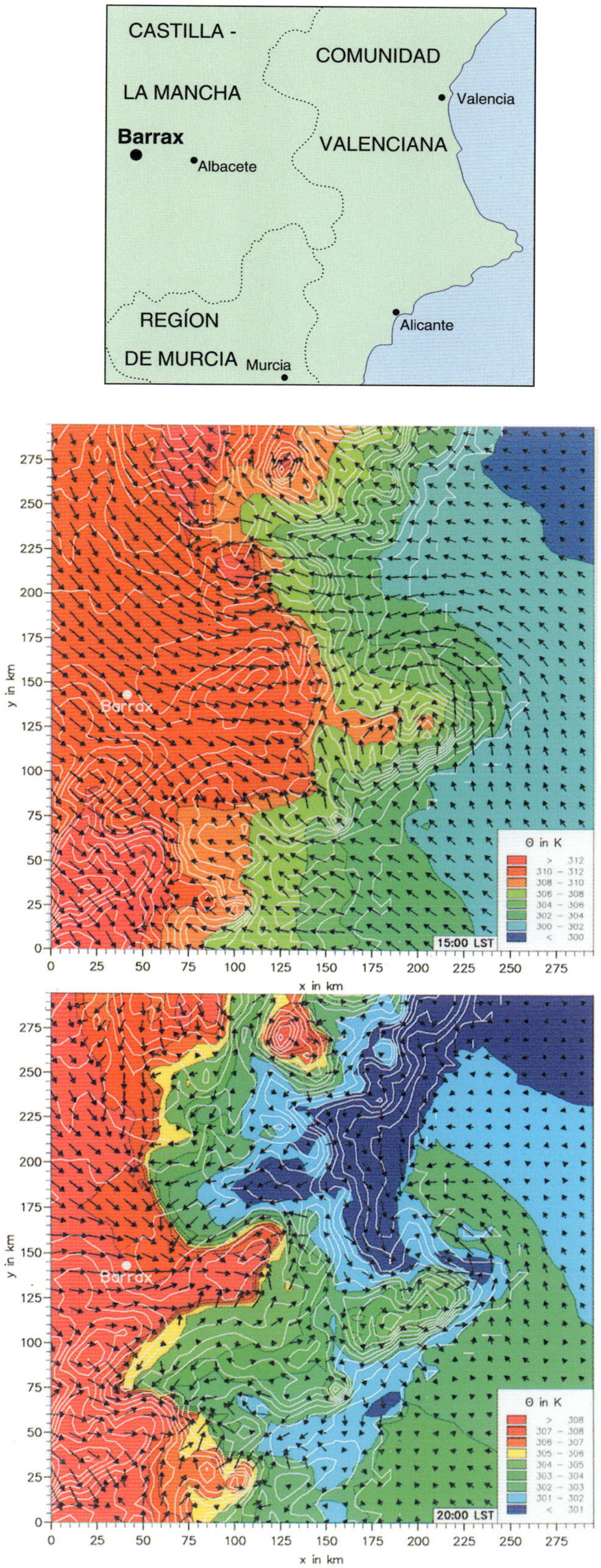

Figure 1.4.5 Result of a simulation of the land–sea circulation at the south-eastern sector of Spain at 15 and 20 hours respectively local standard time, 29 June 1991. The white contour lines represent the relief, arrows mark direction and strength of the winds near the surface, and colours represent the air temperature near the surface (Adrian and Fiedler, 1995). It can be seen that the seabreeze front (yellow) in the afternoon moves close to the experimental area that starts west of Barrax. At the front, maritime and continental air is mixed in an updraft which leads to an exchange between the boundary layer and the free troposphere (Noilhan, 1995).

ecology of the component plant species. Sometimes this is thought of as a reductionistic approach compared to other kinds of ecosystem studies, e.g. on nutrient cycles, energy flows and other ecosystem processes. However, at present there is still inadequate knowledge on most species behaviour and on successional processes taking place in the Mediterranean region, to be able to produce reliable and accurate models for these ecosystems.

A rich literature can be found on functional models for tropical and boreal forests and agricultural systems but not for Mediterranean ecosystems. Problems of scaling up from either point, or patch measurements to larger areas at landscape level are still unsolved. In fact, large gaps exist when comparing the accurate representation of biophysical processes adopted in micro-meteorology, the probabilistic or functional approaches of population models, and the gross approximation of regional-scale models, e.g. the parametrisation of the water balance over large areas. Several MEDALUS projects have been tackling some of these problems by relating climate and soil processes to land use at the regional and local scales. However, links still need to be made between the autoecology of plant species, processes which operate at the population scale, and the ecosystem dynamics at the landscape level. The MODMED research project is specifically designed to fill these gaps: obtaining new knowledge on the autoecology of Mediterranean species and solving the difficulty of spanning several orders of scale in modelling ecosystem dynamics.

The theoretical background to MODMED was the assumption that, despite the apparent dichotomy between the deterministic and the probabilistic/stochastic views of vegetation, there are possibilities of mixed models where some ecological processes (e.g. growth) are modelled deterministically and others (e.g. mortality or establishment of new seedlings) can be viewed in probabilistic terms. These ideas have been the base for the planning and development of a new simulation model for Mediterranean vegetation dynamics.

Participants

Within MODMED a consortium of institutions has been assembled to achieve a balance between the needs of data collection for increasing the knowledge on species behaviour and vegetation processes, and the problems related to the collation of the experimental and observational data and the design of the modelling environment.

In particular, data collection has been organised as follows: the University of Molise (S. Mazzoleni), the Kapodistrian University of Athens (M. Arianoutsou) and the Instituto Superior de Agronomia de Lisboa (F. C. Rego) have been managing the landscape and population analysis in Italy, Greece and Portugal respectively. The University of Pisa (R. Gucci and R. Massai) has been providing ecophysiological analysis, and the Ge.Pro.Ter. of Napoli (A. Di Gennaro) the soil data. The University of Edinburgh (C. Legg and R. Muetzelfeldt) has been in charge of the development of a new modelling environment, collaborating on specific aspects of modelling development with the Lisbon and Molise units.

The study areas, plant species and experimental work

The areas for field studies and data collection were selected within regions characterised by Mediterranean climate and vegetation types representing different successional stages following either disturbances or re-colonisation processes. The chosen areas were, in Italy, the Cilento and Amalfi coasts in

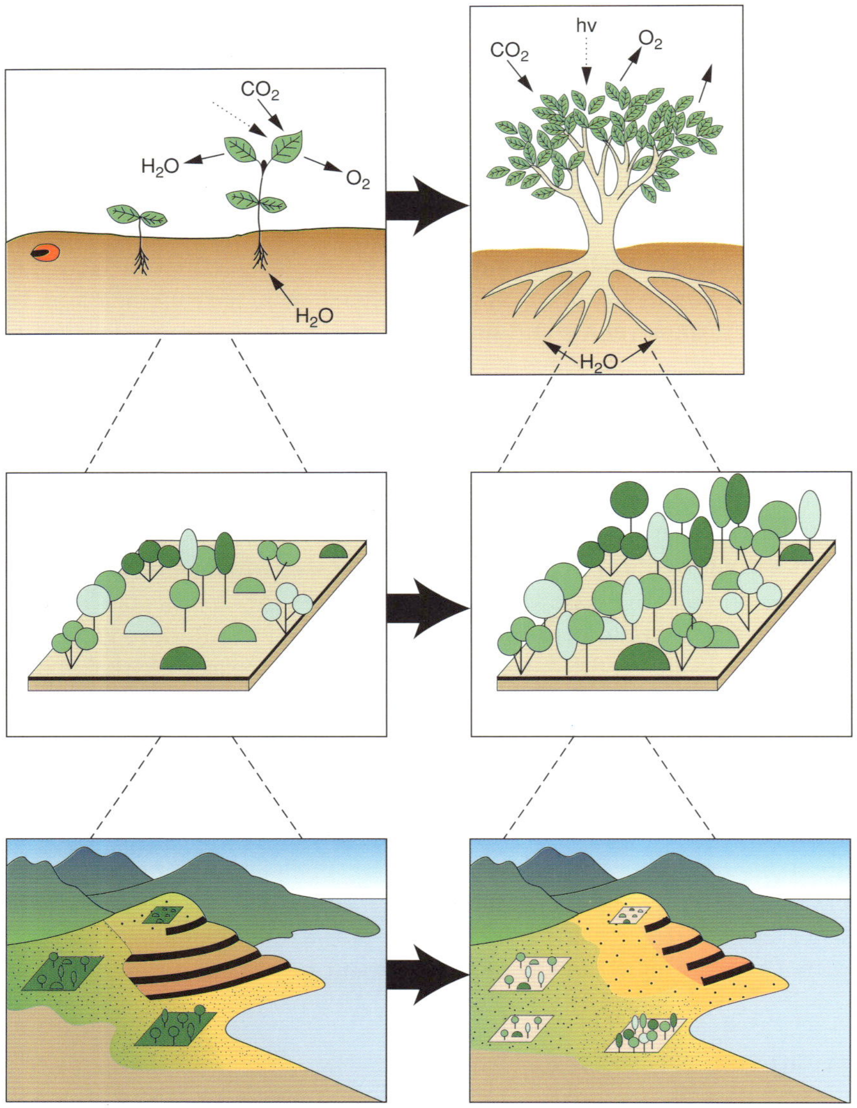

Figure 1.4.6 The MODMED project develops a simulation model for Mediterranean ecosystems spanning several orders of scale, from the behaviour of the individual plants, through the interactions of neighbours at the community level, to the dynamics of land-use units at the landscape level.

Figure 1.4.7 Sampling of post-fire seedlings recruitment in a *Pinus halepensis* forest in Greece. Research work is underway on plant biomass and litter production, architecture, and reproductive biology of species in degrading and aggrading Mediterranean ecosystems.

Campania and the Uccellina Natural Reserve in Tuscany (coastal *macchia* and mixed forests on flysch and limestone substrates); in Portugal, the Serra de Arrabida (*matorral* on limestone soils) and the Serra de Malcata (central interior shrubland on sandstone); in Greece, the Attica (Aleppo pine forests on limestone).

Several species have been selected for detailed ecophysiological analysis (photosynthesis and transpiration behaviour in both field and controlled conditions). The criterion for the selection was the choice of dominant component species characterising the different successional stages at the study sites. This was done to collect sufficient information in order to parametrise the model for simulations of vegetation dynamics and changes of community types. The species list included trees (*Quercus ilex* L., *Q. pubescens* Willd., *Q. faginea* Lam., *Q. rotundifolia* Lam., *Q. coccifera* L., *Ostrya carpinifolia* Scop., *Fraxinus ornus* L., *Olea europaea* L., *Pinus halepensis* Mill., *P. brutia* Ten.), shrubs (*Erica arborea* L., *Phillyrea* L. spp., *Arbutus unedo* L., *Laurus nobilis* L., *Euphorbia dendroides* L., *Myrtus communis* L., *Pistacia lentiscus* L., *Cistus* L. spp.), grasses and herbs (*Festuca drymeja* Mert *et* Koch in Röhling, *F. heterophylla* Lam., *Brachypodium sylvaticum* (Hudson) Beauv., *B. retusum* (Pers.) Beauv., *Ampelodesmos mauritanicus* (Poiret) T. Durand *et* Schinz, *Vicia sativa* L., *Lathyrus cicera* L., *Medicago polymorpha* L., *Hyparrenia hirta* L. Stapf in Oliver).

Experimental works concerned analysis of plant responses to frequency of disturbance (data were analysed from previously established burned and cut permanent quadrats), ecophysiological measurements on light interception, water relations and water use efficiency of available water resources, and gas exchange behaviour of the different species.

The dynamics of land use units at the landscape level has been studied at different sites by searching for evidence of past land cover and use and by remote sensing vegetation mapping and GIS analysis.

The modelling approach

The modelling component of the MODMED project has as its main objective the production of a computer simulation model which can be used to make predictions of successional patterns at both the level of the community (or patch) and at the level of the landscape. By simulating the development of landscapes under a number of different management scenarios, comparative studies will reveal emergent properties of the system which may not be immediately apparent. This will be invaluable for predicting the effects of changes in land use on vegetation, and the effects of changing vegetation structure on biodiversity and conservation value.

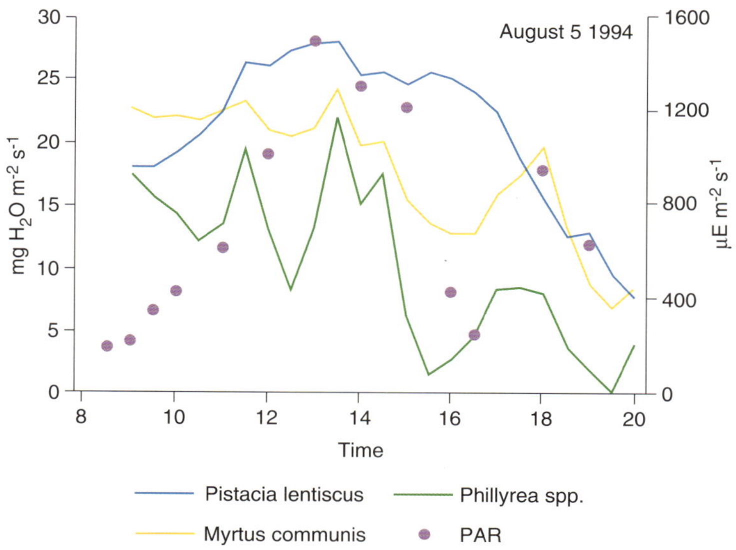

Figure 1.4.8 Field data from the measurements of transpiration for different species and PAR taken at half-hourly intervals during the day. PAR is photosynthetic active radiation. These data are being used for model calibration. The measurements were taken on 5 August 1994 (mean daily air temperature 30.1°C, relative humidity 31%).

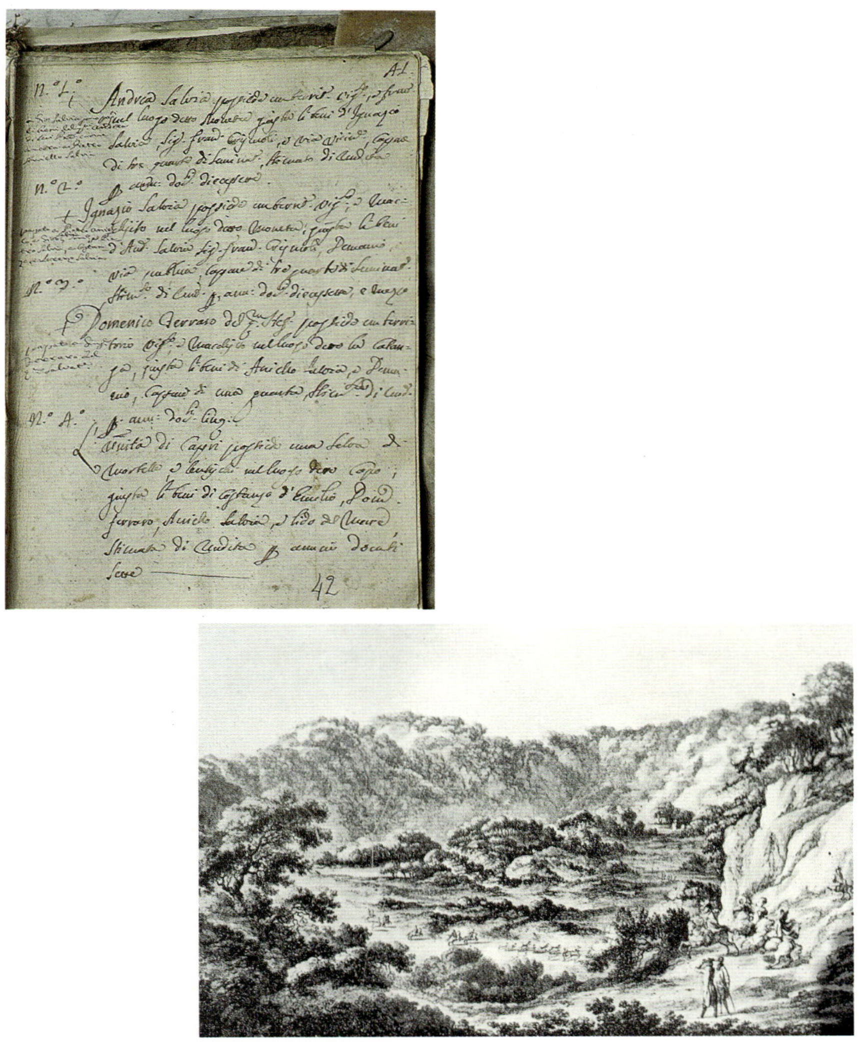

Figure 1.4.9 Archive documents and ancient pictures provide anecdotal and historical evidence of past land uses, and are useful for interpreting present vegetation patterns at the landscape level.

The modelling approach draws on data collected at a wide range of scales, from physiological measurements on individual plants through to long time-scale and large spatial-scale observations from historic records and remote sensing at the landscape level. Data are required for initialisation, parametrisation and validation. Research is also necessary to establish functional relationships between the main state variables within the system. The model will therefore serve as a valuable research tool in providing an integrating framework for planning, executing and co-ordinating data collection and for the representation of functional relationships.

As an approach to producing predictive models, MODMED is exploring the use of hierarchical modular modelling to create a user-friendly modelling environment of potentially wider application in ecology. With the hierarchical modular modelling approach the landscape model may be considered a ‘black box’ using data and making predictions at a single level. However, this box has a modular structure in which each of the modules is itself a model which contains further sub-modules nested within. The concept of hierarchical modular modelling may be illustrated by a diagram.

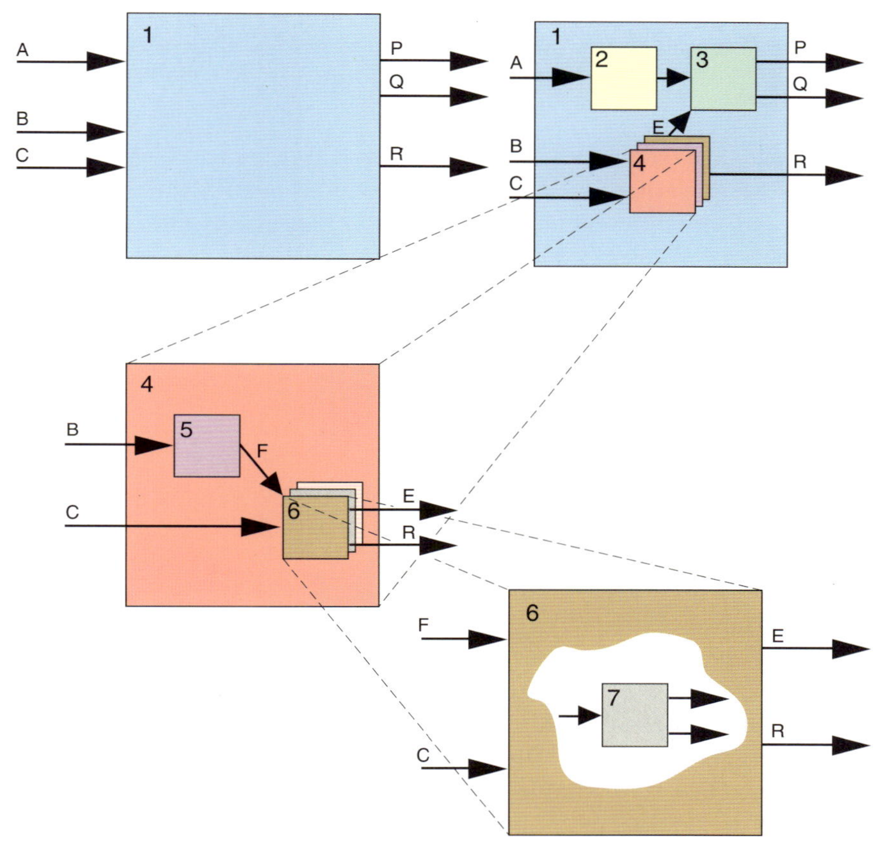

Figure 1.4.10 Schematic diagram to illustrate the concept of hierarchical and modular modelling and the flows of data between modules.

A key feature of the modelling environment is the flexibility to allow the user to specify the structure of the model at run-time. Thus in the full configuration the landscape (represented in a simplified form by box 1 in the diagram) may comprise a number of landscape-level processes (boxes 2 and 3), and a set of communities (box 4), each of which may be represented by a community-level process (box 5) and a spatially explicit population of individuals (box 6). The growth of individuals is then based on established physiological processes of photosynthesis and resource allocation (box 7). Alternatively, the user may choose to represent the community as a population-dynamics model, a

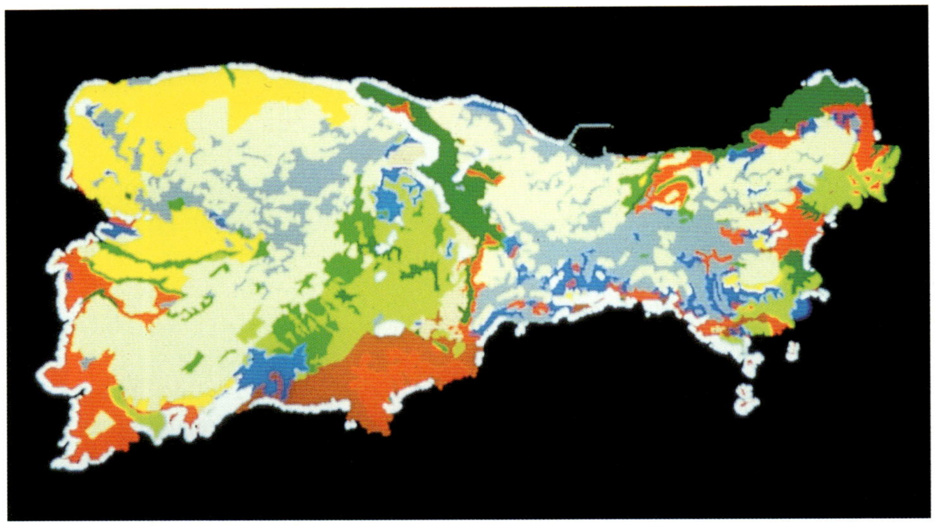

Figure 1.4.11 Vegetation maps of several periods are produced to study the transitions between different vegetation types and land-use units, which can be used for validation of model simulations in similar scenarios. The figure shows a remote sensing and GIS application on the island of Capri study site.

compartment-flow model, or simply as a Markov process in which successional changes are governed by a transition probability matrix (i.e. by replacing boxes 2, 3 and 4 by different modules). The user may also elect to run the community-level module as a complete model in its own right (i.e. by considering one instance of box 4 and its contents in isolation). The appropriate configuration for a particular task will depend on the nature of the problem, the quantity and quality of data available, the precision required and the speed of calculations. Models will be constructed using simple graphical tools within a Windows environment.

The main landscape unit is raster-based and compatible with standard GIS so that the model can be superimposed onto real landscapes. A database contains information on climate, soils, topography and distribution of the landscape units. Landscape units will be susceptible to successional change and disturbance by fire or grazing. Neighbouring patches will interact, for example, through the spread of fire.

In the initial implementation of the model, two community-level modules are being prepared though others may be developed later. The first module is based on a Markov chain with transition probability matrices estimated from actual transitions in vegetation interpreted from historical records and aerial photographs. This very simple descriptive model of succession will be improved by modifying the transition probabilities in response to particular environmental circumstances. The second community-level module comprises a spatially explicit representation of individual plants of a number of species. The micro-environment of individual plants is determined by shading from neighbouring plants and from soil-water availability. Functional modules within the community level include processes such as death and the recruitment of new individuals to the population.

Each individual within the community is represented as a structure with the state variables of crown volume and position, leaf area and a partitioning of resources between different organs. Growth is in response to the micro-environment following established physiological processes. Each species represented within the model is parametrised separately with properties relating to growth, phenology, reproduction, regeneration and response to disturbance.

Finally, the model will have a considerable educational value, both as an ecological tool for demonstrating the interactions between components of the ecosystem and as a modelling environment. The Windows interface will make ecological modelling accessible to a much wider audience than is presently possible and will move the development of models into the hands of those who research in ecology, rather than the minority of computer programmers.

THE ERMES PROGRAMME

C. Boix, A. Calvo-Cases, A. C. Imeson and H. Lavee

The ERMES Project (Erosional Response of Mediterranean Ecosystems) is developing an integrated process-based methodology that can be used to estimate the impact of climate change on ecosystems and land degradation on limestone lithologies under Mediterranean conditions. The ERMES Project mainly studies the physical aspects of land degradation related to climate change and land use. It also contains a socio-economic research component, in order to integrate socio-economic and political information with climatic and biophysical data.

The basic methodology used by ERMES is to undertake research along climatological gradients at sites that form a transect across the Mediterranean. Along this Mediterranean transect, three main research areas were established where transitions from Mediterranean to semi-arid to arid environmental conditions could be studied. These three locations are situated in Judea (Israel), Crete (Greece) and Alicante (Spain). The microclimatic conditions become more arid (increasing temperatures and decreasing precipitation) along the climatological gradients, so that it is possible to evaluate the effect of climate on soil erosion response and land degradation. The selection of the study areas was made taking the climate as the main varying factor within each gradient and along the Mediterranean transect; other factors such as lithology, vegetation and land use were selected to be as similar as possible taking into account the varying histories of human disturbance and policy.

Besides the climate, land-use management influences enormously the behaviour and the evolution of the ecosystems in each area. The Alicante research area is being changed considerably by the effects of fire, land abandonment, agricultural policy and tourism. In Crete the past and present-day history of grazing is the main factor causing land degradation. Israel introduces real desert conditions into the study transect and the behaviour of the ecosystem is also highly affected by the effects of grazing.

Figure 1.4.12 Landcape and grazing in Mishor Adumin (260 mm mean annual rainfall), mid-level site along the Israeli gradient.

Methodology

After the selection of the main study areas along the Mediterranean transect, several locations in each area were selected in order to define local climatological gradients. Three locations were selected in Alicante, with precipitation ranging from 800 to 350 mm; four locations were selected in Crete, with precipitation ranging from 1400 to 400 mm; and seven locations were selected in Judea, with precipitation ranging from 620 to 80 mm.

The ERMES approach to studying desertification and land degradation considers the hierarchical organisation of active processes operating at the study zones, such as trampling and soil compacting due to grazing, consequences of forest fires and aggregation due to biological activity. Thus, the identification of key processes and land degradation indicators was one of the first criteria in the design of the experimental field work. The study of these processes and indicators is being developed at different spatial (i.e. patch, slope, transect) and temporal (i.e. monthly, seasonally, yearly) scales.

In Alicante and Israel, meteorological stations have been set up at each research site to establish the climatological gradients and to provide long-term monitoring.

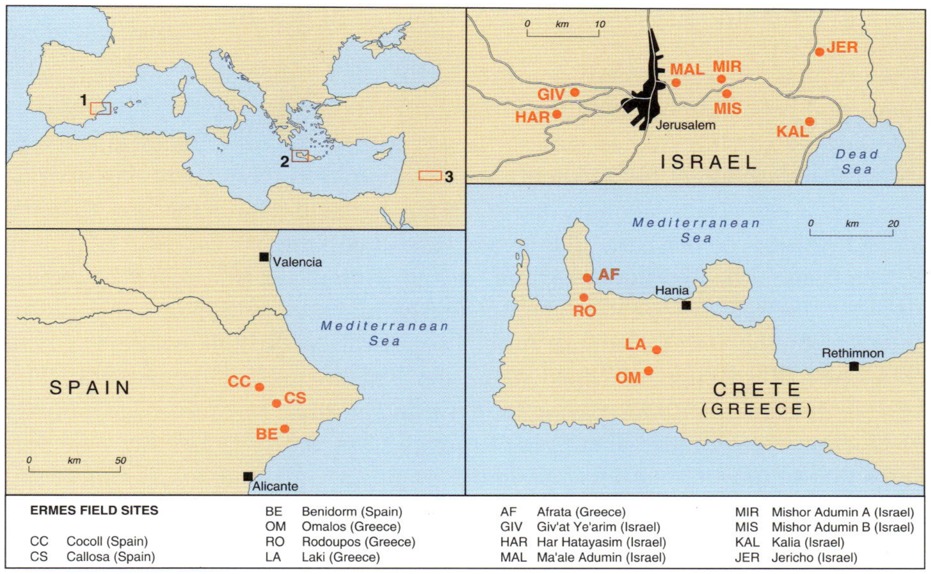

Figure 1.4.13 Location of the ERMES field sites along a Mediterranean transect.

For each site microclimatic conditions (rainfall, air and soil temperatures and soil moisture) have been observed and soil physical as well as soil chemical properties have been investigated in reference to soil profiles. Rainfall simulations have yielded hydrological and soil erosional responses and dynamic soil behaviour, such as infiltration. Vegetation characteristics and other surface properties, such as stoniness, are regularly measured. Hydrological response has been evaluated under both wet and dry soil conditions, as well as changes in organic matter and macro-aggregation. Soil moisture and micro-aggregation have been measured on a monthly basis.

Indicators of soil degradation

The infiltration and aggregation properties of the soil are being used to develop key indicators of soil degradation because the proportion of aggregated particles present in the soil is assumed to reflect the balance between processes of aggregation and processes of destruction. Higher values of micro-aggregated particles have been found when the soil structure becomes better developed. High values of micro-aggregation are also associated with higher values of clay content, biological activity, moderate soil temperatures and water availability, and under plants rather than in bare patches (Boix *et al.*, 1994; Calvo-Cases *et al.*, 1994).

Figure 1.4.14 Benidorm south-facing slope (350 mm mean annual rainfall). The lowest site along the Spanish gradient.

Figure 1.4.15 Setting up a plot to map the vegetation pattern in Rodoupos (350 mm mean annual rainfall), mid-level site along the Cretan gradient.

Infiltration rates vary according to the specific conditions found along each of the local gradients. Rainfall simulation experiments in Crete showed the highest erosion rates to be at the highest and most humid location (Omalos). The opposite occurs in Alicante and Judea where the highest erosion rates were recorded at the lowest and most arid locations. This is because in Crete the grazing pressure is very high in the most humid site due to the high biomass production. In Alicante, the presence of a crust on the soil surface produces the highest erosion rate in the most arid site under very dry soil conditions

Figure 1.4.16 Meteorological station in Callosa (50 mm mean annual rainfall), mid-level site along the Spanish gradient.

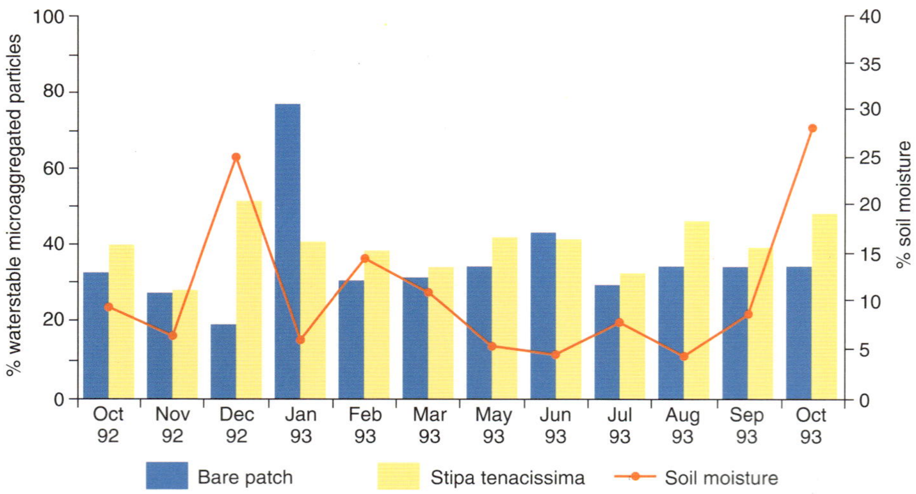

Figure 1.4.17 Evolution of soil moisture and water-stable micro-aggregation in a bare patch and in a vegetated patch on a semi-arid hillslope near Benidorm, Spain. (Data from April 1993 were not available.)

(Benidorm). However, under wet soil conditions the intermediate site in Alicante, having soils with a high clay and stone content, shows the highest erosion rates. In Israel, when the precipitation is less than 300 mm per year, the erosion rates in the most arid site are very high, partly due to the effects of soil salinity.

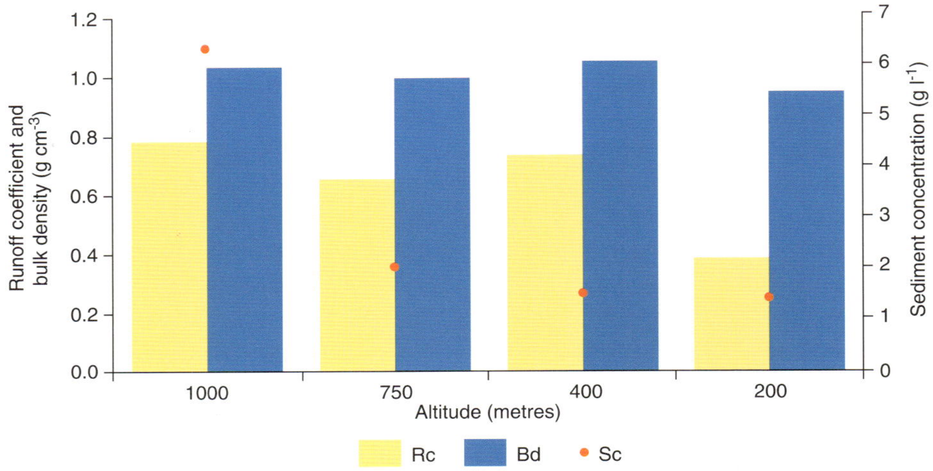

Figure 1.4.18 Relationship of the runoff coefficient (Rc), bulk density (Bd) and sediment concentration with the altitude along the investigated gradient in Crete (Greece).

The erosional response of climatological gradients

The gradients chosen as study areas in the ERMES Project have different environmental behaviours. Although climate provides the driving force for most of the processes being studied (Imeson *et al*., 1994), the differences in the erosional response along the gradients are due mainly to the present-day ecosystem processes. These include those driven by land-use change so that the history of land use in a particular area is very important.

The transect in Israel demonstrates the position of the true desert boundary. In Israel, the risk of land degradation clearly increases with increasing aridity. In Alicante the infiltration rates are, in general, very high, while the high frequency of wildfires is found to be the major cause of degradation. Fire severity increases with increasing biomass and altitude. In Crete, high biomass production means more grazing and trampling of the soil, forming bare surfaces that in moist conditions swell and produce large amounts of runoff and sediment.

Figure 1.4.19 Landscape affected by forest fires in Cocoll (850 mm mean annual rainfall), the highest site along the Spanish gradient.

THE DM2E PROJECT: ASSISTING THE MANAGEMENT OF DEGRADED MEDITERRANEAN AREAS BY RUNOFF AND EROSION MODELLING ON REPRESENTATIVE CATCHMENTS

J. Lavabre

The DM2E project is still in progress and the data are still being analysed so this report is of a limited and partial character. However, the results given are representative of the approach to the processes of desertification in Mediterranean environments which has been defined in the body of this research. The references at the end of the book give fuller details of the issues discussed here.

Objectives

The general objective of the project is to evaluate the risks of perturbations to the hydrological cycle and soil erosion processes created by changes of use of Mediterranean coastal zones. The project links a historical evaluation of erosion and rainfall, with a detailed analysis of contemporary processes. It aims to understand the role of the vegetation cover (or its absence) on the hydrological cycle, flood generation and the processes of soil erosion. The integration of this information in modelling provides predictive capability in relation to the evaluation of desertification in the area under investigation.

The research basins are located on the frontier between France and Spain and in Spanish Cataluñia. Altogether 19 basins, ranging in size from 1.5 ha to 150 km^2, have been studied.

The annual evapotranspiration requirement is of the order of 1000–1200 mm. Average annual rainfall is of this order of magnitude for some of the catchments of the Réal Collobrier, but less on the experimental catchments of Prades (about 550 mm). These data are not *a priori* of concern, but the specific characteristics of the Mediterranean climate (severe summer droughts, strong inter-annual fluctuations) lead to a precarious situation for the use of surface water resources and repeated hydrological stress on the vegetation.

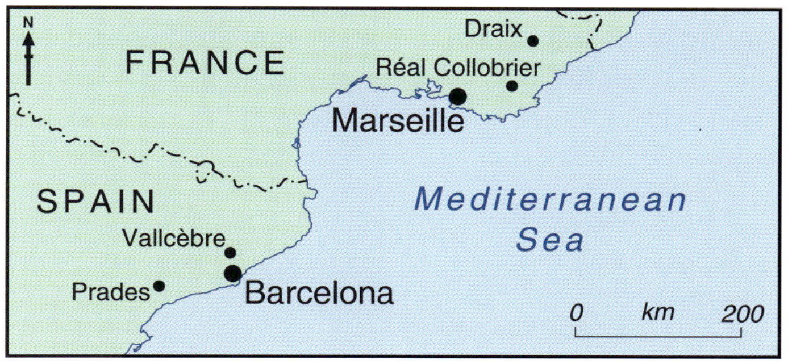

Figure 1.4.20 Location of catchments studied.

Contemporary rainfall: trend or climatic risk?

The examination of annual rainfall over long periods at different French (Perpignan, Montpellier, Marseille) and Spanish (Murcia, Alicante, Barcelona) climatic stations does not substantiate a trend towards drought. However, for the French stations and on the basis of risk assessments of annual rainfall series, the probability of a decade as dry as 1981–1990 is very weak (less than 1 in 1000). The examination of monthly rainfall averages for Marseille indicates that the rainfall deficit of these latter years is a consequence of very dry winter months. The average for 1988–1992 (November–February) is 97 mm compared with 226 mm for the period 1864–1987. The plot has two very distinct maxima in autumn and spring which accentuate the destructiveness of the Mediterranean climate in this region. The perception of the contemporary state of drought is all the more evident because it follows an exceptionally wet period (the averages for the decades 1970–1979 and 1980–1989 are 646 and 454 mm respectively).

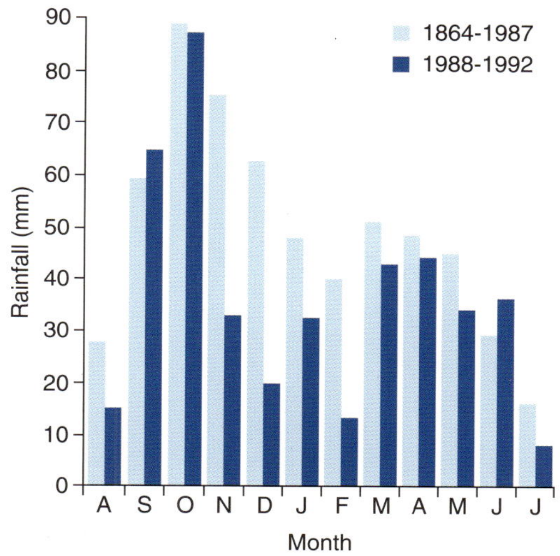

Figure 1.4.21 Average annual rainfall at Marseille.

Catchment runoff

The variations of discharge as a result of rainfall show very clearly and the relation:

$$LA = 7 \cdot \left[\frac{PA - 300}{100}\right]^2$$

explains 80% of the average flow variability. LA is the annual flow and PA the annual average rainfall; 300 mm is the minimum value when flow can be guaranteed in these catchments. This limit is also used as the rainfall limit of the desert in the Sahel.

The general relationship:

$$LA = A \cdot \left[\frac{PA - 300}{100}\right]^2$$

represents the annual flows as a function of annual rainfalls for the catchments investigated through time. For example, the basin of Vaubarnier (Réal Collobrier) with an area of 150 ha has a value of *A* (the slope of the regression line) of 6.4. The strong inter-annual variability in rainfall and the low flows of the last few years are especially notable. The years 1988, 1989 and 1991 were particularly low. Though 1990 and 1992 had more rainfall, they were affected by the earlier very dry years, so that even in a small catchment the effects of drought can persist through two or even three hydrological cycles.

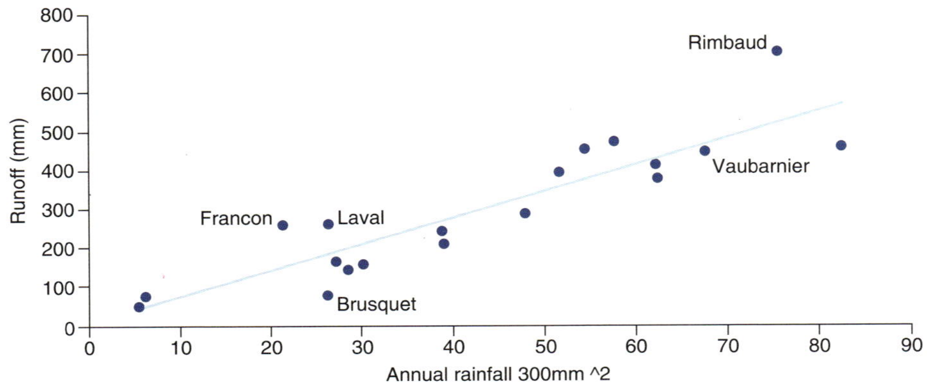

Figure 1.4.22 Average annual runoff plotted against annual average rainfall.

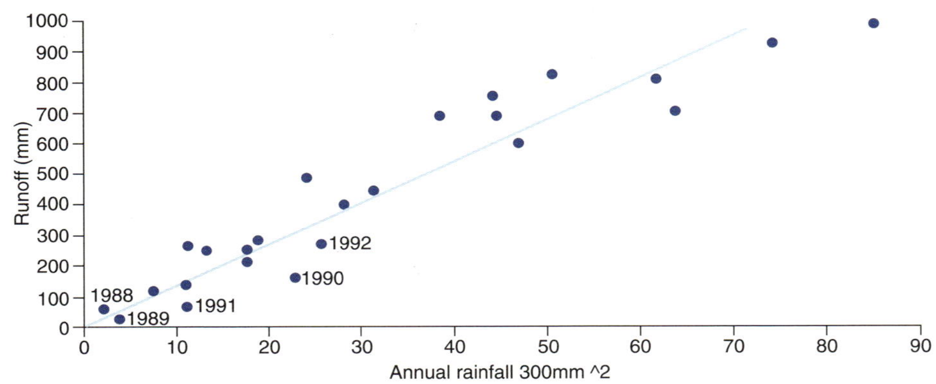

Figure 1.4.23 Vaubarnier Basin: variations in annual runoff against annual rainfall.

The role of vegetation

The role of vegetation is not yet fully worked out but the impact of interception of rainfall on the canopy and the differences between different catchments in the Draix area give some idea of the work being undertaken in DM2E.

About 25% of the rainfall is intercepted by the vegetation but stemflow is negligible in these species. These results mask a great spatial variability, even though over time the values are fairly homogenous. The structure of the trees and the density of the leaf canopy above the collectors control the throughfall, so the same collectors are always under or over the mean value.

Table 1.4.1 Four measurement sites have been instrumented to examine throughfall and stemflow.

Site	Dominant vegetation	Rainfall (mm)	Throughfall (%)	Stemflow (%)
Réal Collobrier Site Boussicaut	Arbutus, cork oak	1094	63	?
Réal Collobrier Site Pounches	Arbutus, cork oak	1200	74	4–9
Vallcebre	Scots pine	845	72	2
Prades	Holm oak	570	74	14–10

Examining the data storm by storm does not suggest an upper limit to interception as rainfall increases. A general expression of the type $I = aP^b$ can be used to describe interception (I) in terms of rainfall P with a between 0.85 and 0.95 and b between 0.6 and 0.7. This accounts for more than 90% of the variance, and can be slightly improved by taking account of the leaf area index (LAI) using $b = 0.6\ (\text{LAI}-1)^{0.2}$ for LAI between 1 and 3.

Comparison of flow in the Draix catchments

Although the structure of the observations did not concentrate on the measurement of the low flows, it is quite remarkable to discover that the difference in annual average flows between the wooded Brusquet catchment (13% deforested) and the Laval and Francon basins (78% and 60% deforested respectively) is in the ratio 1:3 for rainfall and soil conditions that are more or less comparable.

In these same catchments the peak discharges and the coefficient of flow are quite informative. The forest reduces peak discharge by 80%. This is because the forest increases the initial relation before the start of the flood and increases the maximum infiltration capacity. In sediment terms, the Brusquet Basin produces hardly any, while the deforested basins reach 150 ton ha^{-1} year^{-1} on average.

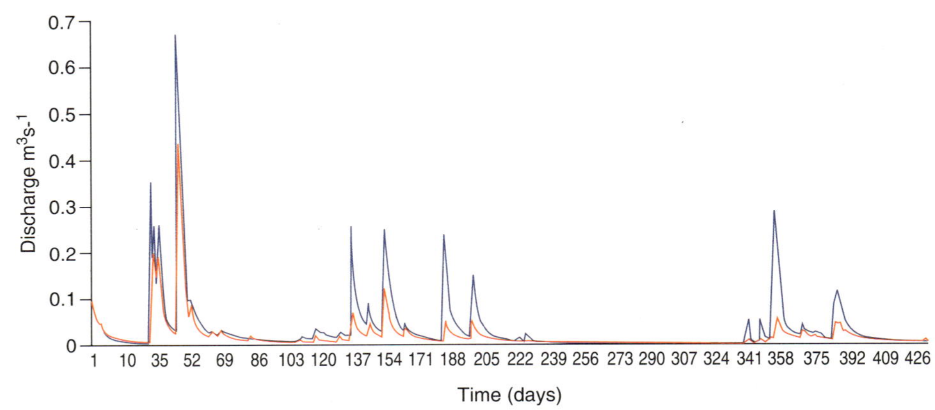

Figure 1.4.24 Modelled pre-fire (red) and observed post-fire (blue) flow conditions for the Rimbaud Basin.

The impact of forest fire: the Rimbaud catchment

In August 1990 the Réal Collobrier catchment was partially burned, destroying 85% of the Rimbaud catchment. This catchment has been under observation since 1968. The impact was an increase of flow in the order of 25% in the first post-fire hydrological cycle, most of it produced in peak flows.

Moreover, the peak storm flows after the fire are higher for a given amount of rainfall in storms of greater than 4 hours duration. Although only estimated for three years, the post-fire peak flows, represented by six events, are all higher than the five-year average peak flows before the fire. The sediment yields, negligible before the fire, are estimated at 6 ton ha^{-1} year^{-1} in the first year after the fire.

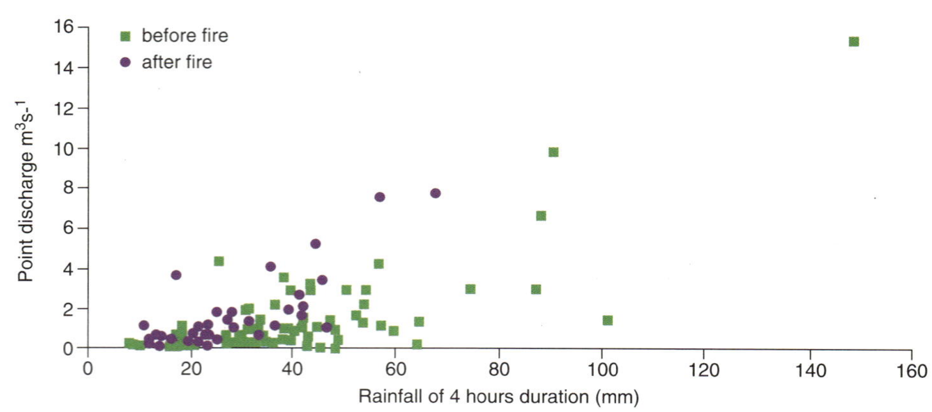

Figure 1.4.25 Discharge at the Rimbaud gauging station in relation to rainfall per 4-hour period.

Conclusions

From the investigations of flows from 20 catchments in the French and Spanish Mediterranean zones, it appears that the surface flow is strongly controlled by climate. Since 1980 the Mediterranean Midi has been experiencing a period of exceptional drought, the probability of which is estimated at 1 in 1000, even though a trend of increasing dryness cannot be demonstrated. As a result, a series of forest fires have affected a number of zones, modifying catchment hydrology in the burnt-over areas. To a first approximation, the increase in flow after a fire is equal to the interception provided by the canopy that has been lost. However, the forest does more than this; it protects the soil and redistributes the rainfall, encouraging vertical rather than lateral movement and further soil erosion.

Figure 1.4.26 Measurement of throughfall below the tree canopy. The throughfall caught by the troughs is stored in reservoirs. In the foreground stemflow is being measured.

Figure 1.4.27 Discharge station in the Rimbaud catchment in the flood of 27 September 1992. The flow is about 8 $m^3 s^{-1}$.

Without being alarmist, even though the phenomena described may be irreversible, it really does seem that desertification is occurring in these areas. Unfavourable climate leads to drought, forest fires and loss of soil and ultimately to desertification.

Figure 1.4.28 Draix catchment. Erosion measuring station for the 86 ha Laval subcatchment. The trench, which contains about 1000 m^3 of sediment, is cleared on average seven times each year.

ARIDIFICATION AND LAND DEGRADATION IN HUNGARY, MEDEAST: THE MEDALUS PROGRAMME IN EASTERN-CENTRAL EUROPE

Á. Kertész, T. Huszár, D. Lóczy, I. Mattyasovszky, J. Mika, K. Molnár, J. Szalay and L. Szalai

Introduction

Aridification, as a consequence of climate change and resulting in land degradation, is a threat not only for the typical Mediterranean region, but also for the neighbouring regions. Hungary, lying in the middle of the Carpathian Basin, is part of one such a region where major transitions occur between Atlantic, continental and Mediterranean climates. However, the last two regimes are the most influential and aridification is becoming an increasing threat for both semi-natural and agro-ecosystems.

The questions arise as to whether semi-aridity or even desert-like conditions are possible scenarios for the future and what would be the consequences of drier climatic conditions on both physical processes (e.g. soil formation and vegetation dynamics) and socio-economic processes.

Within the framework of MEDALUS II, recent and past climates in the Carpathian Basin have been investigated and in particular changes, in precipitation and temperature during the twentieth century. The lowering of the groundwater table in the Danube–Tisza Interfluve has also been studied in relation to the impacts of climatic change on water availability and land use.

Increased soil erosion seems to be a major consequence of changes in the climate regime, particularly in hilly areas, which comprise 25% of the total area of the country, and where approximately 2 287 000 ha are affected by soil erosion. In the lowlands drought is the main problem.

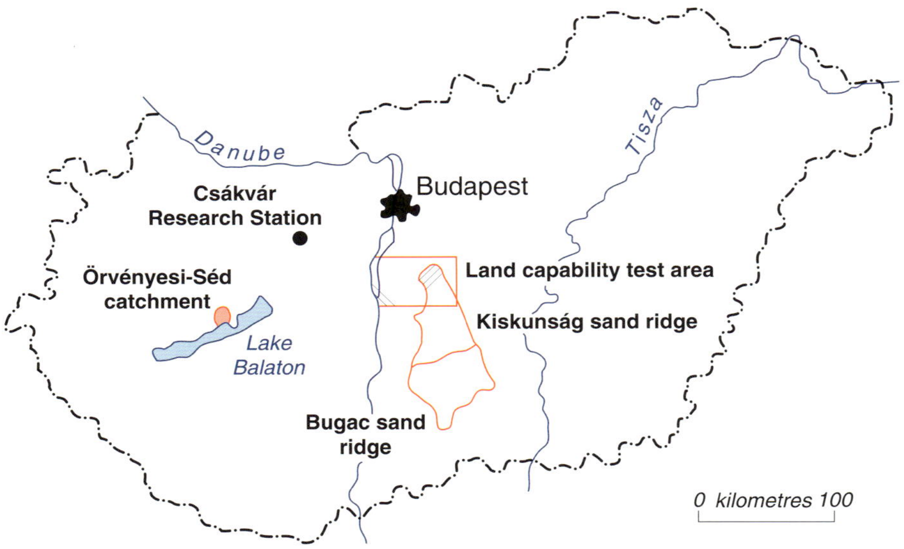

Figure 1.4.29 Location of the study areas of the MEDEAST project in Hungary.

On the hillslopes the dominant lithology is loess, which is very prone to soil erosion. The agricultural practices on the hillslopes also favour soil erosion as the soil remains without vegetation cover for several months during the year. Therefore an increased frequency of high-intensity rainfalls, which is thought to be one of the most important consequences of climate changes in this area, is likely to represent a control factor in increasing soil erosion risk.

The area selected for the study of the effects of climate change on the plain is the Danube–Tisza Interfluve. This is one of the most continental regions of Hungary, with high aridity and soils that locally have low water retention capacity, and therefore represent well the conditions and the problems of the lowlands.

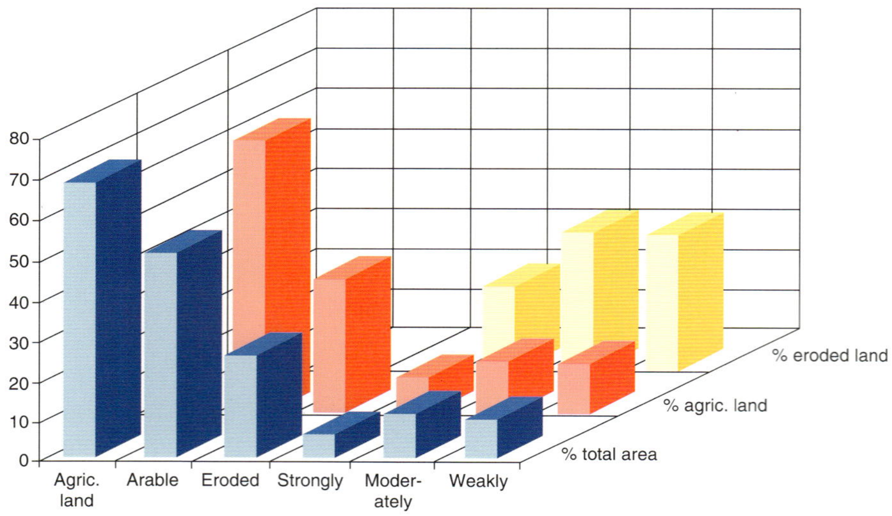

Figure 1.4.30 Importance of agricultural land and erosion in Hungary. Erosion endangers a major natural resource, fertile soil. More than one-third of the agricultural land (2.3 million ha) is affected by fluvial erosion (13.2% weakly, 13.6% moderately and 8.5% strongly). In addition, 1.5 million ha of land are affected by wind erosion. Weakly eroded = loss of 0–30% of the original soil profile; moderately eroded = loss of 30–70% of the original soil profile; strongly eroded = loss of >70% of the original soil profile (Stefanovits and Várallyay, 1992).

Figure 1.4.31 Erosive impact of heavy rainfall on arable land without vegetation in August.

Climate

The recognition that changes in temperature and precipitation could be expected in the short as well as in the long term, provides a new perspective on landscape evaluation. These changes considerably affect the functioning of the landscape system, and global change has local and regional significance. For example, global changes of temperature determine the heat and water budget of each landscape, thereby producing fundamental impacts on agriculture. At the regional scale, the climate of Hungary is characterised by significant inter-annual variability of both the duration of the arid period and the monthly distribution of rainfall.

Linear trends in the annual mean temperature series have been analysed at seven stations for the period 1881–1990, after applying the correction of inhomogeneity suggested by Szentimrey (1994a). After this correction, all series exhibit similar warming trends that are significant at the 95% level and have values of 0.10–0.1°C per decade.

A comparison between the average values recorded in two consecutive periods (1900–1949 and 1950–1989) in 17 meteorological stations, reveals that changes have also occurred in the monthly distribution of annual precipitation. The second period is characterised by a decrease in the average rainfall in spring and autumn months. The largest decreases are those of April, October and September, which show average decreases of 7, 17 and 11 mm respectively, over the 17 stations. For June, conversely, an average increase of the average monthly rainfall is observed (10 mm).

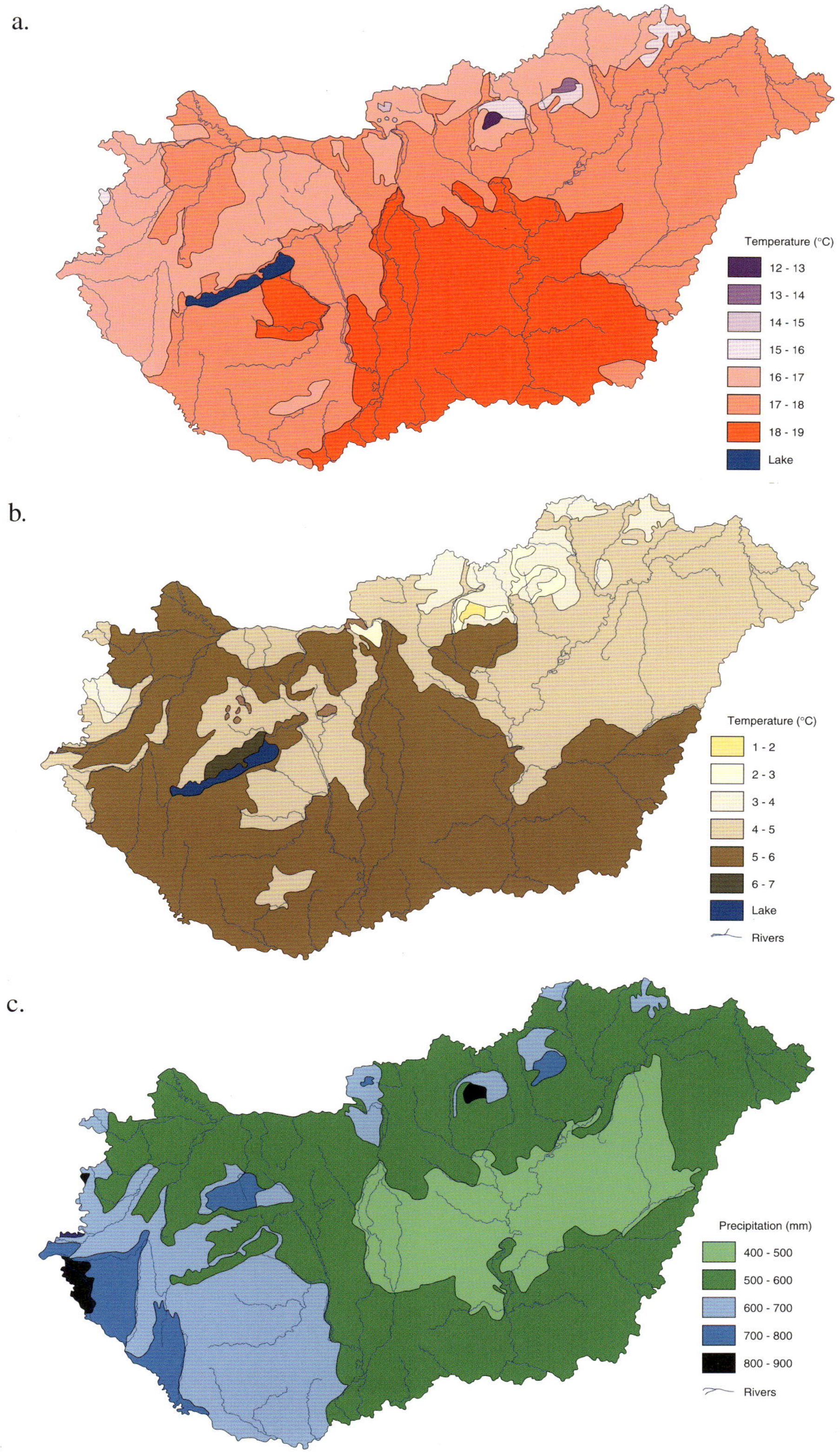

Figure 1.4.32 Assumed temperature and precipitation values under conditons of 1°C rise in hemispherical mean temperature (Mika, 1988, 1991, 1994; Molnár, 1994b). Geographical distributions of the climatic elements in the future are derived by adding values from regional scenarios to the present conditions in the 230 landscape units. The regional climate scenario for 1°C expected hemispherical warming is derived from Mika (1994). Values are mainly taken from different palaeoclimatic reconstructions of the Holocene optimum, taking into account the more detailed scenario for the 0.5°C changes. The latter are based on the statistical relationship between the hemispherical and regional climatic characteristics (but distributed in the country unevenly, which is why the 1°C case has been chosen). (a) Summer half-year (scale 1:2 500 000); (b) winter half-year (scale 1:2 500 000); (c) precipitation (scale 1:2 500 000).

The Örvényesi-Séd catchment

Due to the increasing importance of human activities (mainly related to leisure and recreation) in the surroundings of Lake Balaton, a small watershed, the Örvényesi-Séd catchment (24 km^2), was chosen to repre-

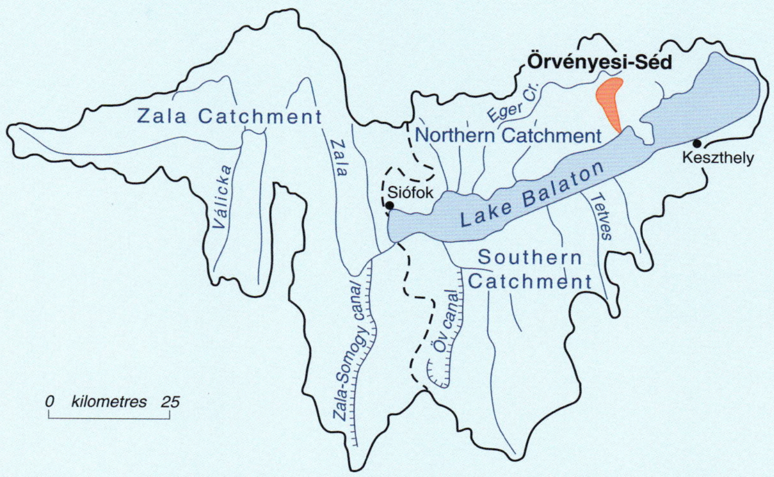

Figure 1.4.33 The Lake Balaton system (55 775 km^2) comprises three subcatchments of different physical conditions. The Zala River subcatchment, the largest of the three (2622 km^2), became part of the Balaton System through a river capture.

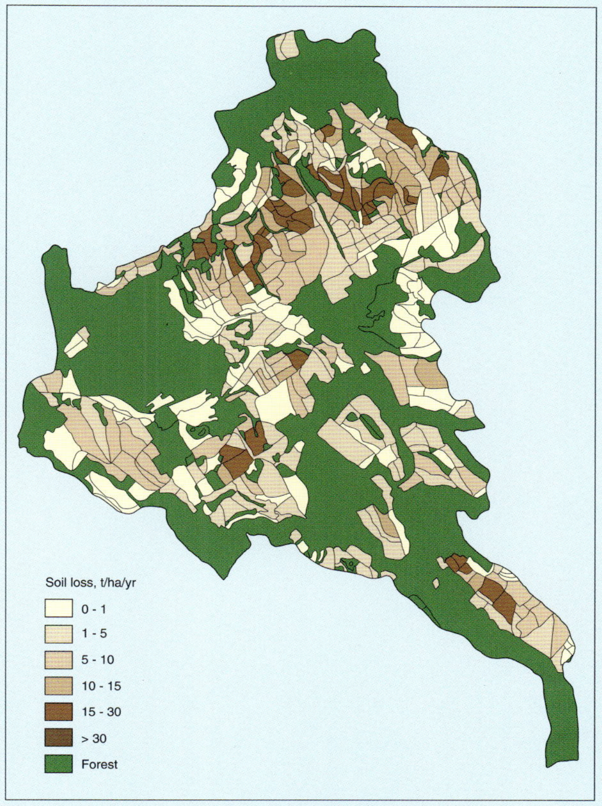

Figure 1.4.34 Assessment of soil erosion in Örvényesi Séd. The calculation was carried out for small topographic units, called 'erotopes'. An erotope is defined by Richter (Kertész *et al.*, 1995) as a unit of land with the same runoff direction and without concentrated flow. Erotopes can be identified using topographic maps (identification of runoff directions). Microrelief and landform maps are also needed because roads, footpaths, ditches and terraces represent the borders of the erotopes. Forests, settlements and flat valley bottoms are not taken into consideration. The soil loss was estimated in tonnes per hectare per year for each erotope.

sent conditions in the northern subcatchment among the three forming the Lake Balaton system (5775 km^2). This research started in 1989 within the framework of a co-operation between Professor Richter of the University of Trier (Germany) and the Geographical Research Institute, financed by the German Research Foundation (Deutsche Forschungsgemeinschaft), the Hungarian Scientific Research Foundation and the Hungarian Academy of Sciences.

Methods

The Universal Soil Loss Equation (USLE) was applied for the assessment of soil loss in the Örvényesi-Séd catchment.

The geographical information system software ARC/INFO was used for spatial data management and manipulation. Runoff directions and slope angles were calculated from the Digital Elevation Model (DEM) so that small units of land (*erotopes*) for the assessment of soil erosion could be identified and mapped. An *erotope* is defined as the smallest mappable unit of land (at the operational scale) with approximately the same runoff direction and without channels or rills (Richter, 1995). The boundaries of erotopes are lines of divergent or convergent runoff direction as well as linear structures such as ditches, streams, roads, field paths and terraces.

Results

For each *erotope* the soil loss was calculated so that erotopes could be assigned to six soil loss classes ranging from 0 to 30 t/ha per year. Most of the erotopes fall within the subrange 0–10 t/ha per year. Soil loss for forested land was not calculated.

Water discharge and sediment load of the Örvényesi-Séd catchment were also measured weekly in Örvényes, near the lakeshore, at the outlet of the stream.

By comparing soil loss calculated for the whole catchment with measurement data for Örvényes, it appears that only 2% (approximately) of the calculated soil loss actually leaves the catchment.

Table 1.4.2 Monthly data series (from 1880 to 1992) of air temperature were statistically analysed for 17 locations in Hungary. Summer temperature strongly decreases, while the winter exhibits statistically non-significant warming. The table shows the trends of the average monthly temperature of July for some meteorological stations in the Great Hungarian Plain.

Location	Rate of change (°C/10 years)
Debrecen	0.11
Kalocsa	0.11
Kecskemét	0.11
Nyíregyháza	0.11
Szarvas	0.10
Szeged	0.10
Túrkeve	0.10

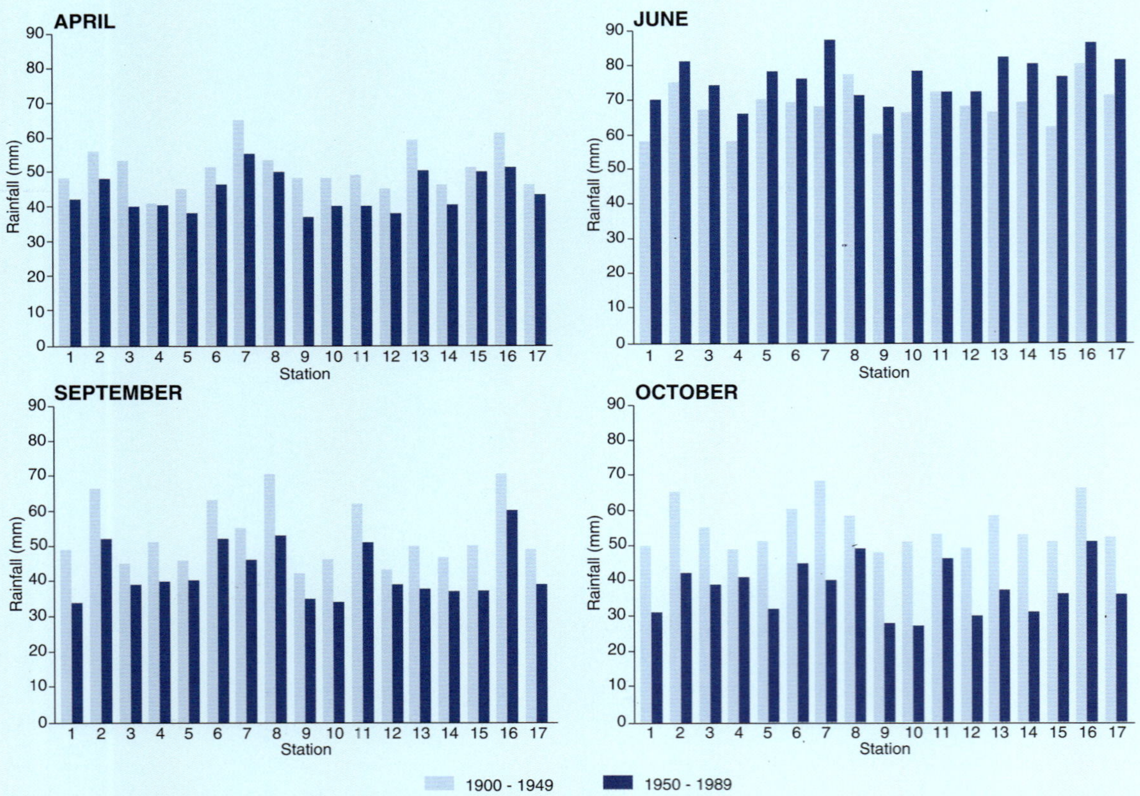

Figure 1.4.35 Precipitation for April, September, October and June (Molnár, 1994b).

Groundwater level changes

Summer drought in Hungary is a typical phenomenon associated with the continental climate regime. In the sand regions of the country in particular, such as the Danube–Tisza Interfluve, low soil water retention capacity and low soil capillarity, mean that water availability is the limiting factor of land capability for agriculture.

One of the areas most sensitive to aridification is the Kiskunság blown-sand region in the Danube–Tisza Interfluve. This is a vast Pleistocene alluvial fan of the Danube where strips of blown-sand dunes alternate with loess-mantled surfaces. Such differences in soil parent materials will be of increasing significance in water storage.

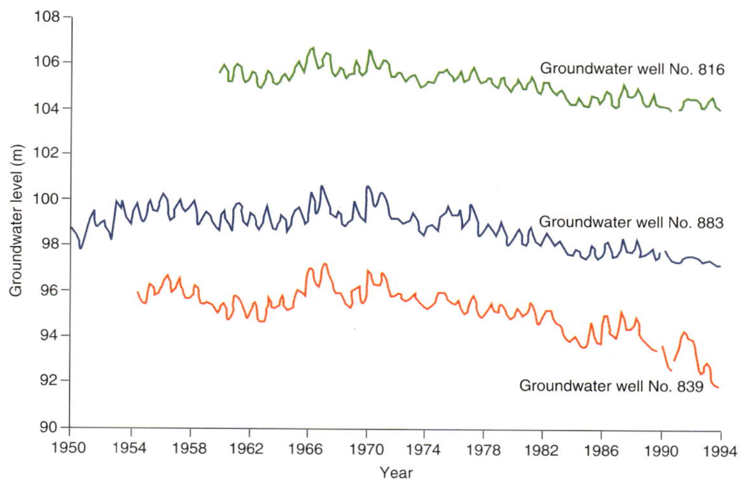

Figure 1.4.36 Groundwater level changes from 1950 to 1994 in the Kiskunság Sand.

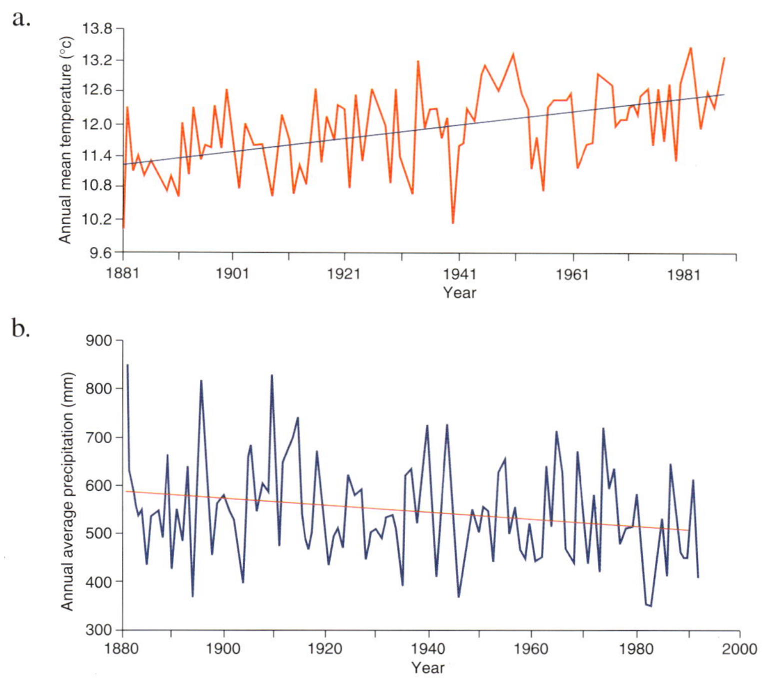

Figure 1.4.37 Temperature and precipitation changes from 1881 to 1990 (Molnár, 1994a) in the Kiskunság Sand region (Meteorological station of Kecskemét, 46°56'N, 19°37'E, 128 m a.s.l.). Rate of change, *m*, for (a) the mean annual temperature and (b) the annual amount of precipitation at Kecskemet (NE), according to the linear trend equation $Y = Y_o + m(t - t_o)$, where *Y* is the meteorological element in time *t*, and Y_o its value at time t_o. The changes are 0.0106°C/year and -0.86 mm/year, respectively.

Regular groundwater level monitoring started in this area in the early 1950s (Liebe, 1994; Szalay, 1994). The data series from wells in various geomorphological locations (e.g. tops of sand ridges, loess-mantled dunes, valley floors) indicate very similar trends for groundwater level changes confirming, incidentally, that the Danube–Tisza Interfluve comprises a distinct regional hydrological system.

The long-term temperature and precipitation curves for the meteorological station of Kecskemét (in the central part of the interfluve), show a clear trend towards aridification. Annual precipitation after 1974 fell below the long-term (1890–1960) annual average of 620 mm; of which 530–570 mm fell during the growing season so that by 1976 the decrease of groundwater level had started. This has been mainly caused by highly irregular winter precipitation producing virtually no infiltration.

The decrease of groundwater levels, however, results from a combination of other factors as well as rainfall. These include the following:

- increased evaporation due to rising summer temperatures,
- increased water extraction from wells,
- reduced groundwater recharge from neighbouring areas,
- afforestation, and
- high water demand for agriculture.

Among these, water demand for agriculture is one of the most important. This is because the central sand region is 20–30 km from the main rivers and 20–30 m above them, so irrigation is only feasible from groundwater.

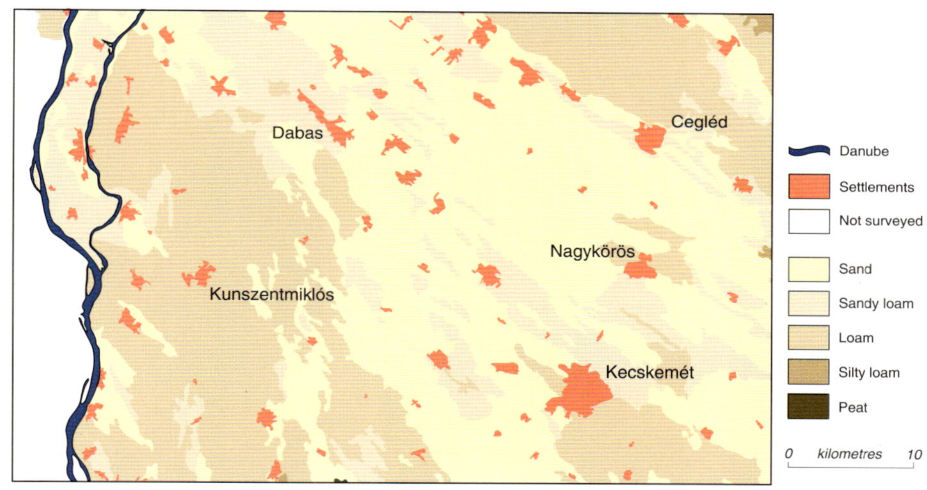

Figure 1.4.38 Soil parent material map for a land capability test area (Lóczy and Szalai, 1993).

Water availability and land capability

In the past (1959–1969), despite the high continentality of the climate and seasonal fluctuations of its levels, the groundwater table was close enough to the surface to supply the root zone of not only the permanent crops (fruit trees and vine, occupying 12% of agricultural land) but also the characteristic arable crops of the region (rye, winter barley, maize and lucerne).

In a recent land capability survey (Lóczy and Szalai, 1993) land capability is assessed using the ARC/INFO GIS, by ranking environmental conditions according to water demands of the most widely grown arable crops in Hungary. The varying sensitivity to water stress of the individual crops, related to either a single parameter or more often to a combination of them, was built into the analysis in the form of weighting.

The findings of the survey highlight the differences in land quality between alluvial soils of the Danubian Plain, the Pleistocene sandy alluvial fan with blown-sand and humic-sandy soils and the loess-mantled surfaces with calcareous chernozems.

When the average groundwater levels of the period 1959–1969 are compared with those of the drought year 1993, decreases of between 2 and 4 m can be observed for most of the geomorphological units of the test area. Dunes and cover sand surfaces are particularly severely affected.

In order to predict the agro-ecological impacts of reduced availability of water, the land evaluation project was re-run for the study area using the same capability indices and replacing the original data set with that for 1993. In relation to soil parent materials, the land suitable for maize appears to be considerably reduced according to soil parent material. In areas with scores as low as 0 or 1, crop cultivation is not feasible without regular irrigation. Such areas are rather extensive over the central sand ridge, the Kiskunság.

If the present trend towards aridification continues, the current pattern of land use cannot be maintained. Crops with a particularly high water demand should be replaced with others (or with drought-tolerant hybrids, as in the case of maize) and the validity of afforestation programmes (aiming at the stabilisation of the movement of sand dunes) should be reconsidered in sand regions.

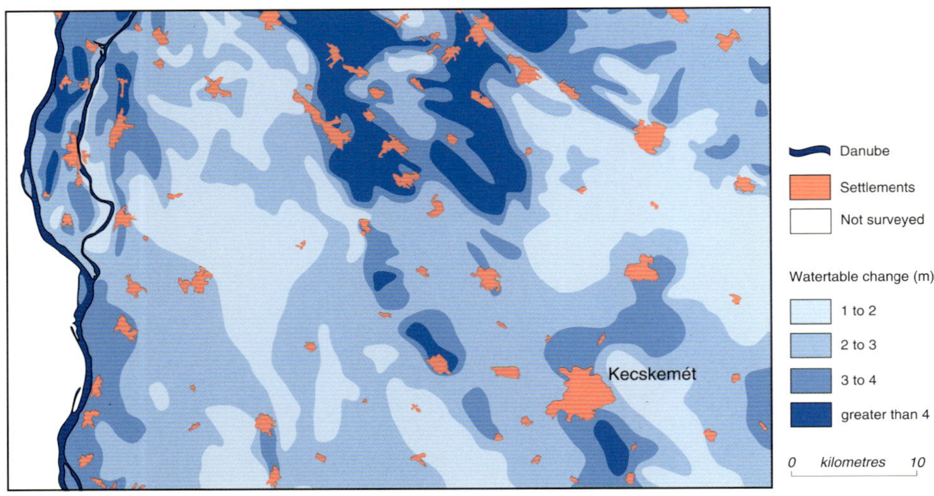

Figure 1.4.39 Average groundwater table (1959–1969) in the land capability test area (Lóczy and Szalai, 1993; Szalay, 1994).

Figure 1.4.41 Classified basic suitability map for maize combining average groundwater table (1959–1969) and soil parent materials in the land capability test area (Lóczy and Szalai, 1993).

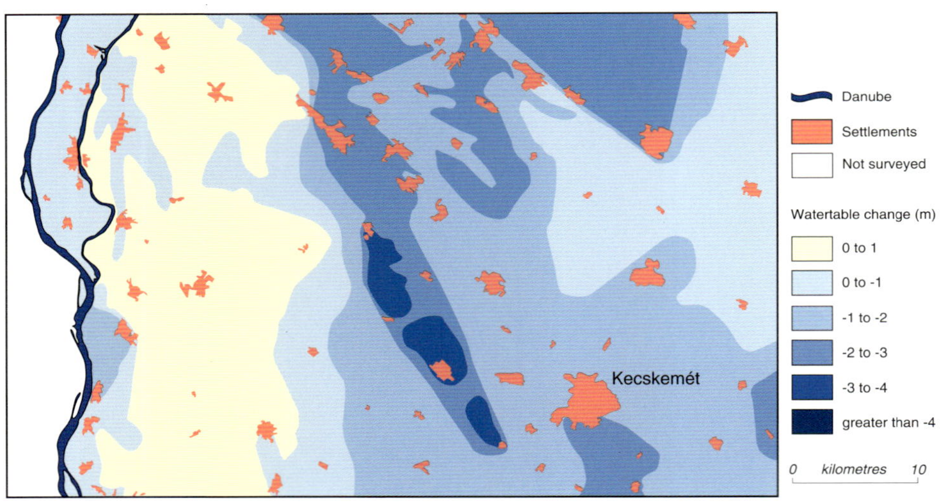

Figure 1.4.40 Average groundwater table change (1969–1993) in the land capability test area (Lóczy and Szalai, 1993; Szalay, 1994).

Figure 1.4.42 Classified current suitability map for maize combining average groundwater table and soil parent materials in the land capability test area (Lóczy and Szalai, 1993).

2

SEMI-NATURAL ENVIRONMENTS AND PROCESSES

2.1 Modelling Change

J. B. Thornes

Why use models?

The usual way of investigating change is to compare places or things through time, for example by comparing old and recent photographs (see Section 3.2). Another is to make careful observations, for example of vegetation cover, or the market price of wheat or rainfall and undertake statistical analysis to estimate the nature and direction of change (as for example in the climate studies in Section 2.3). The third is to actually carry out some laboratory or field experiments, such as those described in Sections 2.6 and 2.7 of this atlas. The advantage here is that the situations can be controlled. However, experiments are expensive to conduct and can only apply to events or processes taking place relatively quickly in time (see Thornes and Brunsden (1977) for other examples).

Desertification is a relatively slow process and although our available records are improving constantly, often the variables or objects measured in the past were not those that now seem critical, or the records have been lost or destroyed, or, simply, the process is so slow that it is very difficult to measure accurately. This especially applies to soil processes, including soil erosion, and as a consequence we have only a very small number of data on erosion rates, under relatively uncontrolled conditions. This is why assertions about the areas undergoing severe erosion or about the rates of erosion must be treated sceptically. By contrast, in floods, the events may occur too quickly or be too large to measure.

Empirical models of change

The simplest types of models are those in which there is a high statistical correlation between events or phenomena so that the occurrence of one normally implies the occurrence of the other. For example the yield of wheat at municipality level in the Province of Murcia, Spain, is directly correlated with rainfall in certain months of the year and the amount of erosion is directly correlated with the vegetation cover and the intensity of rainfall (Francis and Thornes, 1990). There are two important variants on this theme. Firstly we can develop statistical models with more than one controlling variable, such as the Universal Soil Loss Equation (USLE) which is based on the correlation of erosion plot sediment yield against rainfall, land use, cropping type, soil erosivity and a number of other factors. When all the inputs are known, the erosional yield (in tonnes per hectare per year) can be estimated from the equation. The second variant is when two time series are correlated, for example between rainfall at one place through time compared with that at another place through time. This is important in desertification studies, because it helps us to assess whether or not a change in one place (e.g. in soil moisture) is paralleled in another, or if changes at one place lag behind those at another.

There are three major difficulties with such procedures: (1) the data on which the models are based may be small in number or inaccurately measured, so that predictions are subject to wide errors; (2) the models are good for one place but will not work in another, i.e. they are not 'transportable'; (3) if the nature of the relationships between the variables in the model changes, then the model may fail to predict behaviour. For example, the relationship between wheat production and the market price in the previous year may change as result of national or European intervention policies. This problem is called non-stationarity.

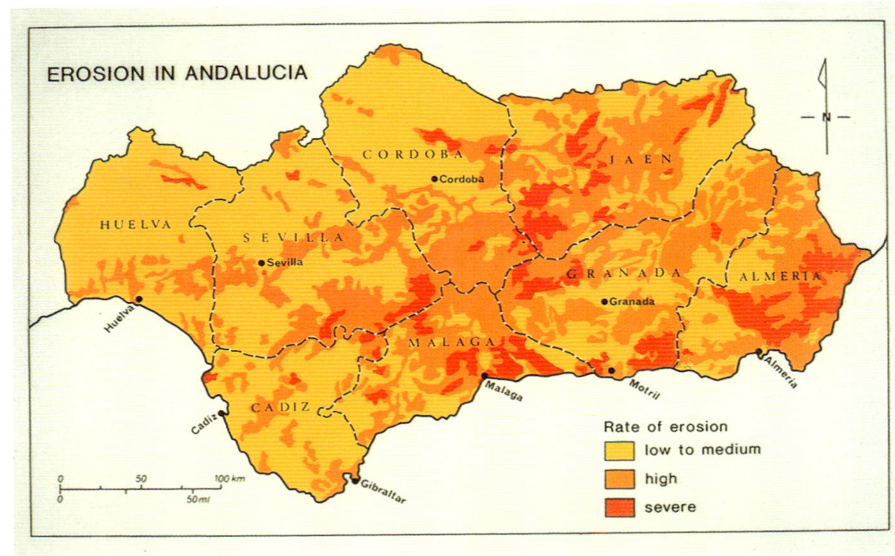

Figure 2.1.1 The USLE is used to model soil erosion. If factors such as slope and rainfall intensity are known, maps of erosion, such as this map of erosion in Andalucia, can be constructed from estimates of erosion at point locations.

Computer simulation models of change

One method that has developed in the last 25 years is to simulate the processes using mathematical models. Here the processes are 'copied' using differential equations or computer algorithms. This means that computer experiments can be carried out that would not be possible in real life perhaps because of scale, cost or time, and also that models can be run for future times under different scenarios. The GCMs referred to in Section 2.4 and the hydrological and erosional models in Section 2.8 are of this type. Equally, socio-economic models can be built to obtain ideas of future populations, future demand for goods and services, or the utilisation of resources (such as water) under different pricing mechanisms.

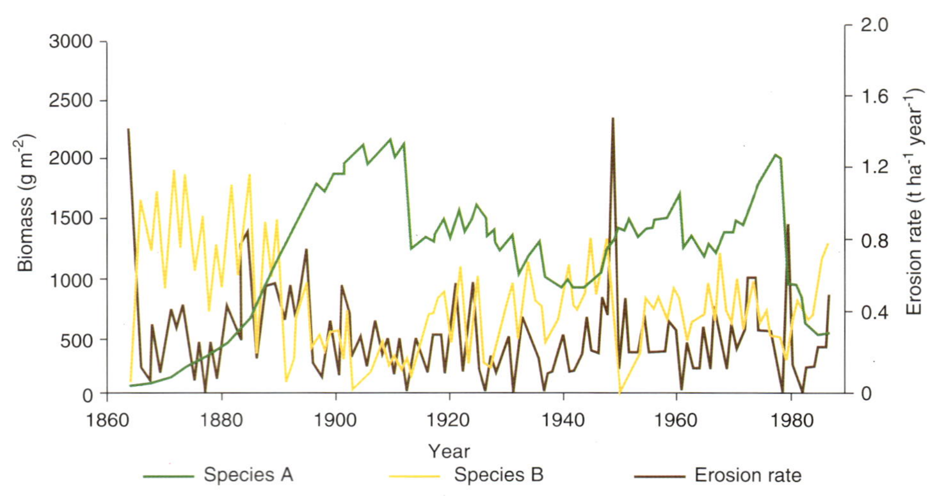

Figure 2.1.2 Computer models can be used to predict vegetation variations over time. Here the competition between two species, controlled by deep soil moisture (Species A) and surface flow (Species B), is modelled to estimate their relative importance over time in Murcia, Spain, over a century.

The development of computer-based models requires a deep working knowledge of the processes to be simulated, so that they can be properly represented by mathematical equations. Moreover the practical computer solution of these equations can be both complicated and difficult, so numerical techniques have to be used. The models have to be calibrated against field parameters, i.e. supplied with realistic values to make them work, and they have to be validated (tested) against changes in the real world (see for example the box in Section 5.2). Finally, they are used to make estimates of future changes. Models are only as good as the 'process rules' on which they operate, and the best models are 'portable', i.e. they can be used in a great variety of situations, such as the SHETRAN-UK and MEDALUS hillslope models described in Subsection 2.8. An example of this type of procedure is the modelling of soil erosion, plant cover and plant functional type undertaken by Thornes and Brandt (1993). Here the model simulates the processes of plant growth, competition between plant types (grass, bushes and trees), and plant control over erosional amounts.

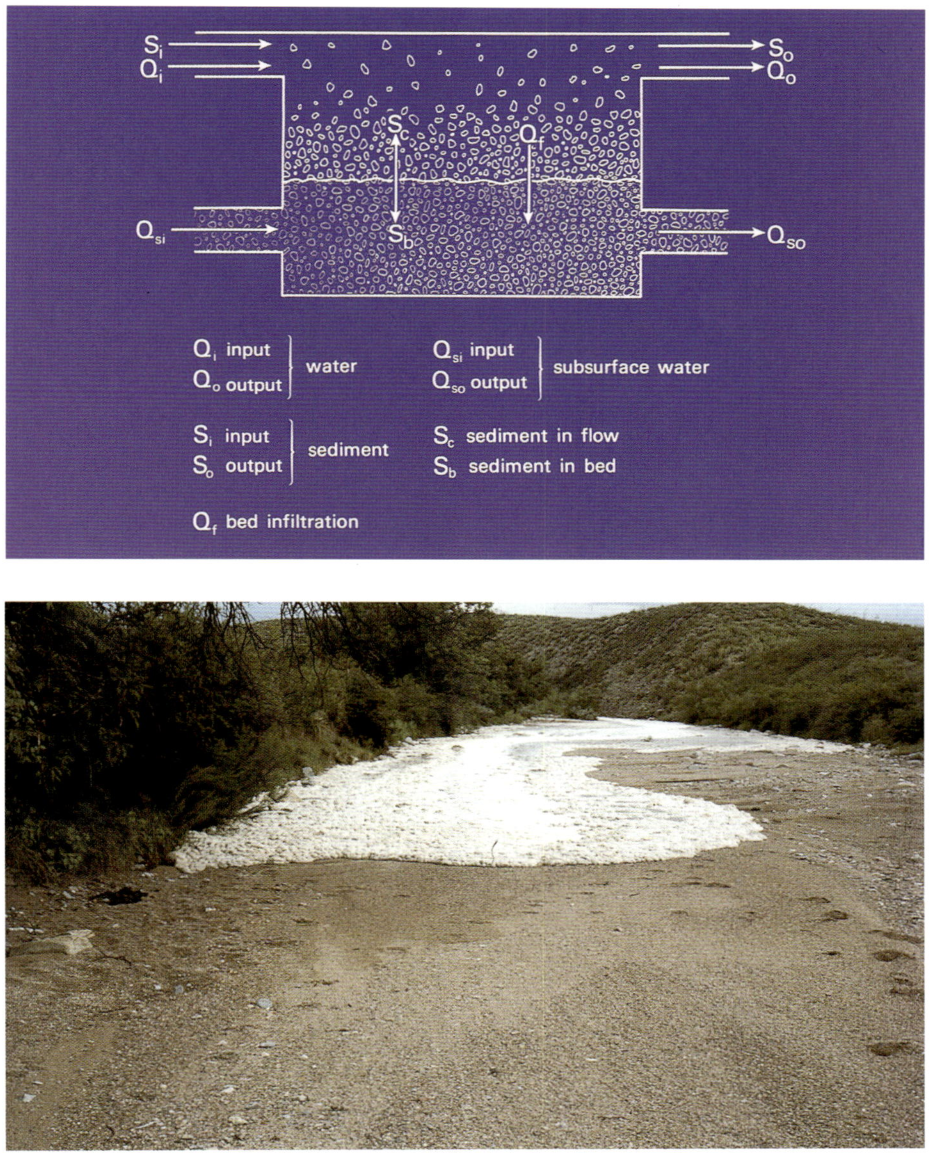

Figure 2.1.3 A simplified representation of a river channel for use in a mathematical model of flooding (Section 3.8). The channel reach is represented as a box of gravel with inflow and outflow to the left and right respectively and with infiltration from the surface into the gravels. The picture shows the real-world process.

There are also problems with this type of modelling. An important one is that sometimes the processes are only poorly understood. Another is that we lack data to describe all the relationships in the model, especially if we wish to apply them over a large area. The third is that the models may be difficult to justify because data to validate the model simply do not exist. It is even sometimes argued that models are not really meant to be applied, rather that they are simply tools that lead to an understanding of complicated processes and their interactions.

Small effects, large changes and butterflies

Not least, models sometimes exhibit quite unpredictable behaviour as they cross certain boundaries or thresholds (Thornes, 1987). This has become a major preoccupation in all types of environmental models, because sudden changes may be very hard to predict if the system of interest is near the threshold. Moreover there are two kinds of change that may occur when a threshold is crossed: one is a simple acceleration of the rate of change, the other is when the processes themselves change. A good example of such a 'bifurcation' is when water starts to pond on the surface instead of infiltrating. Then runoff and erosion begin and the potential consequences change dramatically. In developing our understanding of desertification, especially where the impact of human influence is concerned, we have to recognise that if the soil is near to a critical bifurcation, then a small change, say in agricultural practice, may produce a huge change in soil productivity. Field work, monitoring and modelling can help to identify where such bifurcations exist and help to predict the likelihood of crossing them. This phenomenon is called the 'butterfly effect', because it is sometimes thought that the atmospheric system is so finely balanced that a butterfly flapping its wings in one part of the world might bring about large changes in the weather in another part. One key question is whether or not Mediterranean vegetation is so finely balanced. One view is that the vegetation cover is very fragile and small changes can create large results. Another is that it is very robust, or resiliant, and therefore it can withstand quite large variations in rainfall, grazing pressures or invasions by other plants (see Section 2.2).

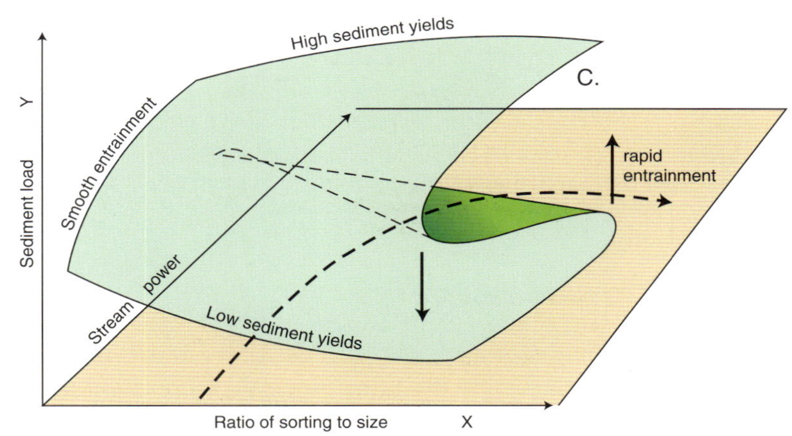

Figure 2.1.4 Processes often cross thresholds, such as the change from meandering to braiding. Near the threshold, a small change in one of the variables can make a big change in the result. Catastrophe theory is one of the ways of handling this. The diagram shows a cusp catastrophe representation of the switch from meandering to braiding. This occurs at point C. When there is a jump from low sediment yields to very high sediment yields, as the process 'switches' from the bottom part of the fold to the top (Thornes, 1980).

2.2 Monitoring Change by Remote Sensing

J. Melía, M. A. Gilabert, M. T. Younis, J. García-Haro, A. Harrison and L. Chiarantini

Introduction

To understand what is happening to the land surface, and establish whether or not it is degrading, it is necessary to monitor present and past changes. Changes can be seasonal and cyclical as a result of vegetation growth, and long term as a result of environmental changes and human impact. The analysis of these changes is complex, but by using remote sensing important aspects of the temporal and spatial dynamics of the land surface can be detected.

The objective of satellite remote-sensing studies in the MEDALUS Project is to provide surface composition parameters at the regional scale and operational microwave remote sensing for soil-moisture measurements. Composition parameters include lithological and vegetation cover maps, particle size distribution of soil and the vegetation fraction obtained from Landsat-5 Thematic Mapper (optical region). Emissivity and reflectance of the surface (bare soil and vegetation) depend on the moisture content, which can be measured in the microwave region, and on the vegetation parameters.

Plant cover is usually estimated using vegetation indices as a surrogate for leaf area index; however, a more meaningful approach is to estimate the proportions of the main cover types within a pixel. This is done through mixture modelling. The procedure is still experimental, and the complete definition of classes necessary to cover the complex nature of the data is difficult.

The use and analysis of remote-sensing data in MEDALUS were carried out by field measurements in the optical and microwave regions and satellite data have been used for the regional studies following validation on target areas situated in Spain (Guadalentín Basin) and Italy (Agri valley, Piomba and Cigno basins).

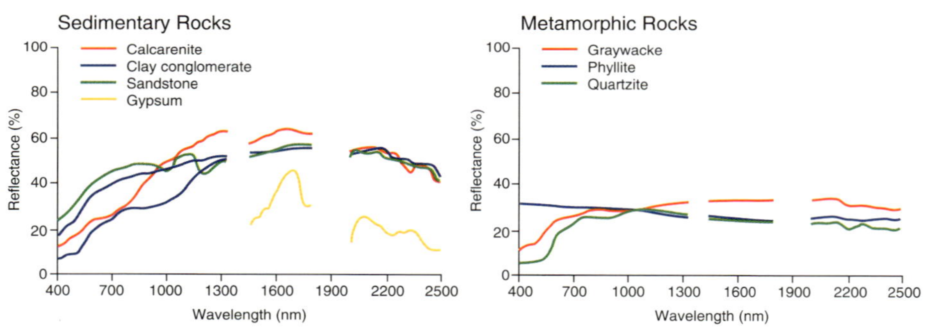

Figure 2.2.1 Spectra of the dominant lithologies in the Guadalentín Basin, including sedimentary and metamorphic units. The spectral measurements were made using the GER SIRIS spectroradiometer in the spectral range 400–2500 nm under clear skies.

Optical and microwave measurements

The role of the *in situ* field radiometry in MEDALUS is to define the distinctive reflectance patterns caused by the mineralogical composition abundance and to select suitable spectral regions (Thematic Mapper bands) that enhance the discrimination between them in image processing. The spectral characterisation of the lithological units in the Guadalentín basin was aided by the mineralogical X-ray diffraction analysis of the most dominant sedimentary, igneous and metamorphic units, which reveals that reflectance derived from the lithological units is mainly controlled by their mineralogical and textural composition. The spectral 'signature' (visible and infra-red regions) of the dominant lithologies shows that there is an appreciable difference in the spectral intensity in the visible region (TM1, TM2 and TM3) due to different iron oxide content, while in the mid-infra-red region, differences are observed in the position of the important absorption features and overall reflectance, as in the case of sedimentary and metamorphic rocks. These spectral differences provide important information for the selection of appropriate bands for image processing and lithological discrimination (Younis, 1993).

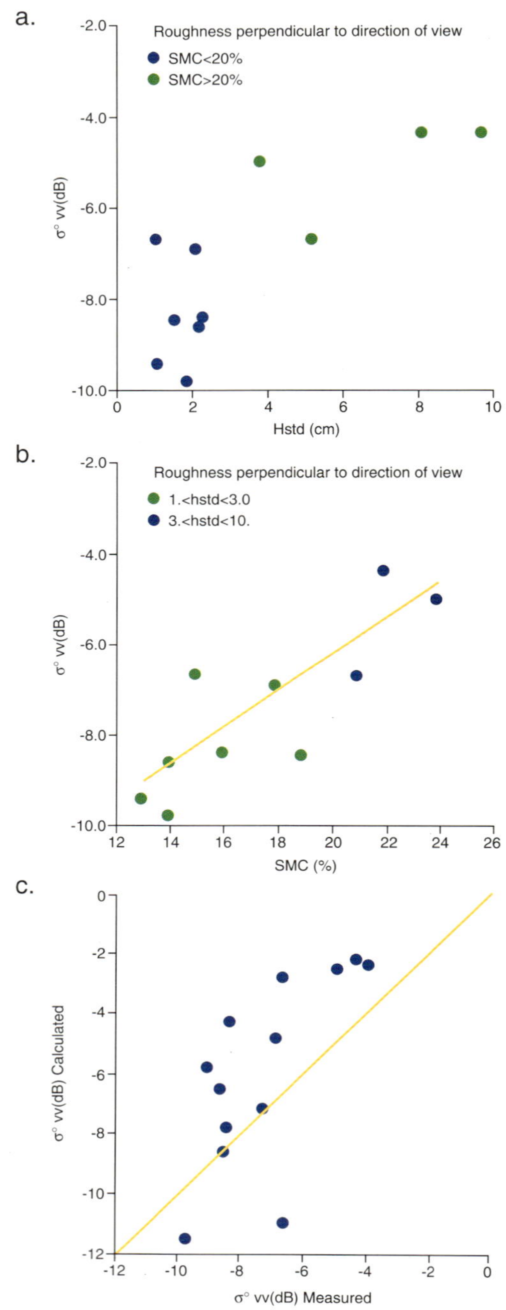

Figure 2.2.2 (a) Measured back scattering coefficient σ^o as a function of surface roughness, height standard deviation (Hstd). SMC = soil moisture content. (b) Measured σ^o as a function of soil moisture content. (c) Calculated versus measured C band σ^o for some bare fields in the Piomba and Cigno basins.

The backscattering coefficient ($\sigma°$), which is analogous to optical reflectance of terrain, describes the scattering properties of terrain along the direction of illumination. These are governed by the geometrical and dielectric properties of the surface in relation to the wave properties of the incident field (wavelength, polarisation and angle of incidence). A comparison between field data, multi-temporal ERS 1 SAR (European Remote-sensing Satellite, Synthetic Aperture Radar) data and electromagnetic models was carried out and the sensitivity of satellite data to soil moisture content and surface roughness have been tested (Chiarantini *et al*., 1995).

The integral equation method has a wider range of validity compared with other electromagnetic models. The calculated backscattering coefficients obtained through calculations based on the Integral Equation Model (IEM) and measured ground truth data, versus measured $\sigma°$ values retrieved from ERS 1 SAR images of the Piomba and Cigno basins, were plotted.

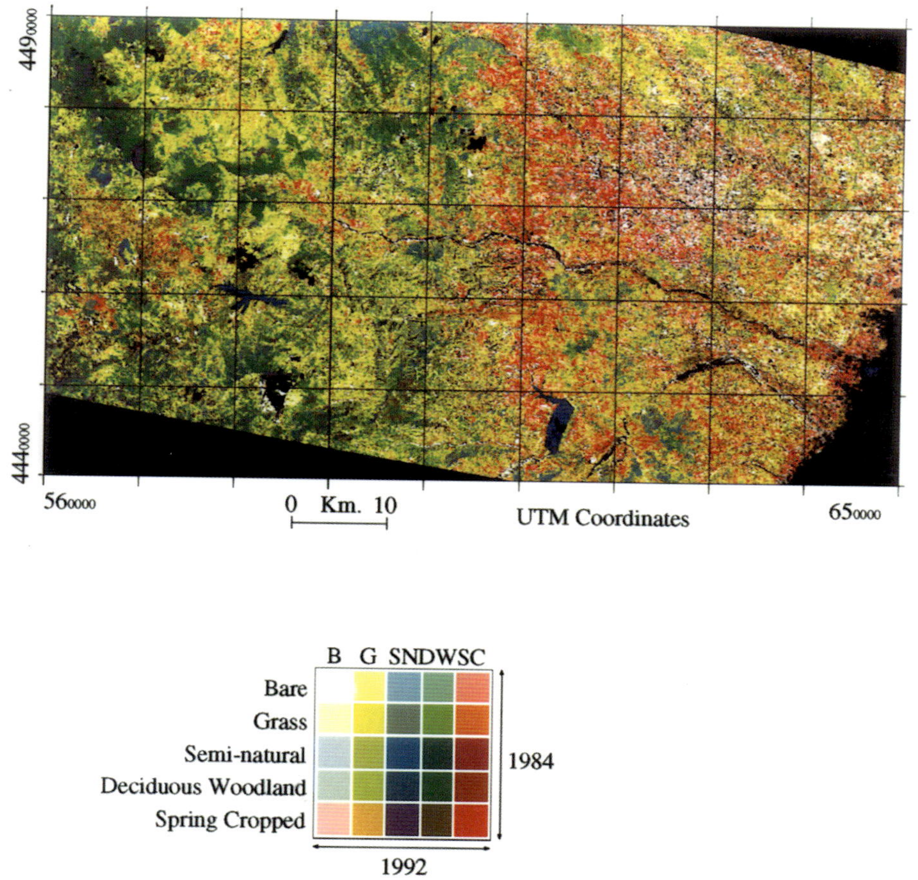

Figure 2.2.3 A multi-temporal classification based on two within-year multi-temporal classifications. The within-year classifications are based on April and August images to maximise vegetation discrimination and land-use practices. This was done for a data set collected in 1992 and, after various corrections and normalisations, for a data set from 1984. The illustration shows the main land-use changes that have occurred over the 8 years.

Multi-temporal classification

The use of within-season multi-temporal data sets allows much finer detail to be extracted from the multispectral classification based on, for example, agricultural practices and differences in the vegetation phenology.

The character of multi-temporal classification between years is sensitive to the timing of the image acquisition within each year. The relative lateness of the season at the time of the spring image acquisition is particularly important otherwise spectral differences will be increased. Ideally, complete within-season monitoring would be used to reduce these errors. The procedure that has so far been developed is based on a multi-temporal classification of recent data, using imagery taken in spring and summer, and tested against ground data or spectral references. Multi-temporal data sets from different years are geometrically registered and corrected for atmospheric effects. They are then classified on the basis of the spectral properties of the reference classes. This procedure has been applied to 1984 and 1992 imagery of the Agri valley target area to determine the main changes in land use.

Spectral mixtures analysis

Spectral mixture analysis has developed in recent years as a suitable remote sensing tool for analysing the biophysical and compositional character of ground surfaces. Spectral mixture modelling assumes that different mixtures of a small number of surface macroscopic materials or components (end-members) are responsible for much of the spectral variation of the observed data, and that these components have different reflectance characteristics so that combining them linearly reproduces the original spectrum. When there is no significant amount of photon multiple scattering between the macroscopic materials, the flux received by the sensor is the sum of all the fluxes from the distinct end-members and the weights are assumed to be proportional to the area covered (end-member fraction).

The constrained least-squares method was applied to generate end-member fractions of three dominant mixture components: soil, crops and pine forest (García-Haro, 1994). Spectral data were obtained from a Landsat-5 TM image taken on 7 April 1993 situated in the south-west of the Guadalentín Basin with an extension of 30 km × 30 km.

Figure 2.2.4 Spectral mixture analysis results of soil, crop and pine forest fractions displayed as red, green and blue respectively. A Landsat-5 thematic mapper image acquired on 7 April 1993 was used.

2.3 Climate Change

J. Corte-Real, R. Sorani and M. Conte

The study of climate change and its impacts is of major concern to scientists, governments and organisations with social responsibilities. Hazardous impacts may result from enhanced climate variability and are more likely in some regions of the world than others. The Mediterranean is one of the areas where the impacts of climate change may be particularly severe.

Potentially hazardous climate changes may be difficult to detect. For example, annual rainfall over Portugal did not change significantly between the two 30-year periods 1932–1960 and 1961–1990. The way in which annual rainfall is distributed between the months has, however, changed dramatically. During the last three decades rainfall has decreased sharply in March. In contrast, rainfall has increased in February and October. These increases are not as statistically significant as the decrease in March. For the same period, the total amount of rainfall in autumn and winter shows a slight increase. The reduction in March rainfall in the three most recent decades is reflected in a decrease in spring rainfall totals. The decrease of spring rainfall reached 19% over the whole country, 27% in the region of Beira Interior and 23% in the Alentejo region.

Thus it appears that, over the last three decades, springs in Portugal are becoming drier, the rainy season is ending earlier and the onset of the dry season is occurring earlier. This has a detrimental effect on the growth of cereal crops. More water therefore has to be stored for longer in reservoirs for irrigation, although the annual rainfall totals have not reduced significantly over the last three decades. A similar analysis for southern Spain also shows decreased spring rainfall in this region. A general decrease in rainfall for the western-central Mediterranean region in recent decades has also been reported.

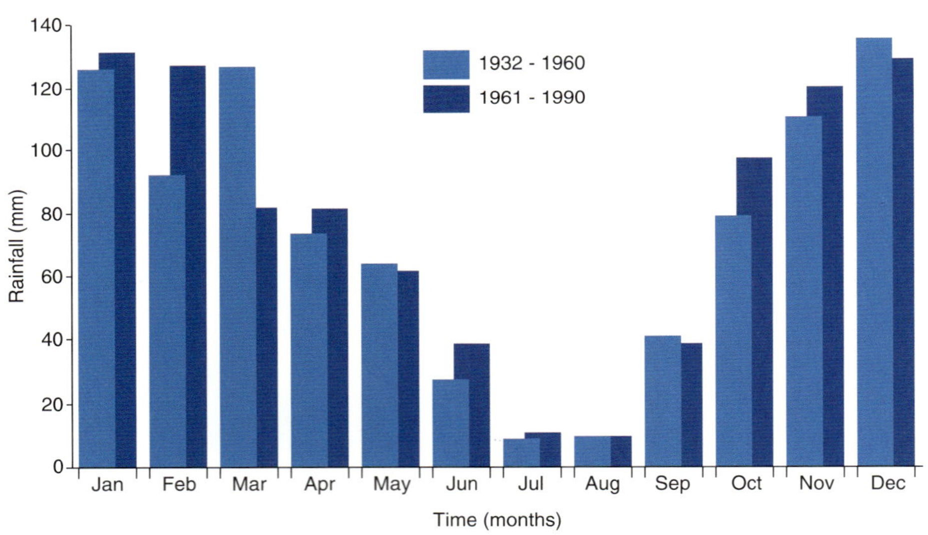

Figure 2.3.1 Changes in the seasonal rainfall cycle over Portugal between the two periods 1932–1960 and 1961–1990. Note the sharp reduction of rainfall in March and the increase of rainfall in February and October.

Climate fluctuations do not always operate in the same direction over the entire Mediterranean area. The intensity, and even the phase, of the anomalies can be quite different in the east and west of the basin. For example, during 1951–1980, the main features of the non-seasonal variation of monthly temperature and precipitation over the eastern and western Mediterranean were different. But why?

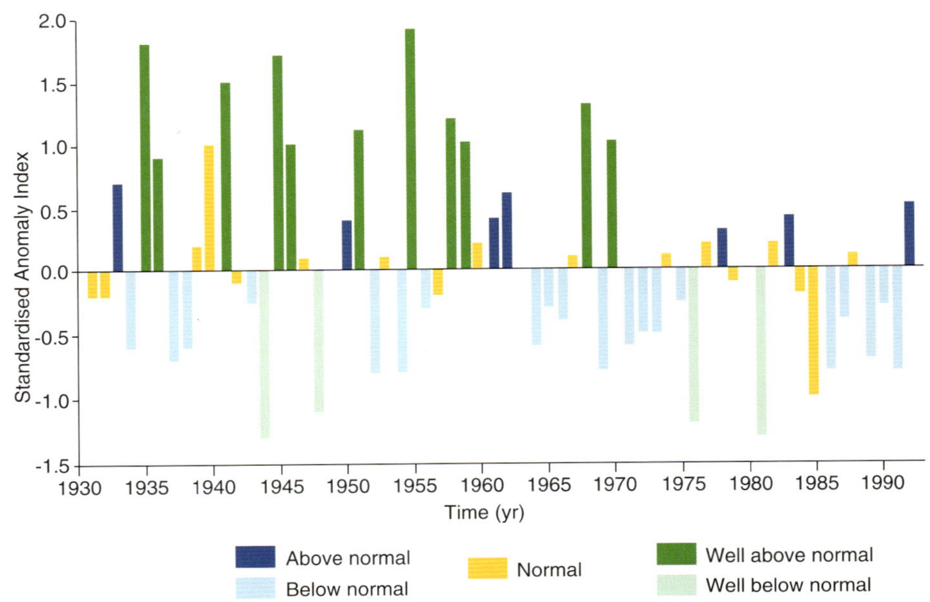

Figure 2.3.2 Standardised anomaly index (SAI) of spring rainfall totals in Portugal. This shows that spring is becoming drier, mainly because of the decrease in rainfall in March. This has important repercussions for the growth of cereal crops.

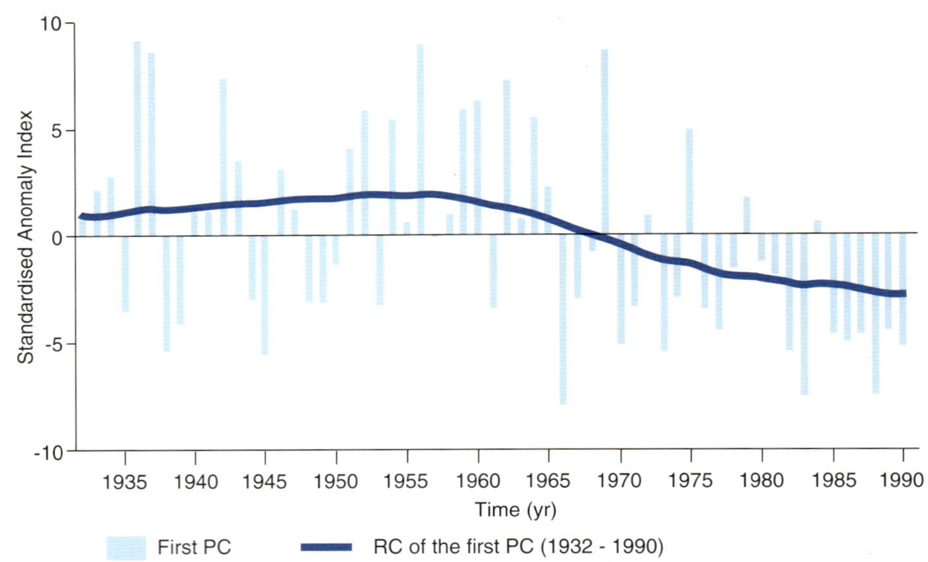

Figure 2.3.3 The first principal component (PC) and its signal component (RC) for March rainfall over Portugal. These show a sharp reduction in March rainfall since the 1960s.

We know that different patterns of atmospheric general circulation can lead to different surface climates. It seems that the different behaviour of precipitation and temperature in the east and west of the Mediterranean may result from particular patterns of the atmospheric general circulation. It is found that a strong sea-level pressure anomaly over the Mediterranean is responsible for different patterns of temperature anomalies across the region. These temperature anomalies are caused by what meteorologists call the Mediterranean Trough, associated with depressions and/or storm activities over the Mediterranean and blocking activities over the Atlantic. This relationship between the surface climate and general circulation is statistically significant and is reasonable from a physical point of view. It can be observed in the real atmosphere as shown in the illustration.

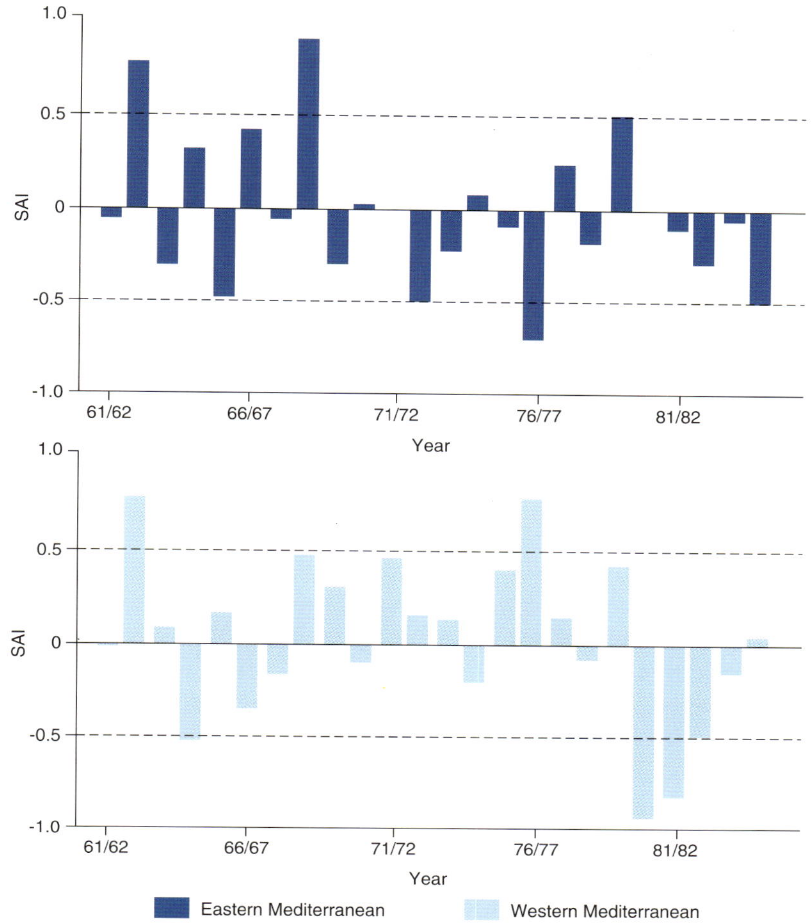

Figure 2.3.4 Standardised anomaly indices (SAIs) of annual rainfall for the eastern (upper graph) and western (lower graph) Mediterranean regions. SAIs allow the incorporation of a number of rainfall time series into a single index, by standardising by the mean and standard deviation. (Courtesy of J. Palutikof)

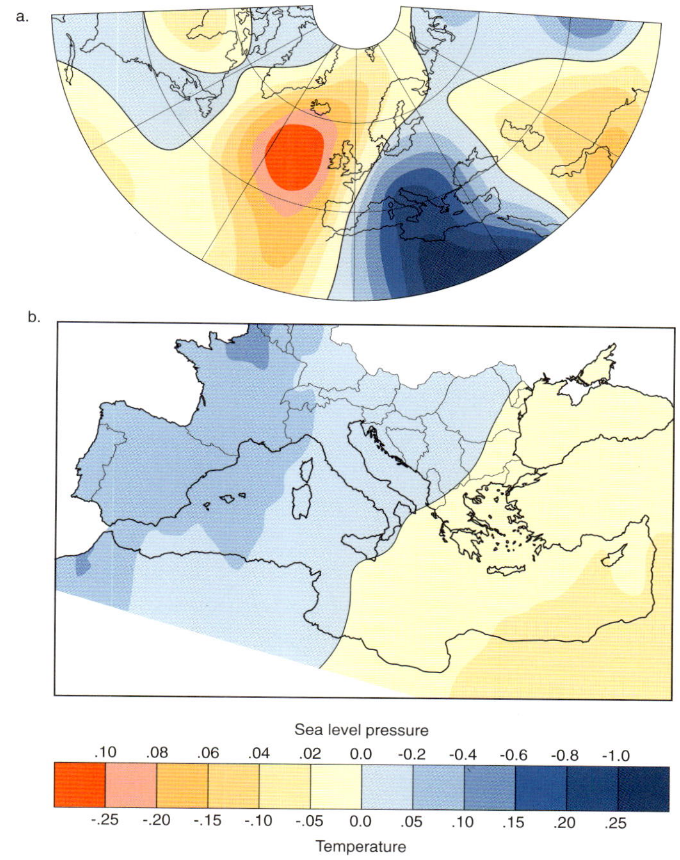

Figure 2.3.5 Anomaly patterns of sea-level pressure and the associated temperature anomalies over the Mediterranean derived by advanced statistical methods. When pressure (a) is below normal over the Mediterranean and above normal over the Atlantic, negative temperature anomalies (b) occur in the west of the region, with positive anomalies in the east. Reversed pressure patterns (above normal over the Mediterranean and below normal over the Atlantic) lead to reversed temperature anomalies.

Links can also be identified between particular patterns of the atmospheric general circulation and rainfall. It is, for example, possible to link rainfall trends in the western/central Mediterranean with a pressure feature known as the Mediterranean Oscillation.

To demonstrate this feature, the pressure field for the Mediterranean is first analysed for the period 1946–1989 using 500 hPa level height data from two guide stations, Algiers and Cairo. These represent the western and eastern sectors of the basin respectively. Changes in the height of the 500 hPa level (which has mean values of 5600–5800 m in the Mediterranean and increases from north to south) are a very good indicator of the pressure behaviour of the whole atmosphere. At this height local and orographic influences are not present and more than half of the Earth's atmosphere is included.

There is a linear increase of the 500 hPa level height over Algiers. An oscillation of about 22 years (the HALE period) is superimposed on this trend. An opposite pattern is evident for Cairo. The geographical distribution of the trend of the 500 hPa height shows that the entire western-central region of the Mediterranean follows the positive trend of Algiers. In contrast, the eastern sector tends to follow Cairo.

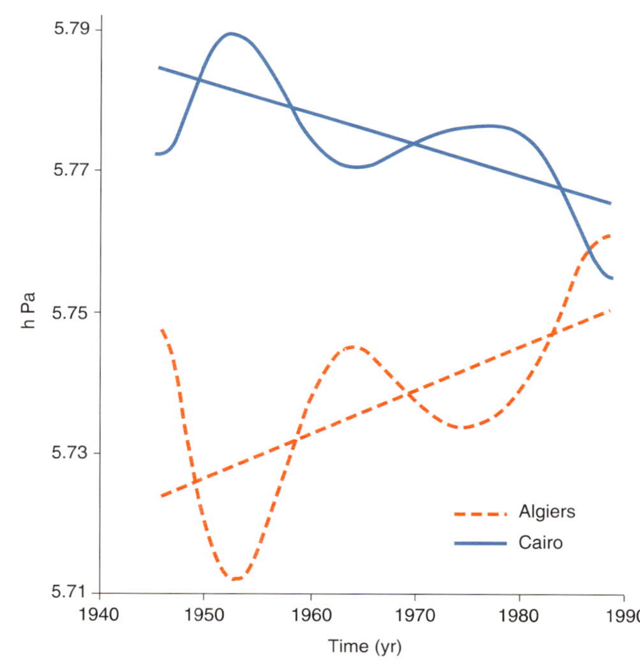

Figure 2.3.6 Change in the height of the 500 hPa level height over Algiers (dashed line) and Cairo (solid line) for the period 1946–1989. A positive trend is seen over Algiers, on which an oscillation with a periodicity of about 22 years is imposed. An opposite trend is observed at Cairo. (The scale of the vertical axis should be multiplied by 1000.)

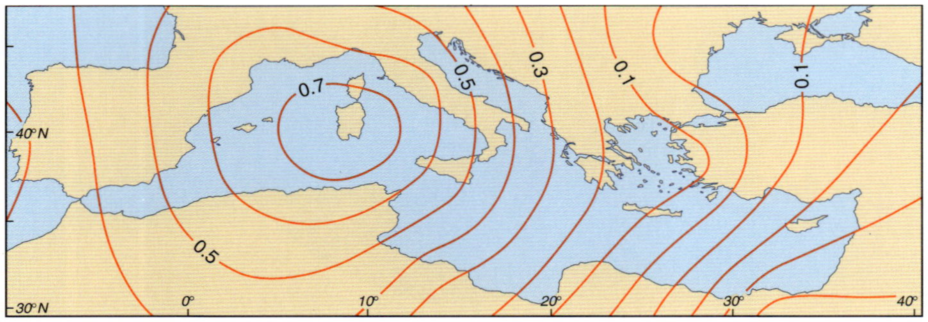

Figure 2.3.7 Spatial distribution of the annual trend of the 500 hPa level height over the Mediterranean for the period 1946–1989 (units in metres). The entire central-western basin behaves similarly to Algiers, with a positive trend indicating increased pressure over this region of the Mediterranean.

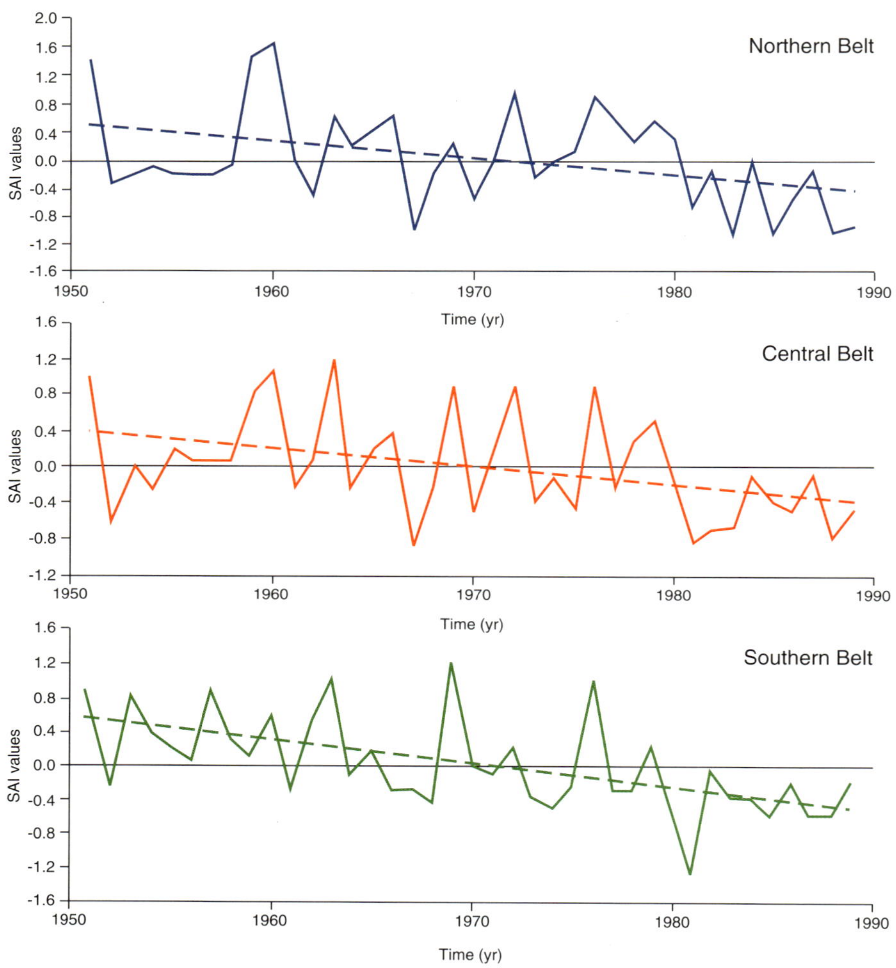

Figure 2.3.8 Standardised anomaly indices of rainfall in three belts of the western-central Mediterranean. Southern belt: south of 38°N. Central belt: 38–42°N. Northern belt: 42–46°N. Negative trends are evident for all belts, indicating decreased rainfall in this part of the basin for the period 1951–1989.

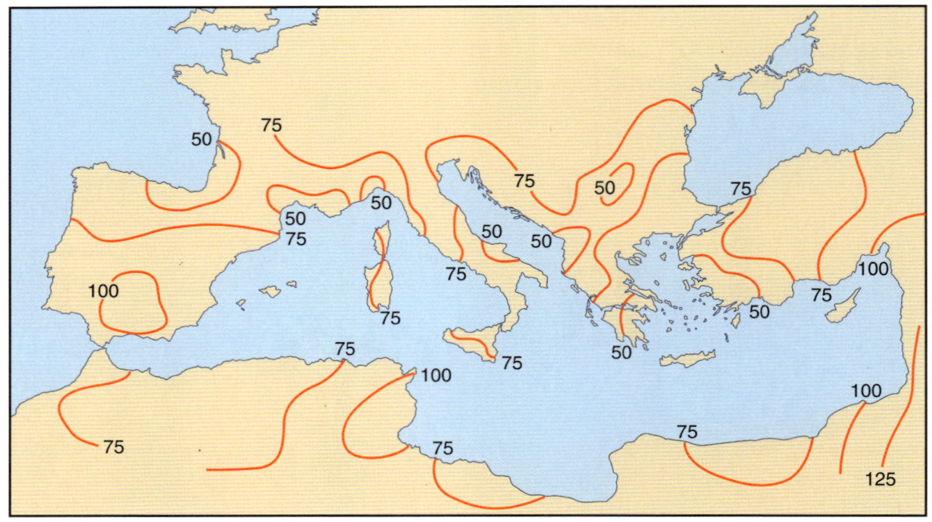

Figure 2.3.9 Percentage of normal (1951–1980) rainfall observed from September 1988 to April 1990.

Standardised Anomaly Indices (SAIs) of precipitation in the western-central part of the basin, for three latitudinal belts, show negative trends, indicating decreasing rainfall over the whole area for the period 1951–1989.

The positive trends of the 500 hPa height level are related to the increased persistence over the western-central Mediterranean region of anticyclonic systems, particularly the Azores High. Since high-pressure anticyclonic patterns inhibit or reduce rainfall, the pressure increase and the rainfall decrease should be physically correlated. In fact, a statistically significant negative correlation does exist between the 500 hPa level height and rainfall in the western-central sector of the basin. This negative correlation is significant at the 95% level. An increased frequency of anticyclonic systems and the probably related decrease in rainfall produced large areas of endemic drought in the basin during the 1980s. This culminated in the very severe event lasting from September 1988 to April 1990. Drought is, of course, a major factor for desertification processes.

Statistical models based on the pressure/rainfall relationships identified above can be used to produce scenarios of future rainfall. These predict a continuing decrease in rainfall for several sites in the western-central Mediterranean from now up until 2025. These decreases are about 10–15% of average rainfall for the period 1951–1980.

2.4 Estimating Future Climates in the Mediterranean

C. M. Goodess, L. Mariani, J. P. Palutikof, V. Menichini and G. P. Minardi

The enhanced greenhouse effect

Greenhouse gases in the atmosphere, such as CO_2, are largely transparent to short-wave solar radiation, but absorb long-wave radiation from the Earth. These gases naturally maintain the Earth at a temperature around 33°C higher than it would be in their absence (the greenhouse effect). Due to human activities, such as the burning of fossil fuels, atmospheric concentrations of the greenhouse gases are rising. This enhanced greenhouse effect should lead to increased atmospheric temperatures. Despite widespread scientific agreement that the effect is real, it is not yet possible to separate out any impact on temperature from the natural variability.

Concern about the potential impacts of enhanced greenhouse warming has given the impetus to a wide range of impact studies. These require detailed regional scenarios of the expected changes in climate. Climate scenarios are not intended as predictions of future climate, but as 'internally consistent pictures of a plausible future climate' (Wigley *et al.*, 1986).

Scenario development

The most widely adopted approach to scenario development is to use output from General Circulation Models (GCMs). GCMs are complex three-dimensional computer-based models of the atmospheric circulation developed from numerical forecasting models. Results are output for a global grid. For a general review of GCMs, see Gates *et al.* (1990) and Goodess and Palutikof (1993).

Until recently, the standard approach has been to perform equilibrium experiments, which require two model runs: first, with the 'pre-industrial' atmospheric CO_2 concentration (the control or $1 \times CO_2$ run) and, second, with doubled CO_2 (the perturbed or $2 \times CO_2$ run), finally comparing results from the two experiments. These simulations do not take into account time-dependent effects, such as the thermal inertia of the oceans, or the transient nature of greenhouse gas forcing, but are less computationally demanding. Advances in computing power are enabling transient-response experiments, in which CO_2 concentrations increase gradually through the perturbed simulation, to be performed. The results are not yet widely available for regional scenario development.

Despite these advances, it is still not possible to model the full complexity of the atmosphere. One reason is that the model grid has a low spatial resolution relative to the scale of processes such as convective precipitation. Such 'sub-grid-scale' processes, therefore, are not satisfactorily reproduced. As a result, our confidence in GCMs is scale-dependent: confidence in their predictions of climate change at the global/hemispheric scale is much higher than it is at the regional/sub-regional scale.

Techniques of regional scenario development, to overcome at least some of these problems, have been developed and applied as part of the MEDALUS research programme. Three examples, involving annual temperature, seasonal moisture availability and daily minimum temperature, are described here. These examples use output from equilibrium-response runs. First, however, we present recent results from a transient-response run.

Climate change in the MEDALUS target areas from a transient GCM experiment

The UK Hadley Centre has performed a 75-year transient experiment with their high-resolution coupled atmosphere–ocean GCM (Hadley Centre, 1992). In this experiment, atmospheric CO_2 concentrations were increased by 1% per year compounded, doubling in year 70. The spatial resolution of the model grid is 2.5° latitude by 3.75° longitude.

Some model results for the grid squares closest to the Guadalentín Basin (Spain) and the Abruzzo/Agri (Italy) target areas illustrate the changes in winter (December to February) temperature and in rainfall during the main rainy season (October to March). The changes are presented for each decade of the perturbed experiment. Although the model has a relatively high spatial resolution, model geography is still not entirely realistic, and the Italian grid square is represented as ocean rather than as land. This may partly explain why the temperature increase is smaller and slower than for the Spanish grid square. The graphs show that the changes in the perturbed experiment, particularly for rainfall, are not necessarily linear with time. This is to be expected with a coupled atmosphere–ocean GCM, which should be able to incorporate the effects of natural variability, as well as greenhouse warming, in the climatology. However, there is a clear trend towards drier conditions for the Spanish grid square and, with the exception of years 45–54, the Italian grid square.

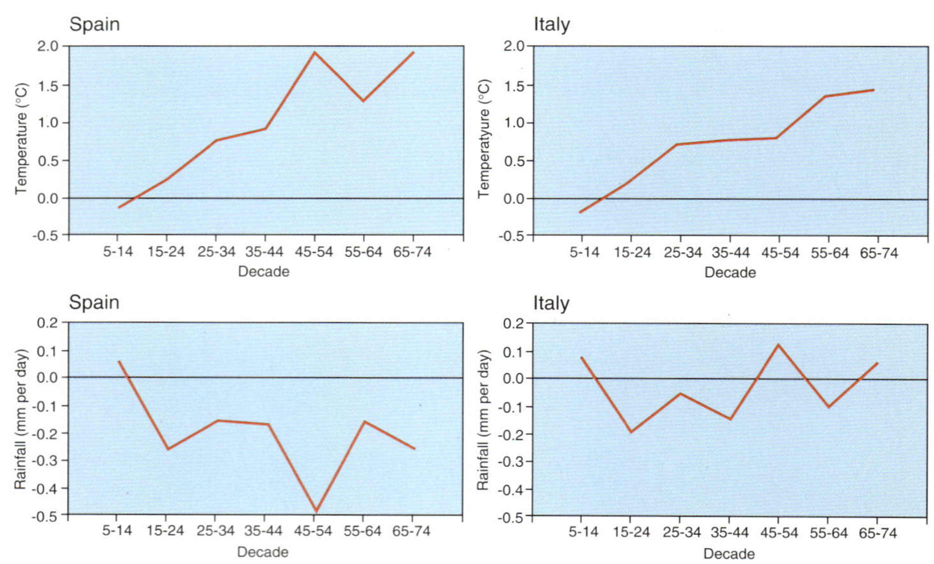

Figure 2.4.1 Changes in winter (December–February) temperature and in rainfall during the main rainy season (October–March) over the perturbed run of the UKTR model for grid squares closest to the Guadalentín Basin (Spain) and the Abruzzo/Agri (Italy) target areas. Changes are expressed as the departure from the control run mean for future decades.

Annual temperature scenarios

In a regional scenario for annual Mediterranean temperature change (from Palutikof *et al.*, 1992) the results from four equilibrium-response GCMs are combined to give a 'mean prediction', since no one model can be shown to outperform the others. The prediction is expressed as the change per 1°C rise in global mean temperature.

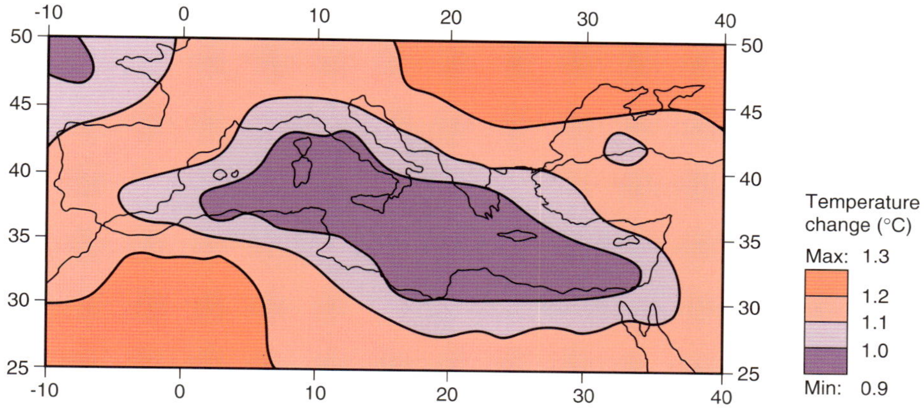

Figure 2.4.2 Annual scenario of the temperature (°C) change per °C change in global mean temperature. Reproduced with permission from Palutikof *et al.* (1992).

The GCMs used to produce this scenario have a coarse spatial resolution, and this is reflected in the very simple pattern of change. In order to obtain information at the subgrid-scale level, GCM results must be combined with statistical models. Taking the example of temperature, one approach is to first construct a regression model to explain the variations in temperature at a meteorological station in terms of observed regionally-averaged climate variables such as temperature, sea-level pressure and precipitation. Then, by replacing the regionally-averaged variables with GCM results, the change in temperature at the station can be estimated. If this method is applied to a dense network of stations, high-resolution scenarios can be constructed. A scenario for Mediterranean annual temperature, interpolated from a network of 248 stations, shows a more complex pattern of change, reflecting, for example, variations due to altitude and proximity to the sea. Changes are again expressed as the change per 1°C global warming.

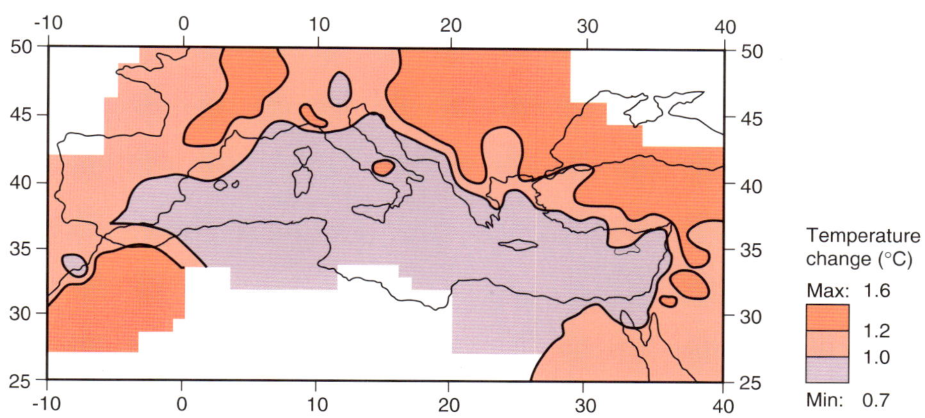

Figure 2.4.3 Annual subgrid-scale scenario of the temperature change (°C) per °C change in global mean temperature.

Scenarios of moisture availability

Scenarios for climate variables such as temperature and precipitation are of limited interest to impact analysts. Effects on agriculture, for example, may be better described by looking at changes in the availability of water for crop growth. We have developed an index of moisture availability, based on the ratio of rainfall (RF) to potential evapotranspiration (PET), and looked at how it may change in a high-CO_2 world.

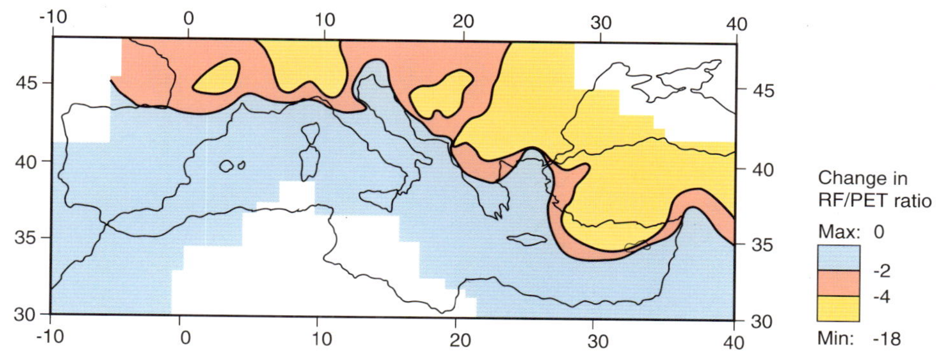

Figure 2.4.4 Winter subgrid-scale scenario of the change in rainfall/potential evapotranspiration (RF/PET) due to a 1°C change in global mean temperature. Reproduced with permission from Palutikof *et al.* (1994).

Reliable scenarios of PET cannot be developed directly from GCM output because the model treatment of ground hydrology is crude. Rather more confidence can be placed in GCM estimates of temperature, which in turn can be related to PET through empirical formulae (Blaney and Criddle, 1950). Using this approach, PET scenarios can be constructed from subgrid-scale temperature scenarios. Subgrid-scale scenarios of precipitation have been constructed using a network of 328 Mediterranean stations and the method described above.

For each site in the network the ratio RF/PET can be calculated for the present day and for a 1°C increase in global mean temperature. The difference between the two ratios can then be calculated and mapped. A negative change in RF/PET ratio indicates a deterioration in the water balance. The map for winter (from Palutikof *et al.*, 1994) shows an unfavourable shift in RF/PET ratio for the whole Mediterranean region. The same pattern is repeated in every season of the year.

Scenarios of this type only provide a rough guide to the implications of climate change for agriculture. They take no account of differences in soil and crop type across the Mediterranean region. However, taken with the scenarios of temperature and precipitation described previously, they represent a set of consistent estimates, based on the same GCM runs, which can be used to evaluate the sensitivity of the economy to future climate change.

Extreme climatic events: the case of temperature

The principal impact of global warming on the rate and extent of desertification in the Mediterranean region is likely to result from changes in the frequency and intensity of extreme climatic events, such as drought and rain storms, rather than from changes in the mean climate. In order to arrive at any estimate of the possible implications of climate change for the occurrence of extremes, we require time series at a daily (or better) time scale.

The ability of one GCM, the Canadian Climate Centre model (McFarlane *et al.*, 1992), to simulate present-day climate at the daily scale has been investigated for Alcantarilla in the Guadalentín Basin. This is a high-resolution equilibrium-response GCM. Ten years of daily temperature data from the $1 \times CO_2$ simulation have been compared with observations. It is found that the model is unable to simulate present-day climate at the daily scale. In the case of minimum temperature, model temperatures are too high and, particularly in winter, do not have the same range of variability as observed temperature. This is clearly seen in the frequency distributions for the winter season. If a GCM is unable to reproduce the present-day features of the climate at the daily scale, this must cast doubt on its ability to simulate future change realistically.

a.

b.

c.

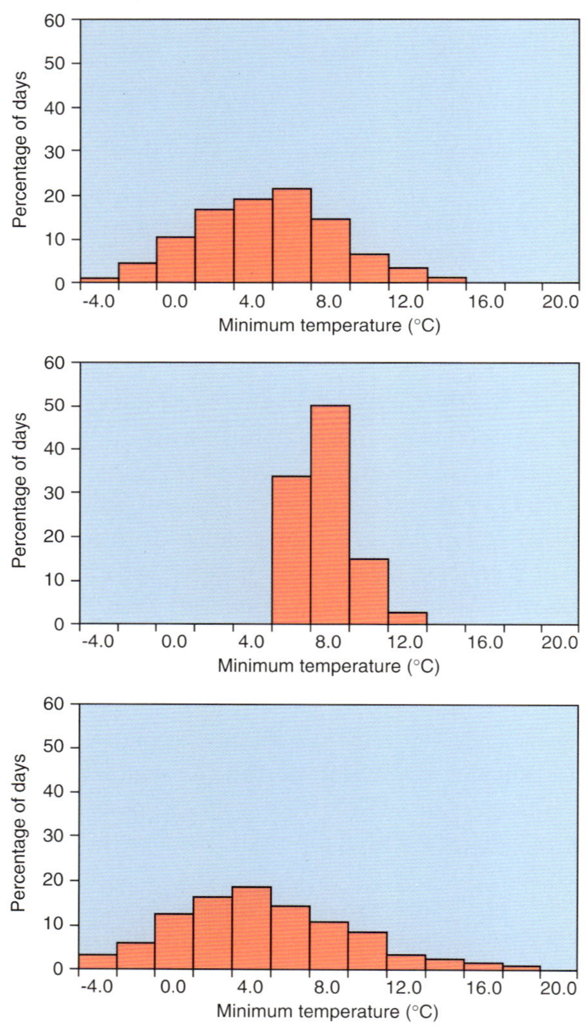

Figure 2.4.5 Frequency distributions of winter minimum temperatures: (a) from the Alcantarilla station record; (b) from the Canadian Climate Centre GCM $1 \times CO_2$ experiment; (c) predicted by free atmosphere variable from the Canadian Climate Centre GCM $1 \times CO_2$ experiment.

The model simulation of free atmospheric variables, such as sea-level pressure, at the daily scale is found to be superior to the simulation of daily temperature. Further analyses indicate that it is possible to use these atmospheric variables to predict surface temperature using a multiple regression model. Model predictions of maximum and minimum temperature, such as the predicted distribution for Alcantarilla winter minimum temperature, are much improved if this technique is used. These improvements allow the occurrence of extreme temperature events to be studied.

Extreme rainfall events

Before future changes in extreme rainfall events can be investigated, the processes that cause such events must be understood. One approach is to see if there are particular patterns of the present-day atmospheric circulation which can be related to their occurrence. An example of this approach, using rainfall data from the Abruzzo region of Italy, is described below. Such methods can eventually be used to predict future changes in the frequency and intensity of extreme rainfall events (Hughes *et al.*, 1993).

Weather types for the Abruzzo

Sequences of daily circulation maps are useful for describing the motions of the air masses at mid-latitudes. Climatologists have created Weather Type Classification Systems (WTCS) which reduce the maps for different heights in the atmosphere to a limited number of patterns. These reflect the variety of patterns produced by turbulent motion of the atmosphere and are a convenient tool for analysing the large amounts of data required for dynamical climatology research studies.

Observations of atmospheric circulation structures at heights which preserve the effects of orography have been used to define weather types affecting Italy, and the Abruzzo region in particular. The aim of this analysis is to study the patterns of surface meteorological variables (rainfall, temperature, wind, etc.) associated with the weather types. The classification system is based on 16 weather types defined at the 850 hPa height (about 1500 m above sea level).

Weather type frequencies in the Abruzzo

The 16 weather types occur in the Abruzzo with quite different annual and seasonal frequencies. Of the four types that each, on average for the period 1983–1987, contributed 10% or more of the total rainfall of the area (types 1, 2, 4 and 6), type 1 is the most frequent of all 16 types. It is most frequent in spring and least frequent in summer. It is characterised by south-westerly upper air flow, perpendicular to the Apennines. It tends to persist for long periods. Weather type 2 is less frequent and occurs mainly in spring and summer. It is a very mobile type, and so exists over the area of interest only briefly and is associated with upper air flow of very variable direction. Type 4 is similar to type 2 in terms of total frequency and persistence. It is least frequent in autumn and is associated with north-easterly upper air flow from the Adriatic Sea, perpendicular to the coast and towards the Apennines. Finally, type 6 is the least frequent of the four types. It is most frequent in autumn and least frequent in summer. It persists for some days and is associated with air flow from the east or south-east.

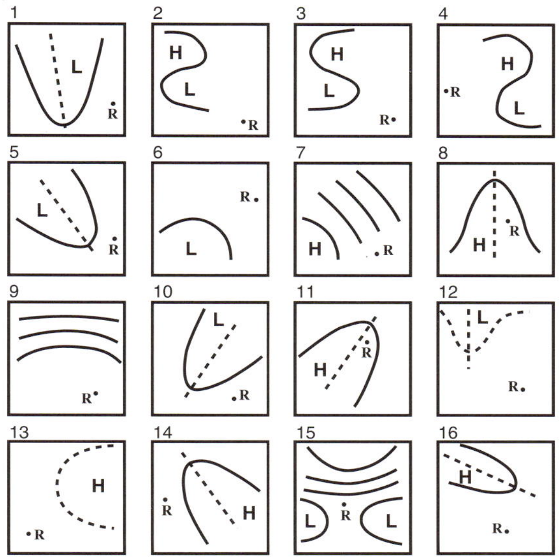

Weather type	Brief description
1	Over the central-western Mediterranean, a trough with N–S axis
2	Over the Mediterranean, 'upside-down S' with depression over the Tyrrhenian Sea
3	Over the Mediterranean, 'S' pattern
4	Over the Mediterranean 'upside-down S' pattern with depression over the Balkans
5	Over the western Mediterranean, trough with axis tilted from north-west to south east
6	African depression over the central-western Mediterranean
7	Over the Mediterranean, north-westerly airstream
8	Over the Mediterranean, North African ridge with N–S (or slightly tilted) axis
9	Over the Mediterranean, high pressure levelled field
10	Over the Mediterranean, a trough with NE–SW axis
11	Over the Mediterranean, ridge from south-west
12	Over the Mediterranean, levelled pressure field; a trough is moving eastwards
13	Over the central-eastern Europe, an anticyclone pushing towards the Mediterranean
14	Over the Mediterranean, ridge from south-east
15	Saddle over the Italian area
16	Ridge located over the north of the Alps, determines easterly airstream

Figure 2.4.6 Schematic representation of the 16 weather types. R = Rome; H = high pressure; L = low pressure.

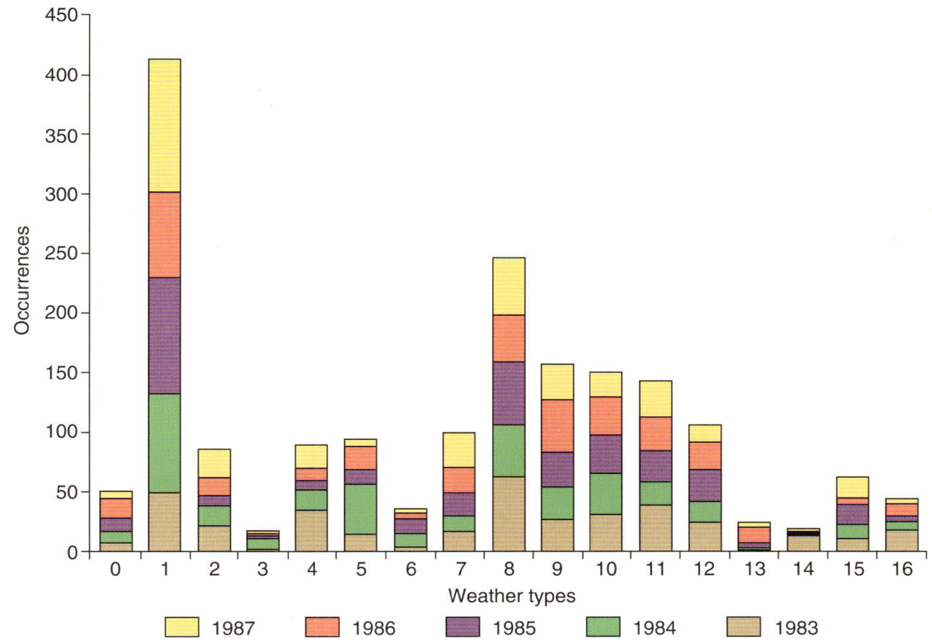

Figure 2.4.7 Annual frequency of the 16 weather types for the period 1983–1987. Type 0 indicates unclassified situations.

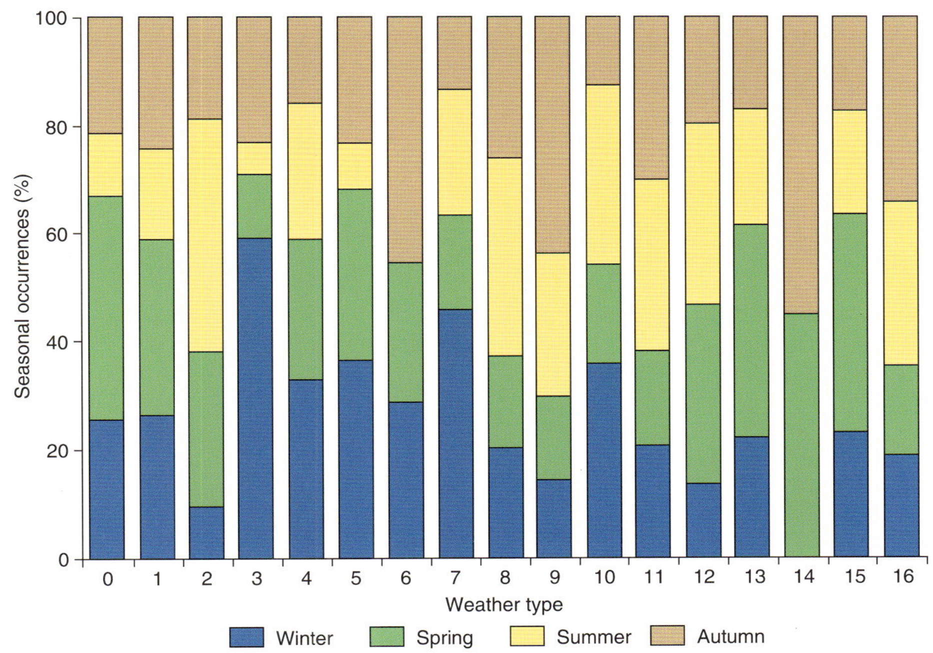

Figure 2.4.8 Seasonal frequencies of the 16 weather types for the period 1983–1987. Type 0 indicates unclassified situations.

Contribution of weather types to rainfall in the Abruzzo

The study of the surface distributions of meteorological variables (temperature, relative humidity, rainfall, etc.) associated with the different weather types provides support for studies of past and present climate. It also provides opportunities to estimate and characterise climate change associated with changes in the frequency and persistence of the weather types. Each of the four weather types 1, 2, 4 and 6, for example, has a very characteristic annual and seasonal contribution and spatial distribution of rainfall.

Weather types and extreme rainfall events in the Abruzzo

The study of rainfall time series for the Abruzzo region reveals the links between weather types and extreme rainfall events. The latter are in turn highly correlated with soil degradation processes. The absolute number of extreme daily rainfall events recorded at 12 key stations in the Abruzzo region during the period 1983–1987 is shown in the tables for each weather type. Extreme events of >100 mm rainfall only occur with types 2 and 5. Two such events are recorded at the Atri station in the Piomba experimental area. For the rainfall class 50–100 mm, types 7, 8, 12, 13 and 14 do not have any events, and type 3 produces only 6 events at one station, Roseto degli Abruzzi.

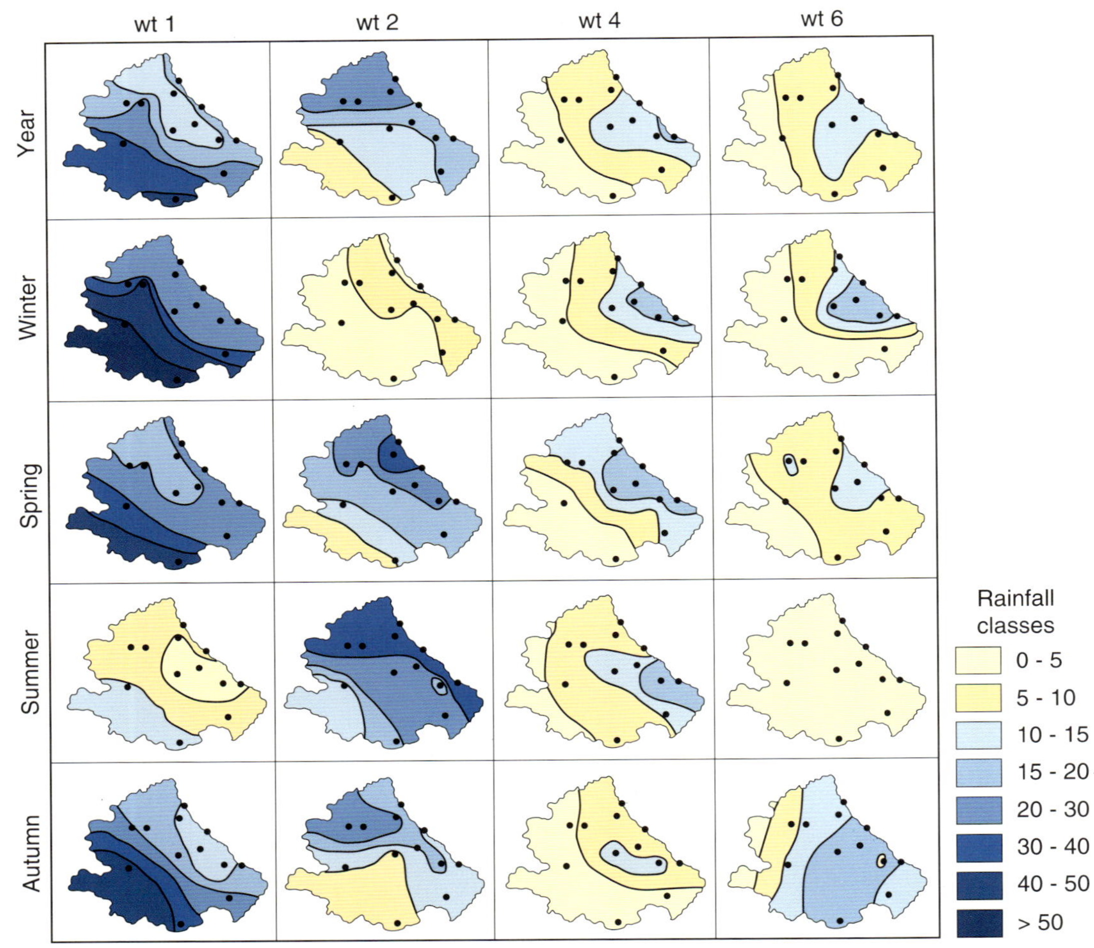

Figure 2.4.9 Annual and seasonal contribution and spatial distribution of rainfall for the four weather types which each contribute 10% or more of the total regional rainfall.

Weather types	Roseto degli Abruzzi	Pescara	Torino di Sangro	Atri	Chieti	Lanciano	Isola del Gran Sasso	Alanno	Pietracamela	Motazzoli	Rocca di mezzo	Barrea
1	0	0	0	0	0	0	0	0	0	0	0	0
2	0	0	0	2	0	0	1	0	1	0	0	0
3	0	0	0	0	0	0	0	0	0	0	0	0
4	0	0	0	0	0	0	0	0	0	0	0	0
5	0	0	0	0	0	0	0	0	1	0	0	0
6	0	0	0	0	0	0	0	0	0	0	0	0
7	0	0	0	0	0	0	0	0	0	0	0	0
8	0	0	0	0	0	0	0	0	0	0	0	0
9	0	0	0	0	0	0	0	0	0	0	0	0
10	0	0	0	0	0	0	0	0	0	0	0	0
11	0	0	0	0	0	0	0	0	0	0	0	0
12	0	0	0	0	0	0	0	0	0	0	0	0
13	0	0	0	0	0	0	0	0	0	0	0	0
14	0	0	0	0	0	0	0	0	0	0	0	0
15	0	0	0	0	0	0	0	0	0	0	0	0
16	0	0	0	0	0	0	0	0	0	0	0	0

Key stations

Colour scale: 0, 1, 2, 3

Figure 2.4.10 Number of extreme daily rainfall events in the class >100 mm recorded at 12 key stations in the Abruzzo region during the period 1983–1987 for each weather type.

Weather types	Roseto degli Abruzzi	Pescara	Torino di Sangro	Atri	Chieti	Lanciano	Isola del Gran Sasso	Alanno	Pietracamela	Motazzoli	Rocca di mezzo	Barrea
1	0	0	0	0	0	0	0	0	0	0	1	0
2	2	3	2	0	3	1	2	0	3	1	1	0
3	6	0	0	0	0	0	0	0	0	0	0	0
4	0	0	2	0	0	4	1	1	0	0	0	0
5	0	0	0	1	0	0	0	0	0	1	0	0
6	0	6	3	0	9	0	3	6	3	0	3	6
7	0	0	0	0	0	0	0	0	0	0	0	0
8	0	0	0	0	0	0	0	0	0	0	0	0
9	0	0	0	1	0	0	0	0	0	0	0	0
10	0	0	0	3	0	0	0	0	0	0	0	0
11	0	0	0	0	1	0	0	0	0	0	0	0
12	0	0	0	0	0	0	0	0	0	0	0	0
13	0	0	0	0	0	0	0	0	0	0	0	0
14	0	0	0	0	0	0	0	0	0	0	0	0
15	0	0	0	2	2	0	2	0	0	0	0	0
16	0	2	0	2	2	2	2	2	2	0	2	2

Colour scale: 0, 1, 2, 3, 4, 5, 6, 7, 8, 9, 10

Figure 2.4.11 Number of extreme daily rainfall events in the class 50–100 mm recorded at 12 key stations in the Abruzzo region during the period 1983–1987 for each weather type.

2.5 Aspects of the Ecology of the Shrub–Winter Annual Communities of the Mediterranean Basin

S. C. Clark, J. Puigdefábregas and I. Woodward

Figure 2.5.1 A dwarf shrub–winter annual community showing the tussock–inter-tussock structure. Tussock grasses as well as dwarf shrubs make up the perennial component.

Origins

There has been tectonic activity and hence instability and erosion in what is now the Mediterranean region for some 200 million years, with dry land habitats available for plant evolution for much of that time (Axelrod, 1973; Thrower and Bradbury, 1973; Windley, 1984). Even in the wetter periods when forest communities were dominant, isolated, open, drought-prone habitats existed in which it is thought the precursors of the present shrub–annual communities of the Mediterranean area evolved (Raven, 1973). A dry season was apparently a feature of the evolving Mediterranean climate, with its timing changing from winter to summer during the Pleistocene. Over the last 1 000 000 years, the destabilising effects of the unsettled Pleistocene climate with extensive erosion and sediment movement in the wetter periods, extended the area available for colonisation (Bradbury, 1981). Human activity during the last 10 000 years has further destabilised the landscape (Pons and Quezel, 1985). This has resulted in large increases in the area occupied by the shrub–annual communities and these are now the predominant type of vegetation in the Mediterranean Basin (Quezel, 1981).

What environment are these communities adapted to?

Climatic features of obvious significance to plant life are a summer drought of varying length and timing, associated high temperatures and most of the rains occurring from October to April, usually as isolated and often heavy storms (Quezel, 1985). The most important feature from the evolutionary point of view is probably the unpredictability within and between years of the amount and timing of the rainfall with no correlation between climatic events. As a consequence of the climatic regime, the soils are unstable, often lack appreciable vegetation cover and are therefore liable to erosion damage (Bradbury, 1981). Heavy grazing and fire, interacting with the climate, are also important evolutionary influences (Aschmann, 1973; Le Houerou, 1981).

The communities have three main components

These are present in varying proportions. One or two may be absent. Occasionally a tussock grass rather than a shrub forms the perennial component, e.g. *Stipa tenacissima* L. in southern Spain. The components are as follows.

Sclerophyllous shrubs

Sclerophyllous shrubs have distinctive 'sclerophyllous' stiff evergreen leaves with thick cuticles, sunken stomata and a dense mesophyll tissue of small thick-walled cells that reduce water loss and maintain leaf shape as the leaves lose water. Broadly, sclerophyllous species are drought endurers. Regeneration is from seed or vegetatively from stools. The importance of each method depends on the species and the environmental conditions at the time of regeneration.

Figure 2.5.2 The evergreen component, *Arbutus andrachne* L., is a typical example of the evergreen (sclerophyllous) shrubs and small trees that are an important constituent of Mediterranean vegetation.

Drought-deciduous shrubs

Drought-deciduous shrubs produce leaves at the start of the wet season that are shed at the onset of the dry season. These 'mesomorphic' leaves, compared with sclerophyllous ones, have large thin-walled cells with mesophyll tissue consisting of loosely packed cells separated by air spaces. These leaves are not drought resistant but in the wet season, compared with sclerophyllous leaves, they are rapid and efficient photosynthesisers. In some species the large wet-season leaves are replaced when they fall by small drought-resistant ones that persist until the start of the following wet season (Orshan, 1964). The main adaptive trait of drought-deciduous shrubs is the avoidance of the high temperatures and drought conditions of summer by a substantial if not complete loss of the wet-season leaves at the onset of the dry season. The type of regeneration, as with evergreen shrubs, depends on the species and current environmental conditions.

Winter annuals

Winter annuals are characterised by autumn germination and completion of the life-cycle, with the production of mature seed before or at the start of the following dry season. Drought avoidance is the main adaptive trait. Dispersal is for the most part within a few centimetres of the parent, leading to a highly patterned distribution (Clark, 1988).

Do the perennial and annual components of the community function independently or interactively?

It is improbable that either the annual or the perennial component is wholly independent of the other if only because the bush and inter-bush 'sub-habitats' are influenced by each other's environments that differ in, for example, the spectral composition and intensity of radiation, in humidity and in the physical and nutritional characteristics of the soils (Patten, 1978). Does competition *sensu stricto* also occur? Shading by dense evergreen canopies excludes annuals from growing beneath them. In the wet season, roots of both annuals and perennials form a dense mat in the infiltration zone. Competition probably occurs in this zone, especially for water in productive years. Conversely, there is evidence that mutualism may operate in water relations and nutrient conservation (Pugnaire *et al.*, 1995). Regeneration of both annuals and perennials is especially affected by the existence of two distinct habitats.

Community resilience and regeneration

Why are these communities so successful in the Mediterranean area, in the face of adverse factors such as fire, heavy grazing, unstable nutrient-poor soils and drought? Vegetation change, given a long enough time-scale, concerns regeneration and hence is cyclical, with building, mature and senescing/old-age stages (Kershaw and Looney, 1985). Cyclical change can be of two kinds: first where the sub-communities exist side by side, each regenerating within its own area, i.e. 'one-dimensional' regeneration, or secondly, 'two-dimensional', where lateral movement occurs, one sub-community replacing the other. One-dimensional regeneration tends to destabilise the community because, for example, soil differences between the sub-communities will increase with time. However, two-dimensional regeneration will enhance stability because differences between the two phases will be reduced as one phase replaces the other. Whichever type occurs, the result is a mosaic of two or more differing components whose proportions will vary depending on the current 'mix' of environmental factors. The stability and resilience of the Mediterranean perennial–annual community is greatly enhanced because the annuals are present

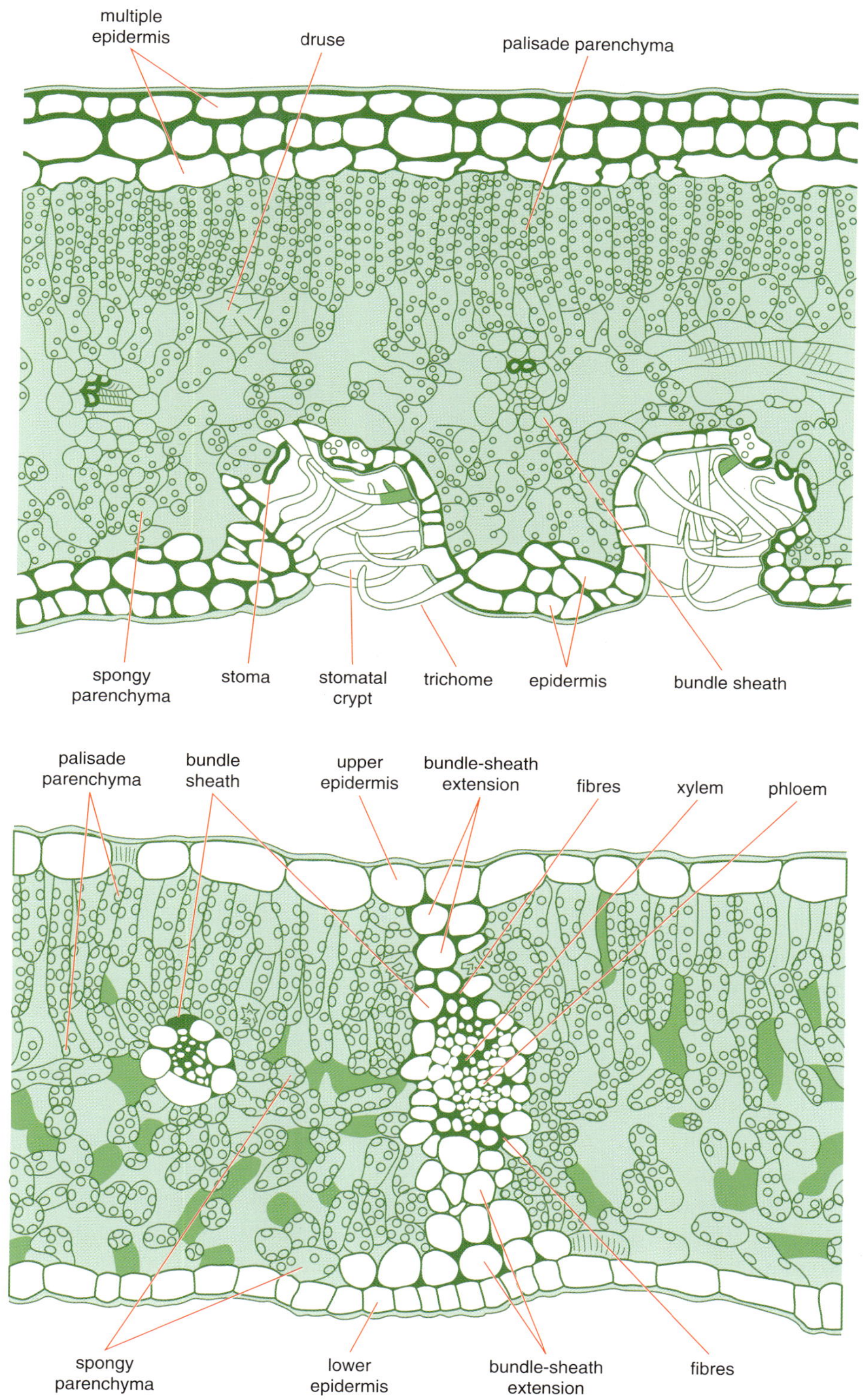

Figure 2.5.3 (a) Transverse section of a sclerophyllous leaf (*Nerium oleander* L.). Note the thick cuticle, multi-layered epidermis of thick-walled cells, tightly packed palisade layer cells and the stomates at the base of pits lined with hairs. (b) Transverse section of a mesomorphic leaf (*Pyrus communis* L.). Compared with the sclerophyllous leaf, note the one-layered epidermis of the thin-walled cells, the loosely packed palisade layer and the absence of sunken stomates. Reproduced with permission from Esau (1965).

Figure 2.5.4 The winter annual component. The annuals occupy the gaps between the perennial dwarf shrubs and tussock grasses. Photographed in April when the cover of annuals is maximum.

Figure 2.5.5 The summer deciduous component. The dwarf shrub *Anthyllis cytisoides* L. in flower in April. Most of the current year's extension growth and all the leaves will be shed at the start of the dry season in late June/early July.

albeit at widely different densities and biomass, at all stages of the community's development. Hence, for example, they are 'on hand' minimising erosion by producing cover rapidly after decimation of the perennial canopy by fire or drought. The three major communities in the Rambla Honda, southern Spain (see also Section 4.2), are examples of these two-component systems, each with different perennial and winter annual elements.

On the upper slopes of the valley, the perennial species is the grass *Stipa tenacissima* L., a slow-growing drought and high-temperature endurer. *S. tenacissima*'s regeneration is mainly vegetative and two-dimensional with slow lateral replacement and only the occasional establishment of seedlings in the inter-tussock areas. The annuals are mainly in the inter-tussock areas. On the fans below the upper slopes, the dwarf shrubs *Anthyllis cytisoides* L. or *Artemisia barrelieri* Besser are the dominant perennials. The community occurs mainly on agricultural land last cultivated 10 or 15 years ago. In dry years, the annuals are largely confined to the areas beneath the canopies of the bushes. In wet years, they also occur, although strongly patterned, in both bush and inter-bush habitats.

In the valley bottom, *Retama sphaerocarpa* (L.) Boiss. is the dominant perennial shrub. Here the annuals occur mainly at the bases of the bushes. Regeneration in *R. sphaerocarpa* is by seed and is usually two-dimensional. The vegetation formed by *S. tenacissima*, *A. cytisoides* and *R. sphaerocarpa* has the typical patchiness of the two-component community just described, with vegetated clumps and areas of bare soil between. The ecological consequences of this patchiness are not yet fully understood. However, there is growing evidence of its importance hydrologically. Thus the less mobile the patches, i.e. the more one-dimensional are the regeneration processes, the greater are the differences in soil properties between vegetated patches and the bare areas. Runoff and sediment yield will therefore be affected by the spatial patterns of vegetated and bare areas. *S. tenacissima* is an ideal species for investigating the effects of patchy vegetation on the hydrology of the associated soils. Modelling its tussock development has produced preliminary information on how to predict the effect of the spatial structure of vegetation on sediment storage, erosion and runoff yield (Sánchez and Puigdefábregas, 1994).

Figure 2.5.6 The tussock grass component. *Stipa tenacissima* L. tussocks, with *Anthyllis cytisoides* L. bushes in the background.

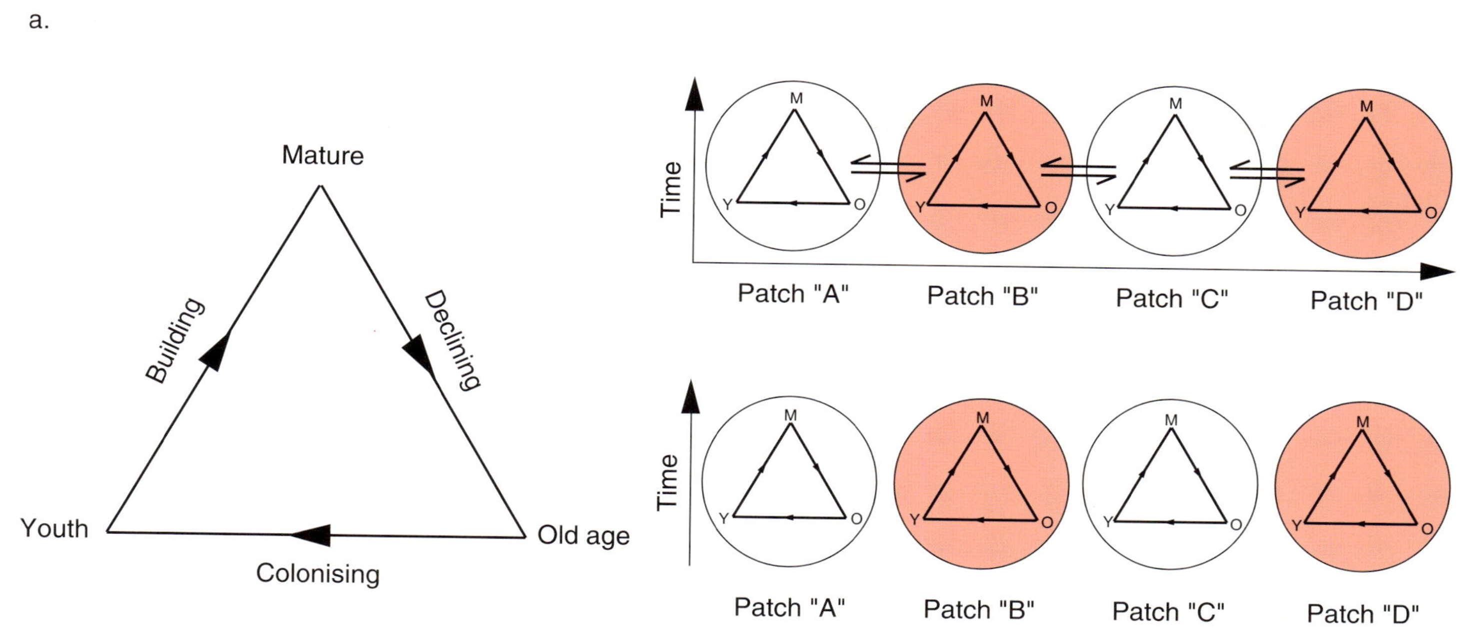

Figure 2.5.7 Types of vegetational change.

(a) Cyclical regeneration. Here 'youth' (Y), 'mature' (M) and 'old age' (O) stages succeed each other, in a closed or nearly closed system, with the changes involving single species or multi-species assemblages. Regeneration always involves some form of cyclical change. Because of the cyclical nature of regeneration, all communities comprise mosaics of different phases of the cycle. The type of mosaic that occurs in a particular community will depend on the species, the habitat and the history of the site. Whether the regeneration is one- or two-dimensional will influence strongly the nature of the mosaic and the community in general. In one-dimensional regeneration, the changes are temporal with no spatial movement during successive cycles. This type tends to be destabilising because differences in the habitats of the components of the mosaic will increase with time. In two-dimensional regeneration change is both temporal and spatial. This type tends to have a stabilising effect as components of the mosaic replace each other, thereby reducing differences between them.

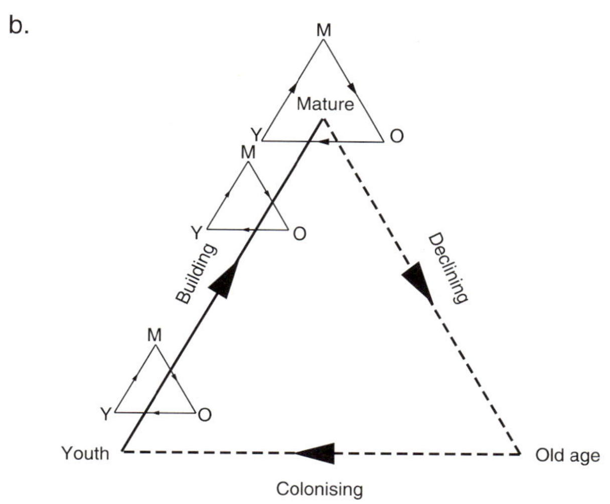

(b) Successional change. This concerns the building phase. Customarily, the term is used for building phases lasting many years that terminate in relatively stable communities. Cyclical regeneration occurs within these terminal or 'climax' communities and there can be gradual non-cyclical changes in climax communities because of climatic changes.

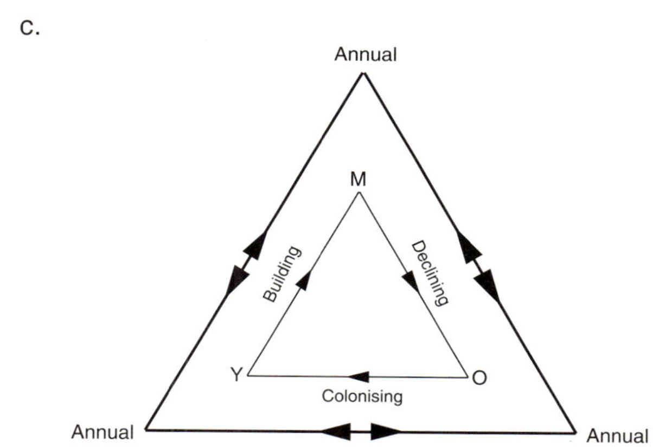

(c) The perennial–winter annual communities of the Mediterranean Basin. Here an irregular form of two-dimensional cyclical regeneration operates. The process is frequently disrupted by climatic irregularities or by perturbations such as fire or heavy grazing. The annuals can germinate and complete their life-cycle from any starting point in the perennials' regenerative cycle, hence they are always present and are often the only component in the community. Their opportunistic behaviour and short life-cycle means that they can take advantage of transiently favourable conditions and produce a cover of vegetation that reduces erosion and provides more favourable conditions for establishment of the perennials than would bare ground.

2.6 Mediterranean Soils and Desertification

Introduction

Soil formation is a relatively slow and complex process in all environments. Materials are gained by weathering from the bedrock, by the incorporation of organic matter from the vegetation cover and the development of animal communities and by transport of materials from upslope. Soil material is lost by erosion or by vertical and lateral (downslope) transport in percolating water. These losses and gains depend on the underlying rock type, the climate and the vegetation cover and are, of course, dramatically affected by human activity. Because these controls vary through time their relative importance in soil formation can be reflected by different horizons or characteristics in the soil profile. So soils are very dynamic living systems continuously changing.

Mediterranean landscapes are enormously varied in rock type, climate, vegetation cover and human activity. So it follows that there is a huge variety of soils. Yassoglou describes this variety below in terms of a soil classification. But, like vegetation cover, the soils have been transformed over thousands of years by human activity, so that there are very few 'natural' soils left. This is why relict soils (natural soils that have survived the onslaught) are of such great interest. There has been a long debate about whether '*terra rossa*' soils found on Mediterranean limestones are ancient or modern, and how much they have been affected by human action.

In the drier areas, soil formation is slower than usual, vegetation poorer and erosion is likely to be higher. So the soil thickness is less and the balance between supply and removal of material more precarious. As a result, small changes in the physical and human factors controlling soil characteristics can have dramatic results.

As a consequence, in the areas below 600 mm of rainfall soils are generally thin, have a high stone content and are subjected to short periods of high soil moisture and long periods of drought. In calcium-rich areas (and especially at the foot of limestone massifs), evaporation and deposition of calcium on or in the soil gives rise to thick indurated crusts called *caliche* or at least to soils rich in calcium-rich horizons. Once formed, these layers act like solid rock to resist both infiltration and erosion. They also make the soil difficult or impossible to cultivate and the *caliche* layers may have to be removed by heavy machinery or even blasting.

Nearly all human activity, cutting vegetation, ploughing, grazing, burning and soil pollution by waste produce conditions in which the loss–gain budget shifts to a net loss. As a result, and through centuries, many hillslopes have been totally denuded of all soils. The transported soil is re-deposited in reservoirs and deltas, or carried out to sea. Therefore the contrast between denuded hillsides and rich, irrigated deposition areas is continually reinforced. This process is desertification, because the formerly productive areas of dryland farming are reduced to barren, unproductive slopes covered only with sparse material which even now may still be heavily grazed.

MEDITERRANEAN SOIL TYPES IN A DESERTIFICATION CONTEXT

N. Yassoglou

Soil typological units (STU) of the north Mediterranean that are threatened by desertification have the following attributes:

- Parent materials are dominantly limestone or dolomite, acid igneous and metamorphic rocks, Tertiary marls and shales.
- Soil depths are shallow for soils on hard bedrock and variable for those on Tertiary marls.
- Textures are loamy or silty.
- Soils are low in organic matter and in biological activity in the surface layer.
- Water infiltration rates are frequently low due to compaction or crusting, except where the surface is nearly vegetated.
- Stoniness is variable, generally low, but in some cases it may be exceptionally high where fines have been removed by erosion.
- Slopes are moderate to steep.
- The climate is thermo-Mediterranean or meso-Mediterranean according to the Bagnouls–Gaussen (1953) classification.
- The electrical conductivity of the saturation extract is greater than 4 dS/m.

Soil associations consisting of the soil mapping units (SMU) and their components (STUs) (FAO-UNESCO, 1974; CEC, 1985) having the above attributes have been ranked in an approximate sequence of decreasing vulnerability to desertification.

The desertification vulnerability list consists of groups of soil associations with the following main attributes (Yassoglou, 1993):

- *Xerosols*. Soils developed under arid conditions on calcareous sandstones, limestones and marl; subjected to extreme water deficits; they occur mainly in Spain.
- *Calcaric lithosols*. Very shallow soils of low productivity; developed on highly permeable limestones; usually on steep slopes; subjected to very high water deficits and erosion risks.
- *Eutric lithosols*. Very shallow soils of low productivity; developed on non-calcareous bedrocks; usually on steep slopes; subjected to high water deficits and erosion risks.
- *Calcaric regosols*. Moderately deep soils of low productivity; on consolidated and unconsolidated parent materials; on moderate to steep slopes; subjected to high water deficits and high erosion risks.
- *Calcic cambisols*. Moderately deep soils of moderate productivity; on moderate to steep slopes; subjected to high water deficits and moderate erosion risks.
- *Eutric cambisols*. Moderately deep soils of moderate productivity; on non-calcareous bedrocks; usually on steep slopes; subjected to moderate water deficits and high erosion risks.
- *Salinisation-threatened soils*. Saline or non-saline soils on deep alluvial deposits in semi-arid zones; the non-saline soils are subjected to salinisation through improper irrigation practices.

Table 2.6.1 North Mediterranean soil mapping units and their components are threatened by desertification, ranked in an approximate sequence of decreasing vulnerability to desertification.

	Associations
	A Xerosols
86	Calcic Xerosol (Xb-3bc), Calcaric Lithosol (Ie), Rendzina (E)
89	Gypsic Xerosol (Xy-4b), Orthic Solonchak (Zo), Gleyic Solonchak (Zg), Calcic Xerosol (Xb-3a), Rhodochromic Luvisol (Lrc)
	B Calcaric Lithosols
51	Calcaric Lithosols (Ic-2/4bd), Rock outcrops (RO)
52	Calcaric Lithosol (Ic-4d), Rock outcrops (RO), Rhodochromic Luvisol (Lcr)
50	Calcaric Lithosol (Ic-4d), Calcic Cambisol (Bk), Calcaric Cambisol (Bc)
	C Eutric Lithosols
47	Eutric Lithosol (Ie-2d), Eutric Cambisol (Be), Rock Outcrops (RO), Chromic Luvisol (Lc)
48	Eutric Lithosol (Ie-2d)
	D Calcaric Regosols
41	Calcaric Regosol (Rc-4d), Calcarochromic Cambisol (Bcc), Rhodochromic Luvisol (Lcr)
42	Calcaric Regosol (Rc-3/4cd), Eutric Regosol (Re), Eutric Cambisol (Be)
43	Calcaric Regosol (Rc-3/4bc), Calcaroeutric Cambisol (Bec), Vertic-Cambisol (Bv), Vertisol (V)
40	Calcaric Regosol (Rc-2/4bc), Calcic Cambisol (Bk), Vertic Luvisol (Lv), Rhodochromic Luvisol (Lcr)
	E Calcic Cambisols
100	Calcic Cambisol (Bk-3cd), Lithosol (I), Chromic Luvisol (Lc)
98	Calcic Cambisol (Bk-3bc), Lithosol (I), Chromic Luvisol (Lc)
99	Calcic Cambisol (Bk-3bc), Calcic Regosol (Rc)
89	Calcic Cambisol (Bk-3cd)
87	Calcic Cambisol (Bk-3ab), Chromic Cambisol (Bc), Calcaric Lithosol (Ic)
102	Calcic Cambisol (Bk-3/4ab), Calcaric Regosol (Rc), Orthic-Solonchak (Zo)
94	Calcic Cambisol (Bk-2cd), Calcaric Phaeozem (Hc), Renzina (E)
	F Eutric Cambisols
17	Eutric Cambisol (Be-2/3cd), Eutric Lithosol (Ie), Orthic Luvisol (Lo)
24	Eutric Cambisol (Be-1/3bd), Ranker (U), Lithosol (I), Rock outcrop (RO)
104	Chromic Cambisol (Bc-2cd), Lithosol (I)
9	Eutric Cambisol (Be-3bc), Eutric Lithosol (Ie)
19	Eutric Cambisol (Be2/3cd), Orthic Luvisol (Lo), Eutric Lithosol (Ie)
21	Eutric Cambisol (Be2/4ad), Chromic Luvisol (Lc), Calcaric Lithosol (Ic-4d)
32	Calcaroeutric Cambisol (Bec-2/4), Calcic Cambisol (Bk)
23	Eutric Cambisol (Be-1bc)
20	Eutric Cambisol (Be-2/3cd), Orthic Luvisol (Lo)
	G Associations threatened with salinisation
9	Fluvic-Calcaric Fluvisol (Jcf-2/4), Orthic Solonchak (Zo)
5	Cleyo-Eutric Fluvisol (Jeg-3a), Gleyic Solonchak (Zg) Eutric Gleysol (Ge)

Classification according to CEC (1985) and FAO-UNESCO (1974).

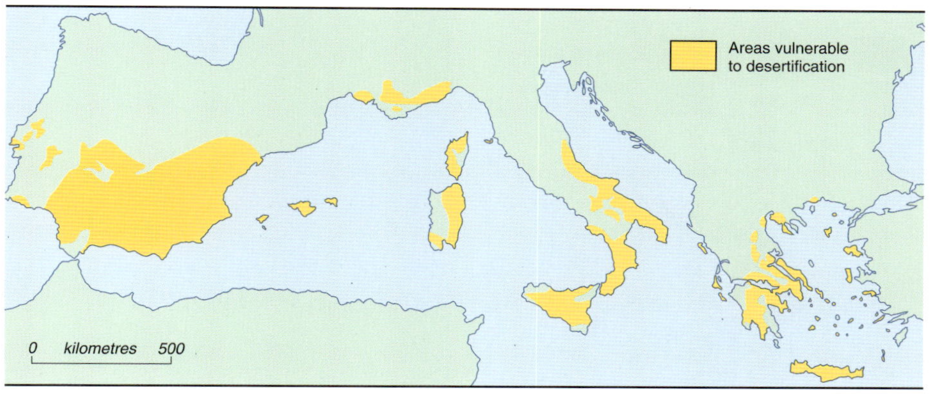

Figure 2.6.1 Map showing areas vulnerable to desertification in the northern Mediterranean region.

Figure 2.6.2 Soil formed on shale formations largely degraded by erosion and classified as eutric lithosol, Sardinia (photograph: C. Kosmas).

Figure 2.6.3 Soil formed on conglomerates largely degraded by erosion and classified as calcaric cambisol, Korinthos (photograph: C. Kosmas).

SOILS CONTAINING ROCK FRAGMENTS AND THEIR RESPONSE TO DESERTIFICATION

J. Poesen, B. van Wesemael and K. Bunte

Spatial distribution of soils containing rock fragments in Mediterranean Europe

Soils containing significant amounts of rock fragments in their top layers are widespread in the Mediterranean belt. Excluding France, they occupy more than 60% of the land surface. Although such soils are most abundant in hilly or mountainous areas, alluvial fans and river terraces can also contain large amounts of rock fragments. Despite their vast extent, these soils have received little attention in relation to desertification processes (degradation of the vegetation cover, physical soil degradation and soil erosion by water). However, it is becoming increasingly evident that such soils are less sensitive to increasing aridity than soils without rock fragments (Poesen and Bunte, 1996).

The fragments come to the surface by selective erosion of fine particles and ploughing. For example, soils in southern Portugal (Alentejo) have been heavily used for wheat production since the last century. This over-exploitation increased the extent of lithosols dramatically compared to similar landscapes in southern Spain.

In smaller landscape units (e.g. individual hillslopes), rock fragment cover often increases nonlinearly with slope angle. Selective erosion of the finer materials on the steeper slope sections and deposition on more gentle slope sections are responsible for this phenomenon. Furthermore, tillage brings rock fragments to the surface where soils are shallow.

Desertification processes affected by the presence of rock fragments

Rock fragments alter the major hydrological processes at or near the soil surface, which in turn determine the intensity of desertification (degradation of vegetative cover, physical soil degradation and soil erosion by water). A distinction should be made between rock fragments at and below the soil surface. Fragments resting on the soil surface or partly incorporated in the top layer affect the partitioning of rainfall into interception, rock flow (i.e. runoff generated by the rock surface), infiltration, overland flow and evaporation.

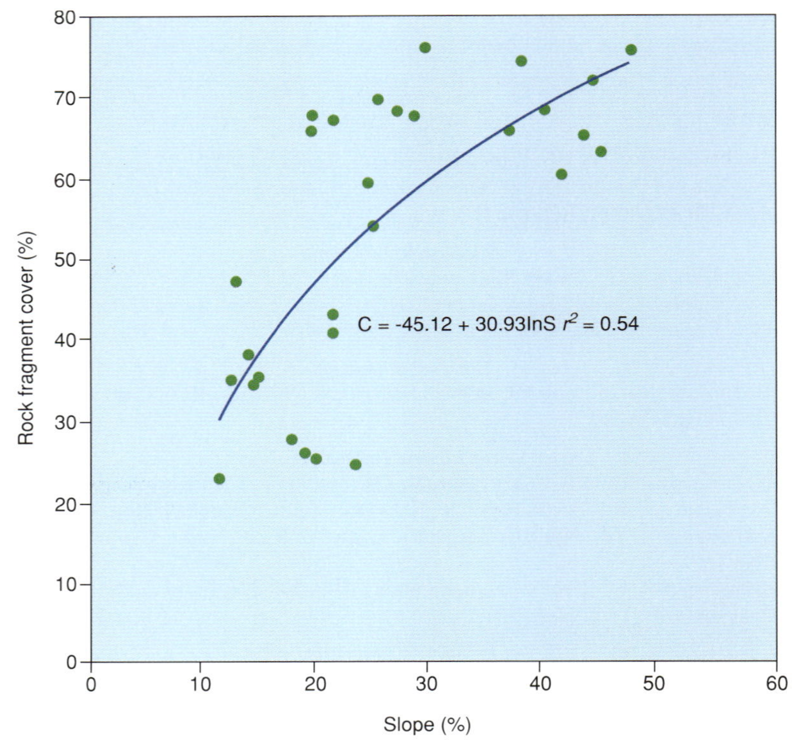

Figure 2.6.5 Relation between slope (S) and surface rock fragment cover (C) along a hillslope in abandoned agricultural land, Almeria, Spain (Poesen and Bunte, 1996).

Fragments below the soil surface affect the water percolation rate and thus also the infiltration rate and runoff generation (Poesen and Lavee, 1994; Poesen and Bunte, 1996).

Fragments below the surface also protect the soil from physical degradation (i.e. an increase of the bulk density) by forming a skeleton which supports the soil structure. In particular, agricultural soils are prone to physical degradation by rainfall. Laboratory experiments showed that small rock fragments (1.7–2.7 cm) were effective in reducing the subsidence rate and in maintaining a low bulk density and a high porosity. These results agree with the lower fine earth bulk densities (high porosity) of agricultural soils containing increasing amounts of rock fragments (van Wesemael *et al.*, 1995).

The way stones at the soil surface influence soil erosion by water varies with the spatial scale. Although at the mesoscale (inter-rill area) rock fragments

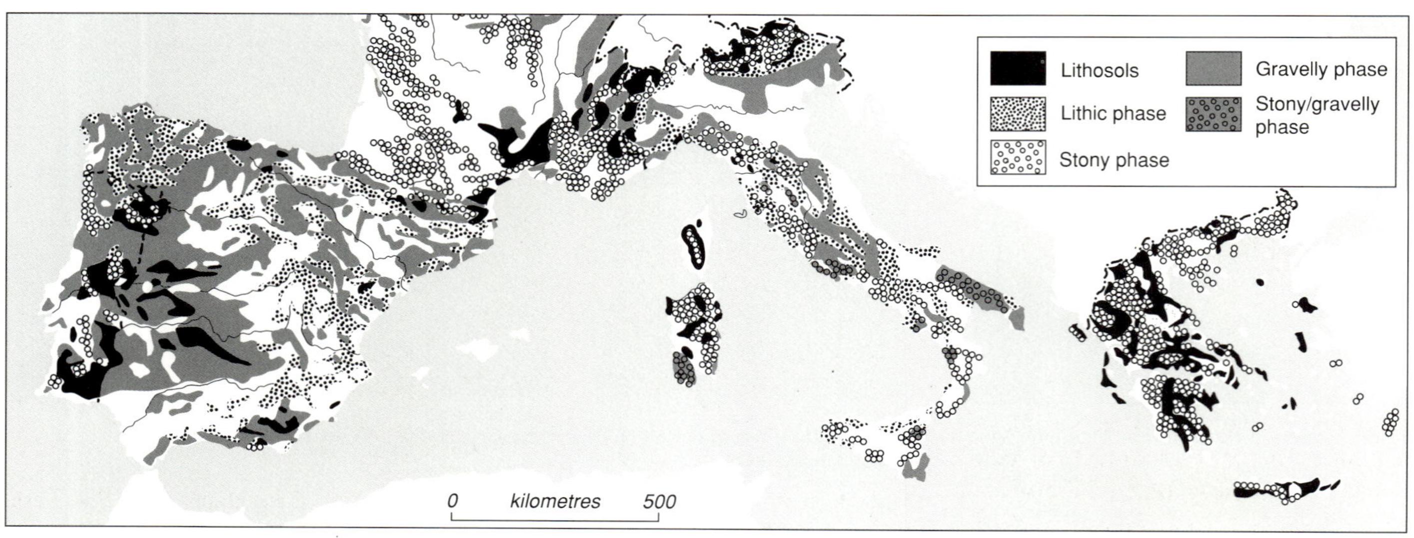

Figure 2.6.4 Thematic map of the European Union depicting soils containing large amounts of rock fragments. Redrawn with permission from Poesen and Lavee (1994).

might activate soil erosion, this trend will often be only temporary, since increased surface erosion will expose more rock fragments, leading to the formation of an erosion pavement. At the micro- and the macroscale, rock fragments act as natural surface stabilisers and relative sediment yield decreases with increasing rock fragment cover (Poesen *et al.*, 1994).

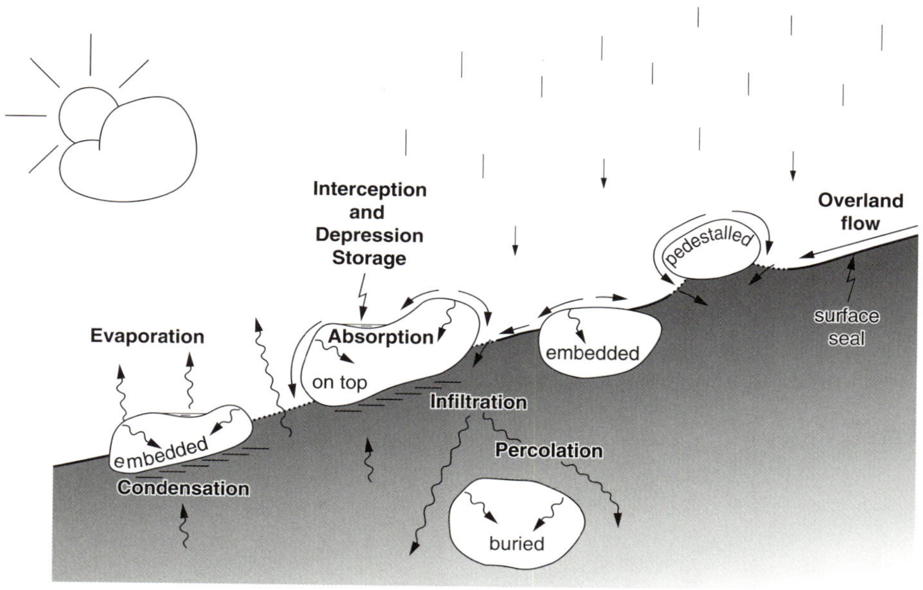

Figure 2.6.6 Hydrological processes which are affected by the presence of rock fragments (Poesen and Bunte, 1996).

Table 2.6.2 Spatial distribution of soils containing rock fragments in the top layers based on the soil map of the European Union and CORINE database (Commission of the European Communities, 1985; Poesen and Bunte, 1996).

Country	Portugal (%)	Spain (%)	France (%)	Italy (%)	Greece (%)
Lithosols	21.1	2.4	3.1	4.2	36.2
Phases:					
Lithic	2.0	20.2	0.4	23.2	—
Lithic/stony	—	0	10.0	8.6	—
Lithic/gravelly	—	—	0	8.4	—
Stony	11.5	—	9.6	7.4	46.2
Stony/gravelly	—	—	—	5.0	—
Gravelly	35.2	37.8	12.7	4.4	0
Concretionary	0	0	—	4.6	—
Total (%)	69.8	60.5	35.8	65.8	82.4
Land area (1000 km^2)	92.1	504.8	544.0	301.3	132.0

Lithosol: hard rock occurs at less than 10 cm depth.
Lithic phase: continuous coherent and hard rock occurs within 50 cm of the surface.
Stony phase: presence of stones (rock fragments with a diameter larger than 7.5 cm), boulders or rock outcrops in the surface layer or at the surface.
Gravelly phase: presence of more than 35% of gravels (rock fragments with a diameter up to 7.5 cm) in the surface layer.
Concretionary phase: presence of more than 35% of pedogenic concretions (with a diameter up to 7.5 cm) in or near the surface layer.

Management of stony soils

From an agricultural point of view, rock fragments are only too often seen as a nuisance, since they damage tillage implements. Therefore, removal of rock fragments is a well-known practice. Considering the mitigation of the most important desertification processes using rock fragments, their removal should be discouraged. Crushing of rock fragments to gravel size, if economically feasible, seems a good alternative, since small stones are generally more effective than large ones. Another practice, which can only be applied in perennial crops or crops with a high return, is stone mulching, where a layer of about 20 cm of gravel often mixed with manure is spread over the fields.

Figure 2.6.7 Simulation of physical degradation by rainfall of gravelly topsoils in the laboratory (van Wesemael *et al.*, 1995).

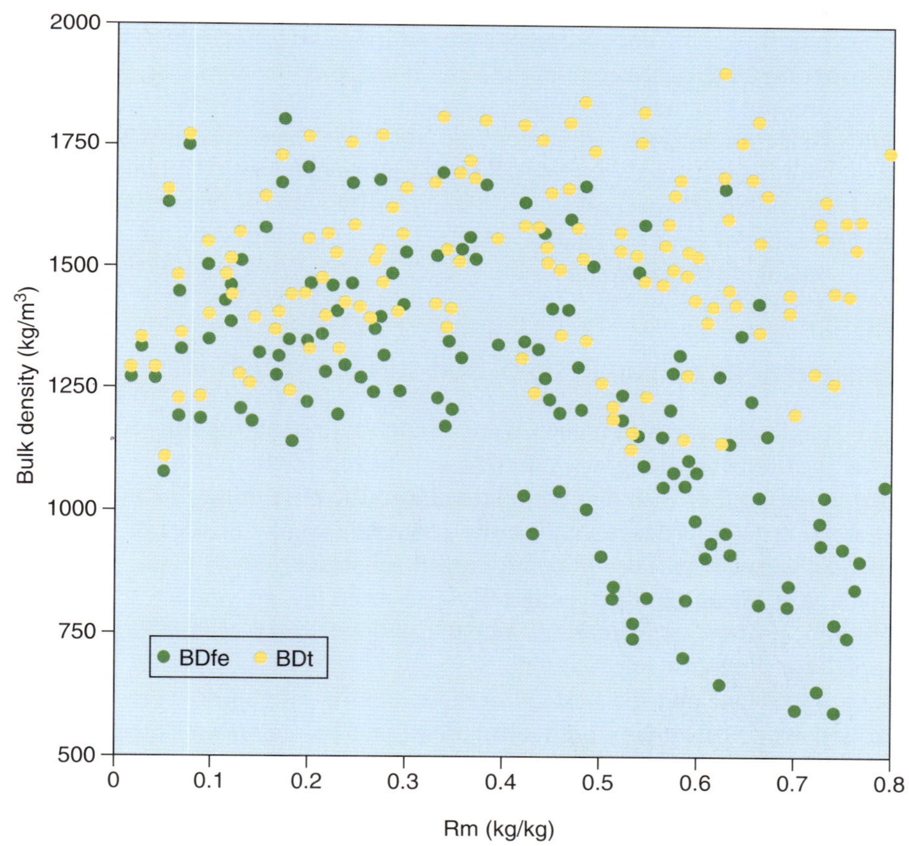

Figure 2.6.8 Fine earth (BD_{fe}) and total bulk density (BD_t) of cultivated soils of the Ebro Basin (Spain) decrease with high rock fragment content by mass (R_m) (data from Alberto, 1971).

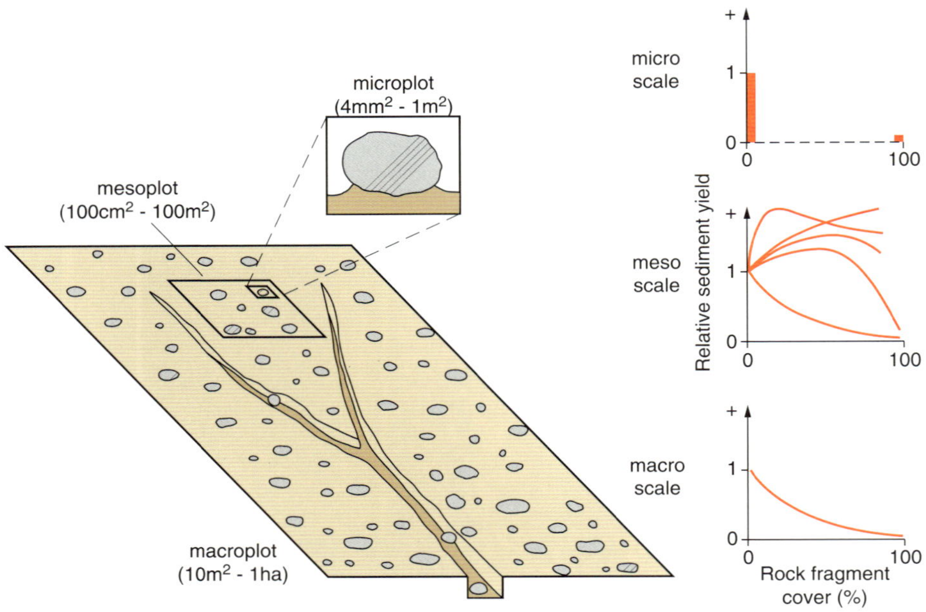

Figure 2.6.9 Surface rock fragments affect soil erosion at various field scales, micro, meso and macro (Poesen *et al.*, 1994).

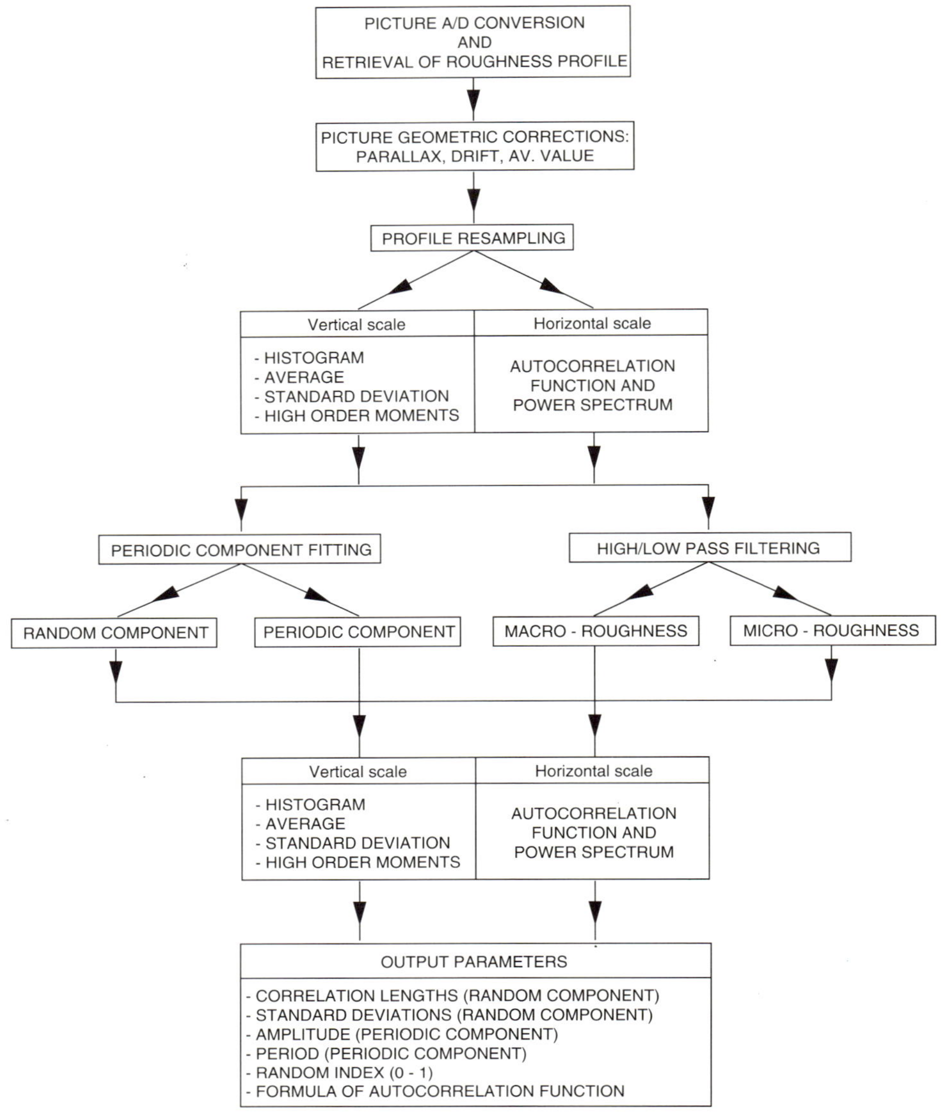

Figure 2.6.10 Flow diagram of the statistical software for the roughness profile processing.

ANALYSIS OF SOIL SURFACE CHARACTERISTICS

L. Chiarantini, P. Coppo, S. Gagliani and G. Luzi

Introduction

Soil micro-relief (roughness) and soil moisture content have a large impact on surface runoff, erosion and sediment transport. This section outlines a procedure for observing these parameters in the field in the Abruzzo test areas. Microwave images provided by satellite remote sensors (i.e. ERS1 and ERS1 SAR) are the result of electromagnetic interactions in which, at characteristic wavelengths, soil moisture and surface roughness play a fundamental role (Chiarantini *et al.*, 1995). The field data was collected simultaneously with ERS1 satellite passages both for data for erosion studies and to verify the sensitivity of microwave remote sensing to these two parameters.

Measurement and analysis of soil surface roughness

Field roughness measurements were made by photographing a 2-m-long needle profilometer along a 4–6 m transect for representative areas of selected fields, both parallel with and perpendicular to the ploughing furrows and erosional features. In order to analyse the geometric characteristics of bare soils, statistical software was developed for profile analysis.

Each roughness picture is digitally converted by an optical scanner, the roughness profile reconstructed, and the geometrically corrected profile resampled for the following analyses.

The horizontal and the vertical components of roughness were separately analysed using histograms, autocorrelation functions and power spectral analyses. The standard deviation of heights (σ) calculated from the histogram defines the vertical scale of roughness, whereas the autocorrelation length (lc) calculated from the autocorrelation function defines the horizontal scale of roughness.

Where the profile shows a row pattern, a periodic component is fitted to separate out the random from the periodic component; otherwise filtering is applied to separate macro-roughness from micro-roughness.

Finally, statistical and autocorrelation functions and power spectral estimates are applied to the retrieved random and possible periodic component of the profile to output roughness data.

As an example, two digitally converted roughness profiles are shown in the figures: a roughness profile with a periodical component perpendicular to tillage direction (the three lines represent the full roughness profile, the periodic and the random roughness components separately), and a random roughness profile parallel to tillage direction. The three lines represent a full profile without obvious periodicities, and the micro-roughness and the macro-roughness components after filtering have been applied to the full profile. An index is applied to establish whether a profile is periodic or random.

The effect of tillage in the Piomba and Cigno basins on surface roughness was examined. Different tools produce different roughness characteristics parallel and perpendicular to the ploughing direction both in the horizontal (lc) and vertical (σ) components. The rougher surfaces have higher σ and lower lc values than smoother surfaces. The horizontal roughness component (lc) is mainly controlled by the tillage method whereas the vertical roughness component (σ) values are closely influenced by the size of the clods.

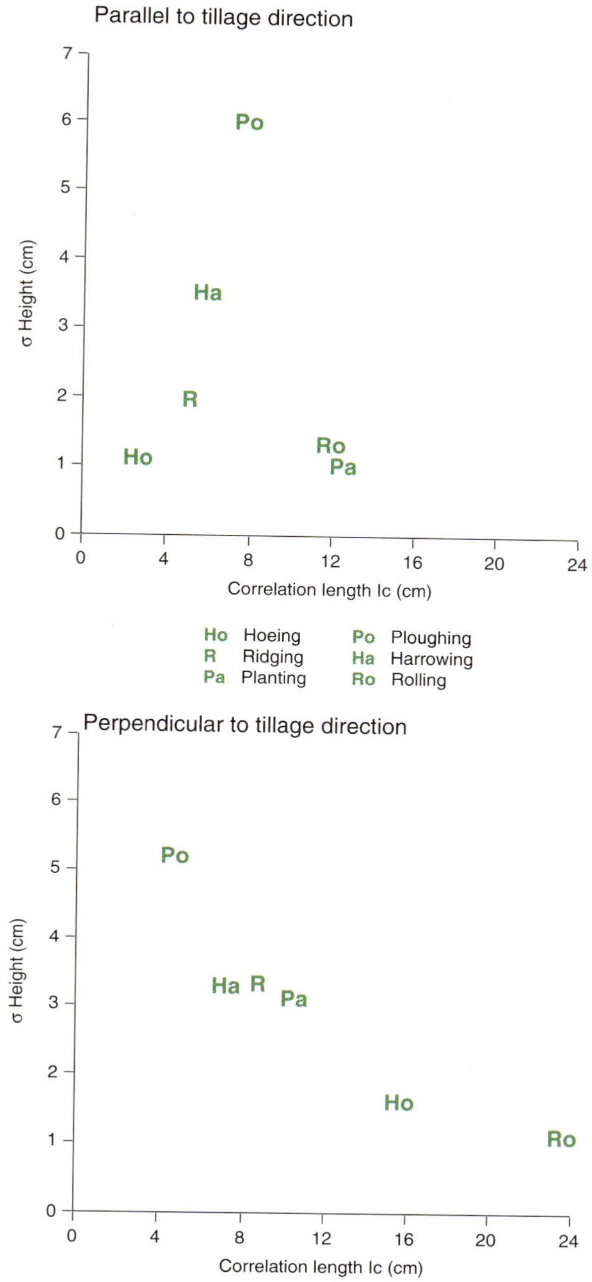

Figure 2.6.11 Influence of some tillage implements on surface roughness parameters (σ, lc).

Table 2.6.3 Statistical parameters of soil roughness profiles (a) perpendicular and (b) parallel to tillage direction.

a	σ	2*A*	*lc*	*P*	*D*
	(cm)	(cm)	(cm)	(cm)	(0<*D*<1)
Full profile	3.3	—	9.1	49.8	0.59
Periodic component	—	9.0	—	49.8	0.66
Random component	1.2	—	2.9	—	0.35

b	σ	2*A*	*lc*	*P*	*D*
	(cm)	(cm)	(cm)	(cm)	(0<*D*<1)
Full profile	1.9	—	5.3	100.5	1.13
Micro component	0.4	—	0.6	3.4	0.68
Macro component	1.9	—	5.5	100.5	0.16

σ = standard deviation of heights (HSTD), *D* = random index (0=random; 1= periodic), *A* = amplitude of periodic component, *lc* = autocorrelation length; *P* = period.

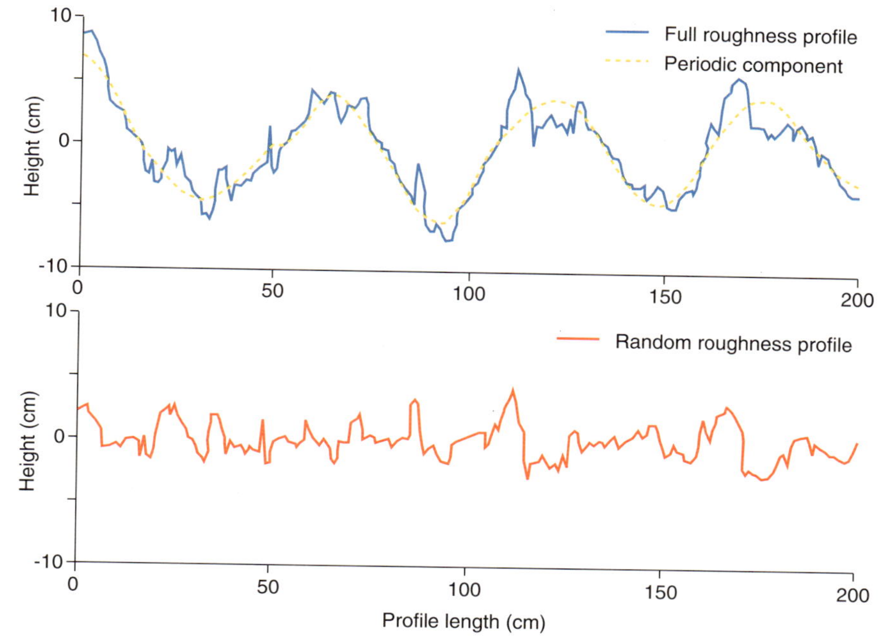

Figure 2.6.12 Example of a roughness profile with a periodic component perpendicular to tillage direction.

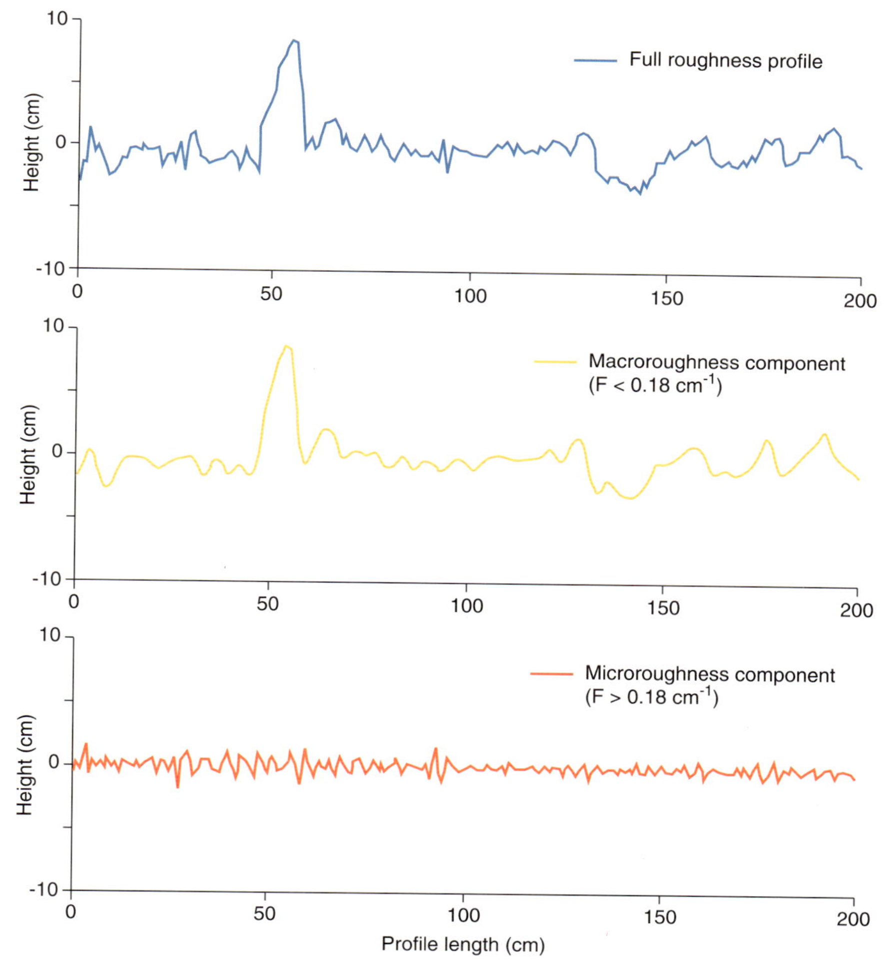

Figure 2.6.13 Example of a roughness profile parallel to tillage direction. F = cutoff frequency (cm^{-1}).

Moisture content measurements by means of a portable microwave probe

Several new soil moisture measurement techniques have been developed in the last few years, including the dielectric constant (DC) method. In the microwave spectrum, the dielectric behaviour of the soil is strongly related to the degree of soil moisture because of the high DC of water.

The response to an incident electromagnetic field of a homogenous and isotropic medium under the influence of an electromagnetic field, can be described by its DC, and is usually referred to the free space permittivity, ε_0. The presence of even a small quantity of soil water can significantly affect its DC. Soils can be considered a mixture of four components: air voids, solid particles, free water and bound water, the last being represented by water molecules tightly held to particles and the lattice structure of the soil. These two different phases of water have dissimilar DC values, and their relative importance is related to soil texture. In short, at a given frequency the DC of soil is affected by its texture, by the bulk density and, for the lower frequencies, by the salinity of the soil. Temperature, specific density, and the presence of organic matter are of less importance.

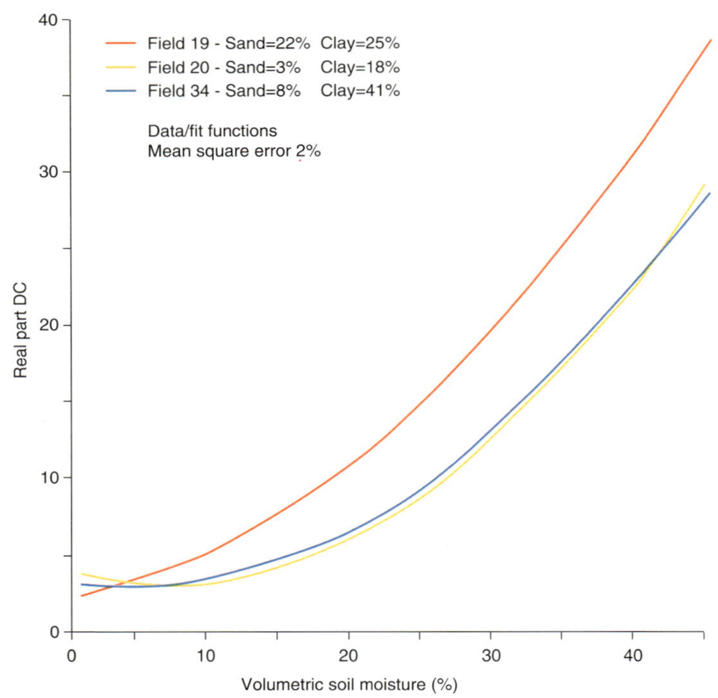

Figure 2.6.14 Laboratory real part DC data of three soil samples corresponding to different textures, as a function of the volumetric soil moisture.

Measuring apparatus and methodology

The dieletric constant measurement system, working at 1.42 GHz, is a two-channel reflectometer that measures the impedance variation of terminated coaxial cable in a dielectric medium on the surface of an open-ended capacitance: the probe's tip. It is battery-operated and easy to use in the field. The instrument measures the DC of the sampled volume (about 4–6 cm^3) and a calibration is used to convert DC data into soil moisture content data. Different approaches were used for this calibration; some based on empirical or semi-empirical formulae, others based on a physically based model. An empirical method similar to the one used by Hallikainen *et al.* (1985), was used here. The coefficients of quadratic functions were obtained over different soil samples at three different frequencies. The inverse relationship was then obtained by means of laboratory measurements of soil samples with different textures.

Data collection

The method has been applied in different field campaigns mainly devoted to remote sensing. For example, Luzi *et al.* (1993) applied the method on agricultural soils along the Pesa River in the Chianti region. In the Abruzzo, within the framework of the MEDALUS Project, DC derived and gravimetrically derived soil moistures were compared (Chiarantini *et al.*, 1995). These samples will be used for the retrieval of SMC from DC data.

SOIL DEVELOPMENT AND TRANSFORMATIONS IN RELATION TO DESERTIFICATION: AN EXAMPLE FROM SARDINIA

A. Aru

Lithological type and landform play major roles in soil formation and its transformation with time. The close relationship between weathering and transformation processes on the one hand and erosional processes on the other (runoff, erosion, sedimentation) currently taking place in the Rio Santa Lucia catchment (south-western Sardinia, Italy) can be clearly recognised in the soil sequences studied. Different soil types, corresponding to the different lithological and morphological situations, can be recognised in the Rio Santa Lucia catchment.

The soil sequences of the upper part of the catchment demonstrate that the degree of soil development depends on slope inclination, slope stability and plant cover. Shallower, less mature soils (Entisols) are found on the upper hillslopes, where the granite parent material weathers very slowly and where erosion is largely characteristically attributable to atmospheric phenomena. More developed and deeper soils (Inceptisols) have been formed in the valley floors as a result of deposition of material as colluvium in stable areas on more stable landforms.

The soil sequences of the flood-plains indicate that (1) soil maturity is related to the age of the parent material, and (2) the ageing or rejuvenation processes depend on morphological and climatic factors (fluvial terracing formation resulting from the climatic fluctuations of the Quaternary). Leached, desaturated and cemented soils (Alfisols–Ultisols) with a high percentage of sesquioxides of iron and aluminium are more common on ancient and stable surfaces, while poorly developed soils (Inceptisols) or those undergoing constant rejuvenation (Entisols) are dominant in the plains where incision by the river is still taking place.

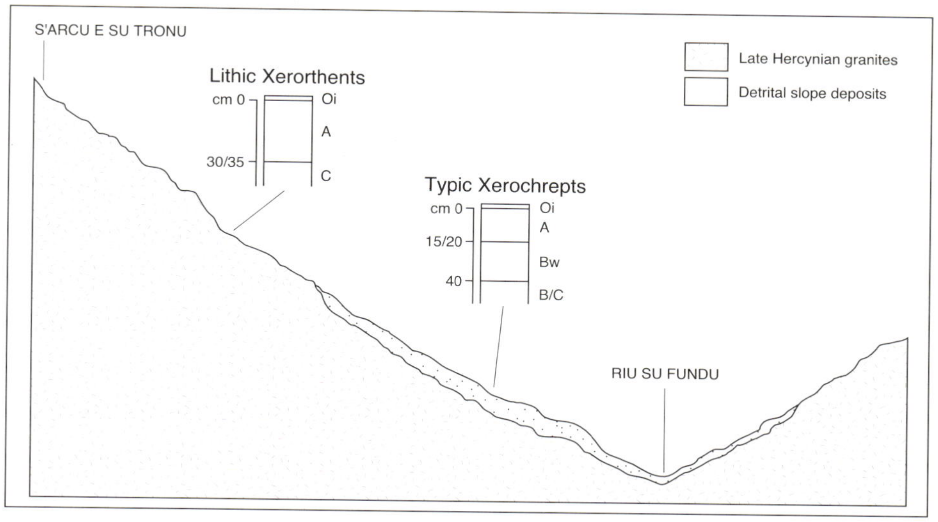

Figure 2.6.15 Schematic diagram of soil sequences in the upper part of the Rio Santa Lucia catchment.

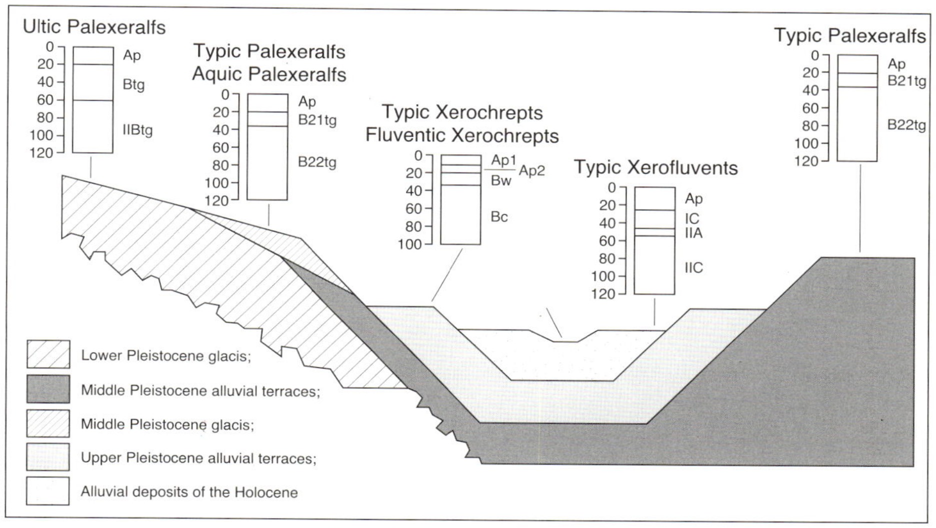

Figure 2.6.16 Schematic diagram of soil sequences in the flood-plain of the Rio Santa Lucia catchment.

The soil degradation taking place in the plain, where widespread impoverishment and deterioration of the soils' physical properties can be clearly recognised, is a result not only of natural causes but also of land use. Particular human activities, such as gravel extraction in the flood-plain and grazing, are geomorphological agents capable of exerting impacts comparable to natural processes on soil development.

Figure 2.6.17 Typic xerofluvent soil.

Figure 2.6.18 Fluventic xerochrept soil.

Figure 2.6.19 Aquic palexeralf soil.

2.7 Runoff and Erosion

L. H. Cammeraat and A. C. Imeson

Runoff and erosion generation

Rainfall that is not intercepted by vegetation cover is mainly divided into two parts: one infiltrates into the soil, the other moves as runoff (overland flow) over the soil surface.

The ratio of infiltration to overland flow depends on the following:

1. the characteristics and type of rainfall, which is variable in intensity, duration and frequency of recurrence, and
2. the soil surface characteristics, including porosity, the occurrence of crusts, texture, physico-chemical properties of the soil material, biological activity of meso/macro fauna, and the presence of lichen and algae.

Runoff generation

During rainfall, runoff can be generated in one of two ways:

1. when the rainfall intensity exceeds the amount of water that can be absorbed by the soil ('Hortonian' overland flow), or
2. when all the pores between the soil particles are filled with water ('saturation excess' overland flow).

If overland flow is generated, the stress provided by the flowing water and impacting raindrops can detach and transport particles present at the soil surface: the process of erosion. Soil material can actually be detached from the soil surface by the impact of falling water drops alone and this also leads to a movement of soil surface material, which is called splash erosion. The sensitivity of soil material to erosion is termed soil erodibility whereas the ability of rainfall or running water to erode soil is known as erosivity.

Figure 2.7.1 Channel forming in a recently ploughed field as a result of rill erosion and rainwash. The photograph was taken directly after the erosive rainfall of May 1994, Halkidiki, Greece.

The different types of erosion are as follows:

1. *Sheet erosion*, where overland flow over a slope carries away fine material from the soil surface.
2. *Pipe and tunnel erosion*, where water is flowing through continuous large pores (pipes) and tunnels within the soil, and where the flowing water erodes the walls of the pipes and tunnels.
3. *Rill erosion*, where overland flow and runoff is concentrated in small channels (a few decimetres wide and deep). Rills are usually ploughed out each year on cultivated land.
4. *Gully erosion*, where overland flow and runoff is concentrated into large channels (gullies), up to several metres deep and wide. These are sometimes formed by the coalescence of several rills, or by the collapse of pipes and tunnels, and cannot be easily removed by ploughing.

Figure 2.7.2 Badlands, near Petrer, Alicante, Spain, developed in white marls, showing clearly developed rills.

Figure 2.7.3 Bounded runoff plots in abandoned olive groves at the MEDALUS field site, Spata, Athens, Greece.

Badlands are areas where gully-like erosion may cover whole hillslopes and may be quite extensive. In the Mediterranean, they often occur on marls or in sediments containing salts, and soil losses from badland areas may be very high.

Vegetation cover above a certain amount protects the soil from raindrop impact and reduces the amount of water that reaches the soil. The soil is also covered and protected by litter from the plants. Often, infiltration around plants is higher than on bare soil because the soil has a more open structure.

Field methods

L. H. Cammeraat and A. C. Imeson

To understand the processes related to runoff and erosion it is necessary to study these either in the field or in the laboratory. In the latter case, only fine-scale processes and soil-water interactions can be studied.

Several methods are in use. Naturally occurring erosion can be measured in the field by means of the following:

- runoff plots where soil loss and runoff following individual rainfall events are measured;
- catchment gauging stations where natural runoff and sediment discharge at the small catchment scale are measured.

Natural processes can also be studied by simulating controlled conditions in the field or laboratory, by means of rainfall simulators. These can be used to generate precipitation that resembles natural rainfall. Every natural rainfall event is different so rainfall simulators have the advantage that the same conditions can be repeated under experimental control.

Studying erosion and runoff processes in detail in the laboratory is achieved by taking soil monoliths or soil material. Then many of the effects present in the natural environment are controlled.

The scale problem is important when undertaking experiments under field conditions. This is especially true in semi-natural conditions, where there is a discontinuous patchy vegetation cover and infiltration, runoff and erosion are very heterogeneous. Erosion rates are strongly related to the vegetation patterns. This explains why, with decreasing scale, the variability of the measured runoff and erosion increases.

Figure 2.7.4 Large rainfall simulator (10 × 2 m) set up on a semi-natural vegetated slope near Belmonte, central Spain (European Field Experiment in a Desertification-threatened Area (EFEDA) field site).

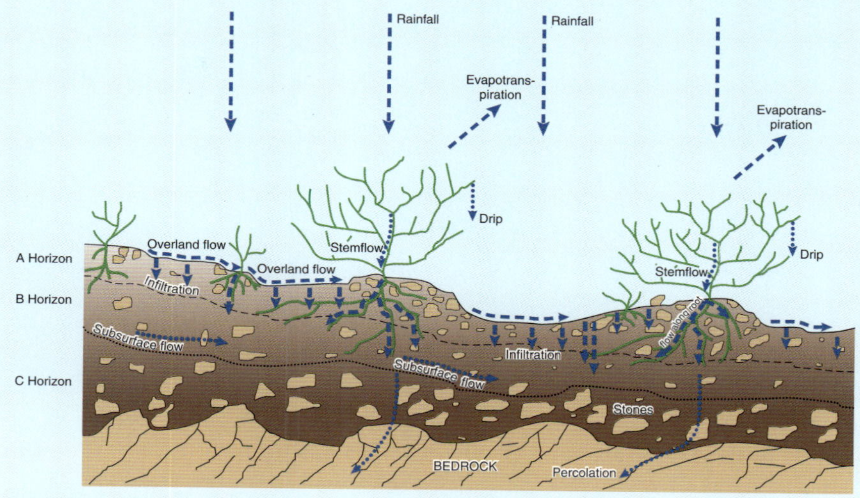

Figure 2.7.5 Flow of water over and through the soil on gently sloping semi-natural open vegetation in semi-arid areas.

On the other hand, in the laboratory, results are often influenced by non-natural boundary conditions, such as limited soil depth and related rapid saturation of the soil. Another problem is the lack of coupling or feedback with the natural environment.

EROSION AND LAND USE: LAND ABANDONMENT, DISTURBANCE AND REHABILITATION

L. H. Cammeraat, A. C. Imeson, J. Quinton and R. Morgan

The very long history of human land use in the Mediterranean region has resulted in changes of the natural vegetation cover. These are forms of ecosystem disturbance as well as causes of soil erosion. In the latter respect, the clearing of natural vegetation for the cultivation of crops and agricultural practices (e.g. tillage), apart from changing the physical characteristics of the soil, exposes it seasonally to rainfall.

In the same respect, intensive grazing, particularly by sheep and goats in semi-natural scrublands, exposes the soil to weathering agents by modifying the structure of the vegetation and influencing its composition. Moreover, they trample the soil surface and reduce the otherwise high soil infiltration rates.

Other forms of more severe disturbance of natural and semi-natural ecosystem disturbance are represented by wildfires, which destroy the protective vegetation and litter cover and also cause severe soil nutrient losses. Timber extraction after fires is a major cause of erosion; so also are desiccation, due to the over-extraction of shallow groundwater for irrigation purposes; and salinisation, as a result of irrigation, with saline or sodic water, or by irrigation mismanagement.

The release of human pressure on rural land, in the form of abandonment of agricultural land, in most cases leads to the re-occurrence and establishment of some form of secondary vegetation, which gradually contributes to the regulation of soil erosion. Even if soil erosion accelerates after abandonment, it gradually decreases as the vegetation cover starts to become established.

Figure 2.7.6 Forested catchment after a high temperature forest fire, Rebalsadores, Valencia, Spain. Note the accumulation of ashes in the channel bed. Many holes are present in the soil surface, where root systems are completely burned out.

Figure 2.7.7 Increased gully erosion caused by overgrazing of semi-natural vegetation (*maquis*) on marl substratum, Eleihora, Halkidiki, Greece.

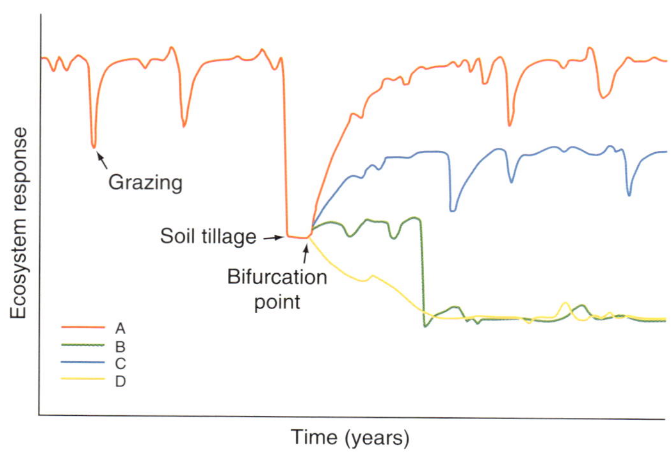

Figure 2.7.8 Resilience of an ecosystem after disturbance by different types of stress. A represents high resilience. B represents low resilience and irreversible degradation after a second major disturbance. C represents low resilience. D represents irreversible degradation.

In other cases, due to the destruction of vegetation and subsequent erosion, some ecosystems have been damaged to the extent that natural vegetation cannot re-establish itself. In such cases the disturbance regime has exceeded the ecosystem resilience and the damage is irreversible.

The resilience and inertia of the ecosystem, of which the vulnerability to erosion is part, determine whether the ecosystem (soils and plants) will recover after the disturbance. Therefore, rehabilitation of a disturbed ecosystem is only possible when the damage is reversible.

Rehabilitation can either be achieved by natural processes (e.g. vegetation succession) or by human intervention (e.g. by planting). Badlands represent highly disturbed ecosystems where rehabilitation is sometimes technically possible but practically infeasible.

Figure 2.7.9 Slope partly covered with semi-natural vegetation. On the left there is undisturbed (unploughed for several years) open cover of *Quercus ilex* L. subsp. *rotundifolia* (Lam.) Schwartz ex Morais trees, with colonisation of annuals and perennials. On the top of the hill is an almond grove. Rambla Chortal, Murcia, Spain.

Both natural and artificial rehabilitation involve the establishment of some form of vegetation cover which represents the only long-term sustainable strategy for controlling soil erosion in degraded ecosystems. In particular, artificial rehabilitation, which is based on the sequence followed by natural vegetation recovery, requires:

- designing a suitable ecological succession;
- creating the necessary soil and water environment to establish the succession;
- providing a long-term management regime to maintain the plant community once established.

A suitable ecological succession comprises:

- coloniser species which will establish quickly in a hostile environment;
- nurse species which will establish next and provide the foundation for other plants to follow;
- long-term species which will eventually form a more evoluted (mature) and relatively stable plant community.

Species must be selected because:

- they are available locally either as seeds or young plants;

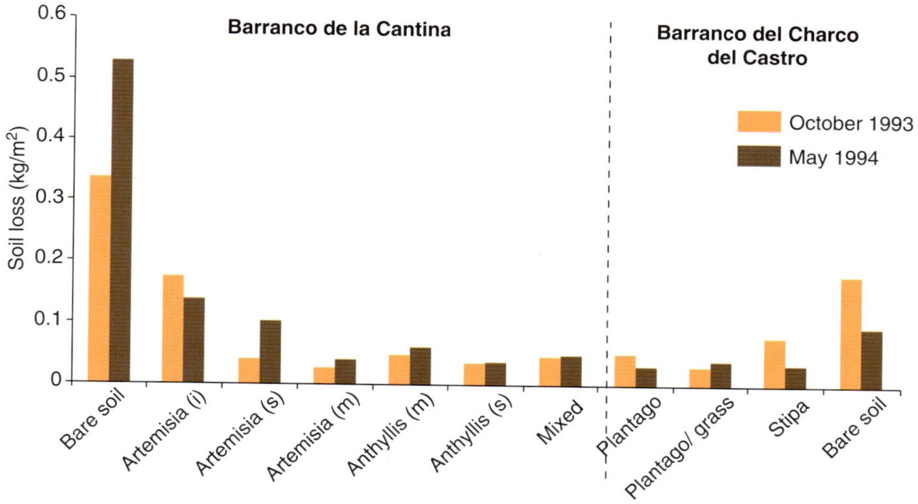

Figure 2.7.10 Soil loss from rainfall simulation experiments carried out in October 1993 and May 1994 on 11 treatments. The experiments were carried out using high intensity rainfall (120 mm/hour) in order to represent soil loss from a big storm. They show that soil erosion is reduced by the presence of vegetation. They also show that soil loss is higher in October than in May. This is likely to be because the more easily erodible material is removed in October, leaving behind very little material to be eroded in May. Immature plants (i): small plants assumed, because of relative size or other individuals, to be in the early stages of growth. Mature plants (m): plants that have achieved the maximum size for the species concerned. Semi-mature plants (s): plants that are intermediate in size to immature and mature individuals.

- they grow well in and are adapted to the local soil and climatic conditions (choice may be limited in saline areas);
- they perform the required 'engineering functions', such as promoting infiltration of water into the soil, increasing soil strength through root reinforcement and protecting the soil against erosion;
- they can withstand disturbance by fire and any proposed light grazing;
- they create a diverse plant community.

Ideally, local rather than exotic species should be chosen and site investigations should be made to establish the composition of the seed-bank remaining in the soil both on site and on adjacent land.

Grasses, herbs and shrubs may provide a more viable approach to re-vegetation than trees. Until a satisfactory litter cover to the soil has developed, trees can enhance erosion through splash from leaf drips, runoff from stemflow, and concentration of infiltrated water in the soil which may feed subsurface pipes. Re-establishment of the original *matorral* cover is, in many instances, a more appropriate and more achievable strategy.

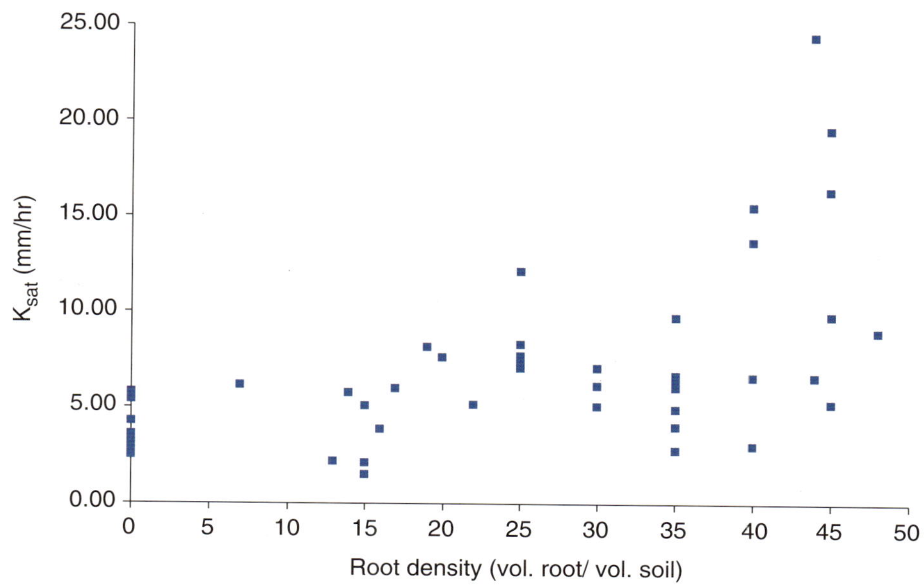

Figure 2.7.11 The effect of root density of *Trifolium pratense* on the saturated hydraulic conductivity of a clay loam soil. Water will move more quickly through a soil with a high saturated hydraulic conductivity than in one with a low saturated hydraulic conductivity.

2.8 Modelling the Physical and Biological Systems

M. J. Kirkby, J. C. Bathurst, I. Woodward and J. B. Thornes

Modelling desertification processes in general

M. J. Kirkby

Desertification is the result of a complex combination of physical, social and economic processes. These include land-use and irrigation changes, influenced by factors which may operate at scales ranging from local to international. The final effects are largely physical, and can be seen through changes in soil properties and their consequences for fertility or stock carrying capacity. The soil may be partly or wholly removed by erosion, usually leading to an increase in stoniness and losses of some nutrients. The soil chemistry may also be changed by irrigation and/or drainage, leading, in some cases, to an increase in salinity. Irreversible desertification in Mediterranean areas most commonly takes the form of extreme soil erosion or salinisation.

Physical modelling within MEDALUS has concentrated on the group of processes that is perhaps best understood, i.e. storm runoff and water erosion. Three approaches have been used to help our understanding of the interactions which are important, both at the present time and in the context of expected changes over the next century. There is likely to be increased pressure on land use for agriculture, increased tourism, and greenhouse warming, all of which may modify conditions for flooding and erosion.

The first of these models, SHETRAN, represents one of the most widely used existing models for routing water and sediment through catchments of 5–500 km^2. Its great strength is that it has an established base of calibration data and experience which gives confidence to its forecasts. Its weakness is that it uses fixed parameter values for soil properties, and so is not able to respond dynamically to changes in land use or vegetation, which can only be accommodated by providing new parameter values from comparable sites with the appropriate land use.

The other two models have been developed within the MEDALUS project. The models are primarily targeted on and validated against Mediterranean landscapes for uncultivated conditions, but the conceptual principles have a wider application. The MEDALUS model simulates changes down the length of a single hillside flow strip, or slope catena, whereas the MEDRUSH model is designed to simulate changes over larger catchments, of up to 2500 km^2. MEDRUSH uses the understanding of slope catenas gained from the MEDALUS model, and applies it, in a somewhat simplified form, to one representative slope catena within each of several hundred subcatchments which make up the total study area. Within the slope catena, these models simulate both the seasonal and longer-term changes in soil and vegetation, so that the runoff and erosion forecasts respond dynamically to these changes in conditions. This allows the models to simulate some of the positive and negative feedbacks which are known to be important, and which can lead to irreversible desertification in some cases and to recoverable degeneration of soil and vegetation in others.

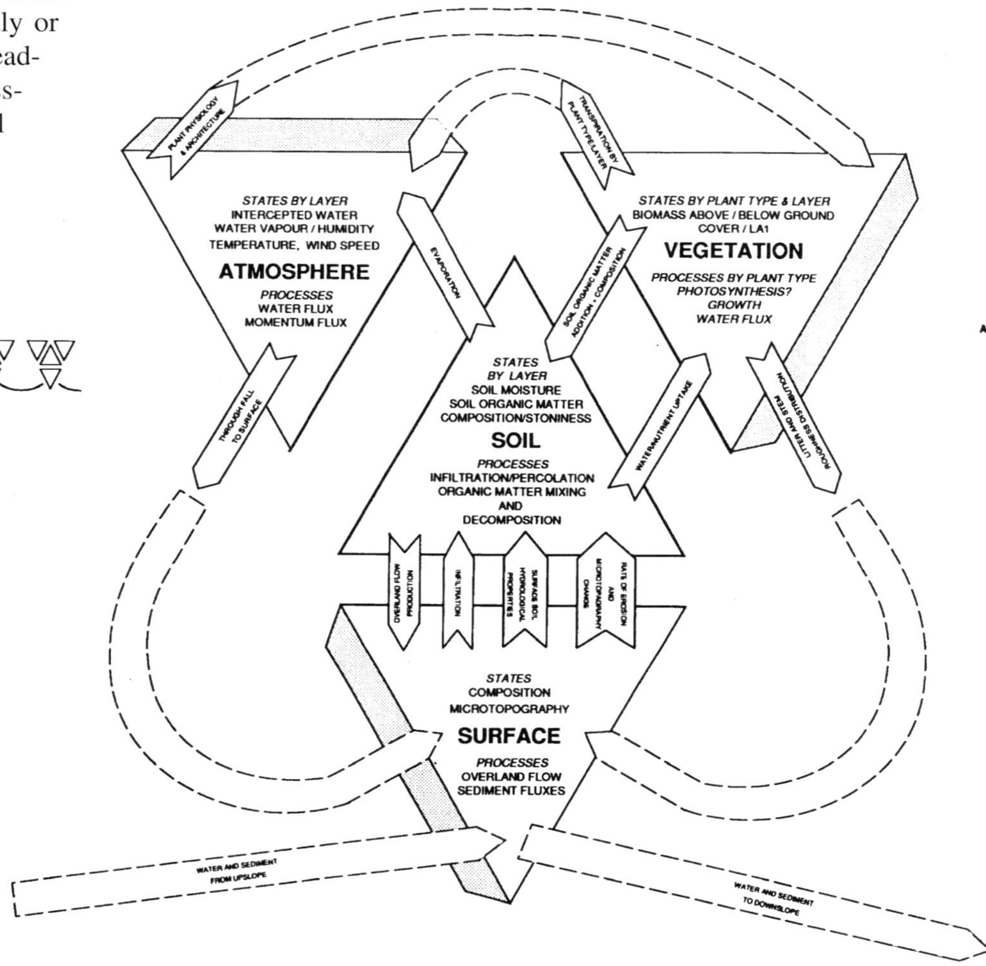

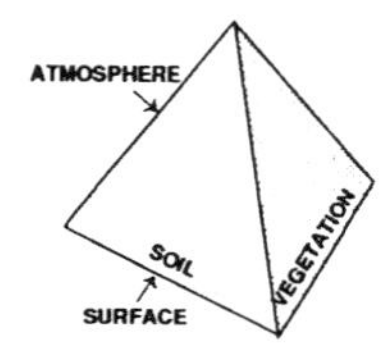

Figure 2.8.1 The structure of the MEDALUS hillslope model. The four sides of the tetrahedron represent the main process domains, and the arrows the flows between them.

The MEDALUS model allows direct comparison with the erosion plot measurements which were set up in Portugal, Spain, Italy and Greece during the first phase of the MEDALUS project in 1991. The model structure directly simulates the standardised plot design, with erosion plots at three positions down the slope catena. Field data have been used to calibrate the model parameters directly. As the understanding gained from these erosion experiments has been applied to larger areas, to provide a regional planning tool, the model has been scaled up to catchments of several thousand square kilometres, retaining the hillslide flow strip as the elementary building block within each subcatchment. At this regional scale the model simulation provides two kinds of forecasts which can be used for planning purposes. First, it simulates the areal distribution of runoff and erosion as it is distributed across the area, with its variety of topography, land use and rock type; and secondly, it provides a forecast of the gross output of water and sediment at various points along the main valleys of the catchment. This accumulated output is important for surface water and groundwater supply, for the frequency distribution of major floods, and for potential sedimentation or incision of the trunk channels, with further implications for land use along valley bottoms.

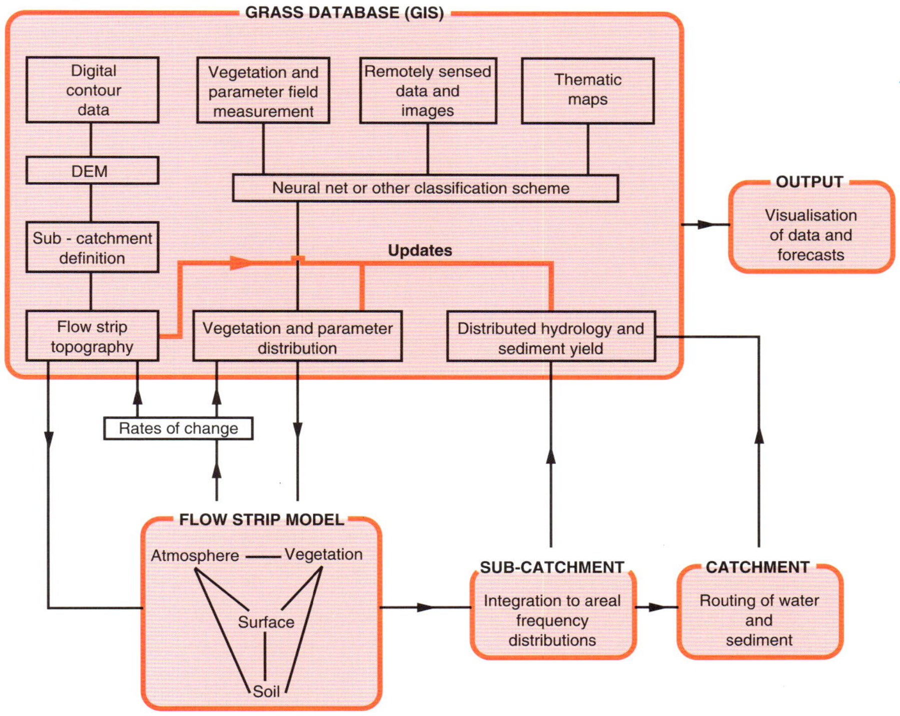

Figure 2.8.2 The structure of the MEDRUSH model.

MEDRUSH basin model

The MEDRUSH model simulates runoff and erosion in catchments of up to 2500 km^2 for periods of up to 100 years. The catchment is divided into a nested hierarchy of up to 250 subcatchments of 1–20 km^2, based on analysis of a Digital Elevation Model (DEM) for the area. Some of these subcatchments cover headwater areas, while others contain a through-flowing main stream. The high specific water and sediment yields from upland areas require a greater density of subcatchments in steep headwater areas. Over the time-spans of interest, it is assumed that there are no changes in the network linking the subcatchments, and that this network is represented by permanent channels (though not necessarily by perennial flows). The subcatchments are connected by routing algorithms for water and sediment which have been modified for use in the MEDRUSH model.

At the scales of interest, it is clear that subcatchments contain many channels. Some of these are permanent, while others may be dynamically extending or infilling in response to changes in climate, land use or vegetation. Their behaviour is generalised from the behaviour of a representative hillslope flow strip, for which a simplified version of the MEDALUS model is used. The total channel flow generated within the subcatchment is then aggregated from a weighted distribution of flow strip responses, using a linear routing algorithm based on the Geomorphological Unit Hydrograph (GUH) approach.

Vegetation growth is an important component within both the MEDALUS and MEDRUSH models. Growth by photosynthesis responds to solar radiation, and is strongly influenced by available water, which is provided by rainfall and stored within the soil. The moisture-holding properties of the soil depend on its texture, including stoniness which is in turn influenced by erosion, and organic matter which is supplied by the vegetation litter. Vegetation not only improves soil structure by supplying soil organic matter but also shields the surface from raindrop impact that can effectively seal the surface. A good vegetation cover therefore encourages infiltration and reduces overland flow runoff and, in consequence, sediment loss.

For example, a field that is abandoned for cultivation may erode rapidly at first. Over several decades it can gradually develop a perennial vegetation cover and a soil that has more stones and more organic matter, both of which partially protect it from further erosion. The MEDALUS and MEDRUSH models are designed specifically to simulate this kind of complex interaction.

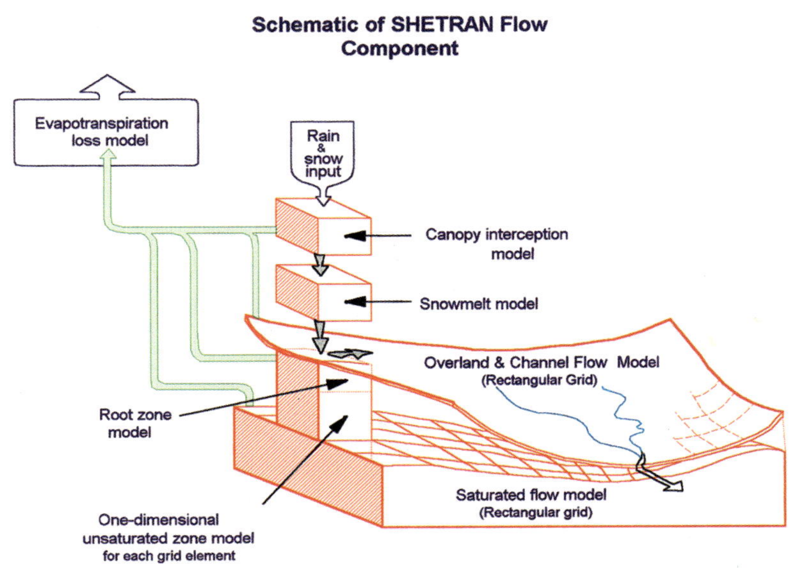

Figure 2.8.3 Schematic diagram of the hydrological component of the SHETRAN modelling system. Spatial variability is represented in the horizontal by a grid square system and in the vertical by a column of horizontal layers at each grid square. Each hydrological process is simulated at each grid square and water is transferred between squares and to the channel system as overland flow or saturated zone flow. (Prepared by R. Adams, WRSRU.)

SHETRAN

SHETRAN is a physically based, spatially distributed modelling system for hydrological, sediment yield and contaminant migration response, applicable at the scale of a river basin. It has been developed by the Water Resource Systems Research Unit, University of Newcastle upon Tyne, UK, from the Système Hydrologique Européen (SHE) hydrological modelling system (Abbott *et al.*, 1986a,b). SHETRAN provides a means by which the understanding of physical desertification processes obtained at the small spatial scale of field studies can be incorporated into environmental assessments at the basin or regional scale at which planning decisions are made. SHETRAN-type models are also the only means of predicting the environmental impacts of future changes in climate and land management. They can therefore make an important contribution to the mitigation of desertification.

SHETRAN FOR THE MEDALUS FIELD SITES

J. C. Bathurst, C. G. Kilsby and S. M. White

The SHETRAN hydrology and soil erosion modelling system was used:

1. To examine the degree to which data and process understanding obtained at the small spatial scale of the MEDALUS field studies can be extrapolated for use in simulations at the larger spatial scales at which planning decisions are made.
2. To examine the response of river basins in areas undergoing desertification to specified changes in climate and land use.

Test data were assembled for the 701 km^2 Cobres Basin in Portugal and the 159 km^2 Mula Basin in Spain. Future climate data were generated from GCM output assuming doubled CO_2 levels. Full catchment simulations were successfully carried out for both basins, and it was found that model parameters derived from field studies could be used in simulations at larger scales either unchanged or with minor adjustment, remaining within the range of observed measurements.

Land-use change impact studies were performed using present-day meteorological input data and model parameters modified to represent the replacement of agriculture by natural vegetation. Reduced sediment yields were found for both basins, due to a reduction in bare ground surface. Climate change impact simulations were performed, using model parameters validated for present conditions and input meteorological data representing a future climate. The mean future climate, although consisting of a substantial reduction in mean annual rainfall, was nevertheless found to lie within the natural variability of the present climate.

TOPOGRAPHY

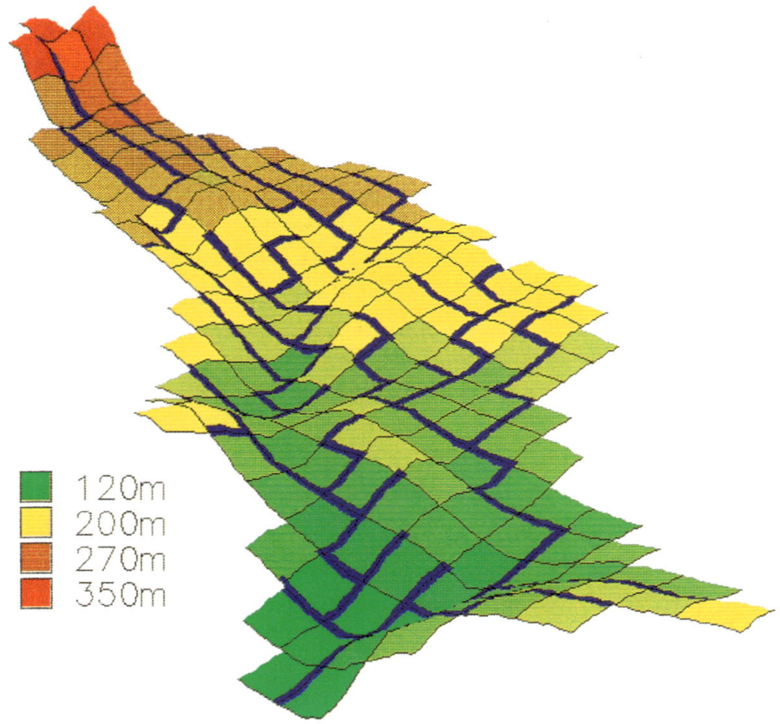

Figure 2.8.4 SHETRAN representation of the Cobres Basin elevation and river network on a 2 km grid.

EROSION

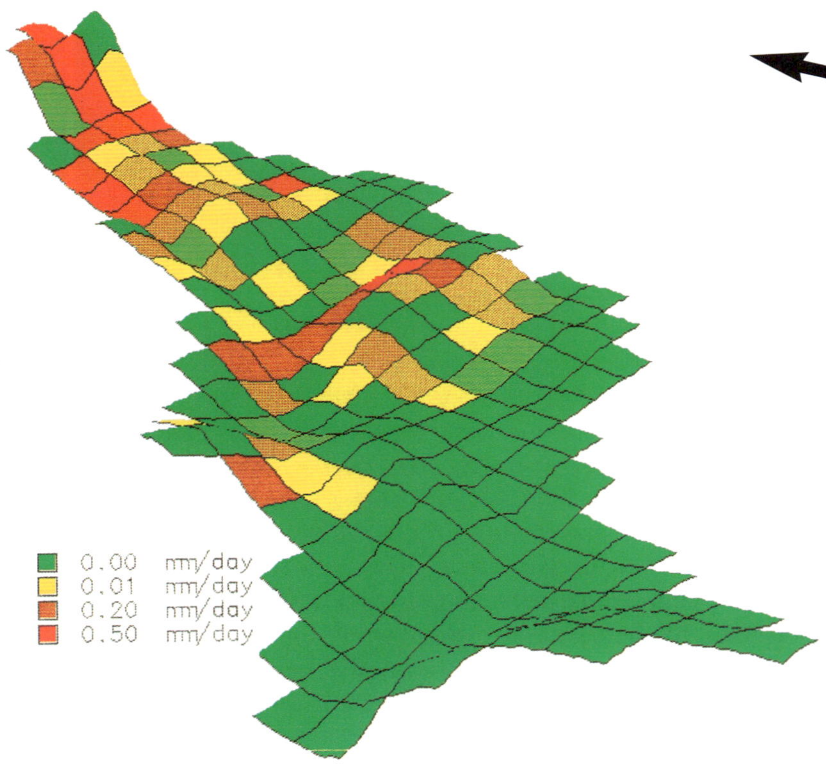

Figure 2.8.5 SHETRAN simulation of instantaneous erosion rates during a storm. High erosion rates are associated with regions with high rainfall rates and steep gradients.

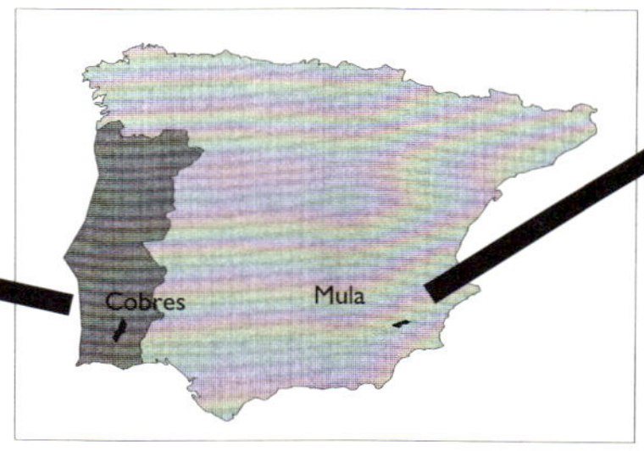

Figure 2.8.6 The location of the Cobres and Mula basins to which the SHETRAN model has been applied.

DISCHARGE TIME SERIES

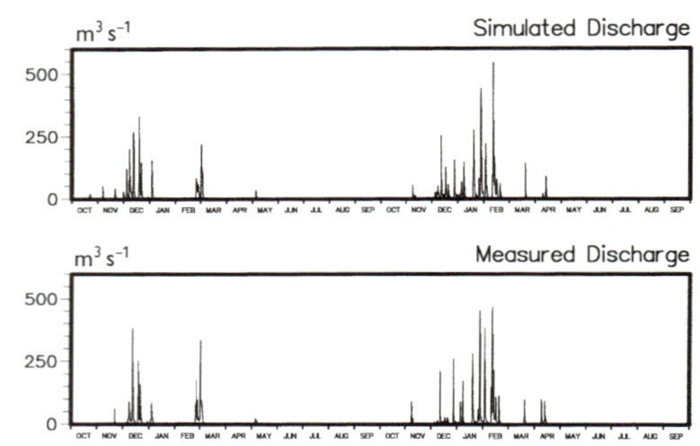

Figure 2.8.7 Measured and simulated discharge at the Monte da Ponte gauging station in the Cobres Basin for the period October 1977 to September 1979.

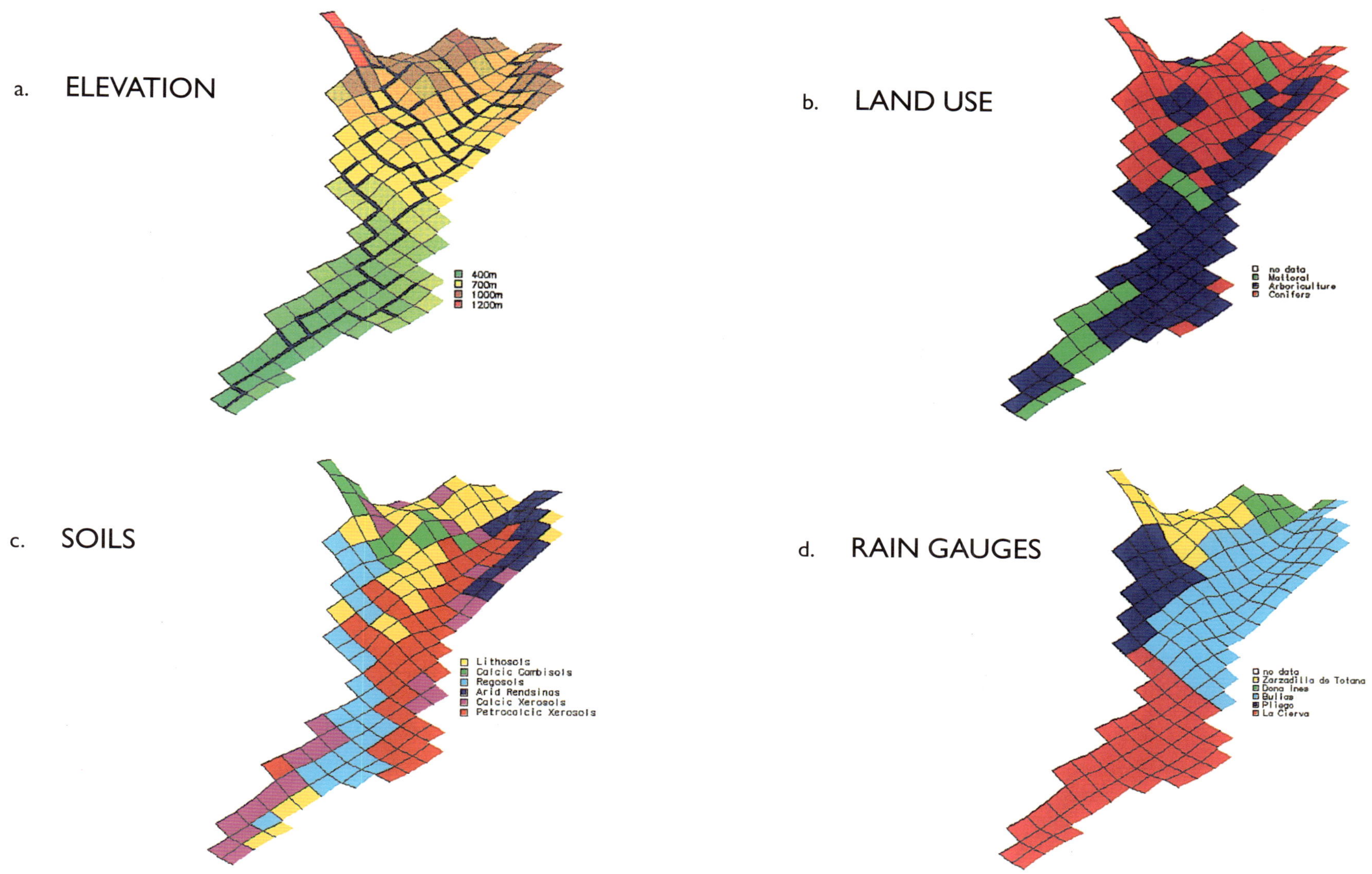

Figure 2.8.8 SHETRAN representation of the Mula Basin on a 1 km grid: (a) elevation and river network; (b) land use; (c) soil types; (d) rain gauge coverage.

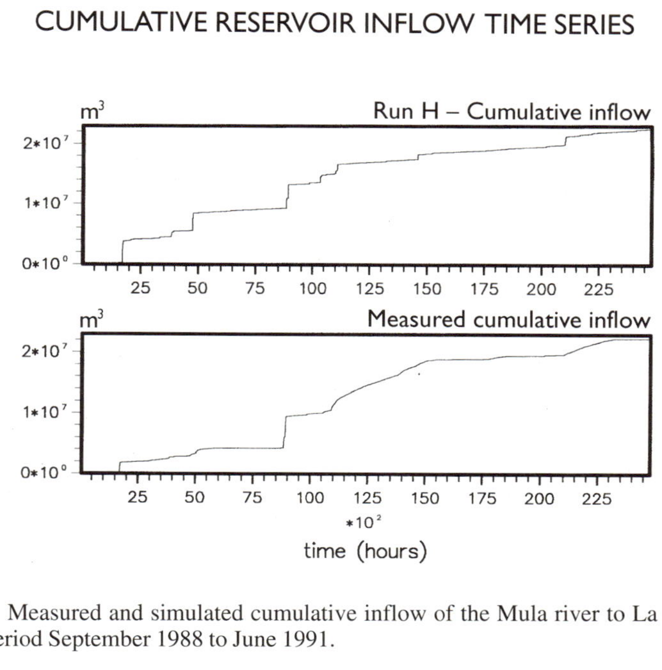

Figure 2.8.9 Measured and simulated cumulative inflow of the Mula river to La Cierva reservoir for the period September 1988 to June 1991.

THE MEDALUS HILLSLOPE MODEL

M. J. Kirkby and M. L. McMahon

The MEDALUS slope catena model represents four interacting submodels for the atmosphere, vegetation, soil and surface systems. These four subsystems are defined at each of a series of points down a hillslope catena, which are connected by routing of overland flow, and where appropriate subsurface flow, and grain-size selective sediment transport. Shrub and grass models simulate vegetation growth and also gradually modify soil organic matter. Evapotranspiration is estimated from a standard sparse vegetation model. Soil water movement is obtained by solving the unsaturated flow equation for up to 10 soil layers. Flow on the surface is explicitly related to micro-topography, which controls the spatial distribution of overland flow, infiltration and sediment transport. The model uses 5–15-minute interval data from automatic weather stations to simulate conditions using a variable time step, to run for periods of up to a decade, using a Windows™-based Turbo-PASCAL™ system running on IBM-compatible PCs.

Field data are generally available at three points along a hillslope profile. This catena represents a flow strip, which may vary in plan width along its length, allowing flow convergence or divergence. Flows of water and sediment are routed between points and erosion is calculated by differencing the sediment storage equation for each grain size.

The long-term behaviour of the model simulations is determined mainly by interactions between the four submodels. Changes in climate and/or vegetation cover tend to produce changes in the amount of overland flow runoff generated. Increases in overland flow lead to increased sediment transport, producing greater erosion upslope and greater deposition at suitable downslope sites. Erosion, as well as being important in its own right, has a strong influence on subsequent hillslope response in at least two ways, through grain size and through micro-topography.

First, sediment transport selects for grain size, leading to the production of an armour layer, particularly in eroded areas, over periods of a few years. Erosion also gradually strips the topsoil, over periods of decades, usually bringing coarser subsoil material to the surface. Armouring generally limits subsequent runoff and erosion, although the exact relationship between stones and the surface may be critical in small plots (Bunte and Poesen, 1993). Surface stones provide some physical protection, and stones within the soil also have a beneficial effect on moisture retention up to concentrations of 30–50%, so that modest increases in stoniness partly counteract the effects of erosion by increasing vegetation cover.

Secondly, overland flow and sediment transport are concentrated in depressions on the surface, particularly depressions running upslope and downslope. Both erosion and deposition are most effective in changing the depth of the deepest depressions. Even where there is not an established or recurring pattern of rills and/or channels, it is normal for roughness to increase systematically downslope in uncultivated eroded areas, and to decrease downslope in areas of deposition.

These long-term interactions have little effect in the short run, although the impact of individual intense storms may be seen on armouring and surface roughness. Over a period, however, they control the positive and negative feedbacks which promote or constrain erosional degradation, and may lead to irreversible desertification of soils and landscapes, through a number of nonlinear feedback loops. Soil organic matter interactions, though not in themselves irreversible, help to drive the erosional interactions. The most important feedback is through soil stoniness. For a while, increased erosion of stony soils creates negative feedbacks through increased water availability in the soil organic matter, and irreversible degradation only occurs when stoniness becomes very high. Nevertheless erosion may be severe if driven by poor security. This buffering effect is, of course, completely absent in deep fine-grained soils, allowing badland development in extreme cases.

TESTING AND USING THE MEDALUS MODEL

J. Shao and J. B. Thornes

In order to apply the MEDALUS model it has to be tested against real data collected in the field. Here we illustrate its application to the El Ardal field site in Murcia. Other applications are illustrated in Brandt and Thornes (1996). The El Ardal field site is described in Section 4.3. Julian day is the number of days since the beginning of the year.

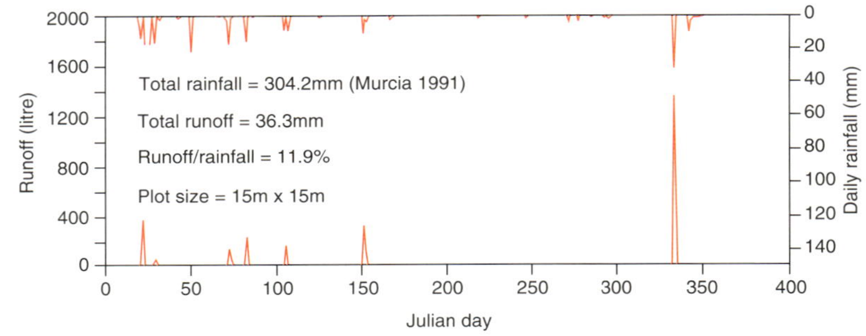

The El Ardal plot runoff is simulated here against the data for 1991. Runoff was closely approximated by the model in both absolute amount and as a proportion of rainfall. Both the long summer drought and the large autumn storm are well simulated.

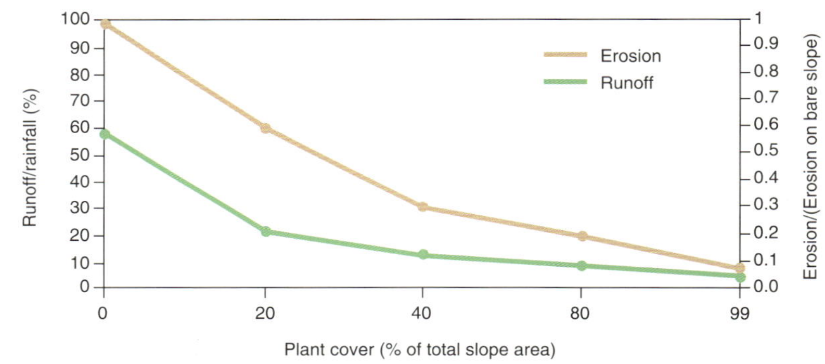

Curves of the impact of vegetation percentage cover on runoff and sediment yield are well replicated by the model.

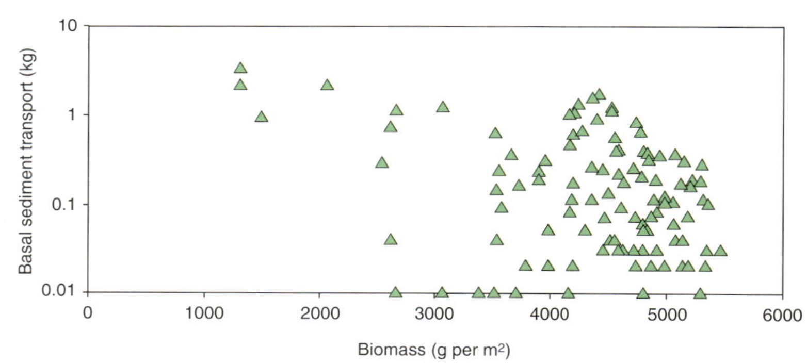

Modelled erosion plotted logarithmically against biomass for individual events shows a large scatter of low values of erosion at high biomass, but a low scatter at low biomass. The overall decrease in erosion with biomass is, however, quite clear.

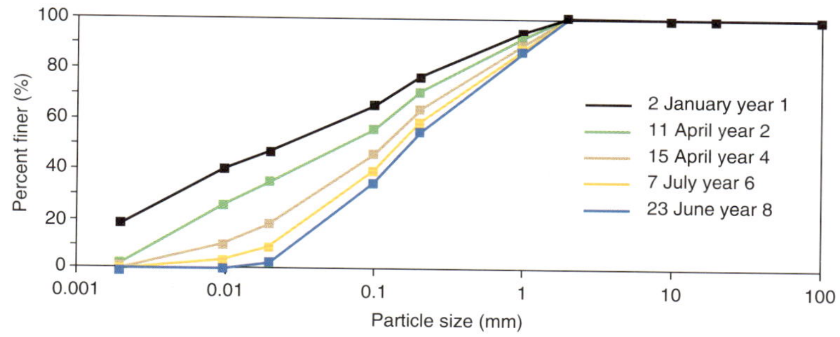

Over time, the slope soil cover becomes progressively coarser in a process called armouring. Here the shift to higher values takes place over a period of eight years of simulated storm events.

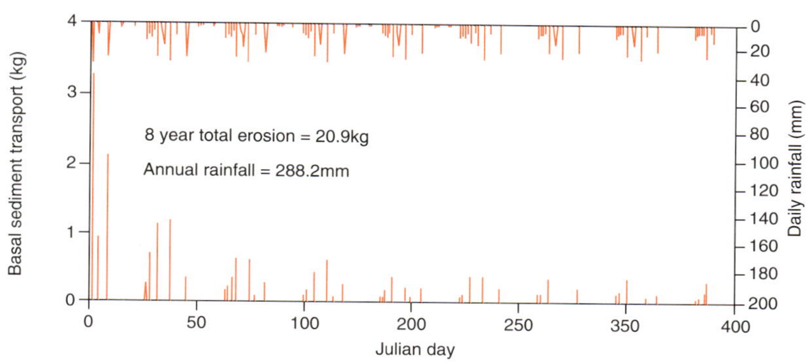

In these simulations the same rainfall (for Murcia, 1991) was repeated for eight years at 288.2 mm until the plant cover stabilised with the rainfall. Then the rainfall was reduced to 149.2 mm for seven years to simulate the drought conditions of the 1980s.

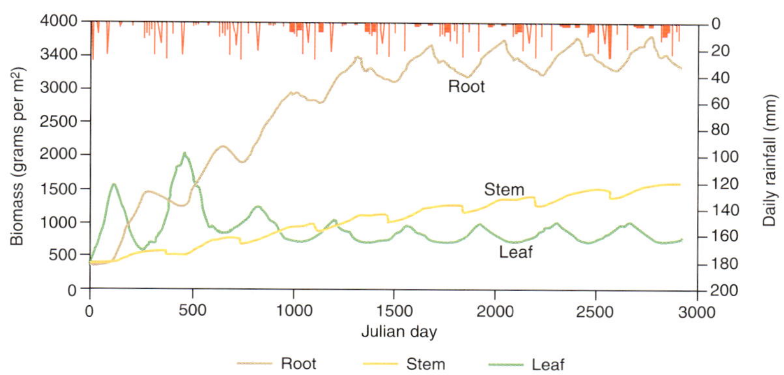

In the first eight years abandoned land quickly generated a substantial plant cover with a total above-ground biomass of 2500 g/m^2, which is fairly typical of low *matorral* in this area. Sediment yield over the eight-year period was higher at the start and then fell as the vegetation grew. In the early period leaf growth biomass dominated but roots gradually accounted for about 60% of the total biomass.

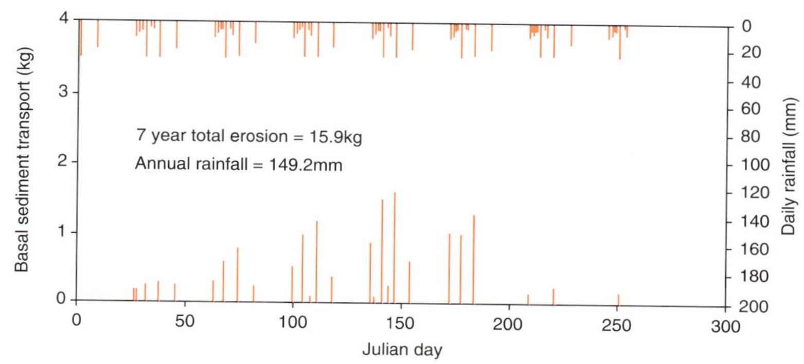

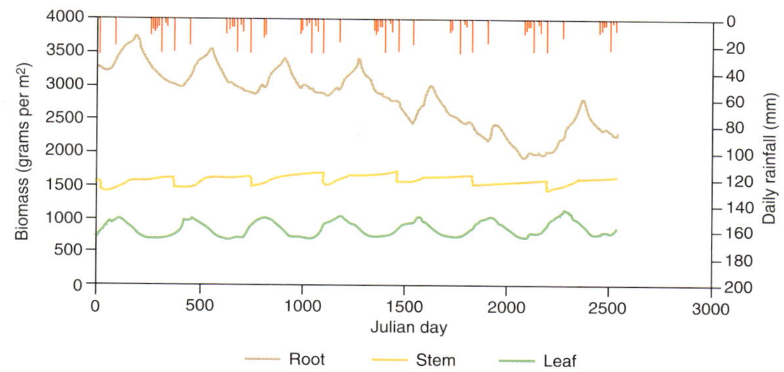

In the next seven years, with recurrent drought, as the plant cover diminished there was root die-back, though above-ground biomass was sustained. A progressive decrease in soil moisture was the main cause of this and eventually all the available soil for erosion was also depleted, so the last two years produced virtually no erosion.

INPUTS FOR THE MEDRUSH MODEL: GUADALENTÍN BASIN

R. Abrahart and J. C. Hawkes

Model parameters are drawn into the MEDRUSH model from data map layers (raster format), held in the integrated geographical information system (GIS), GRASS. These various inputs describe the surface characteristics of the terrain. The GIS provides an efficient data handling tool, in which parameters used to initialise the model are constructed from existing maps, satellite images, published reports, and field data. Changes predicted by the model are stored within the GIS, for subsequent printing or for further analysis.

2. *Sub-basin divisions*. The basic spatial unit, derived from the digital elevation model (DEM), from which representative flow strips are produced. These units are connected via a streamflow network.

3. *Vegetation functional types*. Reclassed land cover map, obtained from satellite remote-sensing images (University of Valencia).

5. *Stone cover*. Fieldwork data are used to produce a truncated soil map.

6. *Micro-topography*. This measure accounts for surface roughness due to surface rock fragments, rills and vegetation mounds.

7. *Infiltration*. Fieldwork data are used to provide estimates of surface runoff and saturated soil hydraulic conductivity for subsequent map production.

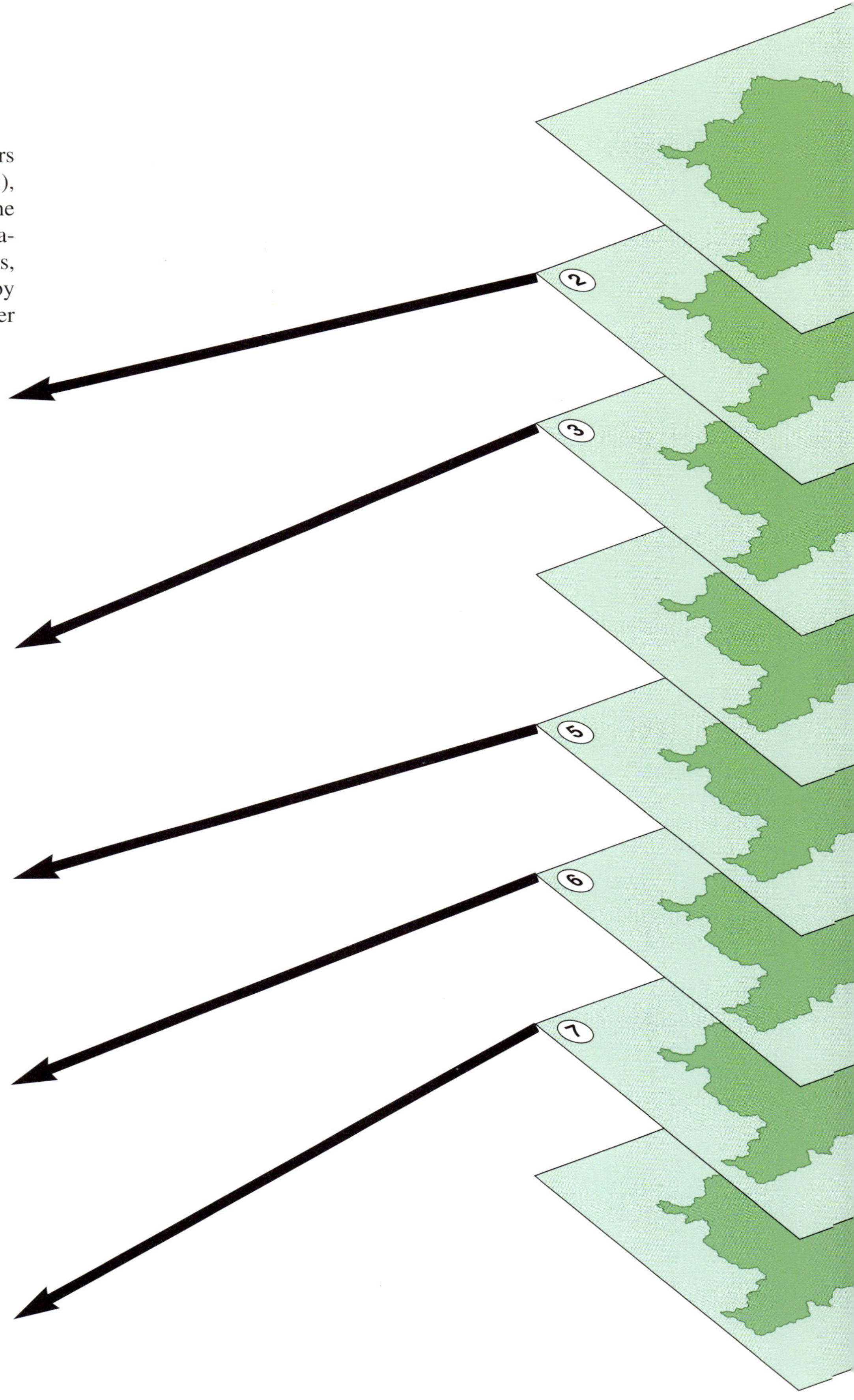

Figure 2.8.10 Data input layers for the M

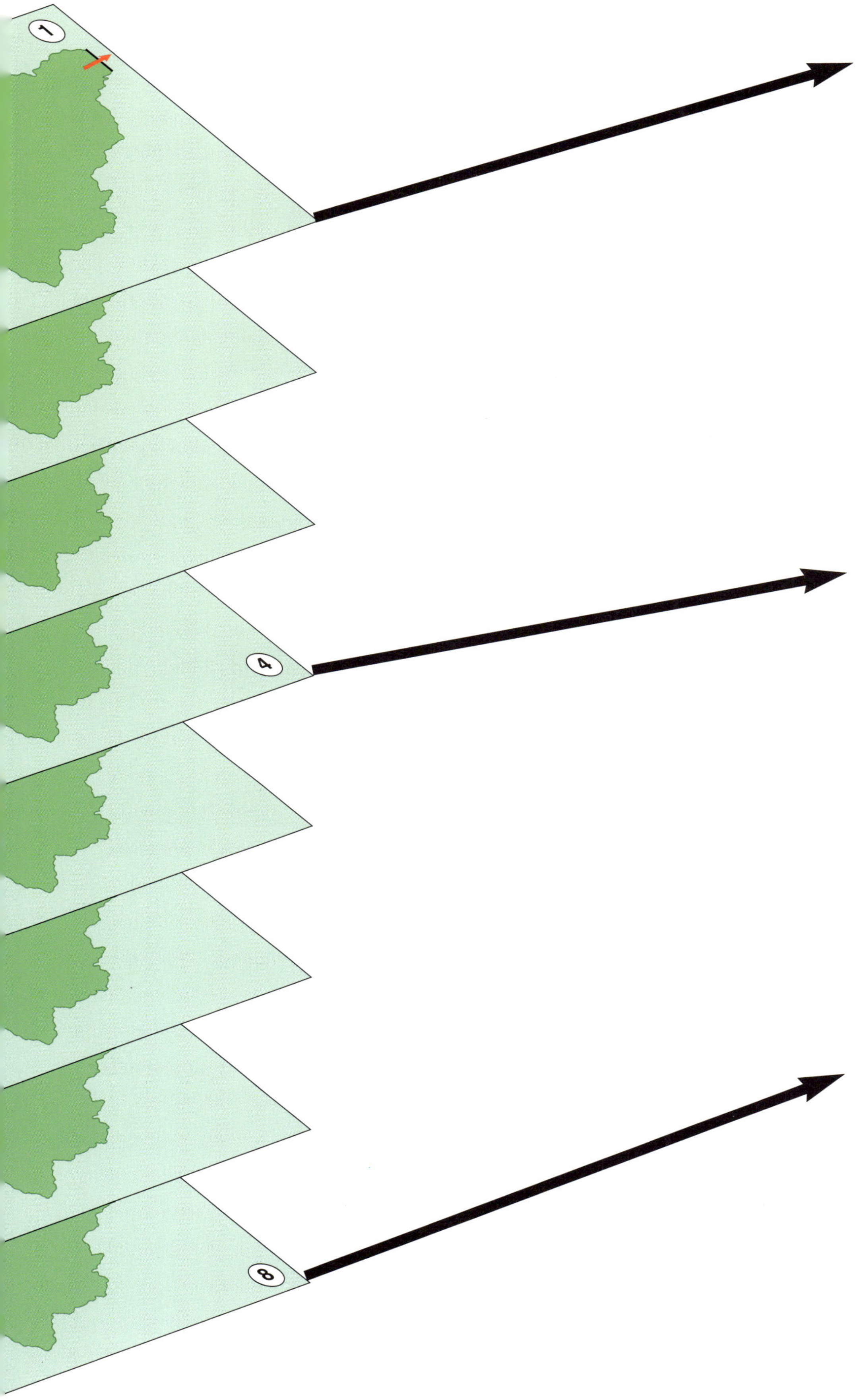

1. *Digital elevation model*. Digital contour data are used to construct an interpolated representation of surface relief, from which all required topographic and drainage attributes are then calculated. The resultant digital elevation model is depicted here as a three-dimensional shaded relief diagram, with appropriate vertical exaggeration, and broad fishnet drape.

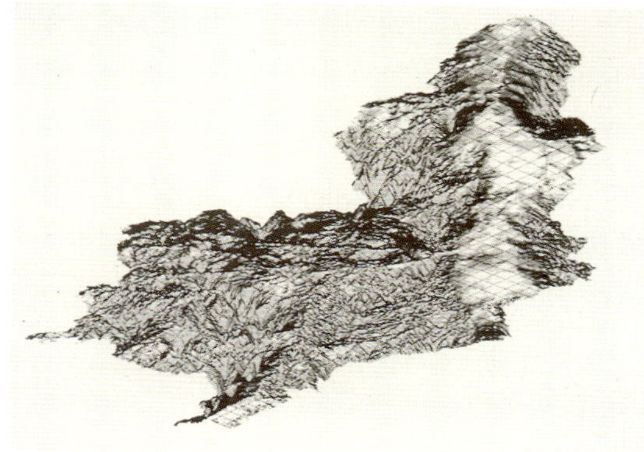

Figure 2.8.11 Guadalentín Basin: Digital elevation model.

4. *Soil particle size distribution*. Percentage sand, silt, and clay in the upper soil layer based on Theissen polygon zonation of published soil pit data (ICONA, Proyecto LUCDEME, Mapa de Suelos). These data are used to determine both the amount of soil that is eroded by surface runoff, which is important in the development of rill networks, and to determine how far eroded material is transported.

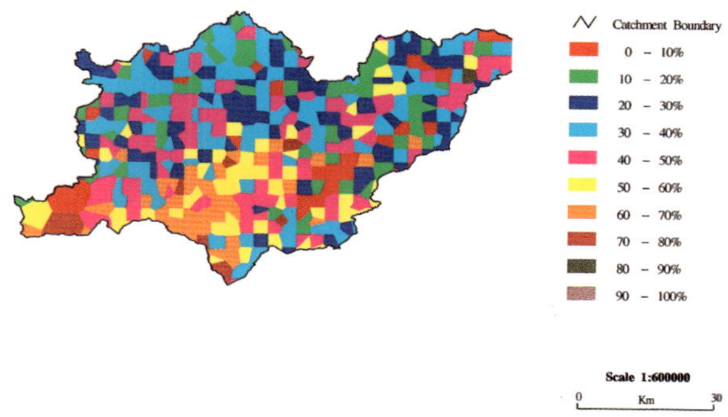

Figure 2.8.12 Guadalentín Basin: Percentage sand in upper soil layer based on Thiessen polygon zonation of available data.

8. *Lithology*. Constructed from existing maps. Used as a base for distributing various data obtained from published reports and fieldwork measurements.

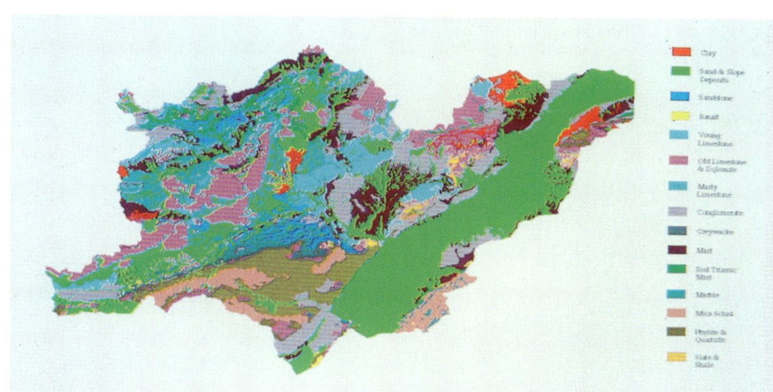

SH model.

Figure 2.8.13 Guadalentíin Basin: Lithological map.

VEGETATION MODELLING

S. Diamond and I. Woodward

Introduction

These models were developed for inclusion within the MEDRUSH model, and therefore require a minimum of computational memory and processing time, and relatively little field-based data. However, they are based on functional relationships describing physiological processes, and so retain the predictive power of mechanistic models.

The sparseness of Mediterranean vegetation presents a challenge for modelling since many modelling concepts assume that vegetation has a closed canopy. Here, we have generalised the diversity of Mediterranean vegetation by models which cover groups of species with similar structure and phenology, which are expected to show common responses to climate change. These functional type models are linked to an evapotranspiration model, and have been run using climate data from Müller (1982) to explore the response of vegetation to climatic change.

Vegetation growth

The main processes simulated are generation of dry matter (which is based on Monteithian-style growth analysis (Russell *et al.*, 1989); partitioning of the matter to plant parts; carrying out growth and maintenance respiration; and producing litter from plant parts. Partitioning and litter production are controlled by pre-set plant phenology; dry matter production and respiration are driven by net solar radiation and temperature respectively.

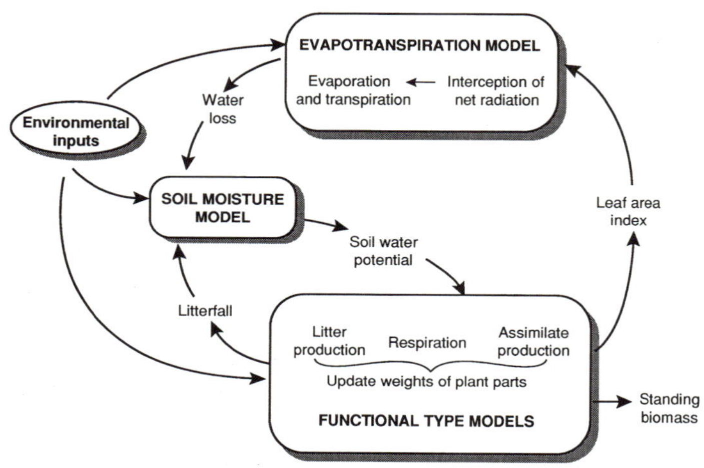

Figure 2.8.14 Schematic diagram of plant–environment processes, showing vegetation and evapotranspiration model links.

Evapotranspiration

The key processes included are as follows. First the net radiation intercepted by each functional type is estimated. This energy load must be dissipated through convection or evapotranspiration, and this is regulated by the Bowen ratio according to soil water potential. Canopy interception of radiation then controls the split between evaporation from the soil and transpiration from vegetation as in the Shuttleworth and Wallace model (1985). Finally a water balance is carried out within the soil, adding precipitation for the time step, and removing water lost by evapotranspiration and drainage.

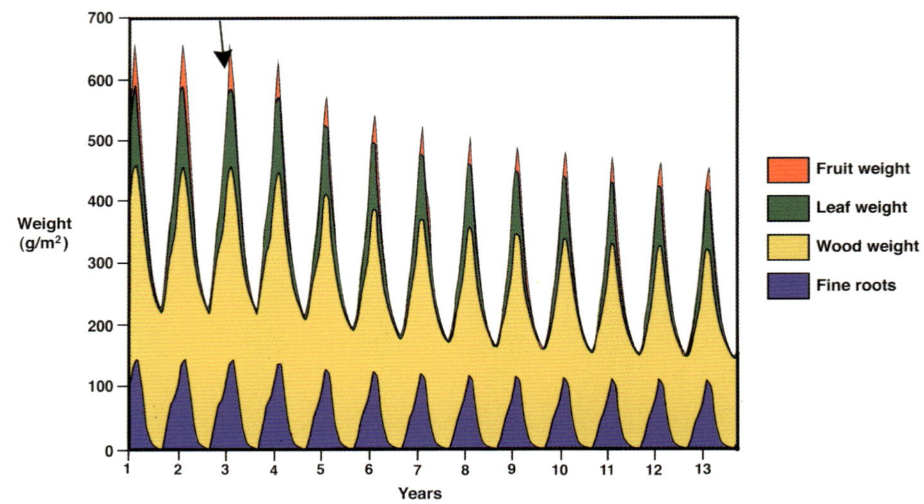

Figure 2.8.15 Simulation for Murcia, Spain, of a rise in temperature of 1.5°C (onset marked by an arrow), showing shrub biomass.

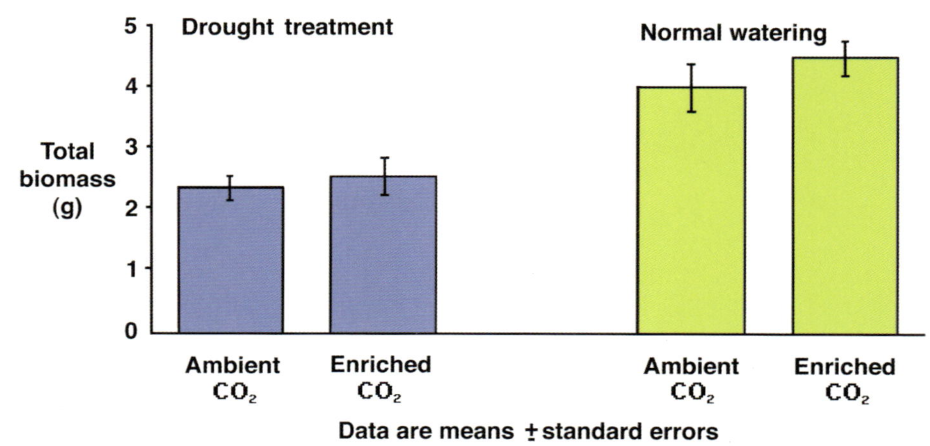

Figure 2.8.16 Response of *Retama sphaerocarpa* to CO_2 and drought.

Outputs

The outputs from the models are those important for understanding desertification processes. For example, soil water is a controlling variable for plant growth; canopy leaf area index controls the interception of precipitation and radiation, and thereby controls evapotranspiration; litter production affects soil erosion; and standing biomass is important for model validation.

Results

The linked models quickly equilibrate on start-up or following perturbation of the climate, through sequential adjustments of internal variables. Thus a rise in temperature causes soil moisture to fall through an increase in total evapotranspiration. Canopy interception decreases due to an increased respiratory burden. There are also indirect effects: as interception reduces, evaporation increases relative to transpiration, and as the soil moisture content decreases, so interception reduces further.

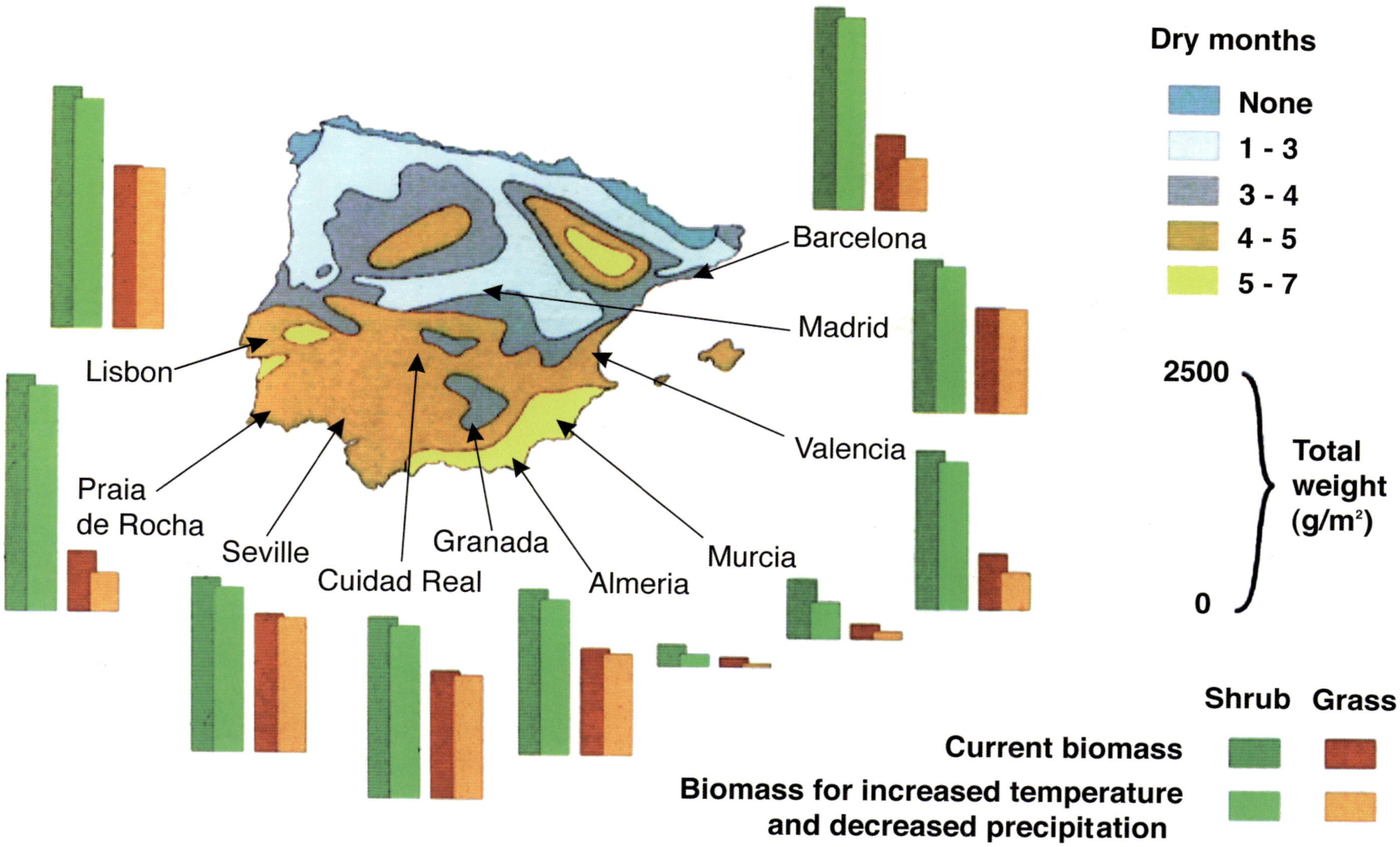

Figure 2.8.17 Predicted biomass of evergreen shrubs and annual herbaceous vegetation for the current climate and for a 1.5-fold increase in temperature (°C) and a 15% fall in precipitation. Dry month = PJ/T < 10; where P = monthly precipitation in millimetres, J = number of rain-days, T = mean monthly temperature (Polunin and Smythies, 1973).

When the models are run to equilibrium, vegetation biomass shows a covariance with the number of dry months for many sites. Simulations of the predicted 15% reduction in precipitation, and the predicted increase in temperature (1.5°C) produced complex changes. Percentage biomass reduction is greatest at sites with most dry months. The models are more responsive to the temperature increase than to the fall in precipitation. Grasses show a greater percentage reduction than evergreen shrubs.

Summary

The dynamic coupling between the vegetation and soil water components allow the models to be used as a tool to explore the interactions between plant productivity and the processes of desertification and regeneration, and the contribution of climate change to these ecological processes. The results presented here indicate that vegetation is more susceptible to decline at sites which are already marginal for plant growth.

3

SOCIO-ECONOMIC PROCESSES AND CHANGE

3.1 Landscape Sustainability

F. Papadimitriou

What is a sustainable landscape?

A rural landscape is defined as sustainable under its contemporary land-management and land-use regime if two conditions hold:

1. the landscape is physically (ecologically, pedologically, geomorphically) viable for the present and the future; and
2. the population acting on and living from the natural resources of the landscape is socio-economically viable in the present and the future.

Unless otherwise stated (Papadimitriou, 1995), a rural landscape is sustainable if its soil degradation rates, soil fertility loss rates and land cover degradation rates are all minimal, whilst its aesthetic, social and economic values are all maximal. The continuation of this set of minima and maxima in the future determines the future sustainability of the landscape.

If a rural landscape is not sustainable, then land degradation may eventually take place, with a wide range of negative repercussions for future generations.

Clearly, there is no singular approach to assessing landscape sustainability and there is a very wide range of factors affecting the status of landscape sustainability or unsustainability (e.g. pollution status, diversity value, traditional land uses, human impact, climate change, use of pesticides).

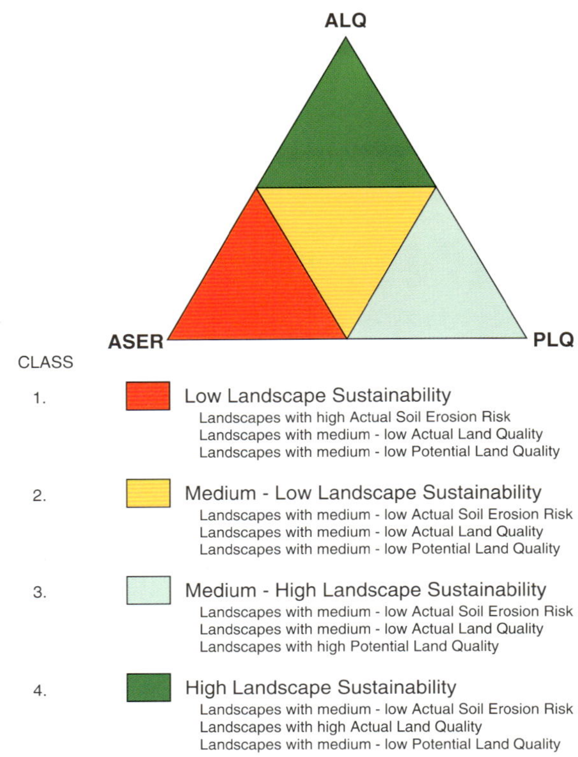

Figure 3.1.1 Evaluation of landscape sustainability for potential land quality (PLQ), actual land quality (ALQ) and actual soil erosion risk (ASER).

Evaluating and mapping landscape sustainability

The procedure outlined here is a *landscape sustainability evaluation* that concludes with the derivation of a *landscape sustainability map* of European Mediterranean regions for two different administrative levels.

In assessing landscape sustainability for some region or country, there must be some restriction to a finite set of interrelated factors that will furnish an adequate picture of the landscape sustainability. Land quality and erosion risk is such a set of characteristic factors for the landscape. The primary environmental asset to be preserved in the case of landscapes is its *land quality*, physical and aesthetic. Land quality is subject to a number of risks (e.g. soil erosion, vegetation degradation, pollution). Any of these risks imposes a threat to landscape sustainability.

Three factors of major importance have been selected for the evaluation procedure: potential land quality, actual land quality and actual soil erosion risk. Consequently, classes of landscape sustainability from these factors and maps for various administrative and policy-making levels were produced.

In this procedure, land quality was considered as a landscape asset and soil erosion as a liability. Other maps and classifications can be made by considering other features of the landscape and defining sustainability thresholds as appropriate.

Sustainability maps will be different from any other existing maps (e.g. natural hazard and land degradation maps) in the sense that, apart from showing the spatial distribution of a certain environmental risk as all other relevant maps do, sustainability maps also show the range of values over which this risk is important, with respect to a predefined set of socio-economic criteria.

Defining sustainability

'Humanity has the ability to make development sustainable—to ensure that it meets the needs of the present without compromising the ability of the future generations to meet their own needs' is expressedly stated in the Bruntland Report (World Commission on Environment and Development, 1987).

In their effort to apply the concept of sustainability to practical problems, certain authors have occasionally stressed the role of different agents affecting the sustainability of farming, forest exploitation, rural systems and the sustainability of a landscape in general. Such often-quoted agents are the soils (Dietz and Hoogervorst, 1991; Hidding, 1993), the ecosystems (Jurgens, 1993) and energy (Hulscher and Hommes, 1992; Jaarsma and Van Lier, 1993; Sharma and Singh, 1994).

Van Mansvelt and Mulder (1993) have given extended accounts of the agents related to land-use sustainability and other authors have used the term 'sustainable landscape' (Barrett and Peles, 1994), but a general definition of landscape sustainability is still missing in the literature.

However, defining the boundaries beyond which a landscape is no longer sustainable is an approach adopted by certain authors (Kessler and Laban, 1994; Perrings and Pearce, 1994). This approach may formulate the basis for landscape evaluation for sustainability following a set of predefined criteria (Werner, 1993).

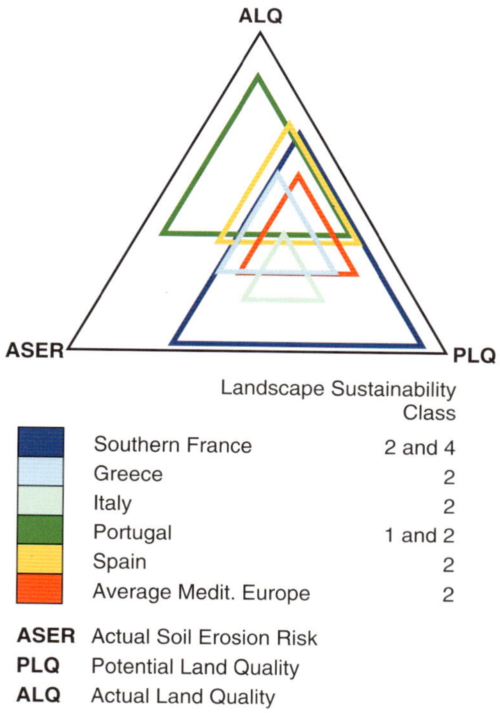

Figure 3.1.2 Range of landscape sustainability at the national level.

The evaluation procedure

Landscape sustainability can be evaluated for southern Europe by considering three of the most relevant and interrelated factors. These are the *potential land quality*, the *actual land quality* and the *actual soil erosion risk* as defined by the European Community CORINE program (CEC, 1992).

- The potential land quality (PLQ) is the 'inherent physical quality of land resources for agriculture, biomass production, and vegetation growth' (CEC, 1992). As much as 52% of the Mediterranean EU area is occupied by lands of low PLQ.
- The actual land quality (ALQ) is the quality of land under present management conditions, including improvements (e.g. terracing, irrigation and artificial drainage). Thus, the actual land quality is usually higher than the potential land quality. Lands with high ALQ are the least likely to be abandoned in the future, since they are highly productive and they are usually well managed.
- The actual soil erosion risk (ASER) is 'the current risk of erosion under present vegetation and land use conditions' (CEC, 1992). If the ASER is high, then the landscapes of the region under consideration are highly unsustainable under their current land-use conditions. Land-use changes and land-management changes may be vital for areas of ASER higher than 50% of the region's territory.

It is assumed that the rural population will remain in the rural sector and will continue to cultivate and take care of the land as long as the ALQ is equal to or higher than the PLQ. Otherwise, people will tend to neglect or, even worse, abandon their lands with unfavourable consequences for the landscape. Also, the landscape has been defined sustainable (as it stands) if the ASER is, at most, medium. The occurrence of soil erosion is but an indication that the landscape is unsustainable.

Thus, an appropriate ranking of these three factors can give a first approximate classification of the physical and socio-economic viability of the landscape and its resources.

Plotting all three variables on a triangle, it is possible to define four regions of landscape sustainability on the triangle (lowest 1 to highest 4).

Notice that the range of ASER, PLQ and ALQ values for a geographical region is not expected to correspond exactly to the range of values in the sustainability classes defined. In such cases, the sustainability class of the region under consideration is defined by the maximum overlap between the triangle of the predefined sustainability class range and the triangle defined by the actual range of ASER, PLQ and ALQ of values of the region (e.g. county, region and province).

Consequently, it is possible to derive a map of landscape sustainability at any administrative level of the European Mediterranean countries by making use of the CORINE program PLQ, ALQ and ASER data.

Observe that most of the provinces belong to the medium–low sustainability class.

An example of a numerical application on the province of the Guadalentín MEDALUS target area (Spain)

Consider the region where the Guadalentín MEDALUS target area is located. The provinces to which the Guadalentín belongs administratively (Almeria and Murcia) have the following ASER, PLQ and ALQ values (from CORINE data, 1992):

$$\text{ASER} = 60\text{–}80\%, \quad \text{PLQ} = 0\text{–}20\%, \quad \text{ALQ} = 0\text{–}20\%$$

These data can be transformed as per their average values as follows.

$$\text{ASER} = 70\%, \quad \text{PLQ} = 10\%, \quad \text{ALQ} = 10\%$$

Now these values can be plotted on the ASER–PLQ–ALQ triangle.

The classification procedure is completed in three steps, as follows.

1. Draw a segment parallel to the line PLQ–ALQ, intersecting the side of the triangle ASER–ALQ at 70% of the length of the triangle side ASER–ALQ. The same segment intersects the side ASER–PLQ of the triangle at 70% of its length.
2. Repeat the same procedure for the values of PLQ (10%) and ALQ (10%).
3. The three intersecting segments define a triangle whose entire area is located within the area of the triangle defining the landscape sustainability class. Hence, it can be concluded that the provinces considered belong to sustainability class 1 (the low-sustainability class).

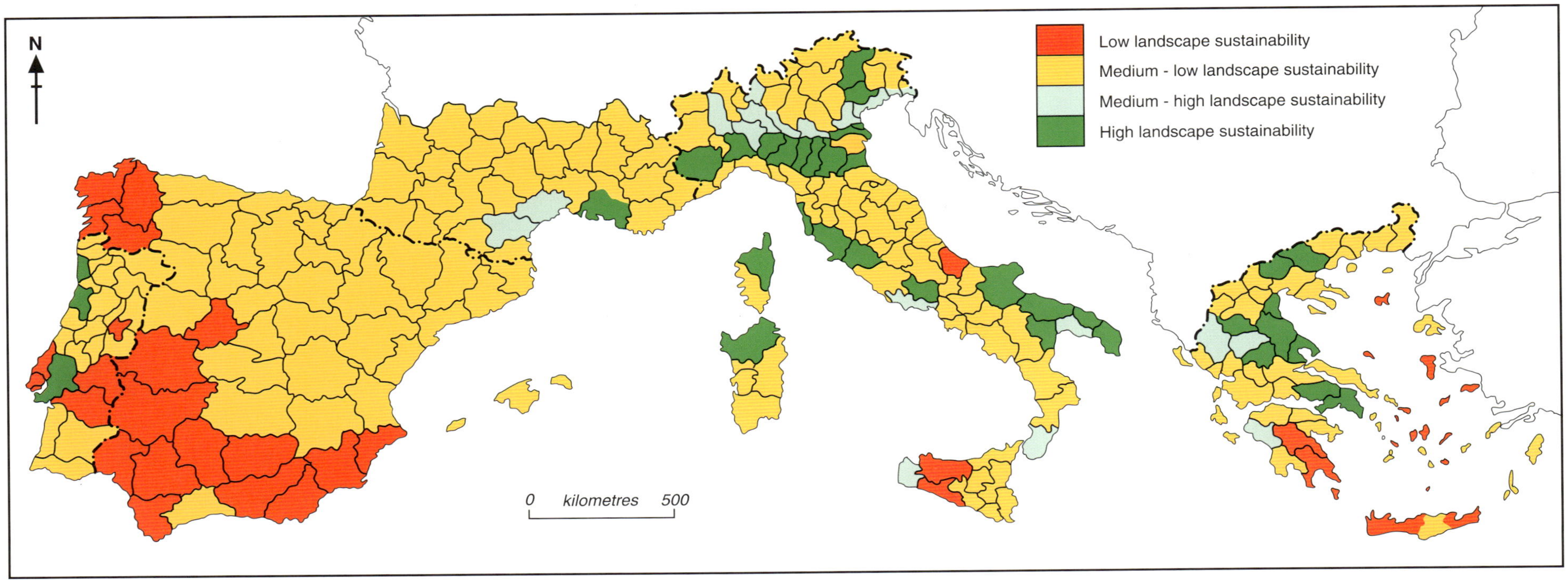

Figure 3.1.3 Map of landscape sustainability of Mediterranean European Union countries at the provincial level. This map has been created solely for the convenience of the reader. The boundaries shown on this map do not imply, on the part of the author, any judgement on the legal status of any territory or any endorsement or acceptance of such boundaries.

Field verification of the classification procedure

The MEDALUS field sites and target areas have been selected in provinces of high desertification risk. Although the occurrence of desertification on a field site can be considered as representative of the provinces' desertification status only up to a limited extent, it is expected that the results of this procedure should be able to verify the reduced landscape sustainability in the sites where desertification has been documented by MEDALUS.

Indeed, this is the case. The sustainability classes of the provinces where MEDALUS target areas are located are 1.75 (Abruzzo), 3 (Agri), and 1 (Guadalentín).

From these values, it can be calculated that the average sustainability class of all these target areas is between 1 and 1.88. The same conclusion would be reached if the average ASER, PLQ and ALQ of all the MEDALUS regions were calculated. Hence, the MEDALUS target areas are located in provinces of low–medium landscape sustainability class.

Table 3.1.1 Landscape sustainability classes for the MEDALUS target areas.

Sustainability	Province	Target area	Target area (average)
1	Pescara	Abruzzo	
2	Chieti		1.75
2	Teramo		
2	L'Aquila		
4	Matera	Agri	
2	Potenza		3.00
1	Murcia	Guadalentín	
1	Alméria		1.00
1.88	average		

3.2 History of Mediterranean Land Use

A. T. Grove and O. Rackham

During the Pleistocene, Mediterranean lands experienced several periods of 'desertification', each lasting tens of thousands of years (Tzedakis, 1993). After fluctuations around 12 000 BC, the climate became somewhat wetter than that of today; vegetation changed to forest in wetter areas and savanna in drier areas (Huntley and Birks, 1983).

The mainland had Palaeolithic populations of hunters and gatherers. Islands such as Cyprus and Crete remained uninhabited until about 6500 BC (Cherry, 1990). About that time, in the Neolithic, when sea level was rising to approach its present level (Pirazzoli and Pluet, 1991) and something very like the present Mediterranean climate had developed, settled civilisation and agriculture began, first in Anatolia and then in Palestine and southern Greece (Trump, 1980). This involved introducing domestic animals and crop plants,

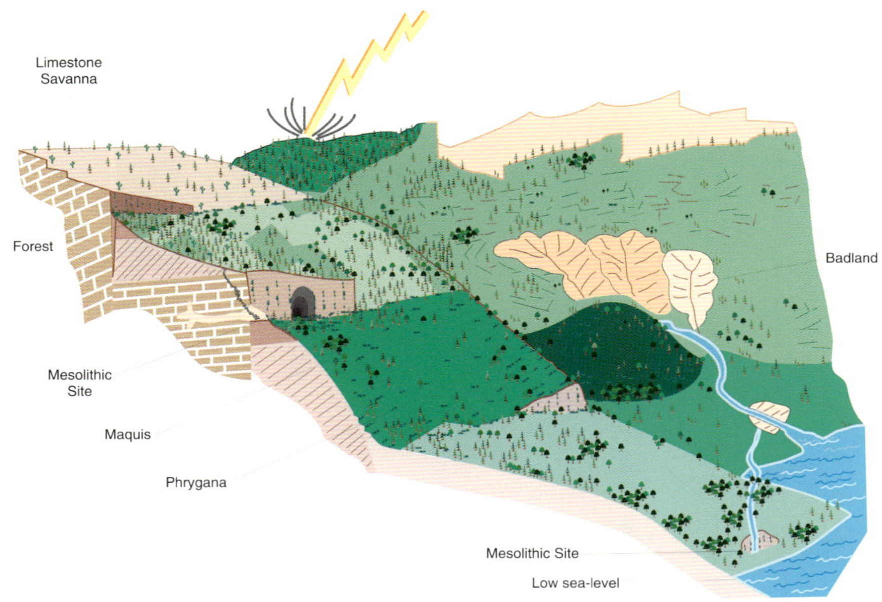

Figure 3.2.1 (a) BC. From the top of the hill to the bottom: limestone savanna, forest, Mesolithic site, maquis, phrygana, and another Mesolithic site. The sea level is low and badlands already occur.

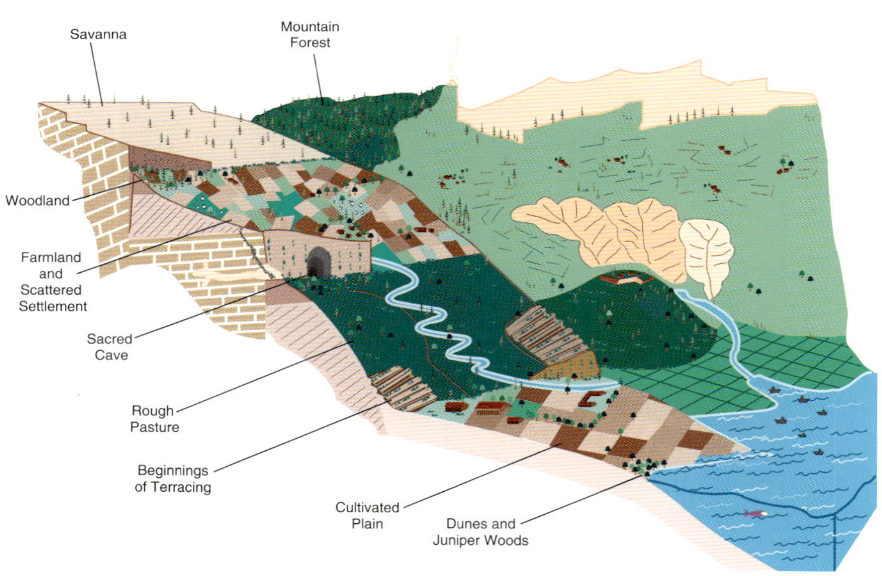

(b) AD 50. From the top of the hill downwards: mountain forest, savanna, woodland, farmland and scattered settlements, a sacred cave, rough pasture, the beginning of terracing, and a cultivated plain. The sea level has increased and there are coastal dunes and juniper woods.

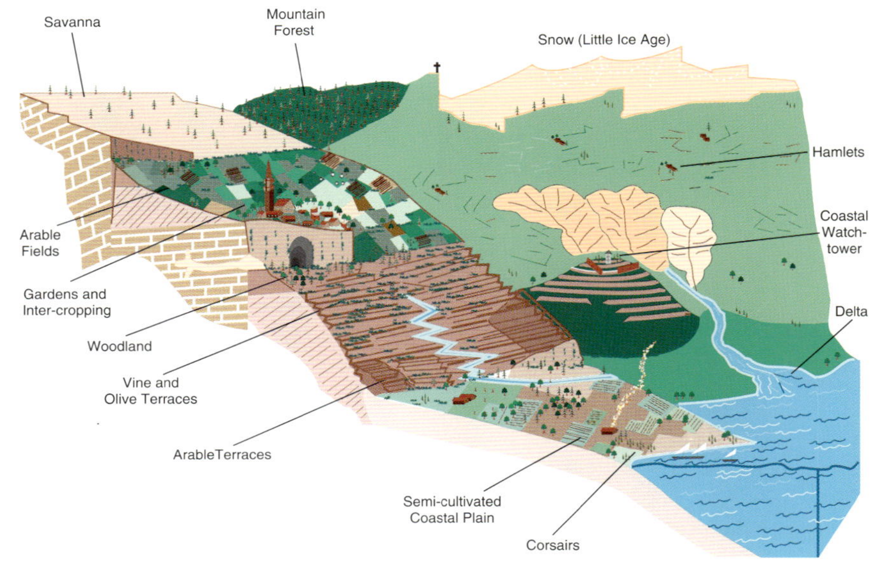

(c) AD 1770. From the top of the hill downwards: mountain forest, savanna, arable fields, gardens and intercropping, woodland, vine and olive terraces, arable terraces, semi-cultivated coastal plain, hamlets. Snow covers most of the landscape (Little Ice Age), a river delta has formed and corsairs infesting the seas determine the need for watch towers.

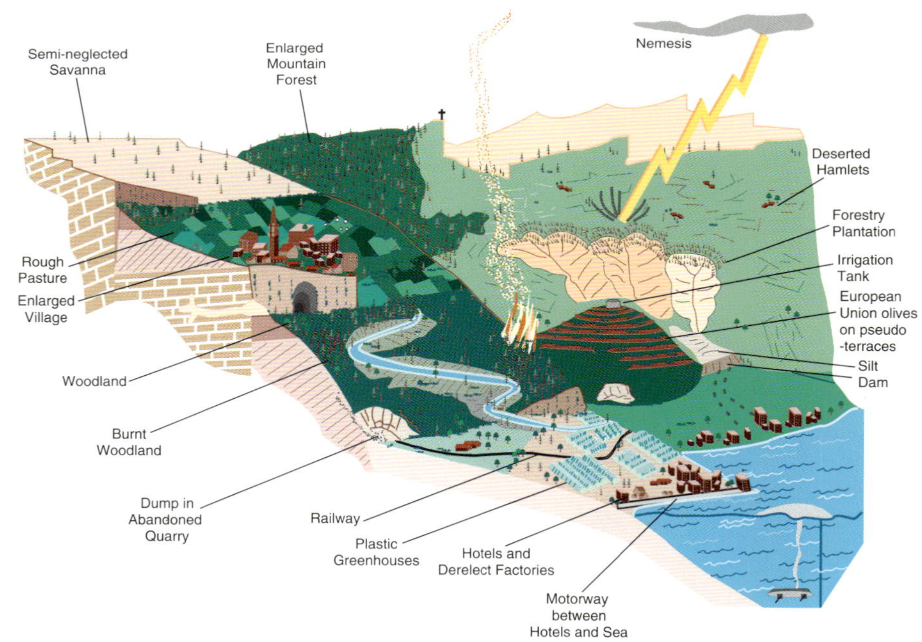

(d) AD 1990. From the top of the hill downwards: enlarged mountain forest, semi-neglected savanna, rough pasture, enlarged village, woodland, burnt woodland, dump in abandoned quarry, railway, plastic greenhouses, hotels and derelict factories. A motorway separates the hotels from the sea, an already silted reservoir has been built on the river. European Community olives have been planted on pseudo-terraces, forest plantations also appear, an irrigation tank replaces the watch tower, and the hamlets are deserted.

exterminating some wild animals, creating fields and pastures, and colonising the islands.

Some gullies and badlands which attract attention today were already in existence 4000 years ago (Gilman and Thornes, 1985; Runnels, 1995). Rivers deposited sediments in their estuaries and the indentations in low coasts were infilled (CNRS, 1987).

By the late Bronze Age, around 1500 BC, terraces and irrigation works were being built and most Mediterranean areas had a 'humanised' landscape with winter wheat, barley, vines and olives, and with cattle, sheep, goats and pigs (Houston, 1964; Bottema *et al.*, 1990). Forests had declined, but wetlands remained; much of the surviving forest and wetland was managed and utilised.

Subsequent history involves repeated episodes of human recession, during which vegetation recovered, as well as human advance, during which it declined (Van Andel and Runnels, 1987). Agricultural terraces are known to have been made on the island of Delos in the Aegean by 400 BC (Brunet, 1990) and it is suspected that they were widely constructed elsewhere, though this is difficult to prove. In Greece, populations peaked in the Hellenistic period, *c.* 200 BC (Sallares, 1991), and in Italy, in the Imperial Roman period, *c.* AD 200 (Delano Smith, 1979). In Crete, rural populations comparable with those of the present day were reached in the Early Byzantine period, *c.* AD 560, and again towards the end of the 16th century (Rackham and Moody, 1995).

New crops were introduced such as carob and mulberry from Asia. The Romans developed irrigation technology, including storage dams and aqueducts; they converted wetlands from pasture to arable. Large-scale shipbuilding allowed Rome to become a mega-city, drawing grain and minerals from all the Mediterranean lands. Then, as now, there were great differences between countries: Greek and Roman authors such as Theophrastus and Pliny contrasted wooded, fertile, populous Italy with relatively arid Greece, and noted the desert-like character of south-east Spain.

Plague came repeatedly during the 6th–8th centuries (Biraben and Le Goff, 1969) and again during the 14th–17th centuries, and could kill one-third or more of the population at each visit (Gottfried, 1986). Malaria made some coastal lowlands and marshy river valleys untenable (Bruce-Chwatt and de Zulueta, 1980). Piracy made many coastlands dangerous, especially from the 14th to the 18th centuries; settlements retreated into the hills, and the coast was left only for winter grazing (Le Lannou, 1941; Braudel, 1973). The Little Ice Age, lasting from 1250 to 1450 and from 1550 to 1850, when Alpine glaciers were in advanced positions, brought sequences of years with droughts, floods, severe winters and famine (Grove, 1988). Often these coincided with wars and epidemics. Deltas extended seawards at river mouths (Hoffmann, 1987), usually draining catchments containing rapidly eroding badlands.

After the Roman period the Islamic world expanded to include Crete, Sicily, Sardinia, and most of Spain and Portugal. The Muslim period lasted longest in south-east Spain, where the Arabs continued and enhanced the civilisation of Roman Baetica (Butzer *et al.*, 1985). They introduced new crops, elaborated waterworks, extended irrigation even onto terraced hillsides, and devised means of trapping sediment washed out of badlands. Much of this was brought to an end by the medieval Christian conquerors, and especially by the expulsion of a quarter of a million Muslims and ex-Muslims from Spain between 1570 and 1614.

In most Mediterranean countries a typical medieval land use was savanna, i.e. big trees scattered in grassland or heath. This may result from forest losing trees, or grassland gaining trees, or may be wholly natural in places too dry for continuous forest. Savanna land uses involve grazing of cattle, sheep, goats and pigs, woodcutting, and sometimes shifting cultivation and cork harvesting. Most savannas in Italy and the Balearic Islands have become forest in the last 200 years. In Greece and Crete, savanna continues. In Spain, where shepherding had been institutionalised in the Middle Ages under the Mesta, a formal kind of savanna greatly increased in the 19th century. In Portugal and Sardinia, savanna has been favoured by the cork industry.

Arab influence brought silkworms (and the mulberry trees on which they feed) to much of southern Europe, and also tropical crops such as sugar-cane, rice, cotton, citrus fruits and egg-plant (aubergine). With the discovery of America came maize, potato, tomato, tobacco and prickly-pear.

Populations again increased, from the 18th century onwards in the west and from the 19th in the east (McNeill, 1992). This was connected with the end of plague and piracy. Although malaria often lasted until the mid-20th century, it became safer to live on coastal plains. Customary restraints on the use of natural resources became unfashionable, and public land passed into private hands (Davis, 1979; Shubert, 1990; Moulin, 1991). Forests were grubbed out and hillsides were terraced to the limits of cultivation (Blanchemanche, 1990).

There had been large-scale trade in cereals, wine, olive oil and olives since before Roman times, which encouraged specialised production in certain areas. From the mid-19th century, railways and steamships made it easier to market perishable crops such as fruit in northern Europe. Conversely, local markets were opened to cereals grown cheaply in the Ukraine and America; this discouraged the cultivation of difficult terrain and the intercropping of cereals with vines and olive trees. In some countries, however, governments imposed import quotas which had the effect, for a time, of encouraging people to grow cereals even beyond the limits of reasonable cultivation (Coccosis, 1991). Some traditional dried products such as figs and raisins declined. Some, but not all, craft products were displaced by factory-made articles and synthetic substitutes. New diseases took toll of crops such as vines and chestnuts.

Most south European countries went through periods of poverty and over-population between 1830 and 1950. Many areas, especially Crete, look desertified in early photographs, compared to their present state. There were programmes of tree-planting in some countries, either to provide employment or to 'restore' a supposed historic forest cover, or in the belief that trees would prevent erosion and floods (Bazire and Gadant, 1991). Military service drew the attention of young men to easier ways of making a living elsewhere. Poverty was relieved by new opportunities for emigration to the cities, to industrial northern Europe, and to north Africa, the Americas and Australia. This began a continuing decline, beginning in France, from the previously over-populated state.

At the end of the Second World War, malaria was finally eradicated. Power-driven agricultural machinery, artificial fertilisers, greenhouses, piped irrigation and pesticides were quickly adopted in the 1960s. They encouraged cultivation on plains and former wetlands. Governments, and later the Economic Community through its Common Agricultural Policy, supported the growing of otherwise uneconomic crops.

Some of the developments are unsustainable. Dams constructed across river valleys to store water for irrigation hold back sediments that would otherwise reach the coast. Sand and gravel extraction from beaches and river beds reduces the natural defences of low shores, allowing coastal erosion to replace progradation (Quélennec, 1990). Excessive extraction of groundwater from coastal aquifers lowers water-tables to levels at which sea-water penetrates inland, causing salinisation and threatening the immediate future of some of the most productive land (Simmers *et al.*, 1992).

International tourism encourages coastal development and provides alternative or additional employment for people no longer bound to agriculture

(Grenon and Batisse, 1989). Terraced hillsides are neglected and being taken over by scrub and woodland (Ambroise *et al.*, 1989).

Over-development affects a limited proportion of the landscape. Mountains and remote areas, where tourism is unfashionable and modern agricultural technology does not work because of steep slopes and shallow soils, have reverted to semi-natural vegetation (Grove and Rackham, 1993). This often consists of flammable trees and shrubs—supplemented by pines and eucalyptus thoughtlessly planted—leading to periodic disastrous fires which have probably not happened before. Landscapes formerly dominated by browsing (e.g. parts of the Portuguese *montado* or savanna) have become dominated by fire; elsewhere, as in the high Alpujarra of southern Spain, there has been a quick progression from terrace cultivation to grazing to forestry to fire. In places, people have built houses, hotels, and holiday and retirement homes scattered in the woods, which adds to the potential for disaster.

Figure 3.2.2 Landscape changes at Omálos, Crete. (a) House of a Lakkiot on Omálos, 1909. (From Trevor-Battye, 1913. *Camping in Crete*. Witherby, London.) (b) The same two hills on the north side of the Omálos Plain on 11 April 1982. As seems to be generally the case in Crete and other parts of Greece, the tree cover has extended over the 73-year interval and the shrubs are taller. In the last 40 years technology has revolutionised Mediterranean environments. This is especially illustrated by the implementation of complex and extensive irrigation systems replacing technologies which had not changed since Arabic times.

3.3 Population Dynamics and Change

L. Leontidou

The boundaries of 'Europe' have shifted through the long history of humankind, and for a long period in ancient times the shores of the Mediterranean were Europe. Population settlements around this 'sea in the midst of land', as its name signifies, were surrounded by preliterate 'barbarians' well into the classical period. Greeks and Romans built glamorous city States and empires 'where town followed town in endless strings, punctuated by great cities' (Braudel, 1975). The Greek *polis* and later city states have been reaffirming their dominance and distinct cultural identity until our times. In fact, the cultural heritage of urbanism is still alive and strong throughout southern Europe, in contrast to the idealisation of the countryside in the countries of northern Europe and Anglo-American cultures (Leontidou, 1990).

Besides being city-oriented, the people of the Mediterranean were cultivators, peasants and landowners, but especially adventurous travellers, merchants, explorers. They moved constantly, first in the seas and later, recently, as migrants to the Americas and to northern Europe. They moved, of course, in widening circles as technology developed, from Odysseus to Columbus. The persistent theme throughout history has been their curiosity about whatever lay beyond their horizon and the familiar Mediterranean Sea, which prompted them to exploration. Ancient Greek scholars developed the theory of a close linkage between environment and civilisation, and Herodotus affirmed that 'soft countries' breed soft men, or that it is impossible for one and the same country to produce good crops and a good army.

This is a very early version of 'environmental determinism', contested by the celebrated geographer Vidal de la Blache (1934); but his 'possibilism' never came to disentangle environmental from cultural formations, at least in the case of the Mediterranean. The same holds for the historians, notably Braudel (1974). Even now, there are few regions in the world where population and its cultures differ so distinctively between coast and inland, plain and mountain, island and mainland, national territory and diaspora; or where population activities are so intensely conditioned by, and have impact upon, the environment. A strong example is deforestation, from the times when the Dalmatian forests were chopped down to provide timber for the Venetian fleet (Pounds, 1990), to the forest arson of the present to make way for new urban and tourist development in Greece and Spain. Such examples underline the important impact which population movements of various sorts have in the formation of the Mediterranean landscape and environment.

When the period of written history began (6th century BC), the Mediterranean had already been inhabited for more than a million years. Middle Eastern cultures entered 'our' Europe through the Balkans. Population levels were low in prehistoric times, but the Neolithic economic development led to a population growth comparable to that of modern times. The Aegean was then the hearth of European civilisation and Scandinavia its preliterate periphery, and this rift was not bridged for a long time. Population growth is difficult to estimate (Pounds, 1990), but was slow. In Italy, for example, the net demographic gain in the entire historic period up to the 1550s is estimated at less than 48%. During Augustus' reign (census AD 14), it was about 7 million, and 10.3 million in the 16th century (Delano Smith, 1979).

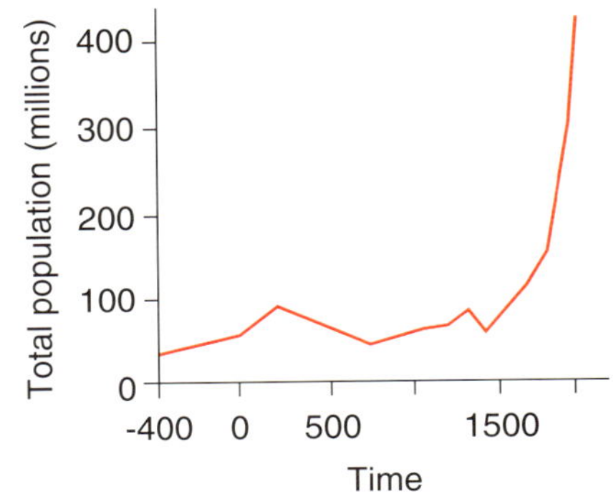

Figure 3.3.1 The standard European population curve for the historic period. It shows the main Roman and late medieval peaks as well as the post-Roman and post-medieval troughs. Reproduced with permission from Delano Smith (1979).

Medieval population development followed a pattern of alternating peaks and depressions, very diverse at local level, following severe storms, crop failures, famines and epidemics. There were four major phases from the 8th to the 17th century, with a 'low' prior to these. The 14th century low is associated with the pan-European Black Death epidemic, but was also due to local economic change with the creation of institutionalised grazing and the abandonment of villages (Delano Smith, 1979). But in general, the demographic curve showed a relatively stable trend with oscillations for almost two millennia.

The explosive demographic increase in the region happened during the 19th and 20th centuries. The total population of Mediterranean Europe has more than doubled since 1850 (Delano Smith, 1979). It also doubled between the 1590s and the mid-19th century (Braudel, 1974). An overall transformation was also underway. Strabo's prosperous *Magna Grecia* had been metamorphosed into the underdeveloped south of Italy by the 18th century, and the Levant was under Ottoman occupation from the mid-15th century. The glorious city states of antiquity in Greece were then but deserted villages: Athens' population shrank from 12 000 in both the 1570s and the 1810s, to the 6000 of 1832, when people fled to the mountains and islands during the Ottoman occupation.

Much has changed since then. The Mediterranean did not see any industrial revolution, but a demographic and urban one, involving higher fertility than the rest of Europe and important rearrangements of the population between rural and urban, inland and coast, national and international levels. Urbanisation was especially rapid during 1950–1970. This was also the period of rapid emigration to northern Europe. Bilateral agreements were signed and northern Europe recruited the productive labourers for its industrialisation from the South. What were thought initially to be Gastarbeiter, guest workers in Fordist factories, became permanent migrants (Castles and Kosack, 1973), who were later joined by their families, even after the 'recruitment stop'. Restrictions to immigration were imposed by northern European countries after the mid-1970s oil crisis, economic restructuring from Fordism to flexible accumulation, de-industrialisation, and creeping social tension or even racism in northern cities (King, 1994).

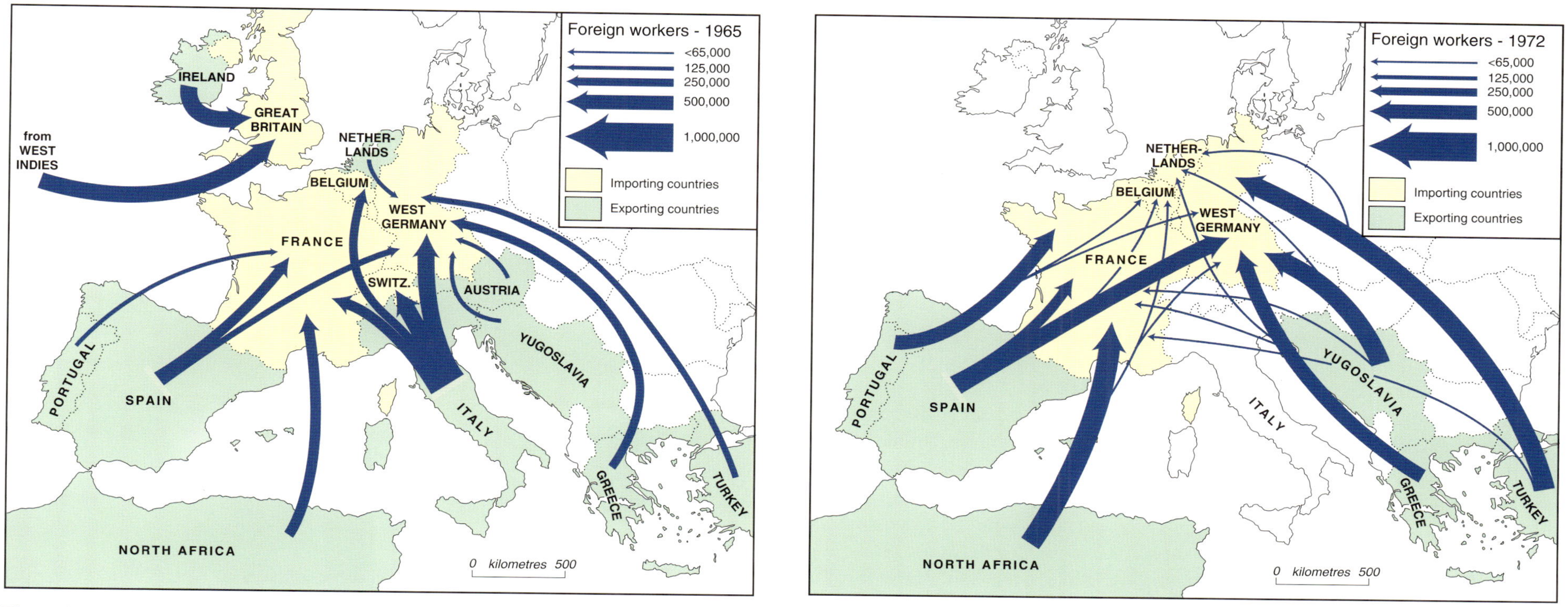

Figure 3.3.2 The destination of foreign workers, 1965 and 1972. Reproduced with permission from Noin and Woods (1993).

This was a major transition for Mediterranean Europe on all levels—from city building to population dynamics and regional structure (Leontidou, 1990). Fertility dropped, halving in Spain within the last decade and relegating the four countries of southern Europe to the bottom of the European Union in this respect. Birth rates were a mere 9.8‰ in Italy, 9.9‰ in Greece, 10‰ in Spain and 11.2‰ in Portugal by 1990—the lowest of all other European countries (Ogden, cited in Noin and Woods, 1993).

Spatial dynamics were reversed. All countries passed from a period of urbanisation, polarisation and centralisation, not to counter-urbanisation, as in the North, but to diffuse urbanisation: the population of larger cities stabilised and the medium-sized towns sprang to life. Their number increased and their rates of growth speeded up. This population turnaround was an outcome of different processes in the various countries, including administrative decentralisation, the decline of Fordist centres, diffuse industrialisation, incentives for regional industrial development, the increase of mass tourism, and the diseconomies of urbanisation in larger cities. In addition, population moved to the coasts, permanently or seasonally, with the consequences of littoralisation discussed in Section 3.6 of this Atlas.

Migration dynamics also changed in the 1970s, on several fronts. First, the period of family reunion in the North was succeeded by repatriation trends. Migrants returned from North to South, towards smaller towns rather than villages. In addition, immigration from the 'Third World' followed: the four countries of southern Europe have been changing from countries of emigration to countries of immigration. It is estimated that 2 million immigrants live there today, half of whom are in Italy, which already has 6 million of its population working abroad (King, 1993, 1994). The informal economy, often illegal or 'black', has always been there, and now develops with illegal migration.

As EC integration engulfed the Iberian Peninsula, retirement and tourism-incited migration also stepped up, and today this is also spreading to Greece. At the same time, skilled migration to northern EU countries from the South involves a brain drain affecting Mediterranean Europe. And finally, just before but especially after the demolition of the Berlin Wall in 1989, and following Near East instability in the 1990s, the trickle of refugees and migrants from Eastern Europe and the Near East is turning into a tide.

These population movements and demographic changes have a diverse impact on Mediterranean regional structures. In general, population stabilises in each country, as fertility rates have dropped. Regional depolarisation does not slow the abandonment of smaller villages and marginal agricultural areas, but is induced by diffuse urbanisation into the middle-sized towns and littoralisation, with population especially congested on the coasts during the peak seasons of tourism.

Today (Noin and Woods, 1993) the four southern European countries are inhabited by 118 million people, or 34% of the European Union's 345 million. In the future, they may grow to represent an even higher percentage, despite their falling birth rates. As the native populations stabilise, eastern European migrants enter north-east Italy and all of Greece, and African migrants enter Portugal, southern Spain and Sicily. An upcoming migration pressure, including refugees and illegal immigrants, is expected across the Mediterranean from population explosion and unrest in the 'Islamic arc' around Europe: the Maghreb states, Egypt, Turkey and Pakistan (King, 1994). After many centuries, the Mediterranean Sea is no longer a barrier. Populations cross it in order to create ethnic and racial mixtures, especially in the countries of southern Europe.

3.4 Agricultural Transformations

N. S. Margaris, E. J. Koutsidou and C. E. Giourga

Traditional land uses

The evolution of Mediterranean systems of management of the agricultural land over the centuries has been driven by the relative isolation of many areas of this region and by the limited availability of natural resources. The original imprint of Mediterranean agriculture, which in general still persists, is given by the co-existence of cultivation of the land and animal husbandry (Braudel, 1974). Both the cropping practises and the plant species used in the Mediterranean region reveal a high degree of adaptation of agriculture to climatic conditions. Typically the Mediterranean has long, hot and dry summers. The limiting factor for vegetation in general, and for agriculture in particular, is the lack of water in coincidence with warm air temperature. Mediterranean landscapes are characterised by the juxtaposition of mountain, hillslopes and flat lands. As a consequence, in the whole of the Mediterranean region, a system of terracing has evolved that varies from country to country. Apart from allowing the cultivation of steep slopes, terracing makes the best possible use of rainfall. Traditional Mediterranean crops are annual, herbaceous crops (e.g. cereals and pulses), and permanent crops (e.g. olive, vine and almonds). The cropping season for annuals typically coincides with the 'wet season' (November–May). Wheat and barley are the most common winter cereals. Traditionally, permanent crops consist of woody plant species which have equivalents in semi-natural vegetation. They were derived from the wild forms by grafting aimed at the production of edible fruits. This is the case for the olive, the cherry and the pear, for example. The wild olive (*Olea europea* L., var. *sylvestris* Brot.) is an important component of *maquis* and it is thought that in many cases Mediterranean olive groves were obtained by clearing *maquis*. Other woody plants typically cropped in the Mediterranean are vine, almonds, plums, pears and figs. Citrus species (e.g. orange, mandarin and lime) are also very common in the region although they were originally from the Far East.

This agro-pastoral system included grazing and was usually organised under the 'rotational' form according to which each year 50% of the land was under cultivation and the other 50% under grazing, and the following year these were rotated (Margaris, 1992). Developed through millennia of experience, this form of management permits a continuous but sustainable exploitation of environmental resources.

At lower elevations under a true Mediterranean climatic regime, three basic systems of cultivated plants developed, phased in time and effort to an annual cycle:

1. wet-season cereals: annuals, green during winter;
2. vineyards: perennial shrub, green during summer;
3. olive groves: perennial evergreen trees.

Wet-season cereals (wheat and barley) planted after autumn rains need the farmer's effort for collection, threshing and processing from June until the end of July or the middle of August. At this time the collection of grapes begins, ending in October, when olive collection takes over as the main activity of the Mediterranean farmer.

In addition to these basic types, other agricultural activities have developed. These include the growth of figs, almonds, carobs and plums which, in the past, covered larger areas of the Mediterranean region. In addition, some special plants were also cultivated in limited areas, such as caper (*Capparis spinosa* L.) on the Italian island of Salina; mastic tree (*Pistacia lentiscus* L.) in the Greek eastern Aegean island of Chios; aromatics in Provence, southern France, and saffron (*Crocus* sp.) in Spain, Corsica and Greece.

Figure 3.4.1 Olive groves, vineyards and cereals were the basic type of cultivation in the Mediterranean.

Figure 3.4.2 Mediterranean-type lands also include extensive flat areas with cereals and vineyards such as those in La Mancha with 'Don Quixote' windmills.

The Mediterranean Basin produces one-fifth of the world's vegetables and the one-third of its fruits. The consumption of fruit in the Mediterranean is the highest in the world, with the Greeks and the Spaniards first (150 kg /cap/yr), Italians second (more than 120 kg /cap/yr) and French third (more than 85 kg per year) (Papagiannakis *et al.*, 1986).

During the last few decades great changes have occurred within the primary sector affecting almost all parameters related to the cultivation of traditional

Mediterranean products. The major changes are connected with mechanisation (introduction of tractors, harvesting machines, etc.), genetics (new varieties and hybrids), chemistry (fertilisers, pesticides), long-distance transport of products, new technology (refrigerators, pasteurised juices, frozen products), new constructions (dams and reservoirs for irrigation) and many others.

Figure 3.4.3 Mechanised agriculture with annuals such as the sunflower means abandonment of olive plantations all over the Mediterranean.

Terrace agriculture and cereals

One of the major changes in cereal production has been the abandonment of cereal cultivation on terraces. This is due to the higher production rates and the ease of production in flatter areas. In wheat, for example, with the introduction of tractors and other machines, production of 4 t/ha can be obtained while 20 years ago production was less than 0.8 t/ha (National Statistics Service of Greece (NSSG), 1961, 1994) and was a painful endeavour involving great human effort.

Abandonment of agricultural land previously cultivated on terraces resulted in the decrease of cultivated areas per capita: from 0.43 ha in the 1960s to 0.34 ha in the 1980s.

Figure 3.4.4 A threshing floor on Tinos Island, Greece, that will never be used again.

With respect to olive groves, almost 98% of world production of olive oil still takes place in the Mediterranean Basin, with Greece, Spain and Italy producing more than 75%.

Table 3.4.1 Olive groves in southern European countries.

Country	Area (thousands ha)	Olive trees (number)
Portugal	500	49 000
Greece	430	83 000
Italy	1200	160 000
Spain	2300	185 000

Source: Tsirtsis (1988).

Table 3.4.2 Production of soya oil (thousand tonnes) in Greece, Italy and Spain.

Country	Year 1980	1987	1988	1990
Greece	0	5	6	22
Italy	0	806	1589	1450
Spain	0	3	4	27

Source: Foundation of Economic and Industrial Research (1991).

The new, so-called intensive agriculture is based on annuals (new species, varieties and hybrids), machinery, and above all, high inputs of water, fertilisers and energy. It has been impossible to apply mechanisation to olive groves, because, up to the stage of collection, no new methods have been developed. In any case, such forms of extensive agriculture have been largely replaced in our time by intensive agriculture.

While olive cultivation has been impossible to mechanise, a substitute for the basic product, olive oil, has recently appeared on the Mediterranean market. During the last ten years in Greece the oil coming from intensive annual-crop agriculture, either directly (sunflower and maize oil) or indirectly (cotton oil), has taken a real share of the market and each year more consumers are turning to these products.

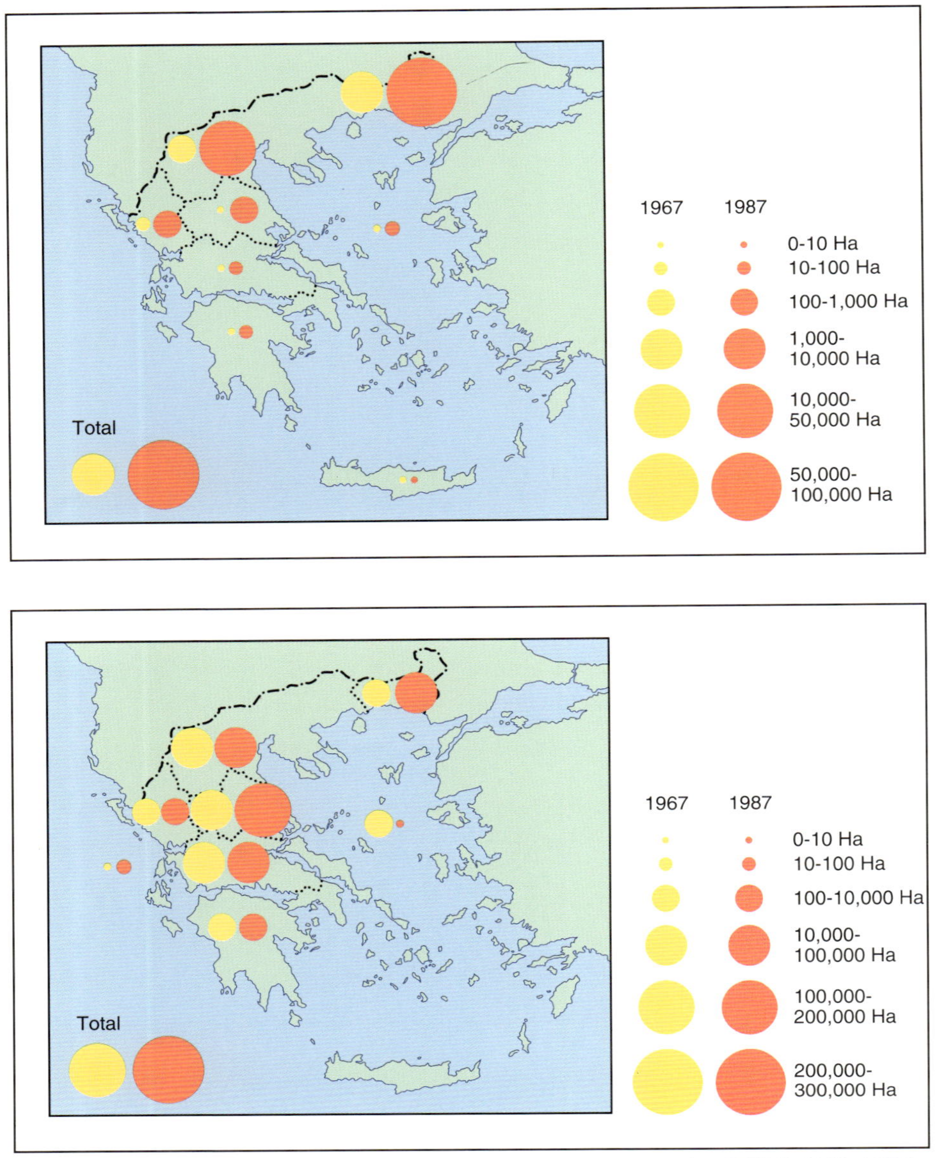

Figure 3.4.5 The spread of (a) sunflower and (b) cotton in recent decades (NSSG, 1969, 1992).

The main driving forces for the Mediterranean primary sector are socio-economic trends related to consumer characteristics. By analysing these trends for different products, such as those involved with the oil market, it is clear that the demand for seed oil, for example, is constantly growing while the use of olive oil is diminishing. A further case is the production of sunflower oil which is increasing constantly in all south European member countries of the Union.

The data for Greece show that the share of the market taken up by 'seed oil' increased from 20% to 36% in the period 1983–1989. In Italy the same share for olive oil fell from 44% to 33% while for 'seed oil' it increased from 41% to 49% in the period 1979–1989. In 1984 the ratio of the price of the sunflower to olive oil was 1/1.37; in 1985 it had moved to 1/1.73 and in 1991 to 1/4.0.

In parallel with these changes the introduction of soya cultivation in the southern states of the EU is another factor affecting the role of olive oil in the market. Due to the above, olive plantations have already started to be abandoned, first in the marginal areas such as in southern France and parts of the Aegean Islands, despite generous economic support given to the producers by the European Union.

Vineyards

Production (as well as the consumption) of wine was of great significance in both economic and social terms to the people of Mediterranean countries. Portugal, Spain, France, Italy and Greece are well known for the production of high-quality wines. Until now the new technological trends applied to cereal and olive production are not very evident in vineyards because the volume production of high-quality wines is low through their restriction to very dry sites.

However, despite the fact that traditional cultivation is generally synonymous with high wine quality, a certain degree of 'modernisation' has had to be introduced. To begin with, the couplings of vine with cereals and olives was broken and, at the same time, the amount of labour input was reduced. In this respect it is worth noting that tourism affected wine production in two ways: negatively by reducing the labour available and positively by increasing the demand for wine (though not necessarily wine of high quality). Moreover, mechanisation in harvesting has been introduced in places like southern France, while elsewhere there was abandonment of the 'marginal' vineyards where mechanisation, at least of harvesting, was impossible.

Figs and other dry fruits against refrigerators and technology

Following the introduction of electricity, new technologies were introduced for preserving fruits such as apples and pears. Today, consumers consider it normal to have apples in April when they are harvested in October. These products are available as fresh fruits until then because of big refrigeration installations, where they remain for months. We have today fresh fruits available all through the winter—something that was not possible three decades ago. This has led to a decrease in the consumption of dry fruits, such as figs, almonds and plums. These come from Mediterranean perennial plant species other than olive and vine, and their cultivation may ultimately be abandoned.

Transition to grazing land and desertification

With the abandonment of terrace cultivation, animal husbandry has begun to dominate and traditional land management systems, which developed over the millennia, have collapsed. Under normal frequencies of the occurrence of fire in Mediterranean climates, herbaceous plants appear in large numbers during the first year after a fire, as a result of the activation of the germination of seeds that lie in the soil seed-bank (Margaris, 1981).

This herbaceous interlude has two major advantages: (1) it helps to combat soil erosion, and (2) although approximately 95% of the nitrogen contained in the above-ground biomass is lost in smoke, leguminous species return atmospheric nitrogen to the soil. Shepherds on the islands know very well that herbaceous plants are abundant after a fire and that leguminous species are an excellent source of food so they have often set fires. However, if intense grazing follows, herbaceous plants are either eaten or trampled by grazing animals, and the whole ecosystem cannot recover so that otherwise-efficient adaptive mechanisms do not work. The only plants that are able to survive are those that are resistant to both fire and grazing. Since most of them are unpalatable, shepherds who wish to get rid of them tend to use the same practice, i.e. they set fires more frequently and the outcome is continuous degradation of both the plant cover and the soil.

3.5 Agriculture

F. Papadimitriou and P. Mairota

Agricultural land use in the Mediterranean EU countries

> *Agriculture is the mother and the nourisher of all other arts: when agriculture is well conducted, all other arts prosper; when agriculture is neglected, all other arts decay, on the land and on the seas.*
>
> (Xenophon, Economicus V-17)

This statement, written more than 2000 years ago, contains elements of truth that are still valid today, because agriculture is a crucial sector of the economy of many countries as a source of income, employment and foreign exchange. Moreover, it also implies that a healthy agricultural sector, with a healthy rural society to drive it, is a precondition for overall prosperity and this is a fundamental argument for integrated and sustainable land management.

The similarities in agricultural production (products and productivities) are well known as these countries specialise in Mediterranean products, having low sufficiency in meat and cereal products. The similarities in land tenure and rural structures are striking. Progress induced by land reforms in these countries (particularly in Greece, Italy and Portugal) is characteristically slow. The cases of Greece and Spain show that although some legal framework preventing the subdivision of plots below a minimum size does exist, there does not seem to be any legislation to prevent re-fragmentation. Thus, the majority of farms are still small in size (e.g. Andalucia's vine-growing *minifundia*).

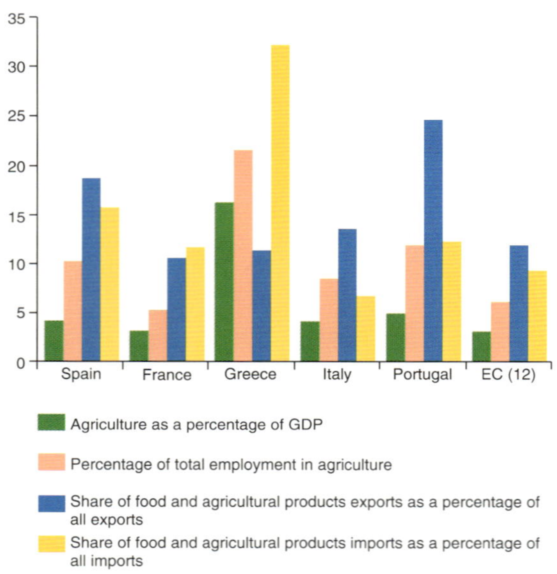

Figure 3.5.1 The importance of agriculture as an economic sector in the Mediterranean countries of the European Union (Spain, Greece, Italy, Portugal and southern France) can be measured by comparing the contribution of agriculture to the gross domestic product (GDP) and the percentage of employment in agriculture in these countries with the average for European Union countries. In addition, food and agricultural products take up a great share of the total export. Traditionally, EU Mediterranean counties export olive oil, wine, oranges, mandarins and clementines, tomatoes, pulses and onions, while importing meat, cereals, potatoes and sugar. (Sources: EUROSTAT, 1989; CEC, 1993.)

The specialisation of production within the agriculture of the EU Mediterranean countries is described (Hecke, 1992) by the relative proportion of five important agricultural production sectors (field crops, horticulture, permanent crops, grazing livestock and granivores), expressed in terms of standard gross margin (SGM). Spain has the most varied production structure, while field crops, permanent crops and grazing livestock are equally important. In Greece, the production structure is dominated by field crops, followed by permanent crops, while in both Italy and Portugal the pattern is exactly the reverse. France as a whole has a production structure more or less similar to that of the EU average.

Regional variations in European Mediterranean agriculture

However similar the production structure of the EU Mediterranean countries may be at the national level, large variations appear when the sub-national administrative level is considered in comparison with the national one (cf. Mairota, 1995a). For example, the production structure of each one of the MEDALUS regions is different to that of the country to which they belong.

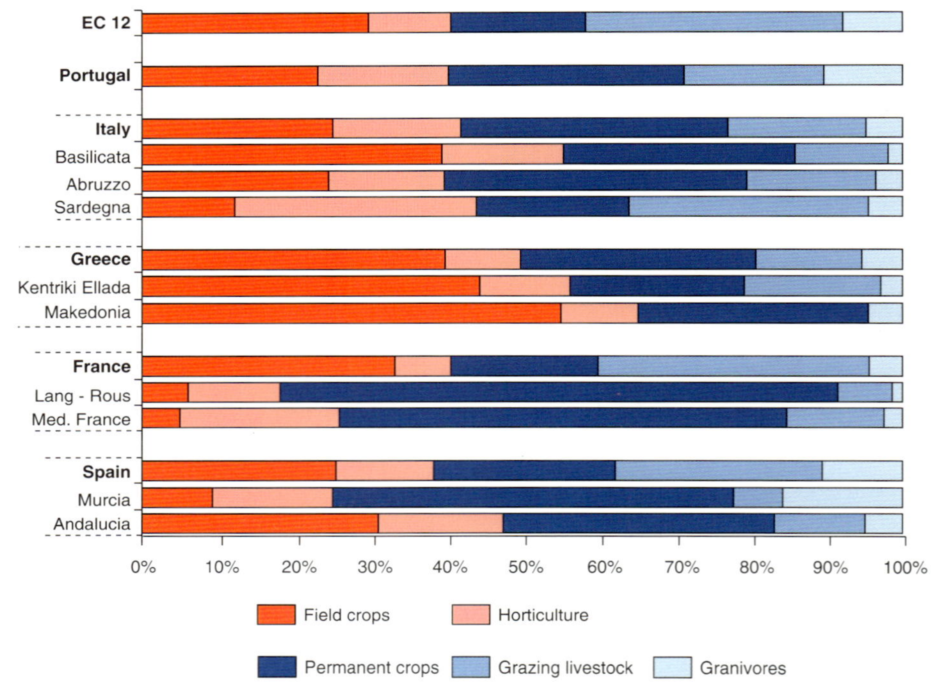

Figure 3.5.2 Production structure of the EU Mediterranean countries and the MEDALUS regions in 1985 (% of SGM). According to Hecke (1992), Languedoc-Roussillon is a 'specialised' region, where the greatest share of the SGM is provided by permanent crops (mostly vine). Kentriki Ellada, Makedonia and Basilicata are 'combined regions' in which the SGM is partially dominated by field crops. Andalucia, Murcia and Abruzzo are also 'combined regions', but the SGMs are mostly provided by crops, among which permanent crops prevail. Sardinia and the whole of Portugal are 'non-specialised' areas, where grazing livestock, horticulture and permanent crops, and field crops, respectively, provide at least 20% of the SGM. (Source: Hecke, 1992.)

The spatial distribution of land-use categories (both at national and sub-national levels) reflects the pattern of the production structure. In all these countries (with the exception of Portugal), the ratio of utilised agricultural land (UAA) to total land is equal to or above 50%. At the regional level, only Languedoc–Roussillon (41%) and Kentriki Ellada (28%) are below the threshold of 50%.

The utilised agricultural area is dominated by arable land in all the EU Mediterranean countries with the exception of Greece, where grasslands prevail instead. In Andalucía, Murcia, Kentriki Ellada, Makedonia and Basilicata, the most important category of the UAA is arable land, with permanent crops and grasslands second and third in order of importance; in Abruzzo and Sardinia, the grasslands are followed by arable land and permanent crops in order of magnitude, and in Languedoc–Roussillon, the grasslands are more important than the permanent crops and the arable land.

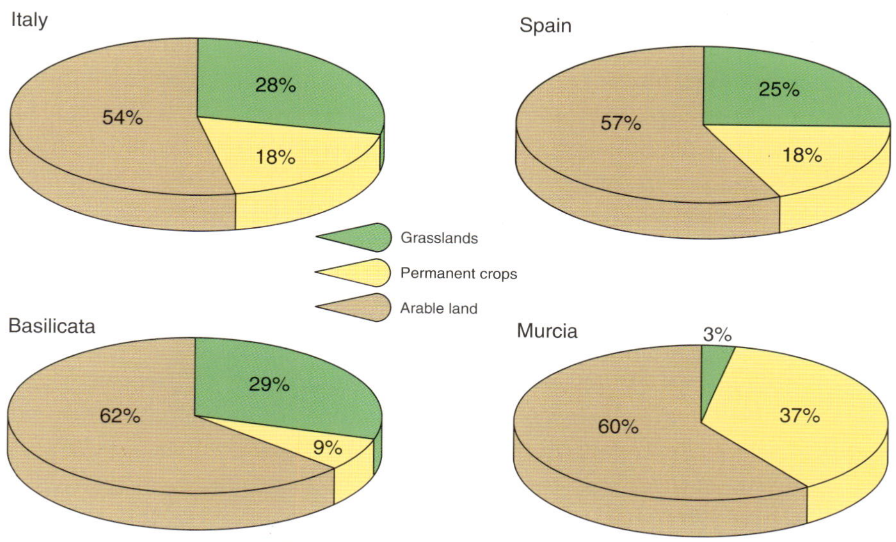

Figure 3.5.3 Spatial distribution of the land-use categories of the utilised agricultural area in Italy, Spain, Basilicata and Murcia. (Sources: EUROSTAT, 1989; ISTAT, 1982.)

The characteristics of agriculture of the EU Mediterranean countries and regions are mainly shaped by the land qualities, climatic conditions and land fragmentation. The combination of these factors (together with the relative degree of socio-economic development of the regions) controls the intensity of the agricultural practices and the level of investment in the agricultural sector, while also determining the economic performances of these regions. On it depend greatly the socio-economic disparities between EU Mediterranean countries and regions, as has been pointed out by a number of studies (e.g. Bartola *et al.*, 1989; Bachtler and Michie, 1993; Dunford, 1993; Gripaios and Mangles, 1993).

The regions of Spain, Greece, southern Italy, Portugal, and both Languedoc–Roussillon and Corsica in France, are among the regions of the EU whose development is lagging behind the EU average (CEC, 1987). Most of these regions specialise in the production of typical Mediterranean products and their agricultural practices are mainly extensive (in terms of investments) and technologically underdeveloped. Such agriculture is 'poor' or 'not efficient' in terms of productivity of labour and/or land productivity (Bartola *et al.*, 1989).

Agricultural structural and socio-economic problems associated with land degradation in Mediterranean Europe

The need to close the gaps with the more developed areas of the EU has led to very important socio-economic changes within the EU Mediterranean regions. A certain degree of rural development has indeed been achieved in these countries within the framework of the Common Policies of the EC (such as the Common Agricultural Policy), as reflected by changes in land use, degree of intensification in agriculture and land-use diversity.

The land degradation that is taking place in many of the agricultural areas of the European Mediterranean countries is closely related to certain structural and socio-economic characters of these countries, particularly increasing intensification, high land fragmentation, low land consolidation rates, and changes within the utilised agricultural area.

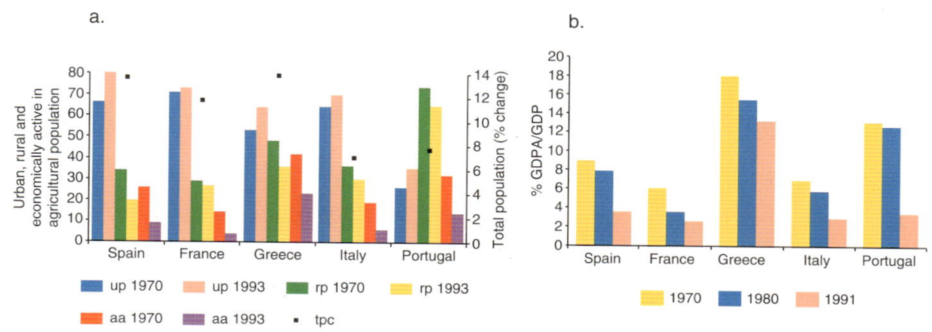

Figure 3.5.4 (a) Population changes in the EU Mediterranean countries: up = urban population, rp = rural population, aa = population active in agriculture, tpc = total population change (1 –tp1970 / tp1993) × 100. (b) Agricultural contribution to GDP in the EU Mediterranean countries. (Source: Centre International de Hautes Etudes Agronomiques Méditerranéens, 1994.)

Increasing intensification

The trends in land-use changes and the increase of intensification in agriculture reveal that the pressure upon the economically productive agricultural land has increased on the one hand, and the less economically viable lands are being devoted to non-agricultural land uses (e.g. urbanisation, infrastructures and forestry) and possibly abandoned, on the other. This has also been determined by demographic growth, increases in the urban population, rural depopulation, decreases in the contribution of agriculture to the GDP, and decreases in the number of the population active in agriculture.

Today's intensification in agriculture results from the increased input of means of production, such as fertilisers, machinery and irrigation per unit of utilised agricultural area. This kind of intensification greatly differs from the intensive use of the land (by means of great labour input per unit of utilised agricultural area) that has been a characteristic of Mediterranean rural environments for centuries.

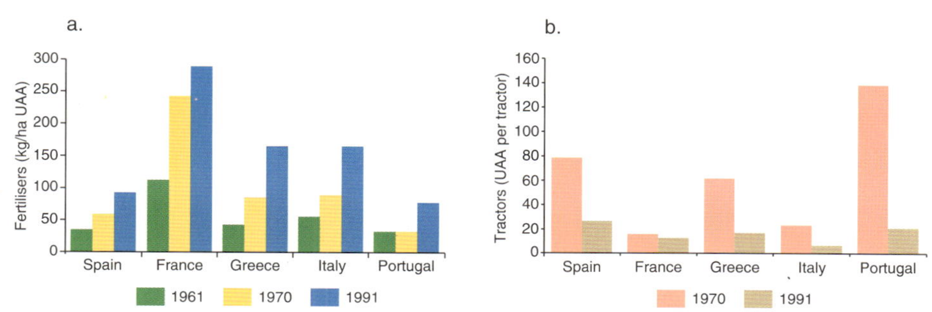

Figure 3.5.5 Increased input of the main means of agricultural production: (a) fertilisers and (b) tractors in the EU Mediterranean countries. (Source: Centre International de Hautes Etudes Agronomiques Méditerranéens, 1994.)

Such a difference has a strong impact upon the physical environment of the Mediterranean EU countries, causing land degradation. One example of the

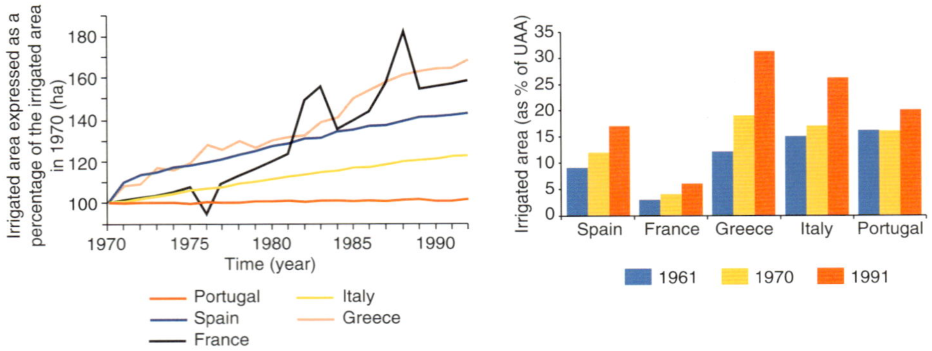

Figure 3.5.6 Irrigated area has notably increased in all countries but Portugal from 1970 to 1992. In all countries a greater percentage of the UAA (in turn slightly decreased in all countries but Greece) is being irrigated.

socio-economic impacts of modern intensification in agriculture (particularly by means of mechanisation) is the decrease of labour per unit of land (hence, the creation of a surplus of available workforce, leading to unemployment) in the areas where intensification of the production is possible. Another example is the economic marginalisation of the agricultural land in the areas where this is not possible. The major constraints to intensification of the agricultural production in Mediterranean Europe, paradoxically, are those very features that have favoured the continuous intensive cultivation of such lands by human endeavour.

Examples of impacts of modern intensification in agriculture upon the physical environment include contamination of soils and waters from the use of fertilisers and animal husbandry wastes, increases in the water supply for agriculture (hence groundwater exploitation) and the need for the construction of reservoirs because of growing irrigation demands, fossil-fuel consumption from mechanisation (hence air pollution and the creation of chemical wastes) and increased erosion risk when mechanisation is applied in geomorphically unfavourable conditions (cf. Mairota, 1995a).

High land fragmentation and low land consolidation rates

Greece displays the highest land fragmentation of all the Mediterranean European countries and France the lowest.

Table 3.5.1 Land fragmentation in the EU Mediterranean countries.

Country	Percentage of holdings by size		
	<20 ha	20–50 ha	>50 ha
Greece	97.8	1.8	0.4
Spain	86.7	8.1	5.2
France	52.7	30.5	16.8
Italy	95.3	3.3	1.4
Portugal	96.7	2.1	1.2

Source: EUROSTAT (1989).

The striking similarity of the Greek agricultural structures with the Portuguese situation is also worth noticing. The small difference between the Greek, Spanish and Italian farm structures is due to the fact that the land-holdings are held by fewer owners in these countries, compared to Greece: 97.8% of the Greek agricultural area is made up of farms of less than 20 ha. This peculiarity can be attributed to the very early agrarian reforms that took place in Greece (1871–1922).

The role of land consolidation schemes in Mediterranean European agriculture has been very important, but the rates of consolidation have been very low.

One of the defects of land consolidation has been the move towards intensive cultivation with high productivity and very advanced mechanisation. Another is the concentration of power by the wealthy farmers with simultaneous impoverishment of the peasants. In most cases, the peasants stayed financially weak even after the land consolidation reforms had taken place, while in the mid-1970s the pressures increased, favouring the enlargement of the holdings. The situation will certainly become worse in the future, particularly if there are changes in the Common Agricultural Policy with respect to subsidies to southern Europe.

The main consolidation measure in Italy, the Land Improvement Act of 1933 (*Legge per la Bonifica Integrale,* 13 February 1933, no. 215), did not produce the expected increase in productivity (OECD, 1965), and the following land reforms of the 1950s (*Riforma Fondiaria*) resulted in the progressive transfer of agricultural activities from the hilly interior areas to the plains, thus deepening the disparities in productivity levels (De Stefano and Lombardi, 1976).

In France, the *Loi sur les Cumuls* defines (according to the country's agricultural legislation) an upper and a lower limit to the enlargement and division of lands. In the effective French consolidation scheme, the incentives provided by the supplementary *Loi d'orientation* (5 August 1960) and *Loi complementaire* (8 August 1962) facilitated the rural exodus (minimising its costs) and gained the advantages of this emigration to improve the farm structures. The French official agencies SAFER (*Société d'aménagement foncier et d'établissement rural*) and FASASA (*Fonds d'adaptation sociale pour l'aménagement des structures agricoles*), established in 1960 and 1962 respectively, bought land which afterwards was used to consolidate fragmented holdings. There are grants available from these agencies to young farmers who wish to settle in areas where the population is below the minimum desirable level (11 inhabitants/km^2) and grants encouraging farmers to retire at the age of 60 or to retrain men after leaving land cultivation for other employment (Minshull, 1990). Such measures would be particularly useful in many Mediterranean European regions, as they would assure a low level of rural depopulation, proper management of the rural human resources and maintenance of the competitive structure of the agricultural holdings.

Changes within the utilised agricultural area

As far as the overall evolution of agricultural land use is concerned, certain observations can be made for the period 1970–1992 using Food and Agriculture Organisation (FAO) Yearbooks.

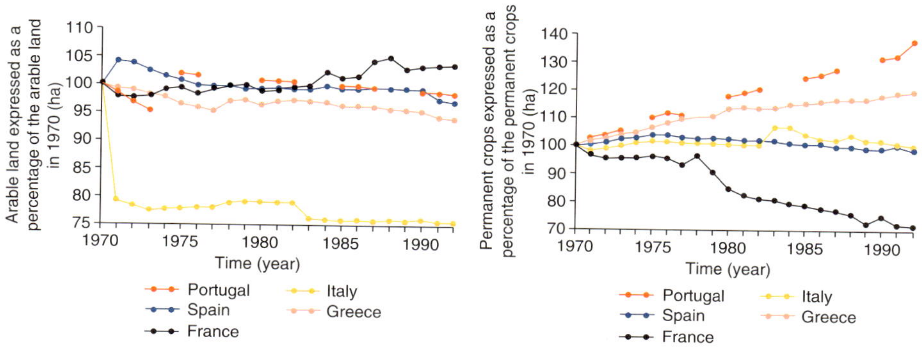

Figure 3.5.7 Evolution of arable land and permanent crops in the EU Mediterranean countries. (Source: *FAO Yearbook*, 1971 onwards). (a) Arable land has decreased in all EU Mediterranean countries but France. This is probably due to the fact that arable farming is increasingly vulnerable to international competition and there has been an established tendency by small farmers either to abandon their arable land or to change its use. (b) The area of permanent crops has increased in all countries (but France) and, apart from Spain and Italy, after 1985. This kind of evolution is explained in terms of relative financial benefits received by farmers cultivating traditional Mediterranean crops (e.g. oranges, olives and vines) until the CAP revisions and reorientation of the mid-1980s.

Future perspectives

Although the Mediterranean European agricultural structures may show increased fragmentation, they are often very productive and they are functioning in a great diversity of purely capitalist modes of production. Yet the perspectives for the agriculture of the European Mediterranean countries are not encouraging (Papadimitriou, 1994), particularly in view of the negative impact of certain international agreements (e.g. CAP, GATT and Lome Conventions).

The increasing competition between the small producers of traditional Mediterranean countries and the large-scale intensive farming of other non-Mediterranean countries is likely to prove ultimately destructive for the Mediterranean farmers. Consequently, larger and larger portions of agricultural land will be set aside from cultivation, abandoned, dedicated to other land uses or to nature conservation. Thus, significant changes in land-use diversity are expected throughout the European Mediterranean countries.

The endemic fragmentation of land ownership and the individualistic attitude of the Mediterranean farmer on the one hand, and the high proportion of moderately steep (slope gradient 15–25%) and steep (>25%) land in southern Europe (Lee, 1990) on the other, are expected to further aggravate the already gloomy picture of European Mediterranean agriculture.

LAND-USE DIVERSITY CHANGES

F. Papadimitriou and P. Mairota

Calculation of land-use diversity

A general description of land-use structure is given by the index of diversity. This index is a 'synthesis' of the relative proportions of land of a landscape under certain land-use types.

An index of land-use diversity (H) for the Mediterranean European countries can be based on five land-use categories (arable land, permanent crops, grasslands, woodlands and other land uses). Shannon's index of diversity (Shannon and Weaver, 1949) was applied to the data published by the Food and Agriculture Organisation of the United Nations[1] (*FAO Yearbooks*).

$$H = -\sum_{i-1}^{n} P_i log P_i$$

where P_i is the proportion of each land-use category over the total and $i = 1, 2, \ldots 5$, with $\Sigma P_i = 1$.

Changes in land-use diversity

Previous research (Papadimitriou and Mairota, 1994) has shown that the land-use diversities of southern European countries changed at significantly different rates during the period 1965–1985. The same study showed that the trends in land-use diversity changes were also different. Hence, a question arises as to how land-use diversity changes with time.

From the present study it emerges that differences in the temporal evolution of land-use diversity amongst the European Mediterranean countries are still significant.

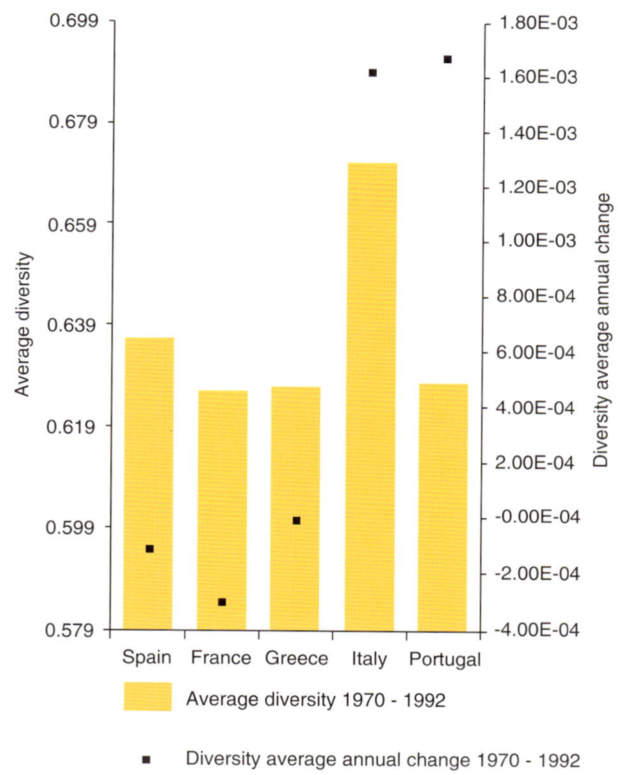

Figure 3.5.8 Changes in landscape diversity in the EU Mediterranean countries. (Source: *FAO Yearbook*, 1971 onwards.)

[1] *FAO Yearbooks* are the only source providing a yearly data set. However, they contain obvious inconsistencies, as the case of Portugal demonstrates. These have been taken into account in the data analysis and the interpretation of the results.

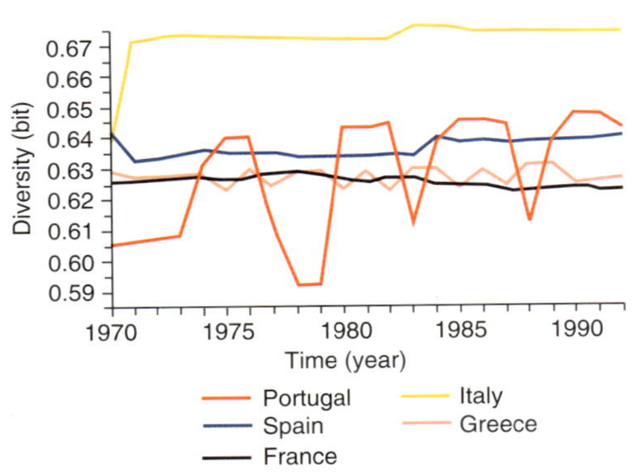

Figure 3.5.9 Evolution of landscape diversity in the EU Mediterranean countries. (Source: *FAO Yearbook*, 1971 onwards.)

For instance, whilst the diversity of Italian land use changes little over time, land-use diversity of Greece varies relatively widely. Also, as confirmed by a simple correlation test, land-use diversity in Spain, Greece, Italy and France is in phase with the land-use category 'other'. In the case of Portugal it seems to be in phase with the grasslands.

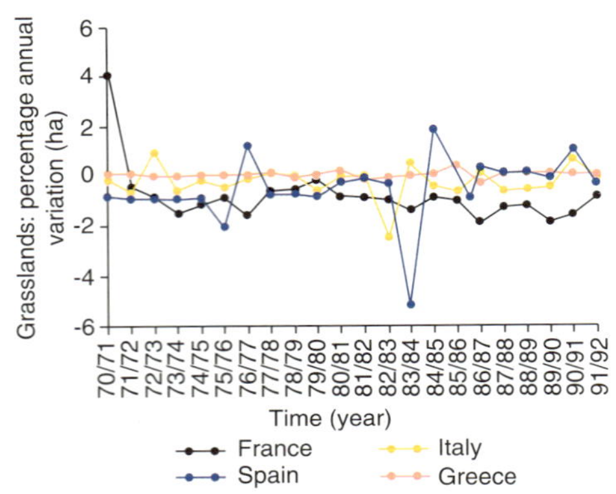

Figure 3.5.10 Inter-annual percentage changes of grasslands in the EU Mediterranean countries. Changes for Portugal are not shown due to inconsistencies in the data set. (Source: *FAO Yearbook*, 1971 onwards.)

As far as the overall evolution of each land-use category is concerned, the following observations can be made for the time interval 1970–1992. The area covered by grasslands and the area designated as 'other' varies from year to year. This may be because of the inter-annual variability in the designation of grasslands and the ease in conversion from any other land-use category to 'other' land use.

The forested areas are increasing in all countries but Greece. This is due both to the increasing tendency of state authorities to reserve larger and larger

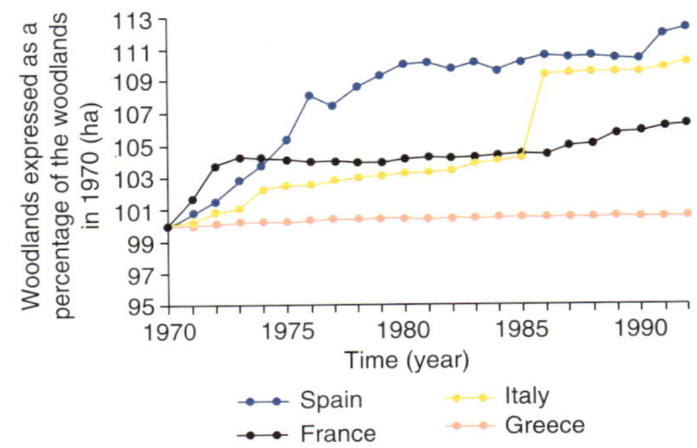

Figure 3.5.11 Evolution of woodlands in the EU Mediterranean countries. Changes for Portugal are not shown due to inconsistencies in the data set. (Source: *FAO Yearbook*, 1971 onwards.)

portions of land for nature conservation purposes, and to the large-scale expansion of shrubby vegetation caused by vegetation degradation.

In view of the European Union's regional policy aims to enhance and increase diversification in the rural areas of the EU (EC, 1994), it is important to take into account the calculated diversity changes in Mediterranean EU countries.

Future perspectives

The Mediterranean European landscapes are diverse and beautiful. As significant changes in land-use diversity have occurred and are expected to continue to occur in the European Mediterranean countries, it is important to put the numerical calculations of land-use diversity change into a landscape protection perspective (Papadimitriou and Mairota, 1995). This is necessary because once the trends in land-use diversity changes are known, it will be possible to discern the regions of countries for which landscape protection is a high priority.

3.6 Urban Expansion and Littoralisation

L. Leontidou, M. L. Gentileschi, A. Aru and G. Pungetti

Introduction

In desertification studies urban expansion, change and development are rarely considered. However, this is the result of a failure to understand that desertification is a cultural as well as a physical phenomenon. Although the evidence for desertification often takes the form of vegetation destruction, loss in soil productivity and of the soil itself, and the accompanying loss of agricultural potential, the causes of desertification are invariably tied up with regional or even national trends in human activity. The large-scale migration from the countryside to the city has led to the abandonment of land, sometimes with beneficial effects, sometimes with the destruction of conservation measures that have been in operation for centuries. Perhaps most importantly, cities create improvements in living standards leading to an almost insatiable demand for water and through their recreative requirements increase the incidence of forest fires. In the Mediterranean at least, urbanisation and the associated littoralisation are key elements in the desertification problem. In this section the trends towards larger urban agglomerations are first outlined, then regional-scale changes are examined, exemplifying the problem from southern Italy, and finally the local level is considered, with examples from small areas near the city of Cagliari in Sardinia.

Urban expansion in Mediterranean Europe

Urban expansion in the Mediterranean is usually unregulated and frequently illegal. In the past, cities were compact and dense, without the squalor which the industrial revolution brought to northern Europe. Their population did not grow very rapidly until the inter-war period, with few notable exceptions; but rapid rural–urban migration was the rule thereafter. Urban expansion stepped up very rapidly during the post-war period, and was inextricably linked with fast urbanisation waves throughout the Mediterranean—in Europe, North Africa and the Near East. Throughout the post-war period, the two largest cities in each southern European country grew at very fast rates, especially in relation to national population figures. A comparison between Athens, Rome and Barcelona in 1971 is very telling, indicating the same population figures (3 million) and rapid annual rates of growth (2.8–3.2%; Leontidou, 1990). Growth also followed technological change—transport and communications, infrastructure development—although the areas gradually 'legalised' and incorporated into the urban tissue they did not necessarily acquire proper infrastructure.

In the largest Mediterranean urban areas, unregulated urban expansion meant polarisation between the inner city and the new suburbs, precariously built, without infrastructure. On the urban fringe, speculative landowners subdivided their land on the basis of rough plans and sold it as plots. This land had been earmarked as agricultural or forest or open land in the urban plans, where it was illegal to build. The new owners did build, however, contravening this regulation, with the tacit acceptance of the State. A standard procedure was then the 'legalisation' of the area settled illegally by the payment of small fines by the settlers, but without any infrastructure provision. Despite European funding, areas of Athens acquired infrastructure only in the late 1980s, and many Portuguese illegal settlements still lack water, sewage, drainage and road networks. In all cases the urban population stabilised throughout Mediterranean Europe (Leontidou, 1995).

Our example below is from a small city, and thus indicative of the relative but not the absolute scale of rapid, uncontrolled urban expansion. The largest cities had reached quite an important urban sprawl by 1981: the built-up area of Rome was then 28 000 ha (compared with Milan at 18 000 ha), equal to that of the second largest Greek city, Thessaloniki (Leontidou, 1990). Expansion contrasts with high densities in the inner city: in the 1980s, the core of Athens, some 4200 ha, accommodated 886 000 inhabitants, and the broader inner city of Barcelona, with 10 000 ha, had a population of 1.6 million (Berry and McGreal, 1995). The figures for urban sprawl do not include the more sparsely populated areas on the urban fringe, which made the area of Rome reach 150 000 ha in 1981; in those areas, land-use succession from agriculture to urban settlement is actively sought by setting fire to forests, and illegal subdivision.

The depletion of physical resources on the rural–urban fringe is clear in our example of a small city which had remained contained in a small area for many decades, but which has grown ninefold in area during the last 60 years. Despite the stabilisation of the urban population in large cities, the change from compact cities to urban sprawl, often illegal, has been a structural feature of urban expansion. Speculation is encouraged, not only by the lack of planning controls, but also because of the fragmentation of land ownership and construction capital, piecemeal urban redevelopment, and the formal/informal economic polarisation in productive activity and the housing sector.

Littoralisation

Trends analysed in Section 3.3 underline the increasing importance of the industrial and tourist sectors rather than agriculture in rural areas of the Mediterranean region. This is very much evident in land-use change: agricultural land is lost and swallowed up by either urban expansion (industry, commerce, roads, housing, quarries, etc.) or tourist development (hotels, apartment complexes, roads, leisure facilities, commerce, etc.). Competition between land uses is especially fierce in tourist localities, not only between tourism and agriculture, but also with industry.

Land-use change on the Mediterranean coast has been partly spontaneous, and partly boosted by government policies since the 1960s (Williams and Shaw, 1991). The first Spanish law to facilitate the creation of new tourist settlements in zones favoured by spontaneous tourism (1963), was followed by continuous emphasis on mass tourism, accommodation, infrastructure and accessibility, but little attention was given to environmental impact, agricultural and forest land. In Italy, tourism was already being discussed in the 1960s as an economic policy to develop the Mezzogiorno, but its effect was not depolarising. The 'Adriatic strip' saw the growth of a continuous unbroken ribbon of tourist resorts, displacing agricultural land use and attracting migrants from the hilly interior as well as international visitors. In Greece, tourism had already started to feature in policy with legislation for incentives by the 1950s, but especially with concentrated legislation after 1969. In fact, coastal development during the dictatorship involved financing, promotion, and special concessions to hotels. There were hardly any environmental considerations, with a lot of damage as a result, especially during the period of the dictatorship. France was rather better targeted in planning the rural landscape in comparison with the other Mediterranean European countries, though not before the mid-1970s. Special development companies were created to oversee plans to expand tourism along the Languedoc–Roussillon and Aquitaine coasts and in Corsica in the early 1960s, and later the *Délégation à l'Aménagement du Territoire et à l'Action Régionale* encouraged tourism. However, the 'concrete walls' along coastlines, restricting seaward access to the coast, and other negative environmental effects were later criticised and this changed official attitudes. Seasonal fluctuations of population on the coast are a challenge to planners. At peak tourist periods, the overwhelming population increase creates a strain on infrastructure, public services, water resources and the environment.

Pollution of the beaches and the sea, forest fires or arson, and concrete walls in place of attractive natural landscapes may result from land-use succession during the present period of economic restructuring. Our examples indicate that even in cases of less active conflict, there can be negative impacts on the environment.

Table 3.6.1 Population (p) and dwellings (d) in per cent in coastal areas .

	1951		1961		1971		1981		1991	
	p	d	p	d	p	d	p	d	p	d
Abruzzo	18.9	22.1	15.0	13.5	25	25	29	28.2	29.5	29
Molise				1.5	12.8	8.8	11.7	13.3	13	15
Campania	47.6	35.6	42.3	40.1	43.6	42.7	41.9	42.1	38.4	39
Basilicata	7.7	8.2	9.0	9.6	10.5	10.2	10.6	11.0	11.0	11.0
Calabria	41.7	40.3	46.9	45.7	49.9	47.6	52.7	54.4	52.1	55
Puglia	40.8	39.4	42.4	41.5	45	43.8	43.8	44.8	42.3	45
Sicilia	57	51.6	57.8	55.5	66.2	59	61	59	60.5	58
Sardegna	37	35.6	43.1	42.3	44	48	50.3	53	49	63
Mezzogiorno	43.8	38.9	43.5	42	47.3	45	46.5	47	45	46

Source: ISTAT (1985 and 1993), data processing: L. Viganoni.

Figure 3.6.1 By the sea, the hillsides still show the scars caused by a wildfire in 1993, two years previously, which destroyed prairies and garigues but could not kill the trees, which are sprouting.

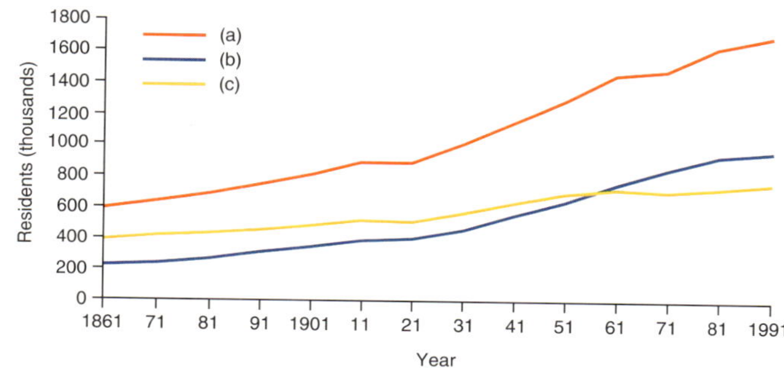

Figure 3.6.2 Resident population (in thousands) in (a) the whole of Sardinia, (b) municipalities, and (c) interior municipalities. (Sources: ISTAT, 1985, 1993. Data processing: M. Jorio. Interpolations for 1891 and 1941.)

Littoralisation in southern Italy: the case of Sardinia

Population and housing in the coastal areas of southern Italy have increased during the last five decades (ISTAT, 1955–1993). Initially the shift to the coast was accompanied by a change from extensive agriculture and grazing on once malarial and sparsely populated land to a market agriculture where horticultural produce and fruit trees were widely grown. Later, however, this shift led to the occupation of prime farmland, also irrigated, to accommodate housing developments and roads. Littoralisation reached its peak in the 1960s and coincided with a phase of notable natural population growth. From the 1970s onward, the trend on the whole started to decline. However, coastline housing continued to expand in the vicinity of the larger cities, e.g. Naples, Bari, Reggio Calabria and Cagliari.

An example of consumption of fertile land due to coastal population growth in southern Italy are the human settlements along the coast. A remarkable growth of settlements has recently occurred along the coasts of Sardinia, once abandoned by the people seeking safer sites inland. Many of the recent settlements are the so called '*marine*'. These are the main form of re-population of the coast in southern Italy (particularly in Basilicata and Calabria) and essentially originate from the remittances of the emigrants abroad or to northern Italy from upland villages.

Coastal areas of Sardinia are particularly prone to land-use conflicts and land degradation caused by human activities and related to land-use changes. Land degradation here takes the form of a loss of non-renewable resources (water and fertile soils), as well as the over-extraction of materials. Sardinian local authorities are continuously faced with the problem of satisfying the conflicting demands of residents, tourists, industry and farmers. As a result, in the last few years coastal settlement growth has been regulated by the enforcement of more strict planning measures. These also aim at the preservation of the coastal landscape and sea-front, and of the fertile arable lands of the delta plains and coastal strips.

Water

Tourism is one of the most important sectors of the economy of the island and is mainly concentrated in July and August. Overnight stays by tourists in these two months account for 57% in Sardinia and 35% in Sicily (the other major Italian Mediterranean island) of all tourist overnight stays.

The expansion of tourism in Sardinia in recent years has by far exceeded what was envisaged by the master water plan of the Regione Autonoma della Sardegna (Regione Autonoma della Sardegna, 1993). This has resulted in the difficulty of coping with increasing demands for water supplies. The increase in the number of tourists has gone far beyond predictions. From 1971 to 1989 an increase of 160% was registered in Sardinia, against 33% for Italy as a whole. The tourists visiting Sardinia each year now total 1 700 000, outnumbering the island's entire population, and forecasts have had to be adjusted accordingly. In 1981 water supplies to tourist resorts accounted for 11% of

Figure 3.6.3 South of Cagliari a mild climate and good soils allow the cultivation of olives, vineyards, and arable plots, enclosed by rows of cypresses which skirt the seashore.

Figure 3.6.4 Rows of olive tree shade the alleys of a residential village by the sea, as living relicts of former groves. The growing housing demand is still causing loss of arable land in the neighbourhood of Cagliari, the main city of Sardinia.

total requirements. This is expected to increase by 7% by the year 2031 (Regione Autonoma della Sardegna, 1993) Furthermore, by then the resident population of Sardinia is unlikely to reach the figure forecast in the master water plan, and so the demand by tourists will have a proportionately greater impact (Regione Autononia della Sardegna, 1993).

As tourists tend to concentrate in a few municipalities by the sea, at the height of the season water shortages become intolerable. Some municipalities (e.g. Carloforte, La Maddalena and Villasimius) have installed plants for the desalination of marine waters to cope with peak demand and in other areas groundwater exploitation is intensified or new reservoirs planned. Public authorities are doing their best to promote low-season tourism with a view to diminishing the peak summer influx, offering attractive packages that encourage tourists to visit the island in off-peak periods.

The municipality of Villasimius, in the province of Cagliari, ranks among

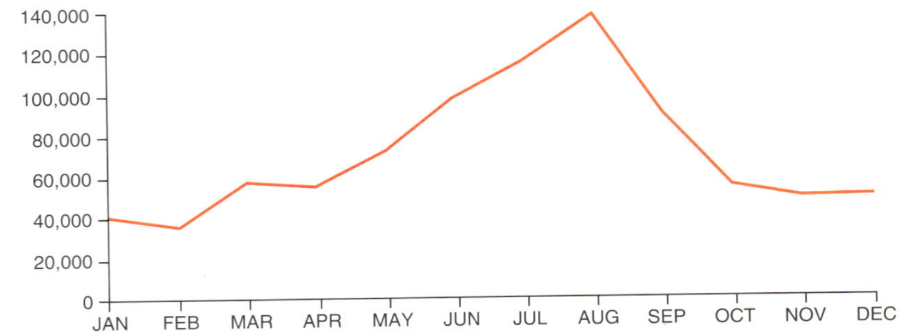

Figure 3.6.5 Monthly drinking water consumption (in cubic metres) for the municipality of Villasimius in 1994. (Source: ESAF Cagliari, private communication.)

the first eight communes in Sardinia for number of tourists. The resident population in the off-season numbers just 2800, but in the two weeks around mid-August, i.e. in the peak season, it swells to 50 000 overnight stays, with an additional 20 000 at weekends. In August (according to data provided by the regional water authority ESAF, which do not include water withdrawn from local artesian wells), water demand increases more than fourfold compared to the lowest monthly demand, registered in February. Domestic water shortage in the summer months has become the local community's most serious problem. To integrate local supplies a sea-water desalination plant has been installed, one of the three currently existing in Sardinia for civil use, despite the high cost of the water obtained with this technique. The recent construction of a long-distance water conveyance system that supplies water from a reservoir in the interior of the island, has alleviated the problem in the short term. In 1993 and 1994, supplies satisfied demand. However, irregular rainfall and the rising demand for water from other tourist resorts along the Sardinian coasts has aggravated the problem once again.

Agricultural land

Good soils are rare in the Mediterranean and in Sardinia in particular, since in the course of its long geological history erosion has stripped bare the hills and uplands, exposing the ancient bedrock. The last 50 years have seen rapid urban expansion in Sardinia and extensive areas of land around the city of Cagliari have been occupied. In the majority of cases the expansion of urban centres has been accompanied by an increase in the number of housing developments along the coast and in the hills.

The coastal plain of the Rio Santa Lucia catchment area is a significant example of this phenomenon. The transformations of the Capoterra area are documented from 1898. Initially (1898–1931) the changes mainly affected the areas around the Cagliari lagoon. Extensive parts of the lagoon were reclaimed for the installations of salt pans (Marchieddu salt works). The introduction of a number of drainage canals broke up the continuity of the water surface. Agriculture remained the main sector of the economy for the village of Capoterra, as its surroundings did not undergo major transformations during this period. From 1931 to 1989, by contrast, the built-up area around Capoterra has increased ninefold, accompanying a population increase of 49.75% from 1951 to 1981 (ISTAT, 1955, 1983). Such urban expansion is the pattern for many other municipalities of Cagliari's hinterland, where mass migration took place in the post-war years.

Urban expansion is by far the major contributor to the loss of agricultural land south of Cagliari. However, alongside the salt pans, industrial agglomerations, such as petrochemical plants and extractive sites, were built on what was once agricultural land.

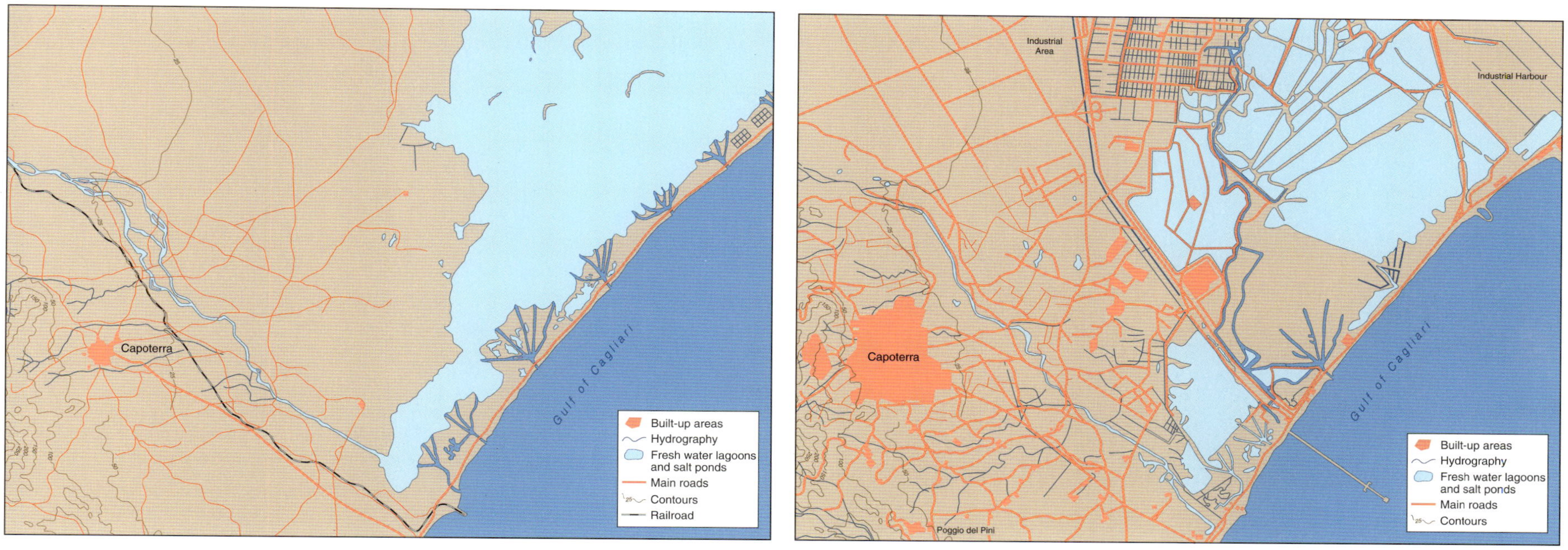

Figure 3.6.6 Land-use change in the municipality of Capoterra (Cagliari) from 1898 to 1989.

Agricultural land loss: the case of Quartu Santa Elena

A. Aru, F. Fantola and M. R. Lai

Among the municipalities affected by extensive conversion of agricultural land for urbanisation-related purposes, Quartu Santa Elena is the most significant. Here both tourist developments and regular urban development have been responsible for the changes.

A cartographic quantitative investigation was used to assess the actual and potential losses of agricultural land, with the following four phases.

1. geomorphological and pedological surveys, as a basis for the study of the physical environment and of land capability evaluation;
2. present land-use survey, contemporary land consumption and its dynamics;
3. assessment of the current municipality's town planning tools in relation to the regional planning policies for future development;
4. land suitability evaluation.

Table 3.6.2 Land loss (ha) due to urban expansion by land unit updated to 1977.

Year	Land units								
	1	2	3	4	5	6	7	8	9
1968	28	477	6137	58	83	66	20	15	0
1977	104	1655	2995	227	464	97	1454	73	1

A map of agricultural land consumption risk was drawn, based on the outcome of the land suitability evaluation.

The results show that such an investigation can provide useful tools for the design and the implementation of plans for future urbanisation without compromising the best soil resources and to prevent irrational uses of the land.

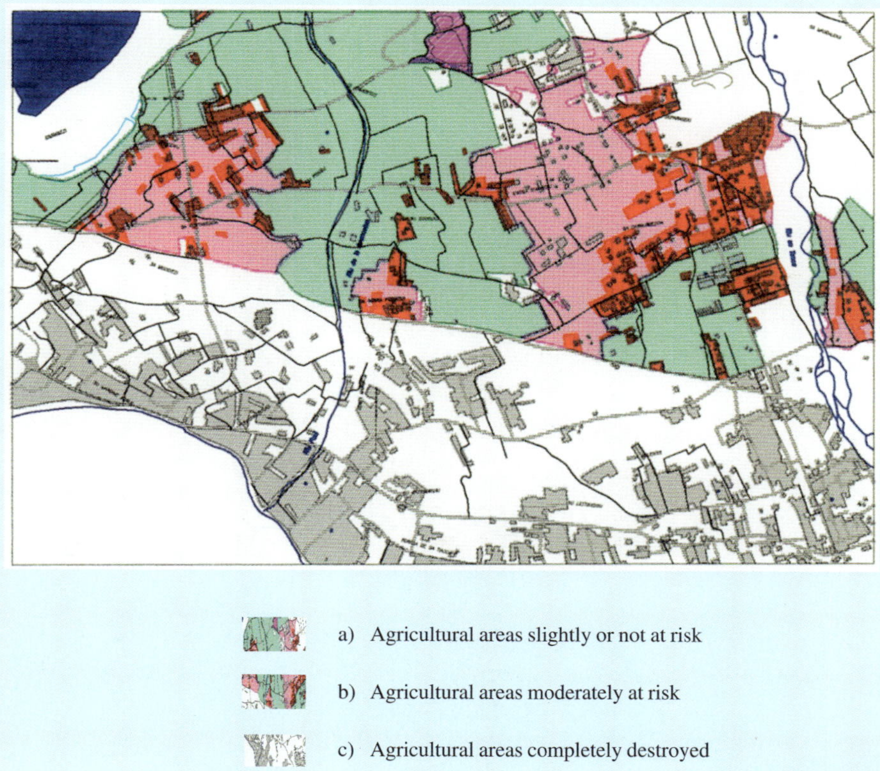

Figure 3.6.7 Agricultural land consumption risk for Quartu Santa Elena municipality. (a) Agricultural areas at low or no risk are those areas where land use is not likely to change and which will continue to be devoted to a more or less intensive agriculture. (b) Agricultural areas moderately at risk are those located at the fringes of built-up areas and are likely to suffer from indirect loss. (c) Agricultural areas at high risk or completely destroyed are those areas where the development of housing, road, quarries, commercial and industrial activities has either already occurred or is more likely to occur, hence where agricultural land has suffered from direct losses.

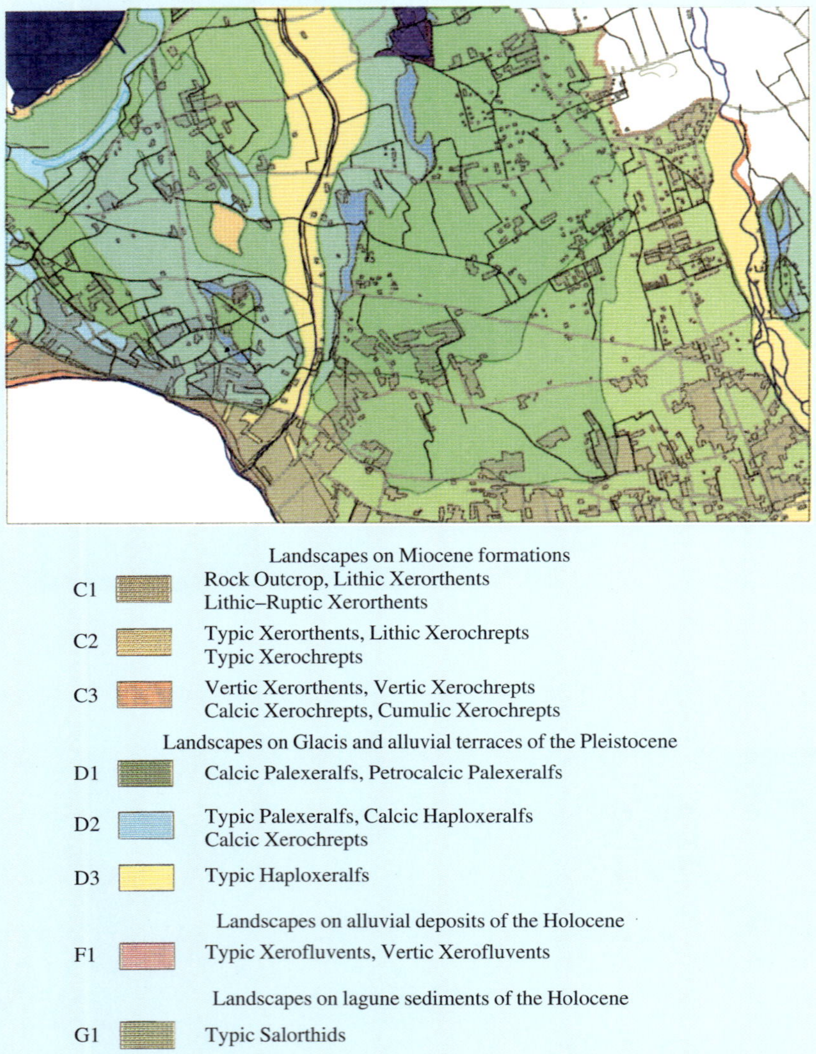

Figure 3.6.8 Pedological classification of Quartu Santa Elena municipality.

Figure 3.6.9 In the early 1960s, the installation of Saris refinery (one of the largest in Europe) and later of the Enichem chemical plant, wiped out approximately 2.5 km of former agricultural land.

Table 3.6.3 Land suitability classification in the municipality of Quartu Santa Elena.

Map units	Agriculture	Grazing	Housing developments	Industrial and commercial developments	Recreation uses and sports facilities
C1	N1–N2	N1	S3–N1	N1–N2	N1
C2	S1–S3	S2	S1–S2	S1–S3	S1–S3
C3	S2	S2–S3	S2	S2	S1–S3
D1	S2–S3	S3–N1	S1–S2	S1–S2	S3–N1
D2	S1–S3	S3	S1–S2	S1–S2	S3
D3	S2–S3	S3–N1	S2–S3	S2–S3	S3–N1
F1	S2–S3	S2–S3	S3–N1	S3–N1	S3–N1
G1	N2	N2	N2	N2	N2

ORDER: S: suitable; N: unsuitable. *CLASS*: S1: suitable land with no constraints; S2: moderately suitable land with some constraints; S3: marginally suitable land with major, but eliminable constraints; N1: land presently unsuitable, but potentially suitable with constraints; currently too severe to be eliminated; N2: land currently and potentially unsuitable.

Table 3.6.4 Percentage consumption of agricultural land.

Map units	Zones a	Zones b	Zones c
C1	38.99	31.14	29.88
C2	58.17	32.50	9.32
C3	55.41	43.48	1.10
D1	28.91	41.68	29.41
D2	86.74	6.19	7.07
D3	53.49	12.81	33.70
F1	9.00	65.10	25.90
G1	100.00	—	—

Urban expansion and quarrying in Sardinia

Among the side-effects of the economic development schemes especially implemented for southern Italy after the Second Word War, are demographic growth and the coalescence of isolated settlements into a sort of network of towns, harbours, industrial sites and agricultural land (Clemente, 1964). This is the case for the Cagliari hinterland, for example, where quarrying, although playing a positive role in economic development (Di Gregorio and Massoli-Novelli, 1986), has led to severe land-use conflicts.

No attention has ever been paid to land-use planning, the impact of extensive extraction works (sand and gravel along river-beds and clay in the alluvial plains) upon the environment, and the need for environmental protection. Therefore quarries are now one of the major causes of degradation in the Sardinian lowlands and river-beds.

The uncontrolled extraction of aggregates has led to profound changes in the landscape, creating an irreversible loss of resources and undermining the overall relationship of the people with their land. Evident transformations have occurred along the bed of the Rio Santa Lucia. The typical landscapes of the fluvial Sardinian countryside, dominated by mixed vegetation of willows, alders, oleanders, tamarisks, elicrisos and lavenders, have now almost vanished. Active quarry sites are characterised by large machinery for sand and gravel extraction, the disused sites by mounds of scrap materials.

An appraisal of the current situation of the quarries in the Rio Santa Lucia was carried out in MEDALUS (Puddu and Lai, 1995) to provide decision-makers with a planning tool to prevent the devastating impacts of quarrying on the landscape. Active and disused quarries were mapped and synthesised in a form suitable for planning purposes. The geomorphologic units on which the quarries are located were also classified and soil losses estimated (Klingebiel and Montgomery, 1961). The landscape impact of each quarry was assessed following Bruschi (1983) and Di Gregorio and Massoli Novelli (1987). The study concludes that fertile areas have been taken away from agriculture, and wet ecological systems annihilated by new quarries. The Santa Gilla lagoon, protected by the Ramsar Convention, is also at risk.

In recent years environmental awareness has increased; regional authorities are being urged to implement the existing legislation on quarrying and to reverse the circumstances that have led to the present conflicts between quarries and environment. These include the lack of thorough enforcement of the regional law on quarrying activities, the short-sightedness and small size of quarrying companies, the lack of resources for proper environmental management and the fact that they are under no obligation to consider the final state of their sites. The complete enforcement of this law (particularly of Article 6) is of special importance and should be accompanied by other specific planning instruments at a more local scale, such as the landscape plan and the quarry reclamation plan. These would provide an opportunity for discussion between regional and local planners, which is also required in order to find the most appropriate ways of handling adverse impacts (Beer, 1990).

An integrated approach to land management, the Regional Plan for Quarrying, would allow regional authorities to harmonise the needs for urban development and nature conservation, for example, by encouraging the relocation of quarries away from areas of ecological value, management and reclamation of the site, and activities contributing to the enhancement of ecological and scenic values. The quarry contractor would have to draw up a redevelopment plan, as opposed to simply camouflaging the excavations. Such plans are widely used in other European countries, for example, in the Netherlands. Likewise, the options for the re-use of sites according to their natural and anthropological potentialities and needs should be taken into account (Bignami *et al.*, 1985).

Table 3.6.5 Quarries impact assessment.

Type of impact	Origin	Effects	Quarries
Geomorphological	Change to landforms; creation of cavities (excavations); creation of fills (dumps of waste rocks)	Accentuation of natural geomorphic processes (sheet, rill and gully erosion); triggering of geomorphic processes on human-made earthworks	1–12
Hydrological	Change in hydrographic network; change in flow rates (drainage, pumping); creation of artificial lakes	Degradation of fluvial aquatic environment; deepening of river bed; change of solid transport regime; probable alteration of accretion	8
Hydrogeological	Quarries in plain that necessitate deflecting water from pit; interception of surface aquifers	Disturbance of flow rates in surface wells; disturbance of underground circulation (aquifers of fluvial terraces); alteration of surface drainage	2, 5–8
Water quality	Streams flowing over dumps and pits; creation of unauthorised waste dumps; salt-water intrusion	Deterioration of quality of surface waters; contamination of waters by dangerous substances; water becomes unusable for drinking and irrigation; effects, some long term, on flora and fauna	1–5
Air quality	Use of mechanical diggers; use of explosives; particular processing techniques (crushers); use of transport vehicles	Salinisation of coastal aquifers; deterioration of appearance of landscape; degradation of vegetation; alteration of chlorophyll photosynthesis function in farm produce growing nearby; possible adverse effects on zooethnical species fed with forage contaminated by dust; alteration of surface water bodies	7–10
Roads	Noise and vibrations; dust; circulation of vehicles	Effects on leisure and for tourist use of land; deterioration of roads; risk to road safety; heavier traffic	7–10
Microclimate	Modification of landscape components; elimination of morphological components; elimination of plant cover	Exposure to wind in formerly protected areas; variations in temperature and thermal excursion	1–12
Fauna and flora	Soil erosion; surface water pollution; noise, vibrations, dust, traffic	Effects on flora–faunistic composition; elimination of spontaneous plant species; introduction of exotic plant species	8–12
Landscape	Removal of grass and shrub cover; removal of agricultural soil; creation of cavities (excavations); creation of fills (waste dumps); plants and laboratories; roads and yards	Changes to appearance of landscape; destruction of landscape features; degradation of panoramic spots; interruption of chromatic and morphological continuity	all quarries, particularly 5, 9–12
Agriculture	Excavation works; powering of surface aquifers; pollution of surface waters and underground waters; emission of dust into the atmosphere; loss of land of high capability	Alteration of chemical and physical features of the soil (texture, structure, organic matter); modification of original land suitability; decrease in soil productivity; abandonment of crops grown in areas nearby the quarries; dumps of solid urban waste or other types of waste on agricultural land near the quarries	all quarries, particularly 1–8

3.7 The Increasing Demand for Water

J. J. Collin, C. Mouvet and G. Barrocu

Throughout the world the natural water cycle has been upset by human water uses in the twentieth century, and in the Mediterranean Basin a large amount of water comes in through river flow from wetter regions, where rainfall is higher or where snowmelt provides peak flows in rivers in the spring. As local (mainly groundwater resources) decrease and demand continues to increase, these inter-basin transfers become more and more critical and more problematic from a social, political and economic point of view.

The water resource demand per person for irrigation, thermal plants and energy production, and drinking and domestic use is a good parameter for representing the socio-economic conditions of the different countries. Water for agriculture is the major demand in almost every country in the Mediterranean, but water demand for thermal plants is essentially limited to the European states.

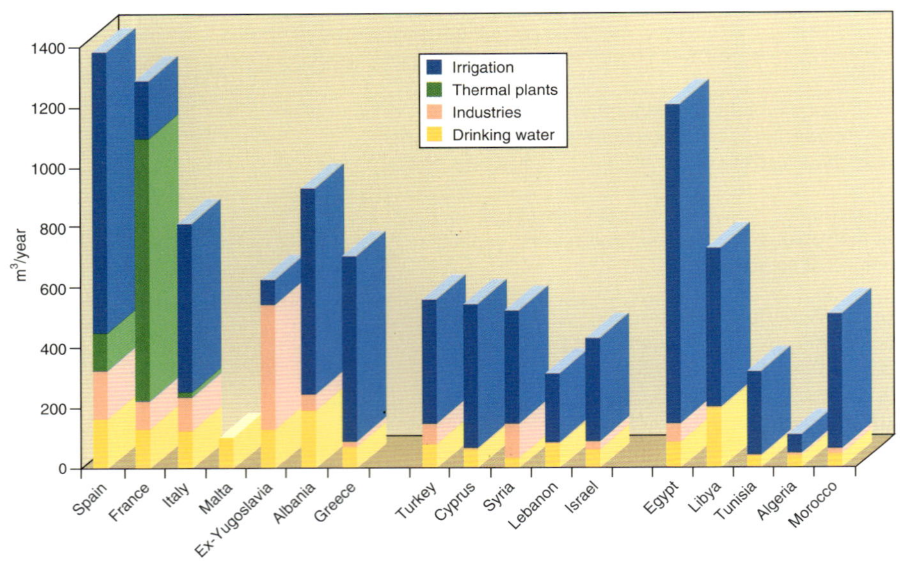

Figure 3.7.1 Water demand per person in the early 1980s. (Source: Margat, 1992).

The available water resources per person are very high in former Yugoslavia, while the demographic pressure on water resources is much higher in the Maghreb countries according to data from the Blue Plan (Margat, 1992). The situation is not irreversible and could change with better management of integrated water resources (surface, ground and treated waters), with close attention to the quantity and quality required for different uses and with a rational development of natural resources.

Water demand will change depending on population growth and socio-economic development, which is likely to be jeopardised by political instability. For the European Mediterranean, some argue that rational exploitation and management of water resources, especially in southern and eastern Mediterranean countries, implies large investments in controlling and storing surface runoff, withdrawing deep groundwaters and building infrastructures. Others believe that better use of existing water might be achieved by a more realistic pricing mechanism. The next decade seems likely to foresee a shift towards such a policy.

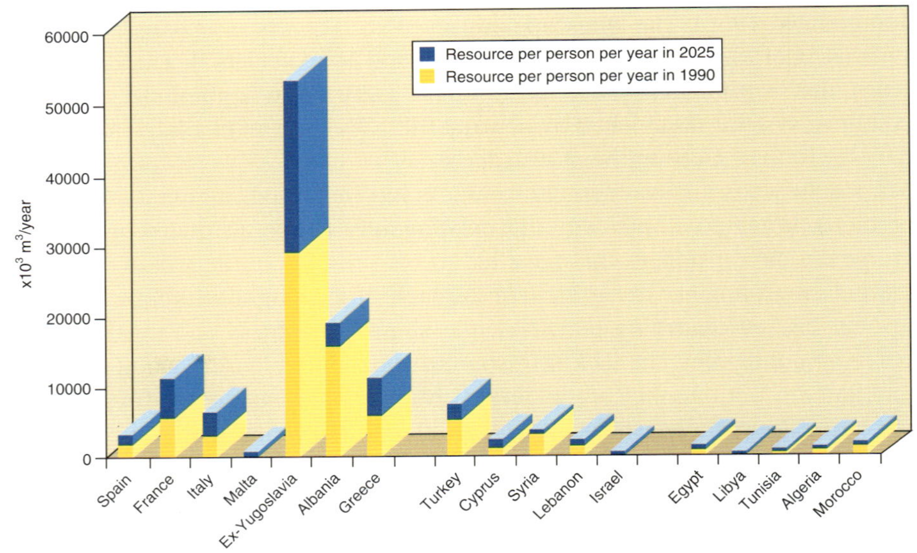

Figure 3.7.2 Water resources per person. (Source: Margat, 1992).

Table 3.7.1 Water consumption index for the period 1992–2025.

Country	Consumption index				
	1992	Year 2000		Year 2025	
		Low hypoth.	High hypoth.	Low hypoth.	High hypoth.
Spain	39.0	48.0	51.0	52.0	60.0
France	2.7	3.0	4.0	4.0	5.0
Greece	6.3	7.0	9.0	9.0	12.0
Italy	8.0	9.0	13.0	10.0	18.0

Water consumption index (WCI) is the ratio of total withdrawal (surface and groundwater) upon total resource not restituted to the water courses. (Source: Margat, 1992).

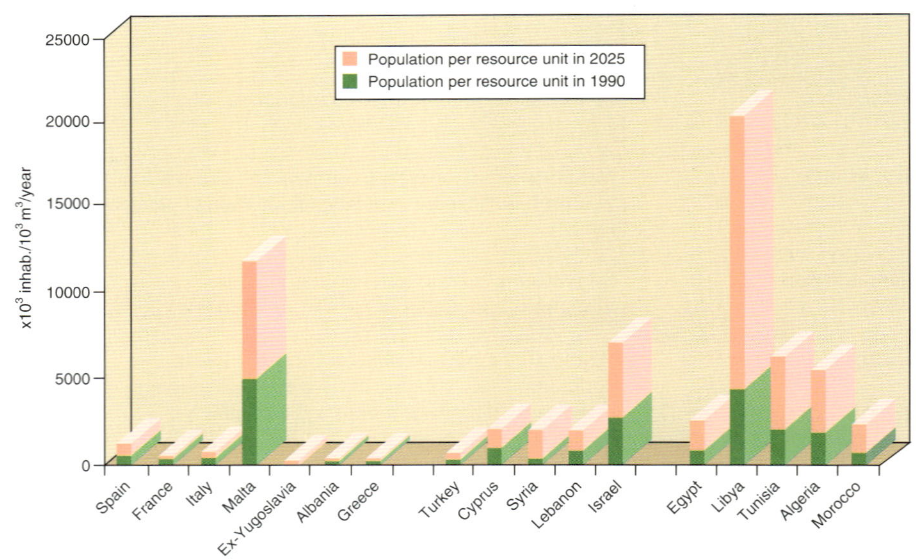

Figure 3.7.3 Population per water resource unit. (Source: Margat, 1992).

Groundwater

In the Mediterranean countries, good aquifer formations are rare. The majority of them are small, located on coastal plains and poorly supplied with rain. The hydroclimatic conditions are half-way between temperate and arid and a further decline of rainfall—observed and forecast—could shift the groundwater resource of these nations towards the pattern observed in deserts.

Of the two schema illustrated, it is easy to compare the respective role of groundwater and surface water in the two different climatic conditions. With greater aridity:

- the main role of groundwater, to feed and regulate river flow, disappears;
- aquifers receive less recharge by direct rain infiltration and are fed mainly by runoff seepage along river channels;
- increasing demand for water generates ‘over-exploitation’ (i.e. withdrawal in excess of recharge) and causes a strong depletion of the water-table;
- as a consequence of deep aquifers (below sea-level), salt-water invades the reservoir. With deep aquifers, many other impacts on groundwater produce dramatic results, such as the effect of pollution, which brings about deterioration of the water at its origin, so long as rainwater dissolves the chemicals applied at the surface.

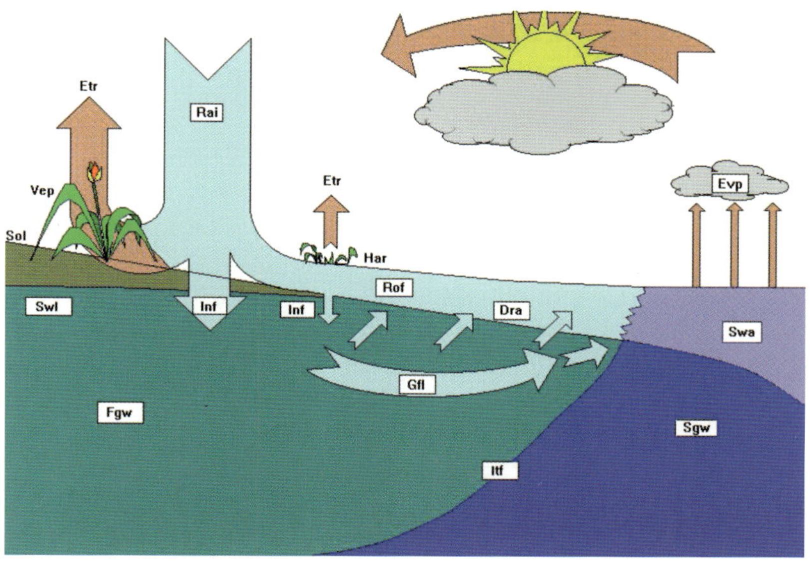

Figure 3.7.4 Consequences of aquifer exploitation on the water cycle in temperate and arid climates: a conceptual scheme of desertification phenomena. Temperate zone, aquifers not used. Abundant rainfall (Rai), well distributed and fed by evaporation (Evp) of sea-water (Swa), maintains vegetation production (Vep), which in itself, is responsible for strong evapotranspiration (Etr). Part of the rainfall infiltrates (Inf) in a diffuse manner towards the aquifer (Fgw) and another part runs off (Rof). Groundwater flow (Gfl) is in turn drained (Dra) by rivers, the flow of which increases downstream. Humid areas (Har) exist as well. Gfl maintains the interface (Itf) between fresh water and the saline groundwater (Sgw) of the coastal area. Swl is the the static groundwater level and Sol is the soil surface.

Salt-water intrusion

Salt-water intrusion in coastal aquifers is encountered quite commonly throughout the Mediterranean and has serious effects on environmental equilibrium. Soils irrigated with groundwater with a high salinity exhaust their natural fertility and vegetation rapidly withers.

The alluvial plain of Capoterra, in southern Sardinia, is a typical example of

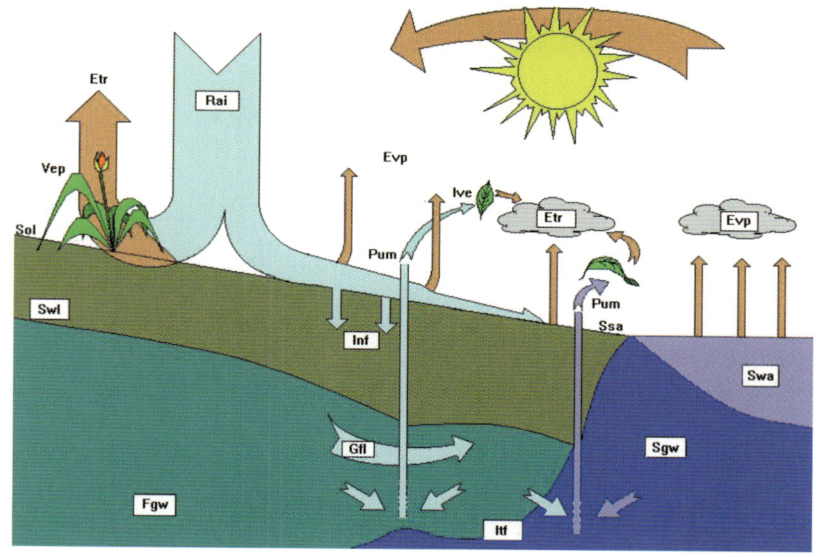

Figure 3.7.5 Consequences of aquifer exploitation on the water cycle in temperate and arid climates: a conceptual scheme of desertification phenomena. Arid zone, aquifers pumped. Pumping (Pum) maintains irrigated vegetation (Ive), strongly lowering the groundwater table (Swl) which receives no recharge because of the absence of durable runoff. Pumping from the aquifer shifts the boundary with saline groundwater (Sgw) inland, thus degrading crops and soil over large areas, by salinisation (Ssa).

that process (Barrocu *et al.*, 1994). The coastal aquifer system is jeopardised by salt-water intrusion in the areas exploited to satisfy the ever-increasing groundwater demand of agriculture, industry and domestic use. Groundwater is withdrawn from the aquifer system through about 300 wells scattered over an area of nearly 60 km^2. The phreatic aquifer occurs in multilayered recent alluvia in the upper part of the plain, and the confined aquifer is to be found at an average depth of 40–50 m in older alluvial deposits.

Continuous monitoring indicates a rise in salinity in those areas where groundwater is over-exploited, particularly in the central part of the plain. The diagrams show the water-table and TDS (total dissolved solids) contour lines in March 1993, for the phreatic and confined aquifers respectively.

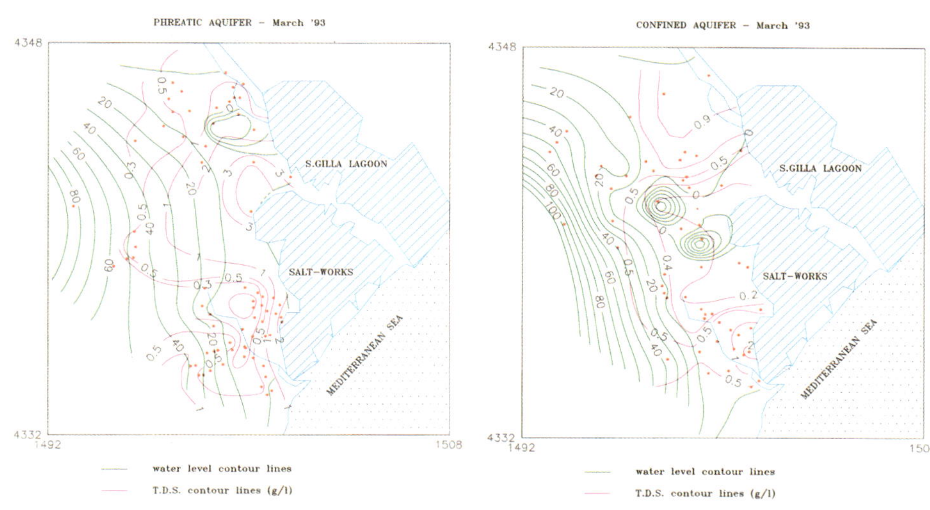

Figure 3.7.6 The alluvial plain of Capoterra in southern Sardinia has suffered salt-water intrusion in areas exploited for agriculture, industry and domestic use. Groundwater is withdrawn from the unconfined aquifer system through about 300 wells scattered over an area of nearly 60 km^2. Continuous monitoring has indicated a rise in salinity, particularly in the central part of the plain. The maps show the depth of the water table and contours for total dissolved solids for the confined aquifer and the phreatic aquifer in March 1993.

The confined aquifer shows a lower salinity than the phreatic one. The salinity is higher near the saltworks and the lagoon than at the edge of the plain. This probably reflects leaching from the lagoonal salt deposits, and spray transported from the salt extraction pans by the wind.

Water quality

The utilisation of foreign biological agents, such as herbicides to control agricultural pests that reduce productivity, is undoubtedly beneficial from an economical viewpoint. However, the environmental risk associated with the remaining residues and their transport to the groundwater after application needs to be assessed in relation to European Union regulations, for which the level of each herbicide or metabolite in drinking water is 0.1 μg/litre.

The control of weeds in vineyards and the production of corn, sorghum and wheat commonly relies on herbicides such as triazines and phenylureas, whose transfer to groundwater has not been sufficiently studied in a Mediterranean context. Therefore, within the MEDALUS Project, BRGM (Bureau de Recherches Géologiques et Minières) monitored herbicide transfer through the soil in an experimental plot of a vineyard at Puits de Férauds, Var, France. The plot was equipped with tensiometers and neutron probe access tubes to determine the relationships between the water content, suction and hydraulic conductivity (characteristic curves) because these relationships govern the water movement through the unsaturated zone. Suction cups installed at different depths enabled soil water to be sampled for later determinations of the herbicide concentrations. The soil had a low organic carbon content (< 0.8%) and a strong prevalence of the sand fraction (50–200 μm; 77%) over the silt (2–50 μm; 15%) and the clay (< 2 μm; 7%).

The results show migration of terbutylazine and diuron, together with degradation products such as hydroxyterbutylazine, to the deepest soil level investigated (60 cm). Very little attenuation is to be expected deeper in the soil profile, except dilution when the contaminant reaches the groundwater. Therefore, the concentrations in the μg/litre range measured in the soil water of this specific plot indicate a potential threat to the underlying groundwater if it were to be used as a drinking water resource. Such a transfer of herbicides through the soil would be lower for soils with a higher pesticide retention capacity, i.e. deeper soils with a higher organic carbon content and lower hydraulic conductivity.

Little is known about the impact of increasing drought on the mobilisation of contaminants. For non-conservative pollutants, longer gaps between events and larger storms could prove beneficial. For conservative contaminants, continued build-up of water in the soil can result in the development of 'chemical time bombs' as large amounts of pollutants attached to soil particles are moved in infrequent but highly erosive rainfalls.

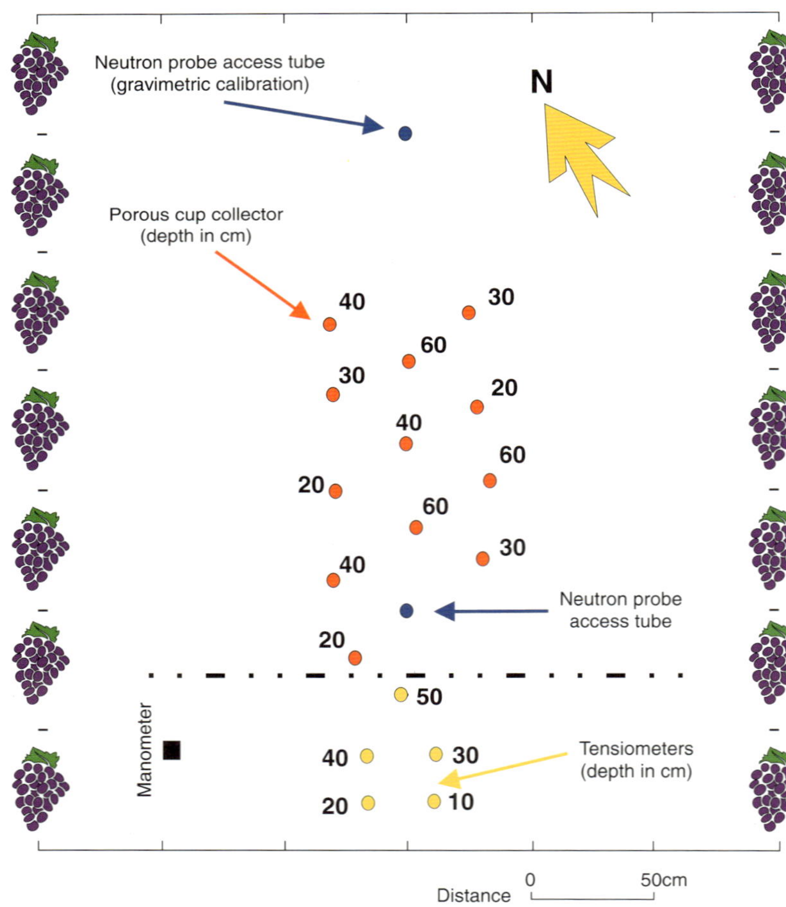

Figure 3.7.7 Detailed location of the various equipment installed at the field site between two vine rows for monitoring water and herbicide transfer through the soil.

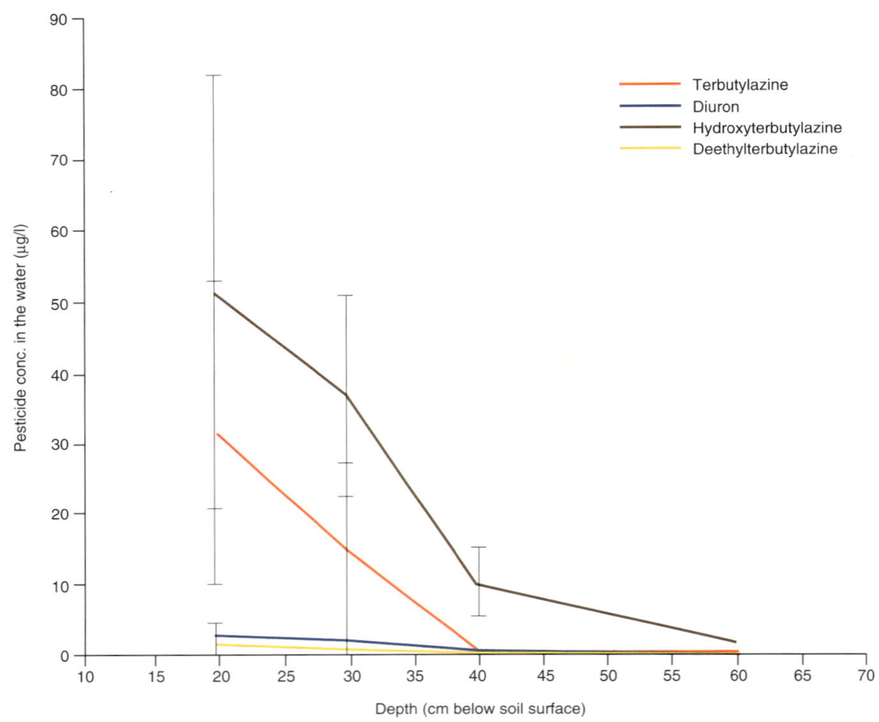

Figure 3.7.8 Concentrations of various herbicides (parent compounds and one degradation product, hydroxyterbutylazine) in the soil water sampled at various depths by suction cups in September 1994.

Groundwater over-exploitation: a case study in Castilla-La Mancha, Spain

S. Burke

A major threat in Mediterranean areas is falling groundwater levels. This is of vital importance in the semi-arid marginal zone where groundwater is the only source of water for domestic and agricultural consumption uses.

An area under great desertification risk is the aquifer system 23, Mancha Occidental, in Castilla-La Mancha, central Spain. Recent declines in groundwater levels can be attributed to the following:

1. Increases in public water consumption for agriculture, especially irrigation. The area was, until recent decades, predominantly an area of dryland farming. Exploitation of previously relatively under-utilised groundwater resources in this area has increased in the last few decades for intensive irrigation schemes, mainly for cereals, vines, alfalfa and sunflowers. The total area under irrigation in Castilla-La Mancha has increased from 159 100 ha in 1970 to 352 452 ha in 1989 (INE, 1990).
2. Reduced inputs into the groundwater system through declining rainfall trends, particularly in the general decline of the 1960s and 1970s and the dry years of the early 1980s. The area has low rainfall (around 450 mm) and high potential evapotranspiration (over 800 mm), which means that direct recharge is likely to be extremely low, and the recovery of the aquifer levels slow. A general rise in temperature has led to increasing evapotranspiration losses. It has been estimated that 90% of the infiltrated rainfall is lost through evaporation and evapotranspiration. This leaves very little to be finally recharged to the aquifer, and as a result the aquifer reserves have not been replenished.
3. Land-use change from dryland species to new varieties. Castilla-La Mancha has traditionally been an area of dryland farming. However, since the early 1970s there have been radical land-use changes. New crops and varieties with higher yields have been introduced.

These three factors have contributed to a reduction of groundwater levels by more than 20 m in some areas, since the 1970s. This has led to changes in surface and groundwater hydrology. In some areas it has led to land degradation, a good example being Las Tablas de Daimiel (Llamas, 1988). This was an important area of scientific interest and wildlife conservation, providing a unique habitat for rare and diverse wetland flora and fauna. It was a wetland area of 20 km^2, at the confluence of two rivers, the Ciguela and the Guadiana, and was one of the main sources of groundwater discharge from the aquifer. During the 1970s water was extracted from the rivers and from the aquifer by borehole extraction to provide irrigation waters for the surrounding agricultural areas. As a result the groundwater levels fell and the area of the wetlands has been reduced to less that 1 km^2. After a large public outcry, the area is now protected and declared a National Park. Nevertheless, many still believe that the action came too late and the area is still under threat of desiccation.

This research was funded by the EFEDA II project, EU Contract EV5V-CT93-0282 (DG 12 DTEE).

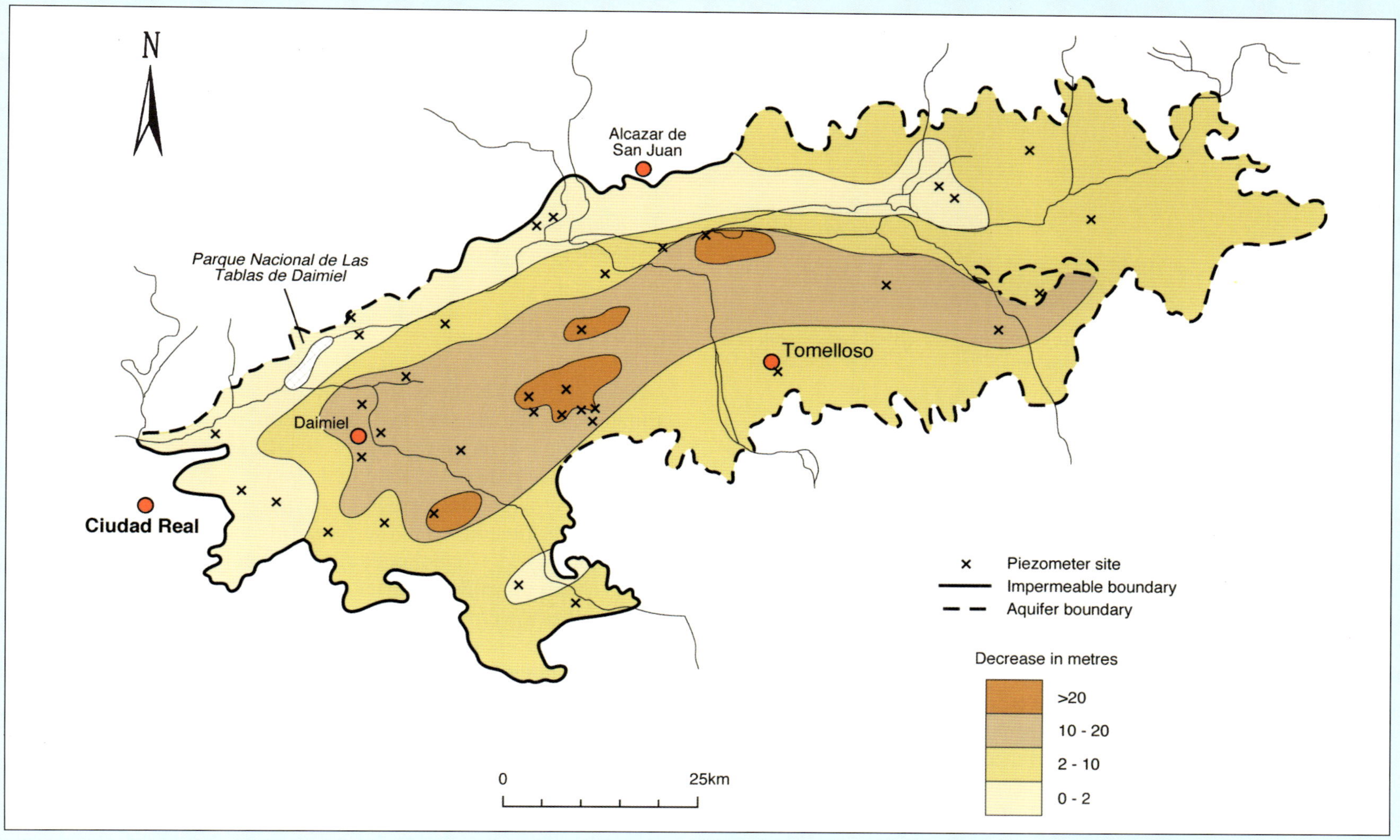

Figure 3.7.9 The aquifer system in Castilla-La Mancha, Spain.

3.8 Floods and River Valley Management

J. B. Thornes

Floods occur when rivers overflow their banks, often causing loss of life and damage to property. In the Mediterranean region, floods are inextricably linked to four major causes, which are themselves related to the desertification problem. First, for rivers arising within the Mediterranean climatic zone, rainfall is often lacking for long periods of time and then when it does arrive comes in exceptionally intense events. Large amounts fall in short periods. Second, rainfall often runs off quickly from the hillsides to the channels because of lack of vegetation, water-resistant soils, or a complete lack of soils and channel systems such as badlands, which with their high drainage density, quickly concentrate the water. Third, the flow of water in the channels themselves, coupled with the channel characteristics, leads to a rapid increase in water depth, and hence higher levels of damage for a given amount of rainfall. Finally, because of the scarcity of both water and good agricultural land of high quality, agricultural activity is concentrated in riparian (river valley floor) zones.

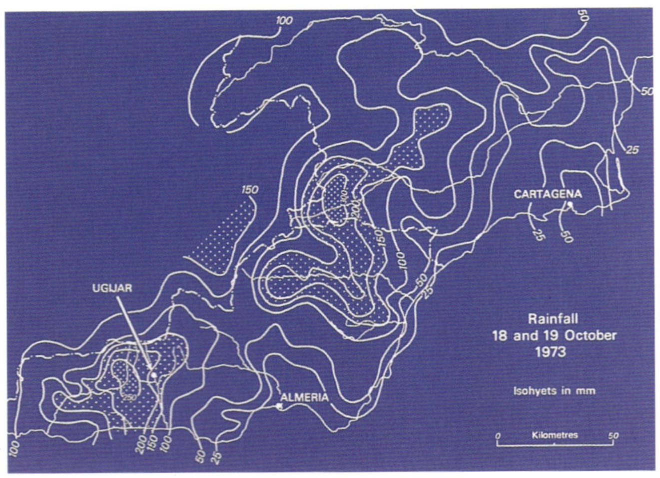

Figure 3.8.1 The October 1973 storm, the largest in living memory, spread across south-east Spain from Malaga to Murcia. Here in the Alpujarras rainfall was locally above 250 mm in 18 hours.

Extreme rainfall events

All of these characteristics can be demonstrated for the famous storm of October 1973 in south-east Spain (Thornes, 1974, 1976, 1993). A deep storm was caused by strong instability where cool moist air coming from the Mediterranean crossed onto land. Such storms almost invariably occur in late September or October and are locally called '*gotas frias*' (literally 'cold drops'). The storm affected a zone stretching from Malaga to Murcia, though the strongest sectors were in the Alpujarras (south-west from Granada) and in northern Almeria–southern Murcia. Here some areas had as much as 300 mm in 18 hours. One of the difficulties in the dry Mediterranean is that the record of large events is too short to estimate the recurrence interval, i.e. the time over which, on average, a storm as big is likely to occur again. Certainly there was nothing to match it in living memory, though historical records tell us of other great floods such as the flood in 16th-century Lorca (Gil Olcina, 1976). As soon as a large flood occurs, the estimate of the probability of its recurrence is changed, so flood estimates have to be continually upgraded, not least because the channel conditions are also changing. The amount of rainfall in the period running up to a large flood may also be important. If the soil and channel storage is already charged, then the likelihood of flooding is higher.

High runoff rates

The 1973 flood fell on summer-parched soils in areas where there was little or no soil as a result of prior severe erosion. For example, in the Alpujarras the rain fell on easily weathered, but also easily eroded, micaschist soils called *launas* whose mica-flakes are oriented parallel to the slope and act like tiles on a roof. Further east, in the Ugijar and Vera basins, the rain fell on marls whose soils turn to a kind of impervious paste, resistant to infiltration and often gullied into badlands meaning that not only is runoff generated quickly, but also that it concentrates quickly because of the very high drainage densities. Slopes are short and channels are steep, so that concentration occurs quickly. Many of these marls are completely lacking in vegetation cover, and although some had been bulldozed into terraces, the poorly constructed terraces were quickly eroded by pipes and gullies, locally called '*carcavas*'. The vegetation produces organic matter, which in turn causes the soils to aggregate and allows water to infiltrate more readily. Removal of the vegetation cover and continuous tillage reduces the soil organic matter and this leads to soil erosion. Deeper soils have lower infiltration capacities because of high bulk densities and therefore also resist infiltration. In floods of this type water and sediment mix at very high concentrations, producing mudflows which move at great speed, destroying all before them. Concentrations of sediments may be as high as 30% of the total volume and the mudflows can move at great speed; in the village of La Rabita, houses and people were destroyed by the moving mud.

Hazardous channels

As the water reaches the channel and begins to flow downstream, some of it infiltrates into the channel bed, and runoff from small storms can easily be absorbed by the wide gravel-filled valley floors called *ramblas*. In these small events (see illustrations in Section 2.1) trapped air escapes through the water to provide a frothy surface, which soon becomes charged with debris, such as organic matter floating on the surface. This together with the roughness in the channel can slow down the water at the advancing wavefront and so water from behind catches up, building up the wave as it propagates down the channel. The speed of the wavefront increases as its depth increases so that the wave becomes a bore, moving down an otherwise dry channel bed. As flows from tributaries join the main channel, they may coincide with the flood wave or be ahead or behind it. Unlike temperate channels, where rainfall is usually widespread, in ephemeral channels it may be very localised so that some areas escape completely. This makes the prediction of flows and their effects, even after a storm has started, very difficult indeed.

In the 1973 flood on the Almanzora River the great iron bridge at Santa Barbara was destroyed and swept off downstream as a flood wave estimated at 18 m high passed through, burying orange groves on high terraces in sand. On the Rambla Nogalte, people setting up market in the rambla bed, in the town of

Figure 3.8.2 Flooding is exacerbated by calcareous crusts on soils (a), which reduce infiltration and by soil piping (b), which speed up the movement of water from terraces to gullies.

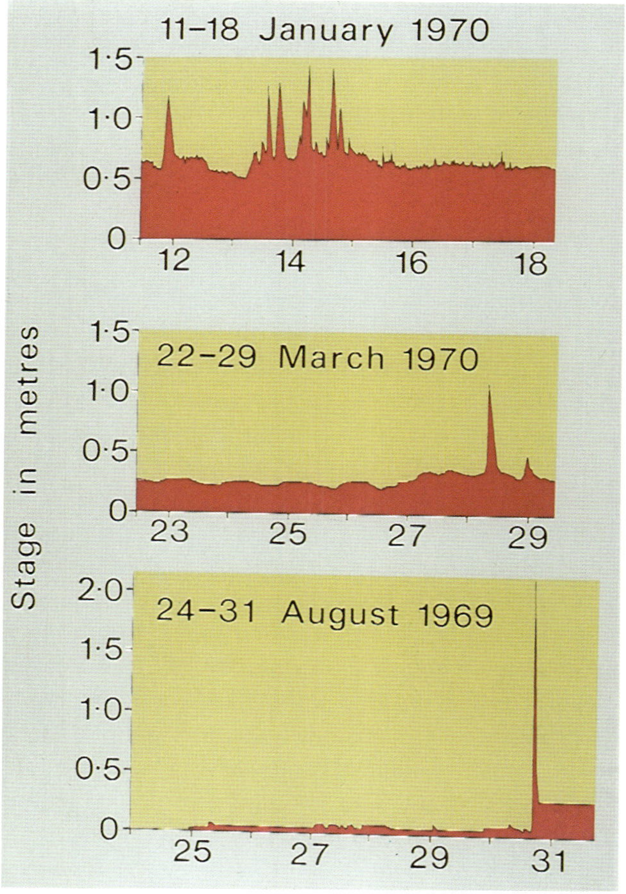

Figure 3.8.3 Flow depths in a *rambla*. In winter, the hydrographs are like runoff in temperate areas and there is groundwater flow. In summer, flood peaks are sharp and short-lived in dry channels.

Puerto Lumbreras, in the Upper Guadalentín valley (see Section 5.1), were unable to escape the great wave of water and many lives were lost.

Bad channel management

One of the great problems in Mediterranean dryland environments is that, as agriculture has intensified, land with soils suitable for intensive cultivation but near to irrigation water, has become more scarce and flood-plain occupancy has increased, despite the high vulnerability of these zones to flooding. In addition, as a result of littoralisation (see Section 3.6) urban development has spilled over onto the rambla beds, and hapless foreign speculators, under slack planning regulations, have been sold terrain that sooner or later will be the subject of intense flooding. Finally, the channels are often treated as dumping grounds for domestic, toxic industrial and occasionally even nuclear waste, and as sources of gravel for urban building projects. This indiscriminate use and neglect of the riparian environments fails to recognise that the sub-alluvial aquifers often supply water downstream. Irrigation return via the channel alone may increase the salt content of underground water to levels that make it unusable.

Mitigating flooding

There are many approaches to mitigating flooding, but they fall into four main categories: (1) prevention of runoff in the headwaters; (2) prevention of the build-up of water in the channel; (3) prevention of damage in the channel; and (4) reduction of the risk of damage.

The 1973 flood in south-east Spain was followed by implementation of some or all of these techniques individually or in combination. Headwater protection means reducing the runoff from the hillslopes. This can be achieved by increasing the vegetation cover, reducing slopes through terracing and contour ploughing, and increasing infiltration by soil treatment, e.g. using special tractor tools. In the watersheds of the Rambla Nogalte and the Rambla de Albunol, two of the most severely affected areas, extensive planting and contour ploughing have been practised. Elsewhere reservoirs have been built to increase in-channel storage. This excess floodwater can also then be used for irrigation. In the valley of the Almanzora and the Guadalfeo rivers, flood walls have been built to protect the land on either side and particularly to stop the floods in the main channels of the rivers from 'backing up' and causing floods in the tributaries. Finally, people living in danger zones on the sides of ramblas have been re-housed and flood warning systems, such as the radar rainfall measuring system in the Province of Murcia, have been installed.

Flooding and desertification

Just as water resources are inextricably linked to desertification through the hydrological cycle, so too is flooding. Degradation of the soil and its biotic cover and content, the characteristic features of desertification, increase the flood risk, so watershed management by increasing the tree cover and terracing seems the obvious solution. This is not as easy as it might first appear, because the establishment of trees in dry climates requires heavy investment in the early years and, with the extreme droughts seen in the last few years in the Mediterranean, may be difficult to achieve. There is, moreover, a balance to be struck between complete absorption of runoff by increased vegetation cover and the relatively high water demand of growing trees. There is also an increased risk of fire to be accounted for in the equation. Severe flooding is caused when rainfall is large in amount and of short duration. If this water can be successfully captured, despite its quick rate of runoff, then it is not lost to the atmosphere. This is the principle of the ancient *boquera* system wherein flood waters were led straight from the channels into the fields where they could be used to irrigate dry farmed crops. The mitigation of desertification and flooding have to be considered together with the conservation of ground and surface water resources, and the problem is one that has to be seen as a whole, rather than in its constituent parts.

Figure 3.8.4 In Puerto Lumbreras, lives and property were lost in the 1973 flood.

4

FIELD STUDIES

4.1 Human impact on land degradation in the inner Alentejo, Mértola, Portugal

M. J. Roxo and P. Cortesão Casimiro

Figure 4.1.1 MEDALUS field sites in Portugal.

Figure 4.1.2 Typical landscape of the Inner Alentejo.

Background conditions

Field studies in the inner, lower Alentejo have been undertaken in two different areas, the Vale Formoso Erosion Centre, north-east of Mértola, and the Almocreva MEDALUS field site, south-west of Beja. These areas are situated on the South Alentejo peneplain, a polygenetic erosion surface, last dissected at the beginning of the Quaternary. The present landscape is characterised by rolling topography and highly impermeable lithology. It has evolved in relation to the regressive erosion resulting from the activities of the main river. As a result, there is a remarkable accordance in the summits of the hills around Almocreva and Vale Formoso at 180–220 m a.s.l.

The landscape is monotonous, with flat summits and gentle slopes. Simple slope profiles never exceed 14°. However, there are occasionally areas with steeper slopes as a result of either lithological contrasts (bedrock outcrops), or deep stream incisions, particularly along the Guadiana, many of which follow fault and fracture structures. Morphology and soils result from the metamorphic nature of the parent material, dominated by schists: Lower Devonian red schists in Vale Formoso and Carboniferous yellow schists in Almocreva (Batista *et al.*, 1976; Texeira and Gonçalves, 1980; Ribeiro *et al.*, 1981).

Soils are generally very shallow (10–30 cm). In Vale Formoso, soil profiles show that the Ap horizon normally lies directly on a transition CB horizon, made of quite weathered bedrock. Recent textural analyses indicate that the Ap horizon is made up, on average, of 30% coarse sand, 30% fine sand, 20% silt and 20% clay (Cardoso, 1965; Dachary, 1976; Roxo, 1994).

The climate is a continentally-enhanced Mediterranean climate, with rainfall concentrated in the autumn and winter, followed by a long hot and dry season which lasts from May to September. The latter creates severe stress conditions for vegetation, which, consequently has sub-xerophytic characteristics. Annual and monthly rainfall variability is extreme, and sometimes concentrated in a few, very violent, thunderstorms. Periods of drought sometimes last several years and, like the less common extremely wet years, have a critical socio-economic impact.

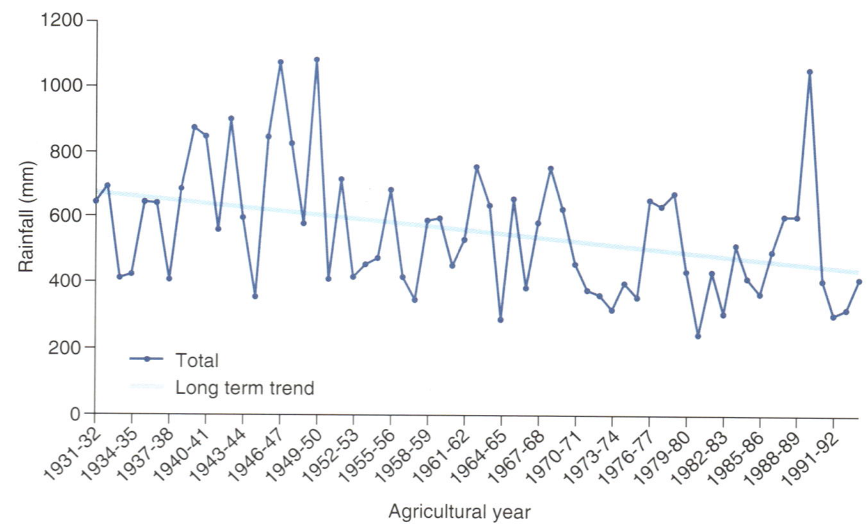

Figure 4.1.3 Average annual rainfall at Vale Formoso (1931–1992).

Land-use history and problems

Land-use changes in the Mértola municipality have occurred in three phases from 1900 to 1985 (Casimiro, 1994).

Table 4.1.1 Land use in Mértola municipality from 1897 to 1985.

Land use	1897 (%)	1950 (%)	1985 (%)
Agricutural land uses	62.95	89.60	53.3
Woodlands	7.13	7.47	13.2
Other semi-natural vegetation	29.17	12.34	31.3
Unproductive area	0.57	1.07	1.1
Total	100.00	100.00	98.9

Sources: 1897 Land-use map, 1950 Carta Agrícola e Florestal, and 1985 Land Use classification based on Landsat TM imagery for April–September. 1.1% of the total of pixels in the TM imagery were not assigned to any class.

Pre-1900

Agriculture was already widespread and comprised cereal cropping and grazing. Only 30% of the municipality was under natural vegetation. Human-induced environmental degradation, especially on areas not suitable for uses other than low-density grazing, commenced before the 19th century.

1900–1950

During this period almost all the remaining naturally vegetated areas (i.e. hill tops and the steepest slopes) were converted into cereal fields. Land clearance was induced by division of communal land and donations. Most of the Inner Lower Alentejo was turned into a tree-free area of cereal monoculture, with some relicts of Azinheira woodlands (*Quercus ilex* L. subsp. *rotundifolia* [Lam.] Schwartz ex T. Morais.) and shrublands (Coutinho, 1913; Pena *et al.*, 1985). The implementation of a governmental policy known as the 'Wheat campaign' also contributed to the clearance of semi-natural vegetation. This supported farmers by means of subsidies for the purchase of seeds, fertilisers and machinery. By the 1950s it was officially recognised that soil degradation had reached serious proportions.

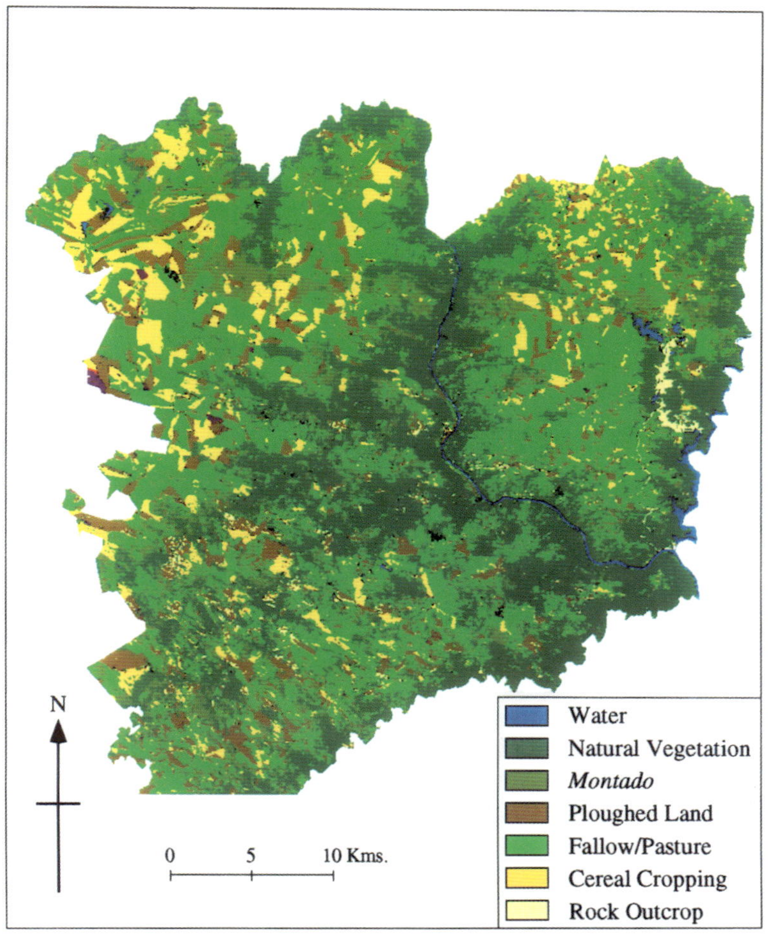

Figure 4.1.4 Land-use classification based on Landsat Thematic Mapper imagery for April 1985.

1950–1985

This period was characterised by the decline of agriculture, due to emigration to the main cities and abroad. Human desertification had started. The process of industrialisation of the economy occurred during the 1950s and early 1960s, and resulted in the loss of importance of agriculture as the main national economic sector, and gave it the final blow. After joining the European Economic Community (EEC) in 1986, agricultural production costs were three to four times higher in Portugal than in northern Europe. Areas dominated by semi-natural vegetation increased to 32% of the total area, and the share of total area taken up by agriculture dropped to 53%. Agricultural land itself is mainly fallow-land, mostly used for grazing, and the current trend is towards land abandonment. Hunting and related enterprises are becoming dominant (using 57% of the area of the municipality), as they are mostly based on semi-natural vegetation.

Problems

Land degradation is the progressive decrease of self-generation capacity in a natural system, which terminates in desertification, considered as the final stage. This degradation works at the soil level, which supports primary activity, both directly and indirectly (through vegetation cover modifications), but ultimately promoting soil erosion by running water.

As far as soil erosion by running water is concerned, land use and land cover are by far the most important factors, along with rainfall amount and intensity. The range between the two extreme land-use situations—continuously ploughed fields and natural pasture—is very high, and substantiates the importance of vegetation cover within the complexity of the processes involved in runoff erosion.

In the Inner Lower Alentejo, with the existing soil type, slope angles and climate, soil erosion is triggered by agriculture, as the soils are clearly unsuitable for the long-practised, fairly intensive, dry cereal cropping. When natural vegetation is allowed to recover, on areas that still hold some regeneration potential (less exhausted soils and some residual vegetation patches), soil loss decreases to what might be considered a normal rate for geomorphologic erosion processes.

Within the framework of the MEDALUS Project, field experiments have been devised to evaluate the impacts of such changes in this area.

Experimental methods

The activities of the Experimental Erosion Centre of Vale Formoso started in 1961. They were based on 15 standardised 20 × 8.33 m Wischmeier erosion plots (Wischmeier *et al.*, 1958; Wischmeier, 1960, 1962), and a 'half' plot (7X) of 20 × 4 m. These are located side by side with aspects ranging from south-west to east. All the plots have had crops and crop rotations, based on cereals and pastures.

The slopes of the erosion plots range from 6° to 11°. During 1989–1990 the land use and rotations of the plots were changed. However, the wheat–fallow rotation on plots 1, 2, 10 and 11 (one year wheat–one year fallow) was kept as a reference. Four plots were subject to a continuous up and down ploughed fallow (plots 6, 7, 14 and 15) in order to estimate extreme soil erodibility conditions. In reality, ploughed fields are common all year long in the region, and this creates a potential situation of maximum erodibility depending on the erosivity of the rainfall events.

In September 1988 a new plot was established, plot 13A, with an average slope of 10°. Its designated land use was natural pasture. The plot was 'abandoned', allowing regeneration without any human or grazing interference. After two years the vegetation cover provided a 100% cover, protecting the soil from direct raindrop impact. In another plot, 7X, several local species were introduced; *Cistus ladanifer* L. and *Cistus salvifolius* L. among others. The objective was to evaluate how forced recovery, with local species, could influence soil erosion by reducing it and increasing soil moisture content.

The amount of soil loss on land uses derived from agriculture, ploughed fields, wheat crops and bare soil is very significantly higher than the sediment yield under semi-natural vegetation, or even under stubble. At these rates, land degradation is a reality and results directly from the inadequacy of the long-practised cereal cropping to the edapho-climatic characteristics of the region.

For Vale Formoso, the available data, for 1961–1994, are in the form of the following:

- daily rainfall and evaporation, rain-gauge graphs, wind speed and direction;
- runoff and sediment yield;
- crop records, type and date of soil tillage;
- weekly soil moisture by plot and monthly wheat biomass (from January 1989).

In the Almocreva standard MEDALUS field site, the data available for the 1992/93 agricultural year are in the form of the following:

- weekly soil moisture: three samples (catena) for the two sub-fields;
- monthly biomass for the two sub-fields;
- runoff and sediment yield per event (suspended load) for six plots along a catena.

Table 4.1.2 Annual average soil erosion by land use in the Vale Formoso erosion plots for the agricultural years 1989/90 to 1991/92 (t/ha^{-1}).

Years	Bare soil	Wheat	Stubble	Ploughed	Natural pasture	*Cistus ladanifer*
1989–92	7.3	7.6	3.1	4.4	0.2	0.2

Discussion of management issues

Major land-use changes, induced by governmental policies, were responsible for the spatial and temporal extent of land degradation. Currently, as a consequence of the Common Agricultural Policy of the EEC, cereal cropping is becoming spatially less important. This matches the edapho-climatic characteristics of the Inner Lower Alentejo, which are inadequate for that type of agriculture.

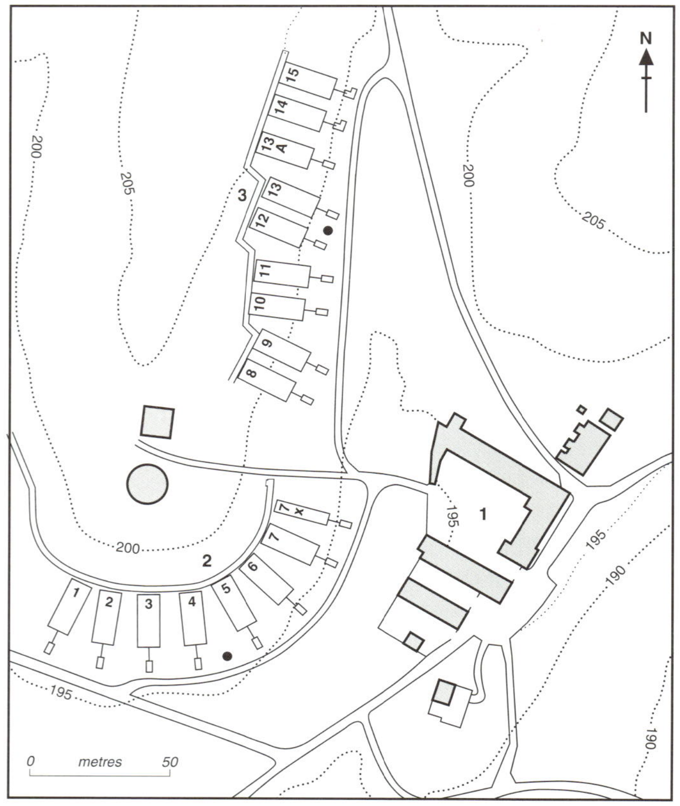

Figure 4.1.5 Experimental Erosion Centre of Vale Formoso: 1, buildings; 2, erosion field plots (group I); 3, erosion field plots (group II); ●, meteorological instrumentation.

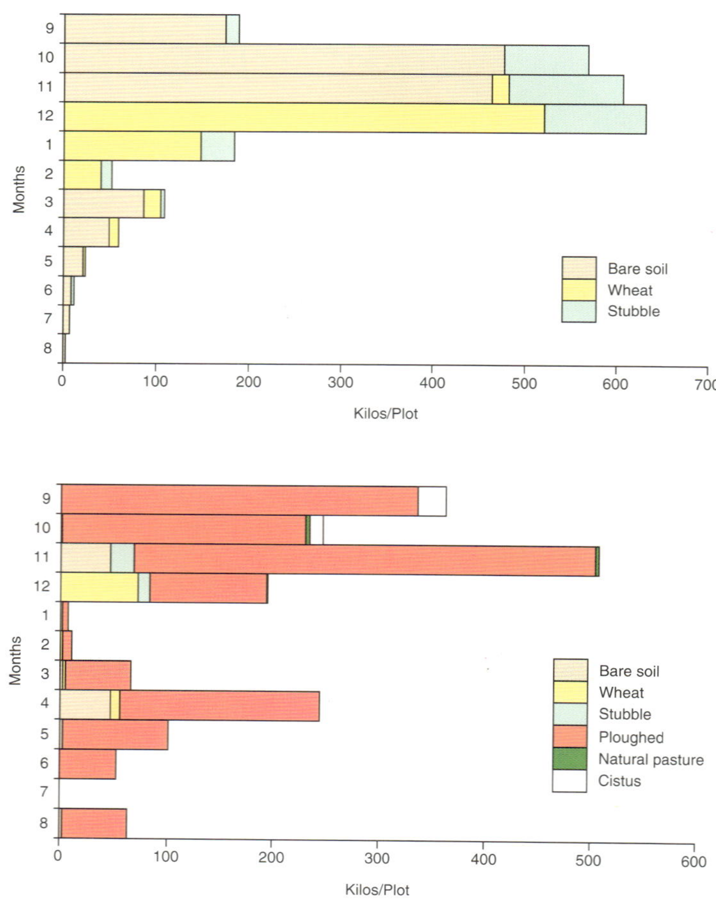

Figure 4.1.6 Average monthly sediment yield under different land uses: (a) 1966–1992; (b) 1989–1992.

The trend towards agricultural extensification, woodland protection and conservation of semi-natural vegetation, is clearly less damaging than the long-practised cereal cropping, and leads to regeneration. Therefore, at a regional scale, and as long as a rational land-use management is undertaken, mainly by adjusting agricultural activity to the soil use capacity, development can be achieved, and the ever-emerging socio-economic problems resulting from the long-term inappropriate human–environment balance will tend to decrease.

The most appropriate measure to be taken would be to progressively decrease cereal cropping, going back to the more appropriate exploitation of the natural indigenous resources: hunting, tourism in areas subject to conservation measures, low-density grazing and apiculture, so as to ensure development in harmony with the environment. The remaining problem, however, is the severe social impact of land abandonment arising from an ageing and unemployed rural population.

4.2 Rambla Honda, Tabernas, Almería, Spain

J. Puigdefábregas, M. Cueto, F. Domingo, L. Gutiérrez, G. Sánchez and A. Solé

Geology

The Rambla Honda is an ephemeral river (*rambla*) that drains a 30.6 km^2 basin on the southern slopes of the 'Sierra de los Filabres', a core of Precambrian to Triassic rocks in the eastern part of the Betic Cordillera. In the lower part of the Rambla Honda, as in most catchments of the mountain range, extensive alluvial fans have developed since the late Pliocene.

Figure 4.2.1 A general downstream view of the lower sector of Rambla Honda, showing the rock-controlled hillslopes and the alluvial fan system that fills the valley bottom.

Climate

The climate is semi-arid with long, hot, dry summers. Most of the rain falls in autumn and spring as isolated, often violent, storms. The amount and temporal distribution of rainfall vary from year to year. The unpredictability of the rainfall regime is an important ecological feature.

Soil

On the steep upper slopes of the valley, bedrock is widely exposed and the soil is restricted to irregularly distributed shallow pockets or to small terraces between the rock outcrops. Soil depth is usually considerably less than 1 m. The texture of these soils is fine sands, with approximately 50% of rock fragments of shaley material. Material washed down from the denuded upper slopes has been deposited midway downslope as alluvial fans, containing much coarse material (>5 mm), irregularly distributed as lenses between silty bands. The soil of the valley floor also consists of irregularly deposited bands of coarse and fine material, reaching a depth of up to 30 m. This material is derived from the valley sides and also from further up the valley, brought down by occasional flash floods.

Land use

Traditionally, this area had dispersed settlements (*cortijos*), each with two or three families running a subsistence agriculture, based on cereal crops; total farm income was supplemented by the husbandry of sheep and goats, the harvest of 'esparto' (*Stipa tenacissima* L.) which was used for cellulose, and by almond and fig picking. Nowadays, in the Rambla Honda, grazing and occasional cereal cropping remain as the only land use.

The catena concept

The soils of the denuded upper slopes, the alluvial fans at midslope position and the valley floor, grade into each other to form a catena of increasing depth and coarseness of texture downslope. Because of irregularities in rainfall, infiltration and runoff, soil moisture does not increase uniformly downslope. However, the presence of deep-seated, relatively moist layers of soil in the valley bottom means that soil moisture is higher than upslope. The increase in soil depth towards the base of the slope is reflected in the physiology adopted by the predominant species in response to the soil environment along their particular part of the catena.

Runoff along the catena

The nature of the rainfall regime is an important factor in affecting both soil movement and soil water relations. The generation of runoff is a complex process, controlled by several factors, e.g. rainfall intensity patterns, soil surface and subsurface hydraulics characteristics, vegetation cover and local micro-relief. For this reason, runoff is not generated evenly along the hillslopes. Thus, some areas tend to act as sources and yield more runoff than others that function as sinks. In large rainfall events in the Rambla Honda, the structural upper hillslopes, on micaschist parent rock, are the main runoff sources, whilst alluvial fans downslope, richer in gravels and sands, tend to act as sinks. In small rainfall events, differentiation between structure-dominated slopes and sedimentary formations becomes obscured by the complexity of the runoff process. The deep (>30 m) alluvial fans at the bases of the slopes can store much more water than is provided by rainfall. This explains one characteristic of the site that is shared with a number of semi-arid hillslopes elsewhere, i.e. that soil moisture does not consistently increase downslope and no permanent acquifer is found within the sedimentary body at the hillslope bases. These results show that, in general, if we wish to influence runoff we should act on the upper hillslopes, but if our goal is to modify the storage, we should consider the sedimentary formations at the slope bases.

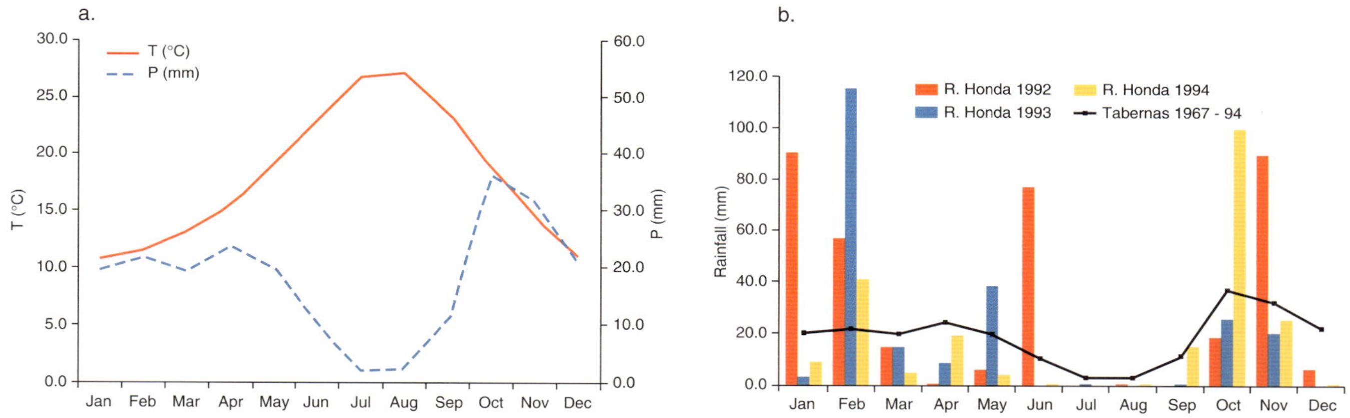

Figure 4.2.2 (a) Climatic diagram (after Walter and Leith, 1960) of the Tabernas weather station, 5 km from the field site. (b) Seasonal and annual variability of rainfall at the Rambla Honda field site, compared to the long-term average of Tabernas.

The perennial vegetation along the catena

There are three different perennial communities along the catena, distributed in a discontinuous, patchy pattern (see Section 3.2). *Stipa tenacissima* L. occurs in the areas with the shallowest soils which have never been cultivated; the community is therefore a long-lived and stable one, with high net primary production (NPP), 326 ± 18 g/m^2, and above-ground biomass (AGB), 356 ± 44 g/m^2. *Anthyllis cytisoides* L., a drought-deciduous shrub, is situated on alluvial fans that were last cultivated years ago, with maximum growth usually in early spring (NPP is about 239 ± 9 g/m^2 and AGB 133 ± 15 g/m^2). *Retama sphaerocarpa* (L.) Boiss, has the largest volume of soil to exploit, the root system reaching up to 30 m in depth, with below-ground biomass four and three times higher than *Anthyllis* and *Stipa* respectively. *Retama* is a drought avoider because it has access to water all the year round. Both the short growth period and disturbance from occasional tillage during wet years result in the lowest values of NPP (188 ± 14 g/m^2) and hence AGB (117 ± 10 g/m^2).

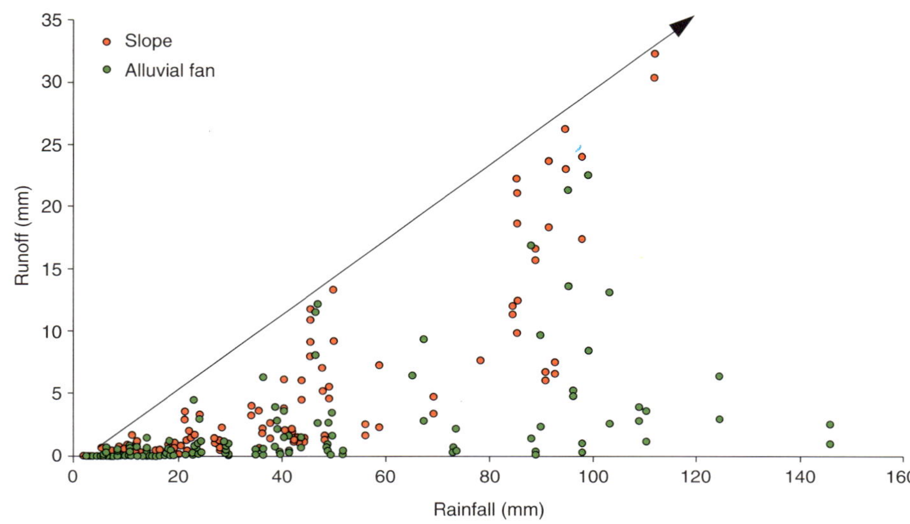

Figure 4.2.3 Surface runoff plotted against rainfall for individual events at the Rambla Honda field site.

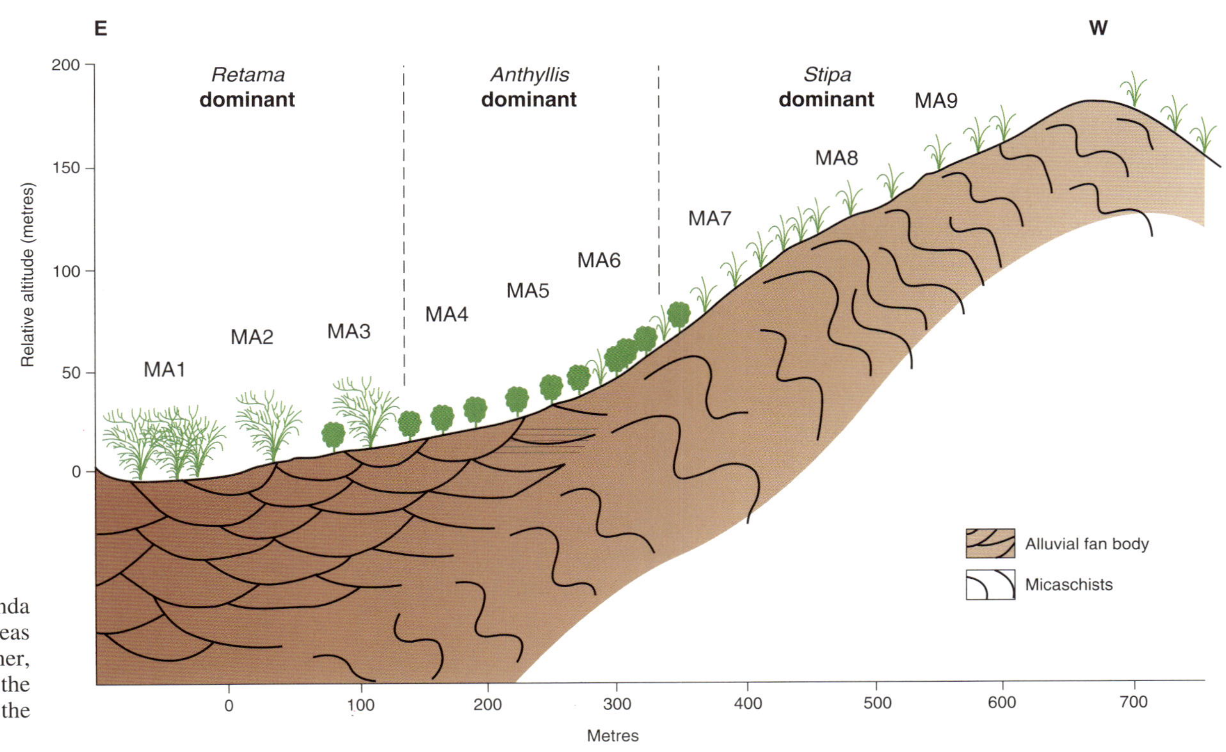

Figure 4.2.4 Schematic cross-section of the Rambla Honda field site, showing the location of the nine measurement areas (MA) along the catena. Systematic observations of weather, runoff, sediment yield, soils and vegetation are made both at the measurement area scale and at the micro-catchment scale on the hillslope.

Figure 4.2.5 *Anthyllis cytisoides* L.

Water relations

Vegetation patchiness affects soil properties and vice versa. Isolated bushes, shrubs and tussocks constitute traps for fine sediments and organic matter deposits. Organic matter content, water-holding capacity and permeability of the soil are significantly higher under perennials than in bare patches. As a consequence, the soil under perennial patches not only acts as fertility islands, but also as sinks for rainfall infiltration and/or runoff.

The three dominant species show different abilities regarding the water uptake and use (Puigdefábregas *et al*., 1995). The relationship between leaf water potential (Ψ) and relative water content (RWC), i.e. the moisture release curves, show that *Stipa* loses turgor later than *Anthyllis* and *Retama* when Ψ decreases. The large change in Ψ for small losses of leaf water is an advantageous mechanism of adaptation in *Stipa*. In *Anthyllis*, small changes in Ψ involve large changes in water content and hence make *Anthyllis* more susceptible to desiccation because harmful levels of dehydration may be reached in order to maintain water uptake. *Retama*, on the contrary, has access to water supplies all year round (Lansac *et al*., 1994; Haase *et al*., 1995) and has a high ratio of Ψ to RWC, which may be advantageous in the valley bottom area of the Rambla Honda. *Stipa tenacissima* L. may therefore be considered as a true drought endurer, while *Anthyllis cytisoides* L. and *Retama sphaerocarpa* L. Boiss. are drought avoiders, the former by selecting time, and the latter by selecting site.

The Rambla Honda as an ongoing study area

The Rambla Honda site is ideal for answering questions arising from the earlier MEDALUS projects as to the mechanisms whereby vegetation responds to climate variability and affects runoff and erosion. Moreover, the Rambla Honda is extremely well placed, both because of its semi-arid Mediterranean climate and because the essential basic information on climate and vegetation has already been acquired and is readily available.

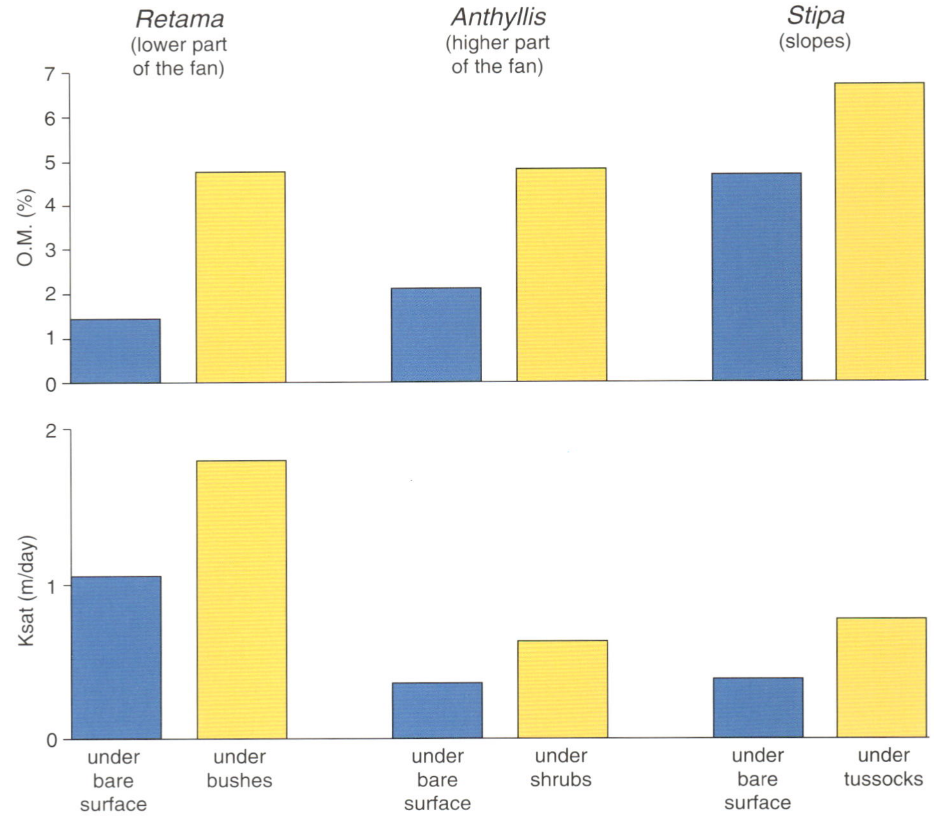

Figure 4.2.6 Average values for organic matter content (OM) and permeability (saturated hydraulic conductivity or Ksat) of soils developed in the three representative plant communities of Rambla Honda. Perennial plants may act as fertility islands and as sinks for rainfall and runoff.

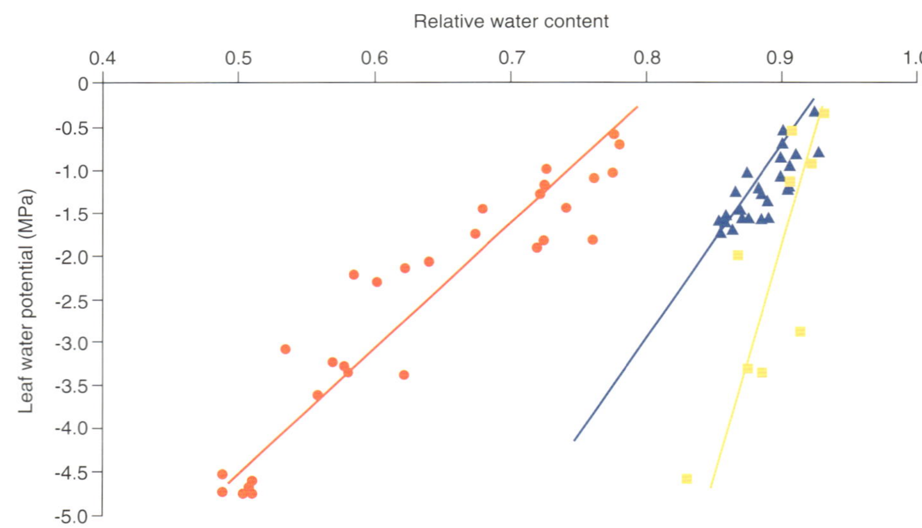

Figure 4.2.7 The relationship between leaf water potential and relative water content (moisture release curves) for *Retama sphaerocarpa* L. Boiss. (▲), *Anthyllis cytisoides* L. (●), and *Stipa tenacissima* L. (■) (from Puigdefábregas *et al*., 1995). Lines show fitted linear regression equations. *Stipa* loses turgor later than *Anthyllis* and *Retama* when leaf water potential decreases.

4.3 El Ardal, Murcia, Spain

F. López-Bermúdez, M. A. Romero-Díaz, José Martínez-Fernandez and Julia Martínez-Fernandez

Background conditions

The El Ardal experimental field site is located on the northern side of the Mula Basin, in the middle of the Region of Murcia. It lies at an average altitude of 550 m a.s.l., has an average slope gradient of 20%, an average annual rainfall of about 320 mm, and is characterised by a large variability because of the alternation of prolonged dry periods and heavy rainfall of short duration, particularly in the autumn. A high insolation (2870 sunshine hours per year) and average air temperature of 14.5°C result in intense evaporation and a potential evapotranspiration of about 1100–1200 mm/year. Limestone lithology, calcareous soil, and Mediterranean scrubland as vegetation cover, complete the representative features of this semi-arid Mediterranean environment.

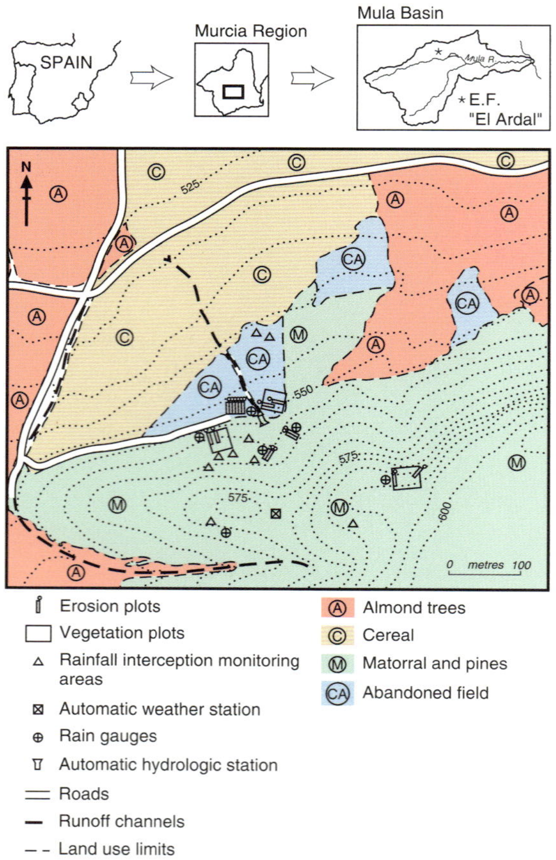

Figure 4.3.1 Location, land use and instrumentation at the El Ardal experimental field site.

Land-use history and problems

It is thought that El Ardal and the whole area of Lomo Herrero in which the site is included, were originally covered by open woodlands (*Pinus halepensis* Mill.) and a dense Mediterranean shrub vegetation. The agricultural use of most of the site, mainly the lower areas with less steep slopes, led to a landscape dominated by cereals and vineyards. In the 1940s, almond trees were introduced and probably the upper areas were cleared to create new agricultural land. In the 1960s, the abandonment of agricultural land, especially of the more marginal and less productive fields, took place. From time to time, coinciding with drought periods, other fields were abandoned. Nowadays, El Ardal has Mediterranean scrubland with isolated pine trees in the Lomo Herrero, abandoned fields in some marginal areas at the bottom of the more elevated and rocky hillslopes, extensive cereal and almond tree crops on the lower slopes and vineyards in the valley bottom.

Figure 4.3.2 Partial view of the El Ardal experimental field site.

Experimental methods

The experimental area has been selected and laid out with the objective of studying different aspects of climate, hydrology, soil erosion and vegetation. Two automatic stations, one meteorological and the other hydrological, were installed. Seventeen erosion/overland flow plots (twelve of 8 × 2 m and five of 10 × 2 m), three plots for plant study (30 × 30 m), and eight plots to control the flow of interception through the vegetation (with different sizes) were set up. Also, a network of 18 Hellman pluviometers were placed throughout the whole experimental field, and 300 ground pluviometers were created.

For the vegetation analysis three areas were selected: (1) an abandoned field; (2) a semi-natural scrubland dominated by rosemary (*Rosmarinus officinalis* L). and juniper (*Juniperus oxycedrus* L.); and (3) an area covered by the semi-natural scrubland dominated by pine. In these plots, species composition, vegetation cover, above-ground and below-ground biomass, litterfall and litter production, leaf area index, net primary production, decomposition cotton rotting rate and shoot growth have been measured. In addition, analyses have been carried out on soil characteristics: aggregate distribution, aggregate stability, water-retention capacity, saturated hydraulic conductivity and moisture content.

Soil losses are being observed on 17 plots with different gradients (7–28%), orientations and soil uses, which have a different percentage of vegetation cover, varying according to the season. In some plots soil fertility loss has also been studied.

Figure 4.3.3 Some experimental plots for soil losses and runoff at El Ardal.

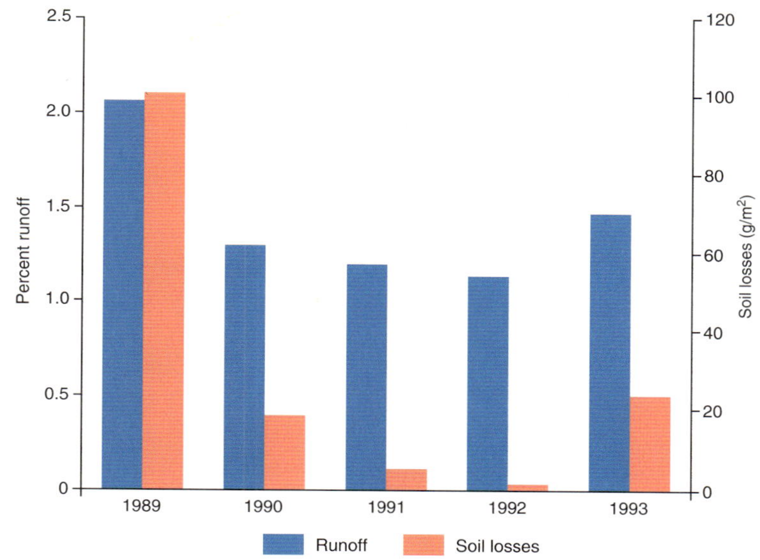

Figure 4.3.5 Soil losses (g/m^2) and runoff coefficient (%) at the El Ardal experimental field site (1989–1993).

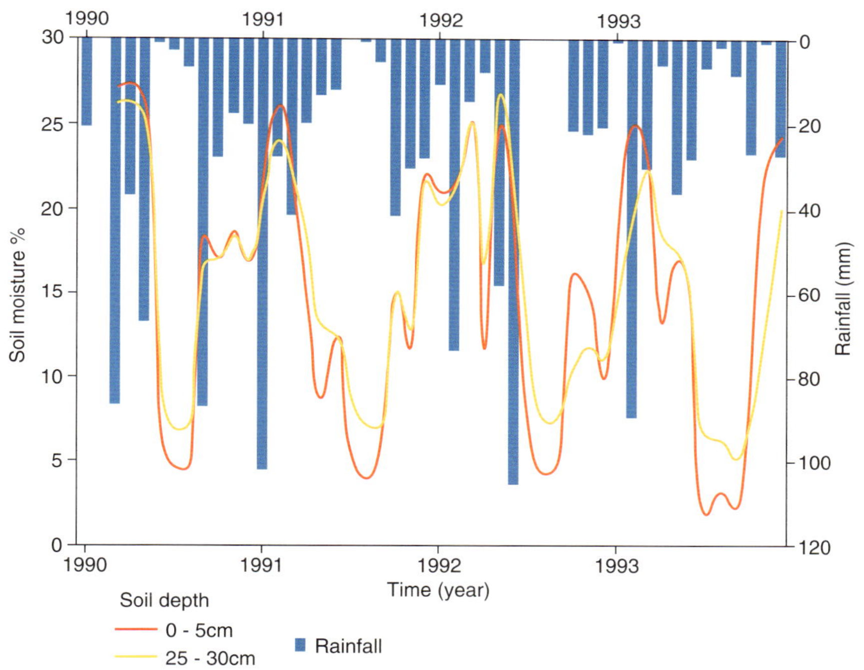

Figure 4.3.4 Monthly rainfall (in millimetres) and soil moisture (%) at 0–5 cm and 25–30 cm depth in El Ardal (1990–1993).

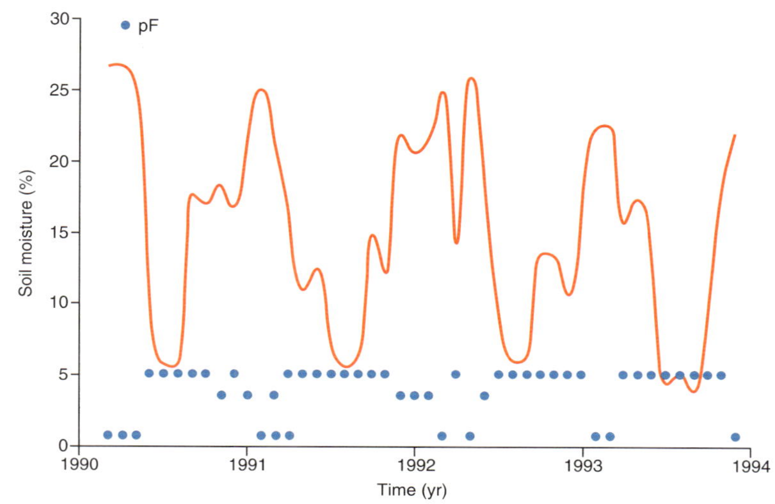

Figure 4.3.6 Monthly soil moisture evolution and soil matrix potential (pF) at the El Ardal experimental field site.

Results

Environmental and soil conditions are substantially different in the hillslope area that has been disturbed by cultivation and in the area left under semi-natural conditions. In the latter, the soil has characteristics that favour good hydro-mechanical behaviour: more organic matter, finer texture, greater stability of aggregates and retention capacity, higher soil moisture content and higher root density. The El Ardal soils have an almost permanently low moisture content, which means that water is well retained by the soil and runoff coefficients are generally low.

In this area, the biological parameters of the semi-natural scrublands have higher values than in the typical semi-arid communities (e.g. high LAI and biomass values). Considering the great capacity for adaptation and resistance which most of the species have to prolonged drought periods, possible changes such as a gradual precipitation decrease would not probably produce more than a local reorganisation of the general vegetation structure. By contrast, different land uses and the differential time of field abandonment may have important effects on the soil properties.

The erosion rates in El Ardal are very low due to soil, climatic and vegetation cover type characteristics. There are very marked differences according to year of study, rain events and land uses on the experimental plots. Soil losses are more related to the rainfall intensity than to total rainfall. The plots with the highest erosion rates and the highest runoff coefficients are those least protected, by vegetation or stone cover.

4.4 Santa Lucia Field Site, Sardinia, Italy

A. Aru, P. Baldaccini, M. R. Lai, R. Puddu, D. Tomasi and A. Vacca

Introduction

Sardinia is the second largest island in the Mediterranean, covering an area of roughly 24 000 km^2. In physiographic terms the region varies greatly. In fact almost all the geological ages are represented, with a broad variety of lithologies of widely ranging composition, which affect morphology, soils, plant cover and past and present land use.

Sardinia is not only a unique environment, an island and one of the less favoured regions of Europe in economic terms, but it also experiences, in its small area, some of the major problems relating to desertification in the Mediterranean region. Water resources, fire, the impact of grazing, soil erosion and land degradation all play their part. Moreover the challenges of urban growth, tourism and industrial development and the introduction of exotic plant species add to the environmental burden.

In the framework of the MEDALUS II Project, two catchments have been studied, the Rio Santa Lucia–Gutturu Mannu in the south and the Rio Astimini–Fiume Santo (Section 4.5) in north-west Sardinia. The two areas have typically Mediterranean climates with mean annual temperatures of around 16°C and variable rainfall with mean annual precipitation of little more than 500 mm, but differ geologically and as far as land use is concerned.

Figure 4.4.1 Location of the MEDALUS field sites in Sardinia.

Environmental context

In the Santa Lucia–Gutturu Mannu catchments the following four problems have been chosen for detailed investigation:

1. erosion in three areas with different land uses at the Is Olias site;
2. loss of agricultural land due to aggregate extraction and urban expansion in the flood and coastal plain of the Rio Santa Lucia catchment;
3. heavy metal pollution from toxic mining waste in the San Leone mine, situated in the mountainous part;
4. salt-water intrusion due to over-exploitation of groundwater with resulting salinisation of the water used for irrigation and domestic supplies.

The Santa Lucia catchment drains north-east and then south-east to meet the sea in the Golfo degli Angeli (Cagliari) just to the west of the city. Within the catchment are the experimental sites at Is Olias.

The catchment comprises four major lithologies (Lai *et al.*, 1995): Hercynian granites and leucogranites in the mountainous Section, mainly in the south-west; metamorphic rocks (quarzites, pellites, metamorphosed sandstones, conglomerates) of the Palaeozoic basement in the middle basin; Quaternary forms and lithologies such as glacis and slope deposits in the piedmont area; and terraced alluvial deposits on the flood-plain.

The soils (Puddu and Lai, 1992), classified according to the USDA Soil Taxonomy (USDA Soil Survey Staff, 1992), are shallow and immature (Xerorthents) on the upper hillslopes of the Palaeozoic basement, where erosion is intense, and more mature and deeper (Xerochrepts) on the valley floors where they have been deposited through gravitational transfer on more stable landforms or where the plant cover is denser. On the Lower Quaternary deposits the soils are leached, desaturated and cemented with a high percentage of Fe and Al sesquioxides (Palexeralfs) on ancient and stable surfaces while poorly developed (Xerochrepts) or constantly rejuvenated soils (Xerofluvents) occur on the plain where incision by the river is still occurring.

The Rio Santa Lucia catchment has a typical Mediterranean climate, with a mean annual rainfall of 551 mm and an average annual temperature of 16.6°C. Typically the main rainfall is in the period October–March, though the rainfall graph shows that during the period 1976–1990 there was appreciably less rainfall than during 1922–1975. During the MEDALUS observation period (1992–1994) the rainfall was above average only in 1993, with heavy falls in October and December 1992, September and October 1993 and September 1994.

The strong contrast between mountain and plain is also reflected in the land use. The mountainous section is mainly wooded with holm oak (*Quercus ilex* L.) and cork oak (*Quercus suber* L.) which in the past was used for the mining industry but is now largely state-owned.

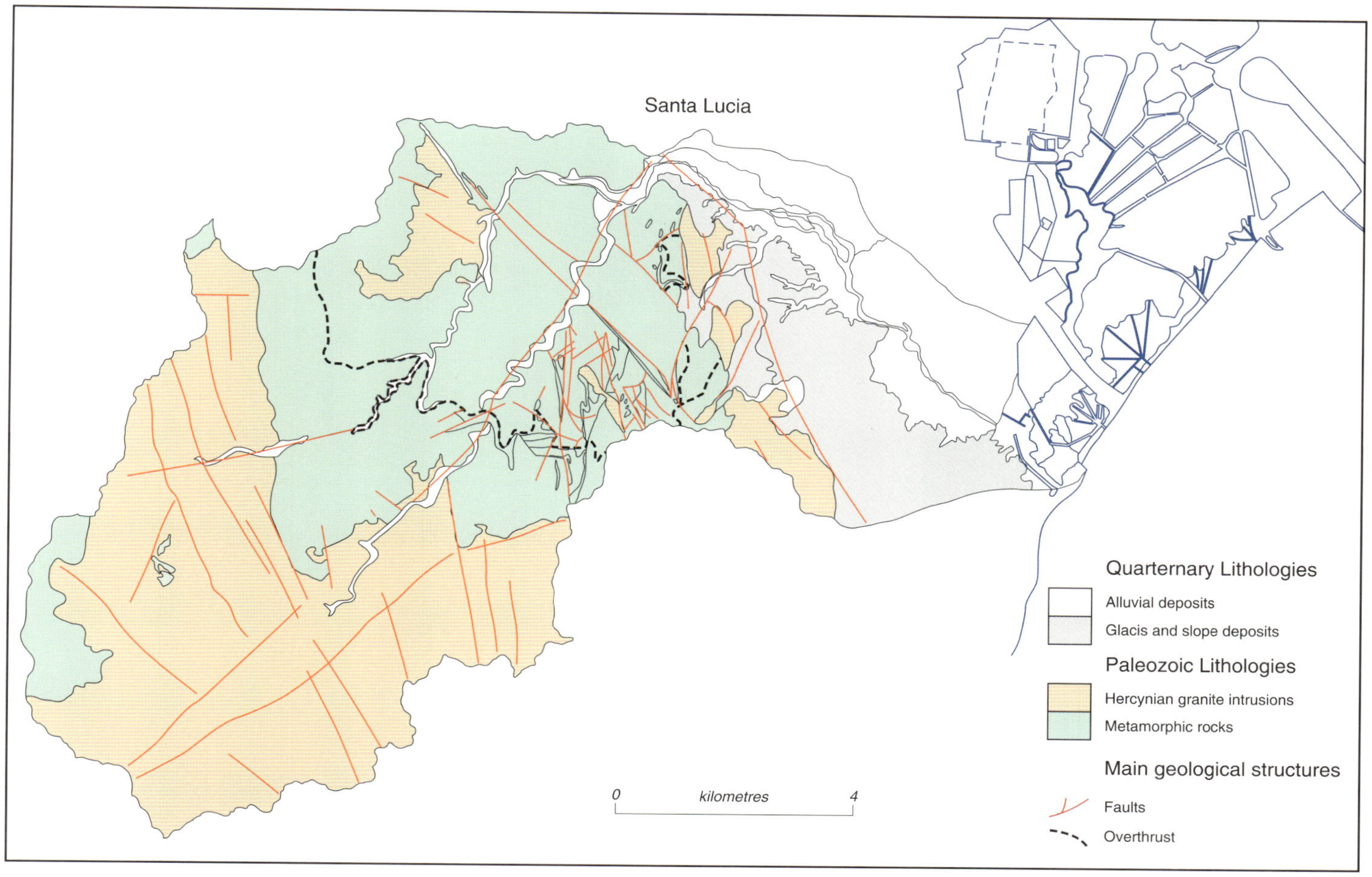

Figure 4.4.2 Geology of the Santa Lucia catchment.

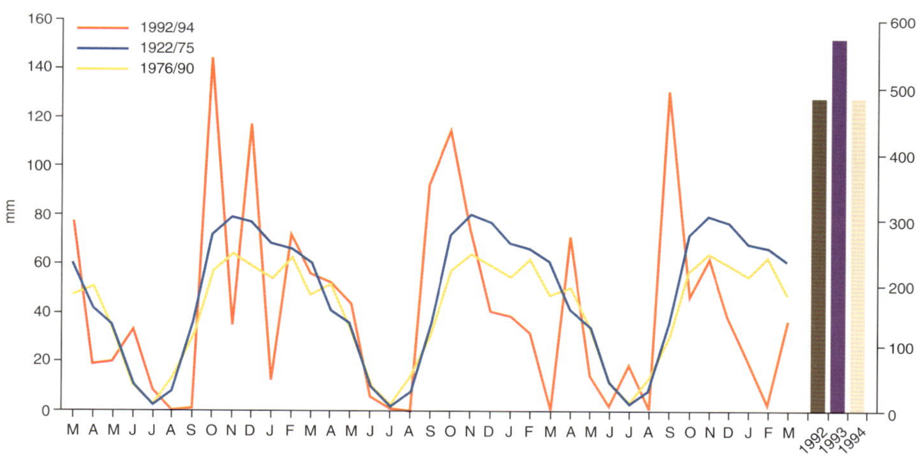

Figure 4.4.3 Rainfall measured during MEDALUS and mean monthly precipitation at Is Olias.

By contrast, urban and industrial expansion in the hills and on the plain, which has resulted in the abandonment of agricultural land, and more generally human pressures and the lack of rational planning, have led to severe forms of land degradation.

Santa Lucia in the MEDALUS Project

In the framework of the MEDALUS Project standard methodologies have been used to carry out geological, soil, land-use and hydrogeological surveys. Detailed maps have been drawn up using the data collected in the surveys of the physical features. Moreover, 18 runoff plots have been set up at the Is Olias site on three catenas with different land uses, namely abandoned grazing land now overgrown with shrubs (catena 1), *macchia* burned at the end of June 1991 (catena 2), and a eucalyptus plantation (catena 3).

The loss of agricultural land due to aggregate extraction and urban expansion has affected soils rated as being of high to very high capability (Puddu and Lai, 1992). The practice used by shepherds of burning land to create pasturage

Figure 4.4.4 Abandoned quarry on former agricultural land.

Figure 4.4.5 Runoff plots under eucalyptus plantation.

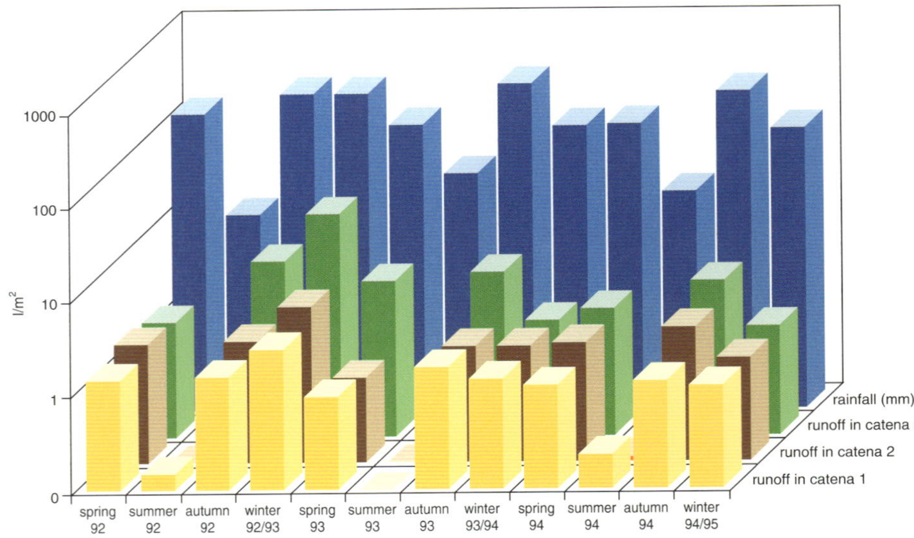

Figure 4.4.6 Runoff (l/m²) against rainfall (mm) in the experimental plots at Is Olias.

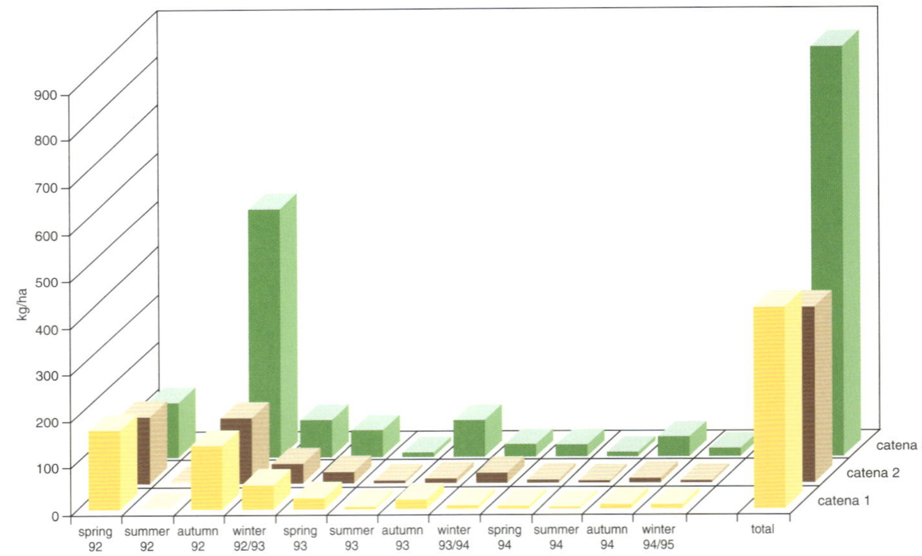

Figure 4.4.7 Sediment yield (kg/ha) in the experimental plots at Is Olias.

has affected extensive areas of *macchia*, encouraging the growth of shrubs. Erosion of flotation tailings from the dumps of the San Leone mine has been estimated for the hydrological year 1992/93 at 22 m^3/km^2 (Lai *et al.*, 1995). Of the heavy metals responsible for pollution, manganese exceeds tolerable limits (Aru, 1993; Aru *et al.*, 1994b; Lai *et al.*, 1995). Over-exploitation of groundwater has led to salt-water intrusion, resulting in the salinisation of water used for domestic supplies and irrigation.

Experimental evidence has shown the important role played by shrub and grass cover in controlling erosion (Aru *et al.*, 1993; Puddu *et al.*, 1995; Tomasi *et al.*, 1995; Vacca *et al.*, 1995). By contrast, it has been found that desertification is exacerbated by accelerated erosion under eucalyptus plantations. Furthermore, the slow growth rates and hence modest productivity of these exotic species raises the question of their suitability and profitability. These findings emphasise the need for proper land resources management to mitigate the desertification process.

4.5 Astimini–Fiume Santo, Sardinia, Italy

G. Pulina, M. d'Angelo, S. Madrau, M. Niedda and C. Porqueddu

Introduction

The loss of soil productivity, strongly related to the present land use and climate, is an important indicator of change preceding more serious deterioration and alterations of the plant–soil–water system, which foreshadows true desertification.

The activity of the Centre for Desertification Research (NRD) of the University of Sassari, as part of the MEDALUS II Project, is aimed at developing an interdisciplinary approach (hydraulics, pedology, agronomy, zootechnics, forestry) for studying land degradation in Mediterranean areas in order to identify the risk factors in the affected areas, to understand the processes leading to erosion and to suggest guidelines for the rational management of these areas.

Background

The area of investigation is the Astimini–Fiume Santo basin (approximately 6000 ha), the largest catchment located in north-western Sardinia (municipalities of Sassari and Porto Torres). The morphological features of the basin vary from hilly to gently undulating. The maximum altitude is 429 m a.s.l. at Punta su Cornu.

Figure 4.5.1 Overview of the Astimini–Fiume Santo Basin.

The characteristics of the basin are dominated by the geological structure, especially the north–south orientation. The structure dates back to the Upper Miocene, when subsidence took place in the central part of the basin, followed by general uplift of the area in the late Miocene and Pliocene.

Different lithologies characterise the basin. Metamorphic rocks are present in the upper part of the basin. In the middle and lower basin, Mesozoic limestones, sandstones and marls occur on the eastern side and Quaternary alluvial deposits on the western side of the basin. Recent alluvial deposits are located in small areas, parallel to the river.

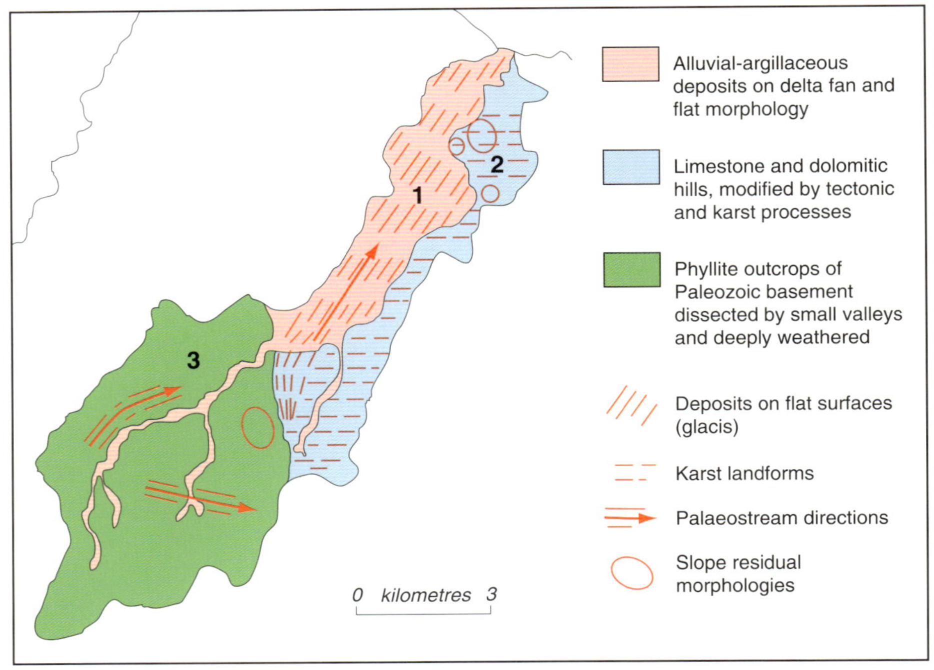

Figure 4.5.2 Geomorphological map of the Astimini–Fiume Santo basin. (1) Geology: alluvial deposits (Quaternary); morphology: delta fan and flat. (2) Geology: marine deposits, mainly limestone (Mesozoic); morphology: limestone and dolomitic hills modified by tectonic and karstic processes. (3) Geology: metamorphic and sedimentary rocks (Palaeozoic); morphology: phyllite outcrops from the Palaeozoic basement dissected by small valleys and deeply weathered.

The main groups of soils, according to the USDA Soil Taxonomy (1975), are Xerorthents, Xerofluvents, Xerochrepts, Rhodoxeralfs, Haploxeralfs and Palexeralfs. The soil moisture regime ranges from an available water content (AWC) of less than 25 mm, for torric and xeric regime soils, to an AWC of 25–300 mm, for xeric soils. The soil temperature regime is thermic.

Land-use history and problems

Agriculture in the Astimini–Fiume Santo river basin dates back to the late Roman Republic, 1st century BC (Caprara, 1991). From the early Middle Ages, north-western Sardinia was totally uninhabited, until the early 17th century when it gradually started to be settled; this process continued up to the late 1950s (Cenami *et al.*, 1991). Currently the land is mainly used for arable crops on flat and gently sloping areas and pasture on the hills; marginal agricultural areas are covered by *maquis* and groves of holm oak (*Quercus ilex* L.), generally degraded by fire and intensively grazed. Abandoned iron mines and an active clay quarry are present and two power stations operate close to the river's mouth. The largest is an oil-burning plant, which will be converted into a coal power station, the other is an experimental wind power plant.

The main causes of land degradation in the basin are:

- sheet and rill erosion of the cultivated land, which are often associated with overgrazing,
- soil and groundwater pollution by heavy metals caused by mine waste dumps and agricultural activities,

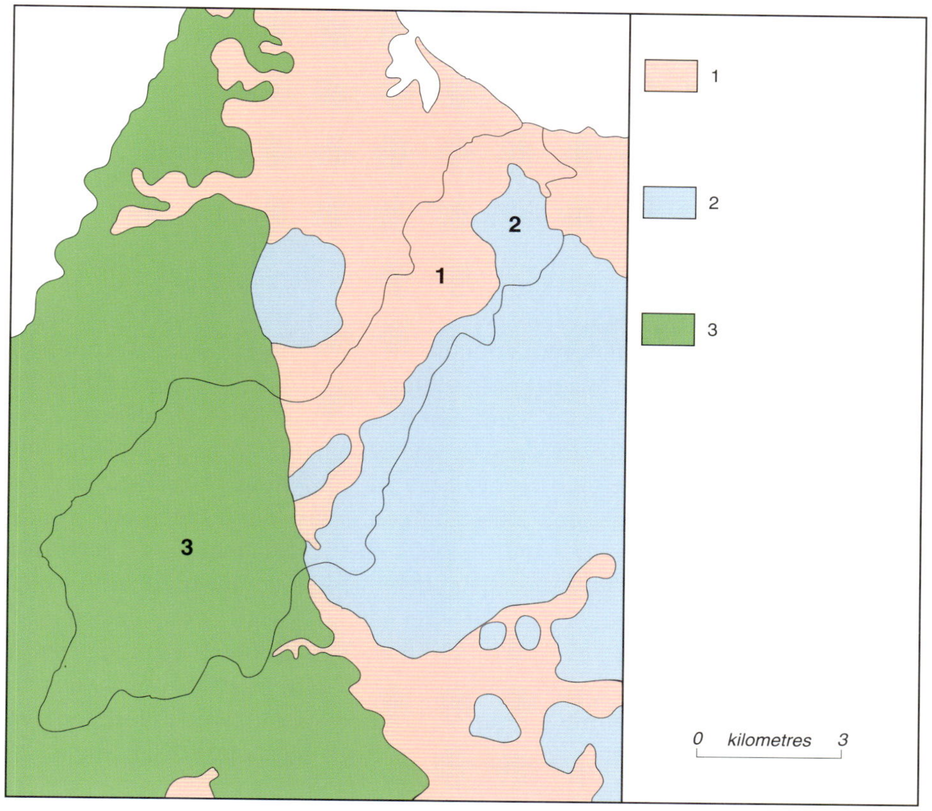

Figure 4.5.3 Geopedological and permeability map. (1) Geology: alluvial deposits (Quaternary); soil: Typic Palexeralfs and Typic Haploxeralfs; permeability: micropore porosity, middle–high. (2) Geology: marine deposits, mainly limestone (Mesozoic); soil: Lithic Xerorthents, Lithic Rhodoxeralfs and Typic Rhodoxeralfs; permeability: micropore porosity, middle–high. (3) Geology: metamorphic and sedimentary rocks (Palaeozoic); soil: Lithic Xerorthents, Lithic Xerochrepts and Dystric Lithic Xerochrepts; permeability: macropore porosity (jointing), low.

Figure 4.5.5 Typic Palexeralf soil

Figure 4.5.4 Lithic Xerorthent soil.

Figure 4.5.6 Frequent fire occurrence with the partial or complete removal of the natural vegetation cover and irrational grazing techniques are recognised as important factors leading to soil erosion in Mediterranean areas.

- industrial and urban developments along the coast and pollution by mineral oils and waste ash from the oil-burning power station dumps,
- wildfires, particularly frequent during the summer, which destroy the natural vegetation cover.

Research activities

The research is being carried out at basin, sub-basin and plot scales. Research activities at basin scale include the following:

- analysis of the main climatic features and climate classification;
- land units and land-use reconnaissance by means of aerial photographs;
- soil survey at 1:10 000 scale: soil types have been identified and mapped, their water storage capacity and compaction index determined, and the proportion of soil eroded evaluated in qualitative and quantitative terms;
- geomorphological survey at 1:10 000 scale: the survey is more detailed in the northern part of the basin where erosional and denudational processes have been stronger;
- groundwater monitoring to evaluate annual and seasonal changes in water-table levels and physical and chemical properties; the hydrological description of the catchment also included the compilation of permeability maps, wells and sources maps and isopiezometric maps for the investigation of fluctuations of the groundwater levels;
- historical investigation on fire occurrence based on existing statistical data to evaluate the role of fires in land-degradation process; remote-sensing data were used to monitor land-cover changes due to fires.

Field studies were carried out in subcatchments falling within two farms representative of main ecological components, land-use regime and management characteristics of north-western Sardinia, and can be summarised as follows:

- soil survey at 1:5000 scale to evaluate the effects of trampling on soil properties;
- agronomic studies on the natural sward, concerning the effect of overstocking and set-aside on the vegetation cover, forage production in qualitative and quantitative terms;
- zootechnical studies to outline the relationship between traditional grazing practices and grazing pressure, forage productivity, nutrition, grazing exploitation rates, and the seasonal arrangements for livestock management;
- erosion studies to understand the effects of rainfall on erosion and surface runoff for different land-use and land-cover typologies.

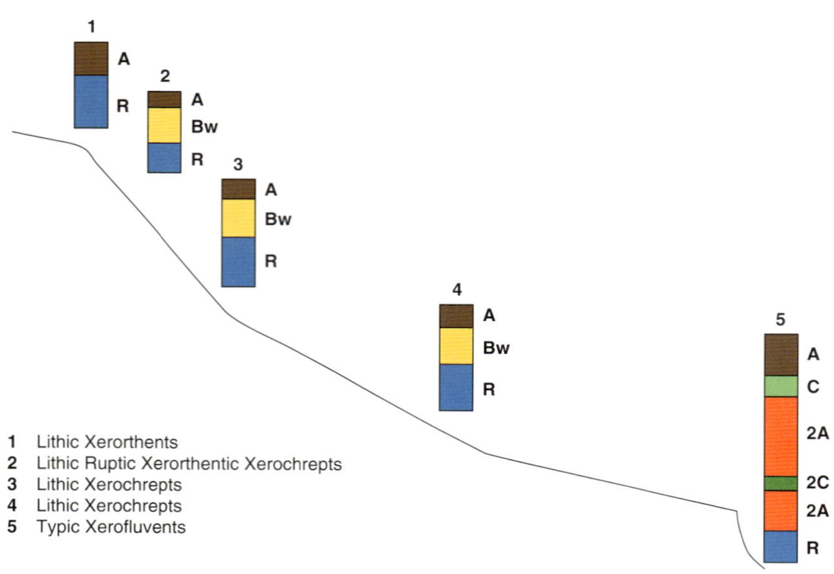

Figure 4.5.7 Experimental area B (Nonnis Farm). Soil profiles along a catena in the ungrazed sub-basin.

Results

Agronomic studies carried out on natural sward, Mediterranean *maquis* and winter cereal crops, highlighted the relationship between agricultural practices and soil erosion. Dry matter production and the qualitative variation (biomass composition and value for grazing) of natural sward in overstocked conditions was comparable to that of other Sardinian marginal areas. Nevertheless, over 100 herbaceous species and 15 shrubby species were identified. Overgrazing increased the maximum proportion of bare soil (22% in ungrazed pasture versus 38% in overstocked pasture). Due to animal selection in the grazed condition, the percentage of grasses and legumes was lower than in the ungrazed condition (Ledda *et al.*, 1995). The sustainable stocking rate in relation to the productive and qualitative aspects of the natural sward were determined. Studies were carried out to evaluate the relationship between the mean grass height and the available forage dry matter of the natural sward. A lower percentage cover of cereals during the autumn–winter season, when rainfall intensity is higher, increases the soil erosion risk on the slopes. The impact of grazers under high stocking rate conditions was studied, pointing out modification of the soil physical properties due to compaction, structure alteration, creation of terracettes and paths, and direct removal of soil (Pulina *et al.*, 1994). A biomass balance finalised to the quantification of biomass consumption (per animal per year), indicated that the variation of unitary removal (from 0.25 to 1.74 t ha^{-1}) was strictly related to differences in farm management, meteorological conditions, animal species and behaviour of grazing animals (Pulina *et al.*, 1995). All the data were processed using a dynamic model developed during the research for the determination of a sustainable stocking rate for extensive cattle production in Mediterranean areas (Masala *et al.*, 1995).

The erosion risk for Sardinia has been mapped and a soil loss model based on climatic data developed (Niedda and Greppi, 1995) according to the available data (43 rain-gauge stations, over the period 1930–1980). At the basin scale (experimental subcatchments equipped with weather stations) during 1994 a total rainfall of 300 mm was recorded and neither runoff nor erosion was found; this was one of the driest years of the century. In the study of the hydrological response of the catchments, the stream network obtained by the digital elevation model (DEM) was used to study the rainfall–discharge transformation using distributed conceptual models (Niedda and Greppi, 1995). The daily rainfall process was studied as a discrete-time point process including sequences of dry and wet days. A stochastic model of daily rainfall as a function of wind direction and cloud cover was developed using historical data and validated (Niedda and Greppi, 1994).

The occurrence and frequency of fires, related to soil erosion risk, and the affected surfaces were monitored by means of remote-sensed data (Landsat TM). The methodology, based on standardised multi-temporal digital change-detection techniques, was validated on different test areas (D'Angelo and Podda, 1994) and provides a classification accuracy of about 90% for the mapping of burned areas.

Finally, pedological, hydrogeological and geomorphological surveys as well as the findings of the other studies have enabled the development of a Land Capability System for extensive grazing which can be applied to north-western

Figure 4.5.8 Erosion in action.

Figure 4.5.9 Irrational agricultural tillage on marginal Sardinian areas (frequent ploughing along the maximum gradient) can result in a severe soil erosion.

Sardinia. A land evaluation for forestry, both for soil conservation and timber production, has been set up and management and conservation plans for groundwater resources have been designed.

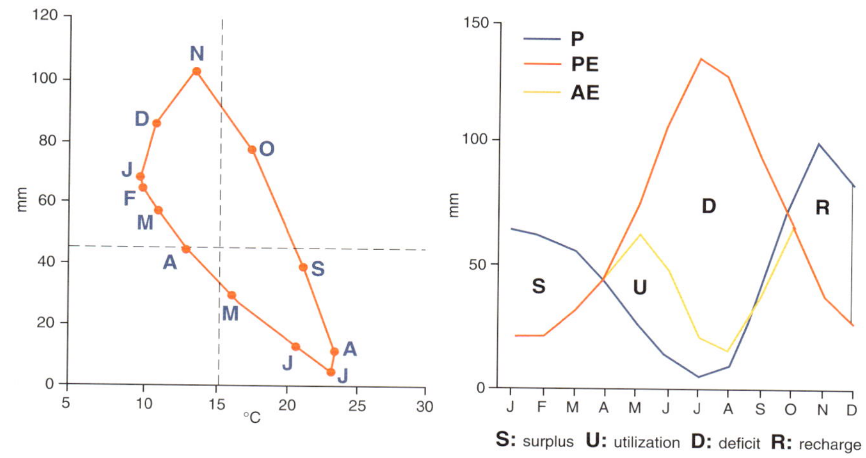

Figure 4.5.10 (a) Climogram of mean monthly rainfall and temperature for Fertilia calculated over the period 1958–1987. The dashed lines indicate type boundaries for the Thornthwaite (1948) classification system. Climate types are defined using a precipitation effectiveness index (precipitation minus potential evapotranspiration) and thermal efficiency index based on temperature). According to the Thornthwaite scheme Fertilia is classified as mesothermal (B2), dry subhumid (C1) to semi-arid (D). (b) Mean monthly water balance for Fertilia calculated over the period 1958–1990. Differences between precipitation (P), potential evapotranspiration (PE) and actual evapotranspiration (AE) indicate soil-water utilisation (U), surplus (S), deficit and recharge (R). PE and AE are calculated using the Thornthwaite method which is based on monthly temperature and station latitude (Thornthwaite, 1948).

Discussion and management issues

The research confirmed the effects of irrational agronomic practices, overgrazing and natural vegetation removal on soil degradation. On this basis the NRD group can suggest some guidelines for rational land management in areas similar to the Astimini–Fiume Santo Basin.

In hilly landscapes with steep slopes the following practices should be avoided: the elimination of natural vegetation, e.g. by tillages and fire; land drainage works; frequent ploughing along the maximum gradient; high stocking rates; and high impact works on slopes (e.g. roads and water-pipes). In these areas the introduction of minimum tillage techniques, the conservation of natural vegetation in areas already covered, the restoration of vegetation in the most exploited and degraded areas and a drastic reduction of stocking rate through the employment of rational grazing techniques, is recommended.

In morphologies varying from flat to gently undulating the following practices should be avoided: the adoption of crops requiring deep tillage and with high hydrological requirements, overgrazing on remarkably clayey soils, the excessive exploitation of groundwater. In these areas it is recommended that there should be recourse to agronomic minimum tillage techniques and the introduction of rational breeding systems.

4.6 The Case Studies at the Petralona and Hortiatis Field Sites, Northern Greece

G. P. Stamou, J. Diamantopoulos, J. Pantis, S. Sgardelis, A. Dalaka, E. Papatheodorou, S. Pirintsos, Th. A. Mardiris, T. Lanaras, C. M. Cook and G. Iatrou

Introduction

Overgrazing is the most important factor inducing degradation in the Mediterranean-type ecosystems of northern Greece. The impact of grazing pressure can be observed mainly on vegetation physiognomy, plant physiology, plant growth and nutrient turnover grounds.

Fieldwork has been carried out in the Petralona and Mount Hortiatis MEDALUS sites.

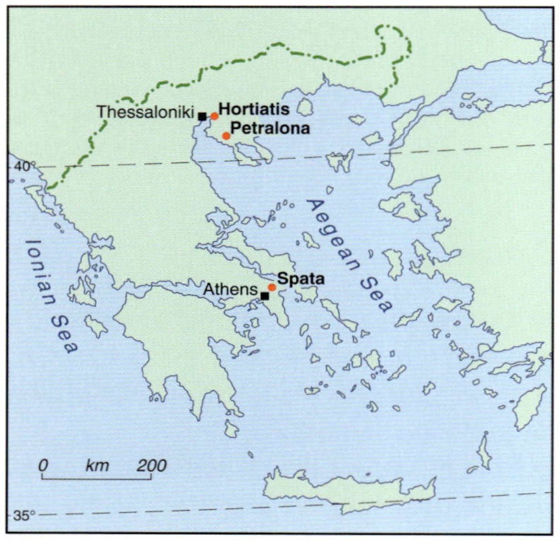

Figure 4.6.1 MEDALUS field sites in Greece.

The main results achieved within the MEDALUS Project show the following:

- Grazing induces changes in species composition and the architecture of vegetation. It affects both the age of different layers in the canopies and the structure of canopies themselves. Younger plant modules (new shoots and leaves) and higher rates of plant growth were recorded in overgrazed areas.
- The direct effect of grazing on plant photosynthetic activity is of lesser significance than that of the indirect effect. By changing plant architecture, grazing increases exposed photosynthetic units and results in a positive response of plant photosynthesis.
- Grazing also affects nutrient dynamics in already degraded ecosystems. For instance, nitrogen (N) content in overgrazed upper soil level remains more or less stable throughout the year, while strong oscillations in the N content of leaves and litter can be observed. Analogous conclusions can be derived for all nutrients.

The data suggest that, in general, vegetation responds positively to grazing pressure and is the most important component of the ecosystem preventing further degradation.

On the other hand, re-afforestation, which is usually used as a land degradation mitigation measure in this kind of area, has unpredictable impacts on ecosystem components. It changes both the compositional and structural characteristics of the *maquis* communities. 'Natural' areas (non-reforested), richer in species and more homogenous, are located towards the middle of the Canonical Component Analysis (CCA) plane. Reforested areas, with low species richness and higher species dominance, are found towards the edges of the CCA plane. Re-afforestation also affects nutrient availability in the humus layer.

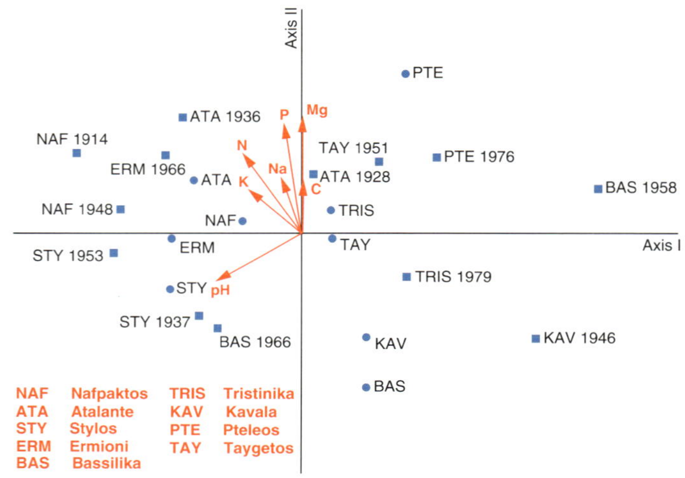

Figure 4.6.2 Ordination of the 13 reforested areas (■) and the 'natural' *maquis* communities (●) on the first two axes of CCA. Vegetational data of the species used in re-afforestations have been removed.

Petralona

The Petralona site is located near the village of the same name that has the famous cave where the only Neanderthal remains in Greece are to be found. Its approximate co-ordinates are 40°20'N and 23°10'E at a height of 400 m a.s.l. and 52 km from Thessaloniki. The site lies in the central-western part of the Halkidiki Peninsula.

Physiographically it is a Triassic limestone hill emerging from Miocene–Pliocene deposits. The landscape configuration imposes the two land-use types in the area, i.e. grazing on limestone and wheat cultivation on the Mio-Pliocene deposits. These land-use categories show a distinct response to soil erosion, with total sediment yield higher in the *maquis* and runoff higher in the wheat sites.

The *maquis* vegetation consists mainly of four evergreen species: *Quercus coccifera* L., *Juniperus oxycedrus* L., *Phillyrea latifolia* L. and *Pistacia lentiscus* L.

Hortiatis

The site lies on the north-east facing slope of the Hortiatis mountain (approximate co-ordinates 40°37'N, 23°03'E). This land has been selected for urbanisation purposes. However, for many years it has been used for grazing.

This has resulted in a vegetation dominated by *Q. coccifera* shrubs with randomly dispersed *J. oxycedrus* shrubs. Isolated *Q. coccifera* stands are formed by

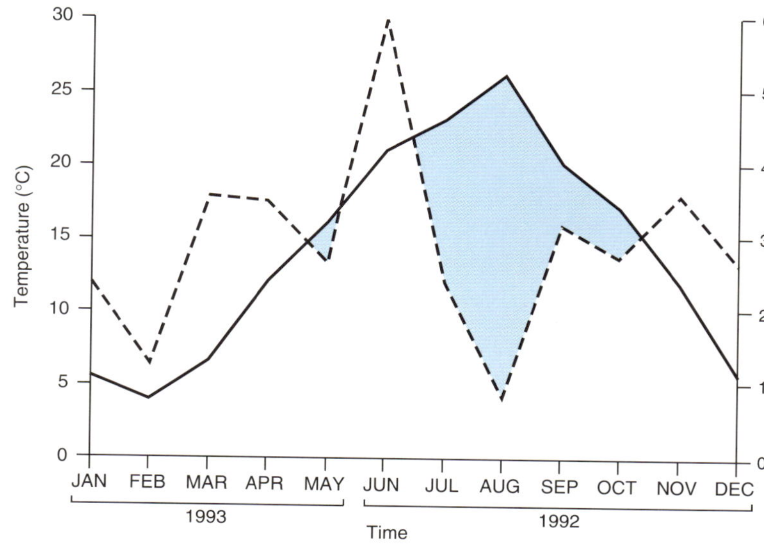

Figure 4.6.3 Climogram (according to Walter and Leith, 1960) showing the regime of mean monthly air temperature (—) and precipitation (– –) for the experimental period. Shaded area is period of water deficit.

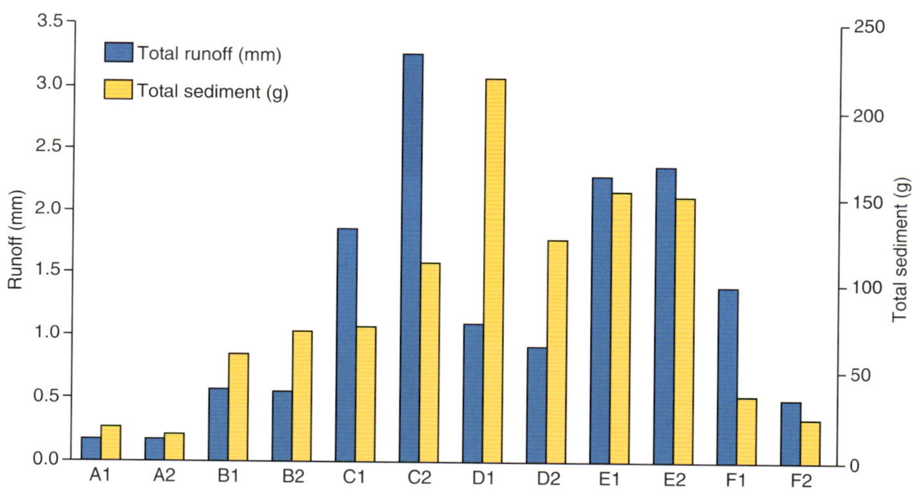

Figure 4.6.4 Runoff and sediment yield in the *maquis* and wheat field plots at Petralona. A, B and C are under wheat cultivation and their slope gradients are 22, 37 and 31°, respectively. D, E and F are under *maquis* and their slope gradients 35, 45 and 45°, repectively.

specimens of different heights. Consequently, low shrubs that can be easily reached by grazing animals are browsed and distorted, while higher shrubs escape grazing. This is an ideal situation for the study of physiological responses of heavily grazed and ungrazed shrubs of which photosynthesis was the main parameter studied due to its contribution to productivity. Soil meso-fauna has also been studied extensively in this site.

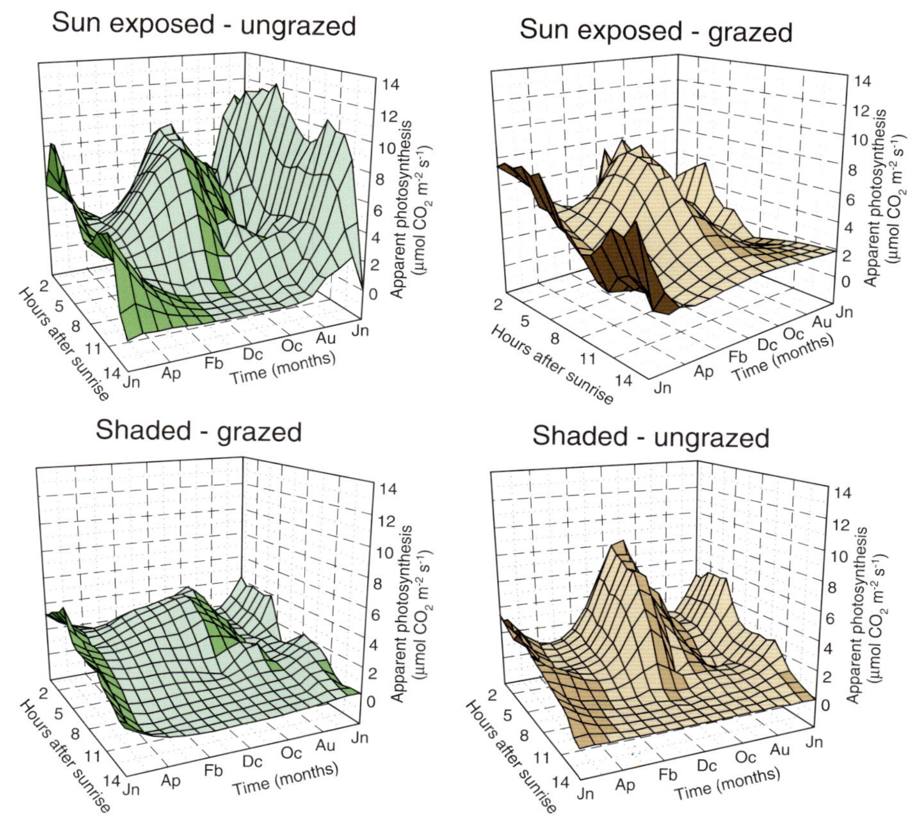

Figure 4.6.5 Variations in the net photosynthesis of *Quercus coccifera* L. by month (July 1993 to June 1994), time of day (hours after sunrise), leaf position (sun exposed, shaded) and site condition (ungrazed grazed) on Mount Hortiatis. Net photosynthesis is equal to the gross rate of photosynthesis (uptake of CO_2) minus the rate of respiration ('dark' respiration and photorespiration, loss of CO_2).

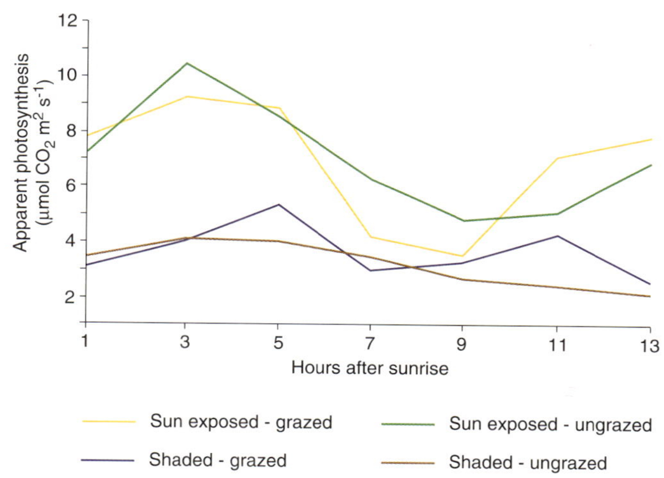

Figure 4.6.6 Variations in the net photosynthesis of *Quercus coccifera* L. for June 1994, by time of day (hours after sunrise), leaf position (sun-exposed, shaded) and site condition (ungrazed grazed) on Mount Hortiatis.

4.7 Land-use Change and Soil Properties Affecting Desertification at the Spata Field Site, Athens, Greece

C. S. Kosmas and N. G. Danalatos

Introduction

Large areas of the Mediterranean region cultivated with vines and olives are mainly confined to hilly lands with shallow soils prone to erosion.

In an effort to study the effect of land use on soil properties and erosion, an experimental site was selected comprising two transects, one under olives and one under vines, extending from the top to the bottom of a hillslope catena located about 30 km east of Athens.

The site is characterised by a Thermo-Mediterranean climate (Bagnouls and Gaussen, 1953). The mean annual air temperature is 17.8°C and the annual precipitation is 496 mm, with more than two-thirds falling in the period November–April.

The major land use of the general area is olive groves. These have been grown under semi-natural conditions in the study area for more than 160 years and have an understorey semi-natural vegetation of annual plants that largely protect the soil from erosion. In part of this area, the entire vegetation was removed, and vines have been cultivated since 1979. Due to frequent weed control, the soil in the vine fields remains bare for most of the time, favouring runoff and erosion.

Figure 4.7.1 Hilly area cultivated with vines and olives.

Soil properties

In the 12 years since the shift from olives to vines, the land and soil characteristics have also undergone important changes. There has been a tenfold decrease in the mean aggregate size of soil particles on the transect under vines as compared to that under olives, associated with a significant increase in the erodibility of the soils (under vines) and a 40% decrease in their organic carbon content.

The decrease in the organic carbon content resulted from increased erosion and reduced biomass production and was associated with an increase in the reflectance in the visible region. Following the erosion of the topsoil, the underlying B-horizon was exposed on the surface (Kosmas *et al*., 1996), and the soil surface became progressively lighter, as reflected by the increased Munsell soil colour value.

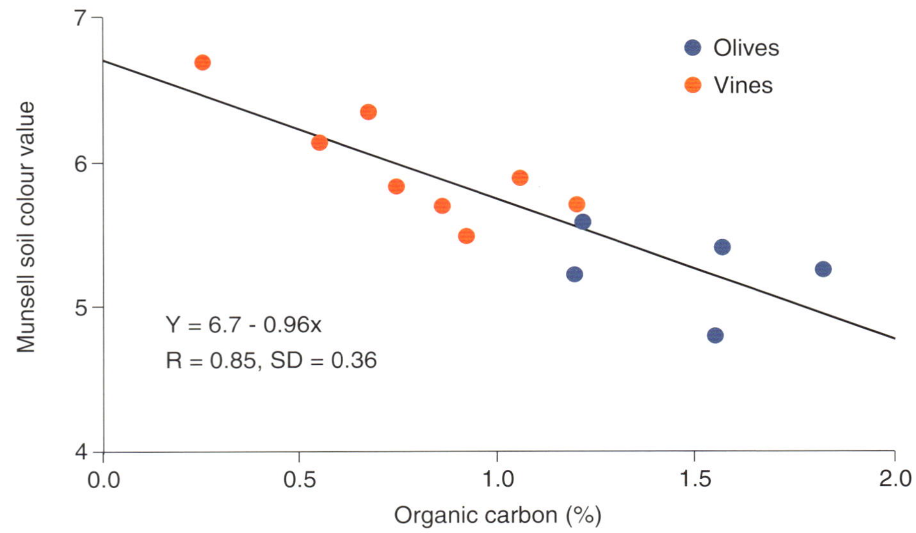

Figure 4.7.2 Relationship between Munsell soil colour value and organic carbon in selected sites along the transects under vines and olives.

Water runoff and erosion

In a 21-month observation period, the total runoff ranged from 0.32 to 2.62 mm in the sites under olives, and from 4.87 to 12.25 mm in the sites under vines.

Soil survey data obtained in 1979 demonstrated that the downslope deposition of new materials through erosion was responsible for an increase in the depth of the Ap-horizon by 25 cm, giving a deposition of at least 216 kg/m^2 in a period of 12 years in the transect under vines.

It is therefore clear that the impact of land-use change on environmental conditions, and particularly on land degradation, in hilly Mediterranean areas should be seriously considered in future land-use planning. Vine cultivation creates favourable conditions for overland flow, erosion and desertification.

Biomass production under cultivation in semi-arid hilly areas vulnerable to desertification

Wheat biomass production along catenas

Rainfall amount and distribution are the major determinants of cereal production on hilly lands in the semi-arid climatic zone of Greece. The continuous cultivation of these lands with rain-fed cereals in the last century resulted in accelerated erosion and the removal of most of the soil, and the formation of the present shallow stony soils.

A study of the total above-ground biomass production (TAGBP) of wheat was carried out in a hilly area (Thivas, 100 km north of Athens) along catenas

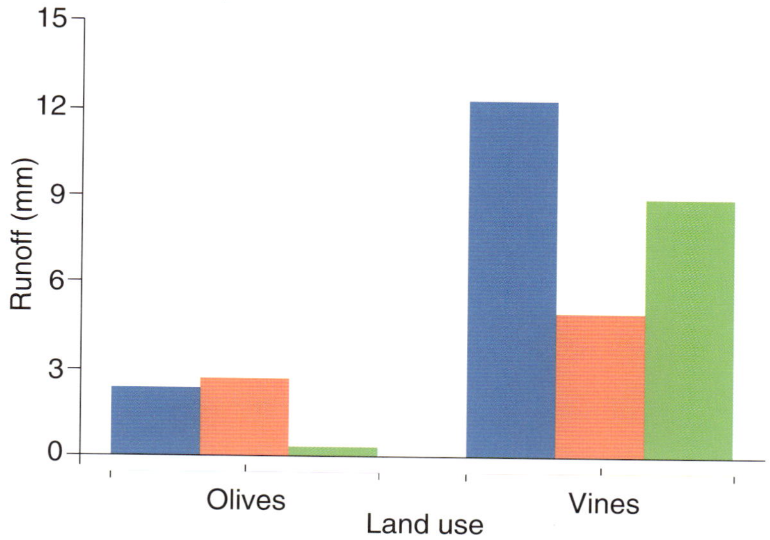

Figure 4.7.3 Total runoff measured in plots installed along two transects under olives and under vines over a period of 21 months.

with soils formed on Tertiary and Quaternary deposits of marl, conglomerates, and shale–sandstones for two successive growing periods, 1989/90 and 1990/91.

The first year (1989/90) was extremely dry (95 mm rainfall) while the following year was exceptionally wet (663 mm, versus 370 mm for an average year). Only at the end of the second growing period did a slight water deficit occur.

In the wet period (1990/91), the soils on marl were more productive than their stony counterparts on shale–sandstone and conglomerates. Since water availability was practically optimal, this is attributed to the relatively high fertility status of the soils on marl.

Under dry conditions (1989/90), the catenas on marl were almost bare, whereas a significant biomass production (TAGBP) was measured at the various locations of the shale–sandstone and conglomerate sites.

Rock fragments on the soil surface appeared to be extremely important in dry years by conserving appreciable amounts of soil water from evaporation

Figure 4.7.4 Experimental field for studying the effect of rock fragments on soil moisture conservation and erosion.

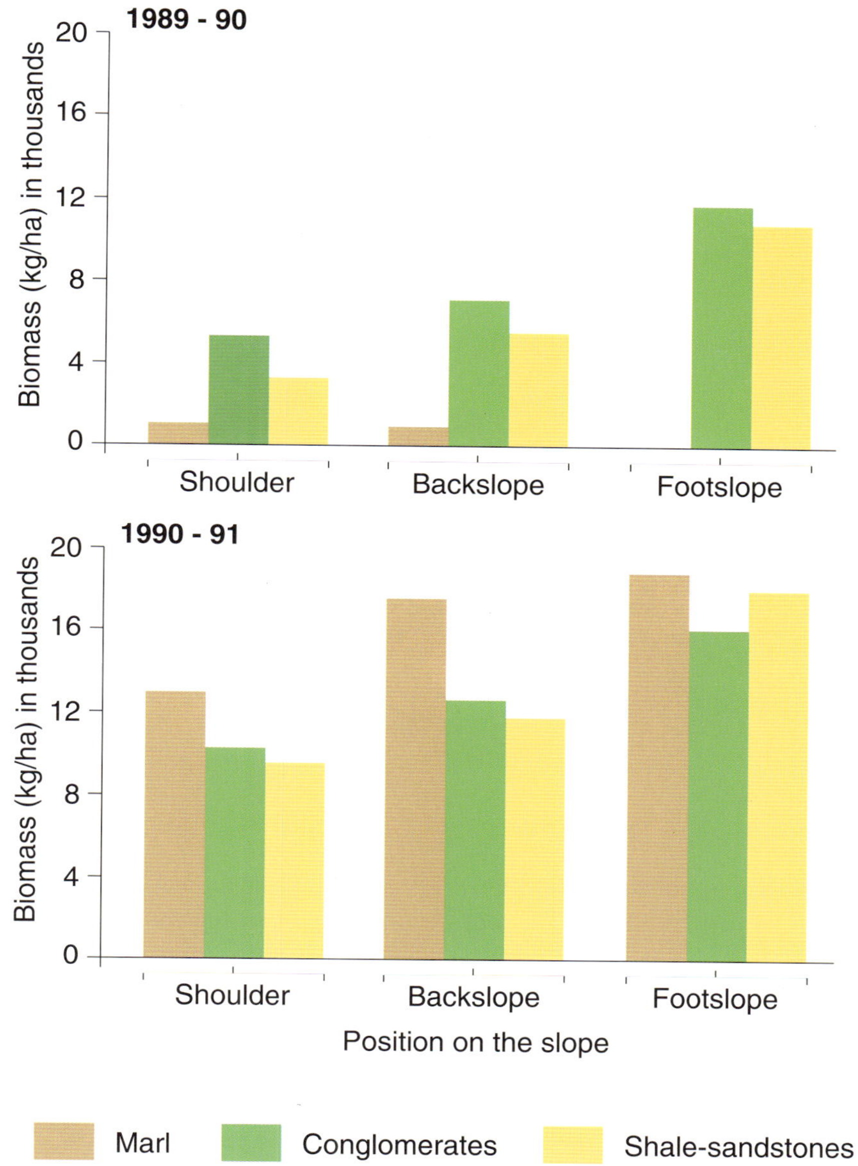

Figure 4.7.5 The effect of parent material and landscape position on biomass production (TAGBP) of rain-fed wheat growing under very dry (1989–1990) and very wet (1990–1991) weather conditions.

through surface mulching and preventing large areas from desertification (Kosmas *et al.*, 1993a). Stony hilly soils along catenas of conglomerates and shale–sandstones, despite their normally low productivity, may supply appreciable amounts of previously stored water to the stressed plants and produce significant biomass even in extremely dry years. Soils formed on marl are free of coarse fragments and despite their considerable depth and high productivity in normal and wet years, are very susceptible to desertification, because they are unable to support any vegetation in particularly dry years as a result of the absence of rock fragments.

Rock fragments

Given that during dry years wheat yields on stony soils are higher than on non-stony soils, a field study on the effects of rock fragments on hydrology and soil erosion was carried out. Thirty plots with soils containing rock fragments of different size, amount and placement were installed on a hillslope. The results

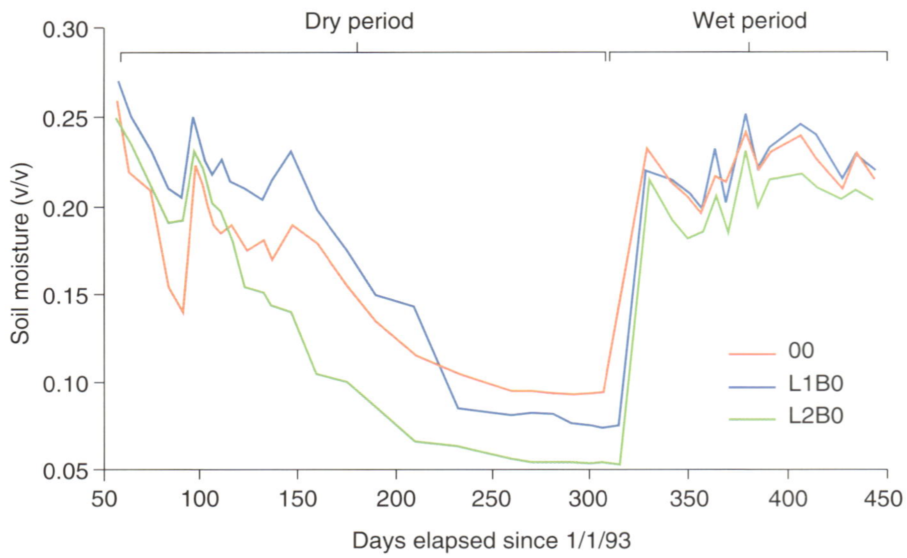

Figure 4.7.6 Changes in volumetric soil moisture content with time measured in plots free of rock fragments (00), with cobbles partially embedded in the soil surface (L1B0), and with cobbles incorporated in the upper 15 cm soil layer (L2B0).

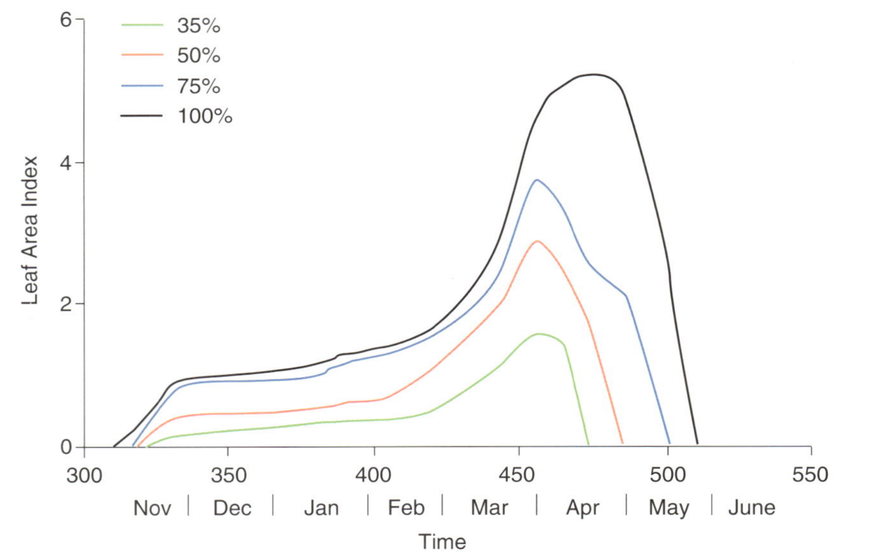

Figure 4.7.7 Change in the leaf area index (LAI) of wheat grown in plots receiving 35%, 50%, 75% and 100% of the incoming rainfall in 1991/1992.

indicated that during dry years, rock fragments can be very valuable in conserving appreciable amounts of water which is supplied to the vegetation, so protecting large areas from desertification (Danalatos *et al*., 1995).

Rainfall exclusion

The effects of diminishing soil moisture on wheat biomass production were also investigated in a rainfall exclusion experiment involving four treatments of rain interception. Three out of four experimental plots were partially shielded with Plexiglas covers, so that of the total rainfall (361 mm = 100%) that occurred during the growing season only 256 mm (75%), 177 mm (50%) and 123 mm (35%) were allowed through. As expected, the reduction in soil moisture affected the leaf area index (LAI), which was measured at 5.2, 3.7, 2.9 and 1.6 respectively (Kosmas *et al.*, 1996). The plants receiving 35% of the total rain were completely dry 37 days earlier than the control plants in the open field.

A very high correlation was found between (the logarithm of) the total aboveground biomass production (TAGBP) and the ratio of actual over maximum crop evapotranspiration (ET_{act}/ET_{max}). Such data can be used to reliably assess the reduction of TAGBP under different weather conditions and to predict future changes in biomass productivity in areas vulnerable to desertification.

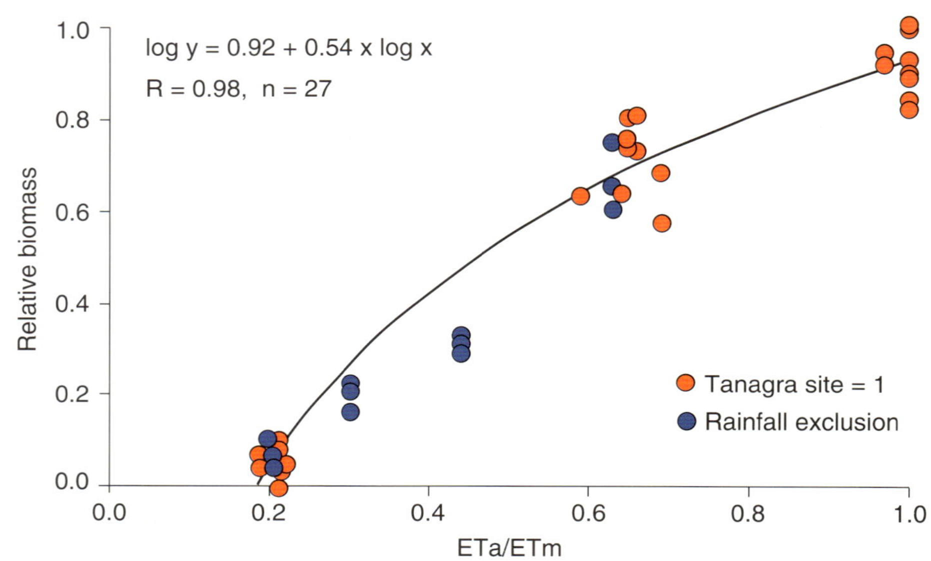

Figure 4.7.8 Relation between the calculated relative evapotranspiration rate (ET_{act}/ET_{max}) and the relative wheat biomass production for different sites and for three growing periods.

5

REGIONAL STUDIES

5.1 The Guadalentín Basin, Murcia, Spain

F. López-Bermúdez, A. Romero-Díaz, F. Cabezas-Calvo-Rubio, L. Rojo-Serrano, José Martínez-Fernandez, Julia Martínez-Fernandez, M. Böer and G. Del Barrio

Introduction

The Guadalentín Basin is a territorial unit in which physical factors and human actions converge to create one of the severest cases of desertification in Europe. It is located in the south-eastern part of the Iberian Peninsula, with an area of 3300 km^2. The headwaters of the river are located in the north-eastern part of the Almería Province (eastern Andalucia) and meet the Segura River, of which the Guadalentín is the main tributary, near the city of Murcia. The basin is a transition zone between the arid and semi-arid climatic domains of Mediterranean climate type. The combination of a whole set of degradation factors (climatic, geological, geomorphological, hydrological and socio-economic) led to the selection of the basin as one of the MEDALUS target areas for the monitoring and investigation of desertification processes in the Mediterranean at the regional scale.

Geological characteristics

The Guadalentín Basin lies within the domain of the Betic Cordillera. This is the biggest and most complex mountain system created by the Alpine orogeny, extending from Cadiz in the west to Alicante in the east—more than 500 km.

The rocks of the basin range from Palaeozoic to Quaternary. The oldest are metamorphic and above these in the sequence are clays and limestones, dolomites, conglomerates, sandstones and gypsiferous marls. In the Pliocene and Quaternary sequences, which are widespread, sands, loams, colluvium, alluvium and basal conglomerates occur. There are also volcanic outcrops of Pliocene and early Quaternary ages.

The most characteristic structural feature is a large graben which forms a depression, along which the river flows. This runs from SW to NE for more than 90 km, from the borders of Almería Province to the city of Murcia. The boundaries are defined by two of the most important faults in the Betic System and comprise faulted mountain fronts, with important alluvial fans, which provide the principle Quaternary sediments filling the graben.

The Lorca–Alhama Fault on the north side of the graben is very active, with many earthquake epicentres. In recent years a number of earthquakes have occurred which have reached 4.5 on the Richter scale (e.g. those of 18 February 1986, 6 June 1977, 14 May 1946, 4 April 1932, 24 October 1927, 21 January 1921, 16 April 1907).

The activity of this and other faults is reflected in the hydrological network. River courses, deflections in *ramblas* (ephemeral water courses) and cracks in irrigation channels and buildings mark the fault. The Guadalentín Basin is therefore an area of significant seismic risk.

Figure 5.1.2 Air photograph of the Guadalentín tectonic trough in the Lorca-Totana sector (Source: Servicio Geografico del Ejercito, Ministerio de Defensa, Madrid, 1996).

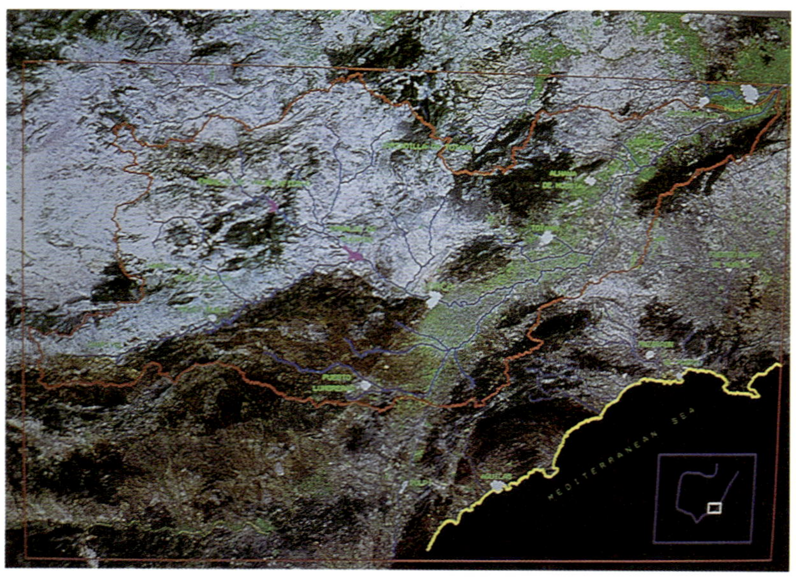

Figure 5.1.1 Landsat image of the Guadalentín river basin.

Topography

The highest lands are found in the Sierra María range to the south-west and the uplands in the north and west. The main peaks are Espuña (1579 m a.s.l.) and Cambrón (1525 m a.s.l.) in the north; Gigante (1493 m a.s.l.) and María (2045 m a.s.l.) in the west; and Almenara (888 m a.s.l.) and Carrascoy (1066 m a.s.l.) in the south and east respectively. The lowest area is the confluence of the Segura and the Guadalentín near Murcia at an altitude of 40 m a.s.l. Of the surface, 63% is at 500–1300 m and 35% is below 500 m, while 55% of the area has a slope gradient of less than 8%; 36% a slope of 8–24%; 8% a slope of 25–45%; and only 1% has a slope over 45%.

Climatic characteristics

The Guadalentín Basin has a Mediterranean climate characterised by dry summers and mild winters with rare torrential rainfalls. The scarcity of precipitation (less than 300 mm/year), the strong insolation (about 2900 hours/year) and

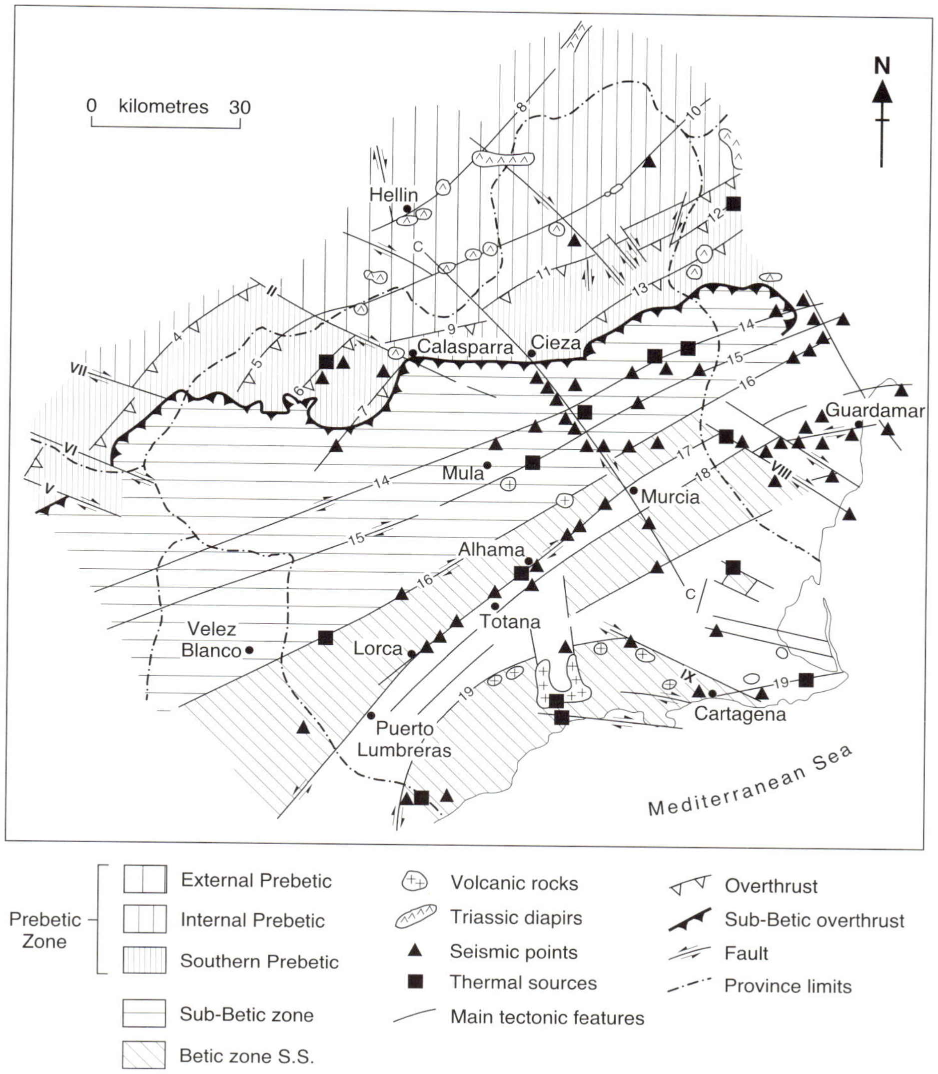

Figure 5.1.3 The main tectonic elements in the Murcia region affecting the Guadalentín Basin (Rodriguez-Estrella, 1979).

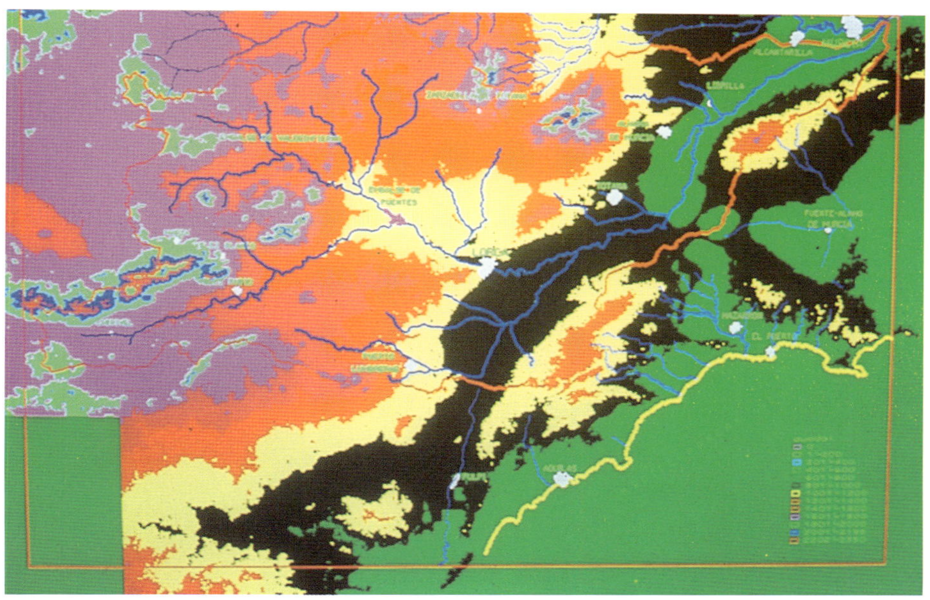

Figure 5.1.4 Digital terrain model of the Guadalentín river basin.

Figure 5.1.5 Tectonic graben of the Guadalentín valley.

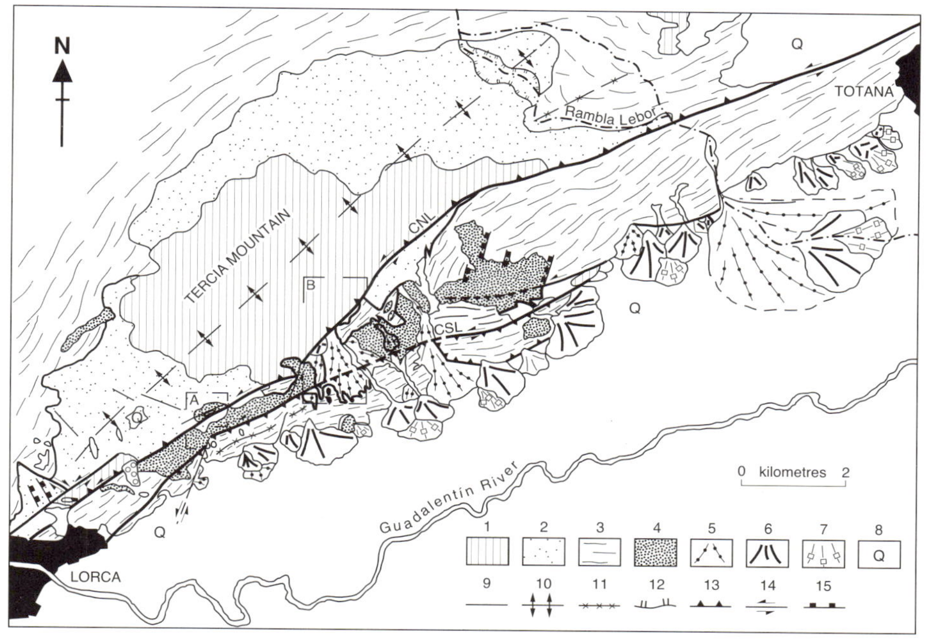

Figure 5.1.6 Geological cartography of the Lorca-Totana sector (Martinez-Diaz and Hernandez-Enrile, 1992). 1, Base of the internal Betic; 2, deposits of early Serravallian–upper Tortonian; 3, deposits of upper Tortonian–Messinian; 4, upper Pliocene–lower Pleistocene; 5, pediment and alluvial fans of middle Pleistocene; 6 and 7, alluvial fans of upper Pleistocene–Holocene; 8, Quaternary; 9, main vestiges of the north passage (CNL) and the south passage (CSL) of Lorca; 10, anticlinal axis; 11, synclinal axis; 12, escarpment; 13–15, faults.

the high potential evaporation, produce a great moisture deficit and give an extremely arid character to the basin.

Mean annual rainfall is around 350 mm in the semi-arid eastern sector, and 500 mm in the subhumid sectors at higher altitudes in the north-west. The rainfall is in autumn and spring, and the *gotas frias* coming from the Mediterranean in autumn can yield 300 mm in 24 hours and cause violent floods.

Temperature decreases from an average of 12°C in the mountains of the north and west to an average of 18°C in the east.

Evapotranspiration ranges between 900 and 1200 mm and so three climatic domains can be distinguished (Navarro Hervás, 1991):

- a *subhumid domain* from the headwaters of the river to the reservoir of Puentes, including the Espuña range, with low average temperatures and rainfalls averaging over 500 mm/year;
- a *semi-arid domain* (covering 50% of the area), in the eastern part of the basin and interior uplands, with moderate temperatures and rainfalls between 300 and 500 mm;
- an *arid domain* that occupies more than 40% of area of the basin, in the central and lower parts, with the highest temperatures and rainfalls less than 300 mm in many places.

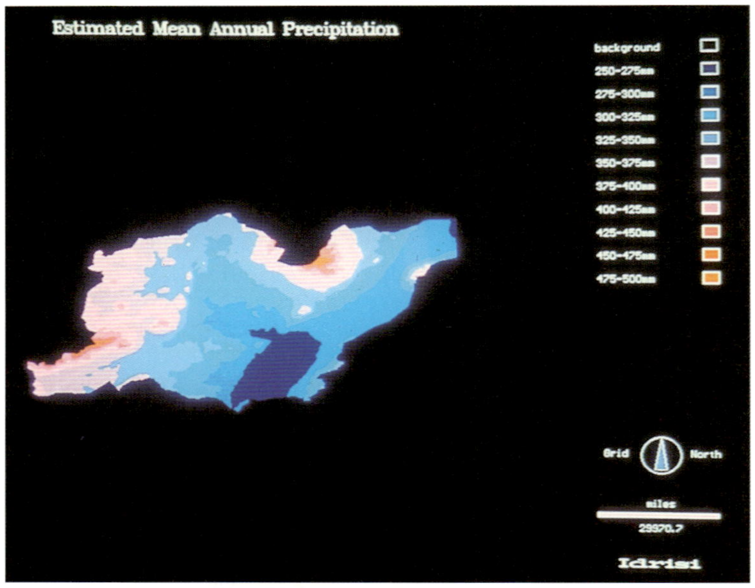

Figure 5.1.7 Estimated mean annual precipitation. Orographic effects produce relatively high precipitation amounts in the main mountain ranges, especially on the eastern sides, and relatively low amounts in the main valley and on lee sides. This map was produced (by Böer) in two steps. First, the general trend of precipitation increasing with altitude was quantified by a linear regression equation and distributed by renumbering a digital elevation map. Secondly, the residuals of the regression equation were interpolated using kriging, and then added to the values obtained from the regression equation to produce the final map.

Surface hydrology

The Guadalentín Basin is 17.5% of the Segura Basin. Starting from Puentes, it narrows where the Velez and Luchena rivers join, and then flows for 14 km towards the south-east, before joining the depression near Lorca, turning towards the north-east and continuing along the bottom of the valley. It is canalised in the lowest reach and joins the Segura near Murcia. The slope of the channel is 3.05% above the Puentes reservoir, 0.70% between Puentes and Lorca, and 0.46% between Lorca and Murcia.

Gullies and ephemeral water courses (*ramblas*) are of fundamental importance in the network, and the Guadalentín receives many ephemeral inputs along its 121 km course. The *ramblas* of Torrealvilla, Lebor, Algeciras, Librilla and Belen come in from the left bank and the Viznaga from the right bank.

In the south-west sector there is an endorreic zone that receives water from the *ramblas* of Bejar and Nogalte. This zone occasionally overflows in floods via the Viznaga gully into the endorreic basin. The ephemeral flows of the Nogalte *rambla* constitute one of the most dynamic elements of the Guadalentín Basin even though it is only 4.2% (139 km^2) of the area. Its loamy soil and schist lithology have favoured strong headward erosion, high transport rates for solid materials, and a badland landscape. Soils are generally undeveloped as a consequence of the arid character of the climate, the scarcity of vegetation cover and the erosive impact of intense rains.

The Velez and Luchena rivers, main contributors to the Guadalentín, acquire great importance in times of intense rains because of the steep headwaters of slopes and the shortness of their main channels. Where important changes of slope occur, extensive deposits are formed, giving rise to a mixture of pediments and alluvial fans.

Groundwater hydrology

The valley of the Guadalentín River is formed in a tectonic trough underlain by Tertiary deposits resting on a Betic basement of Precambrian to Triassic age.

There are two aquifers in the trough, the Alto and Bajo Guadalentín, with a combined areal extent of 596 km^2, whose separation causes them to be strongly over-exploited. Inputs from rainfall, flood transmission losses and base flows

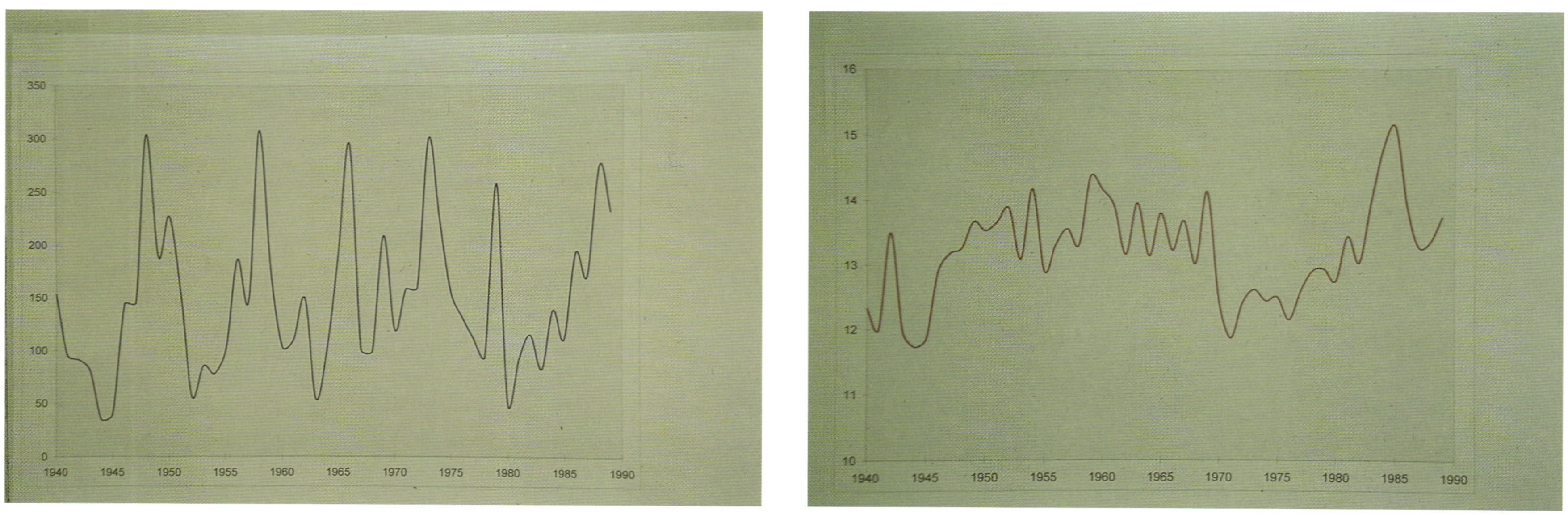

Figure 5.1.8 Annual rainfall and mean temperatures of the Lorca C.H.S. station (1940–1990).

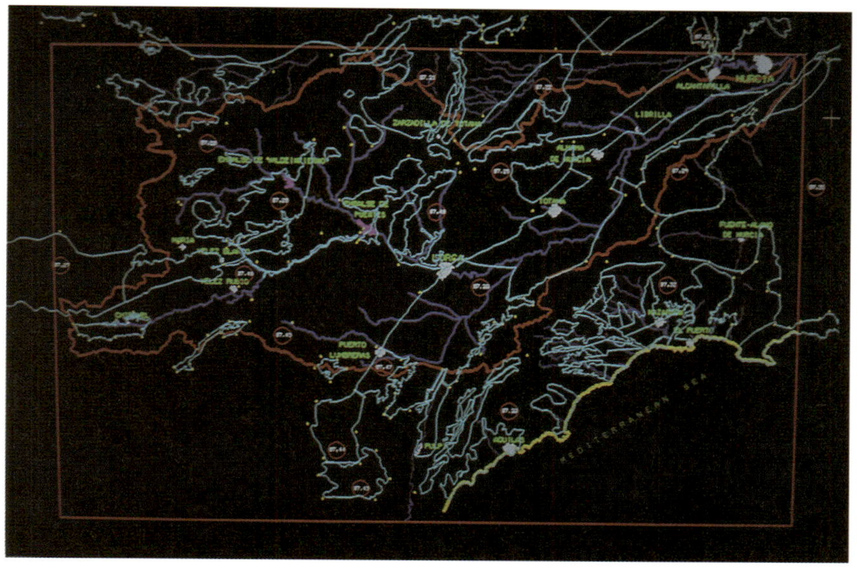

Figure 5.1.9 Inventory of aquifers in the Guadalentín Basin.

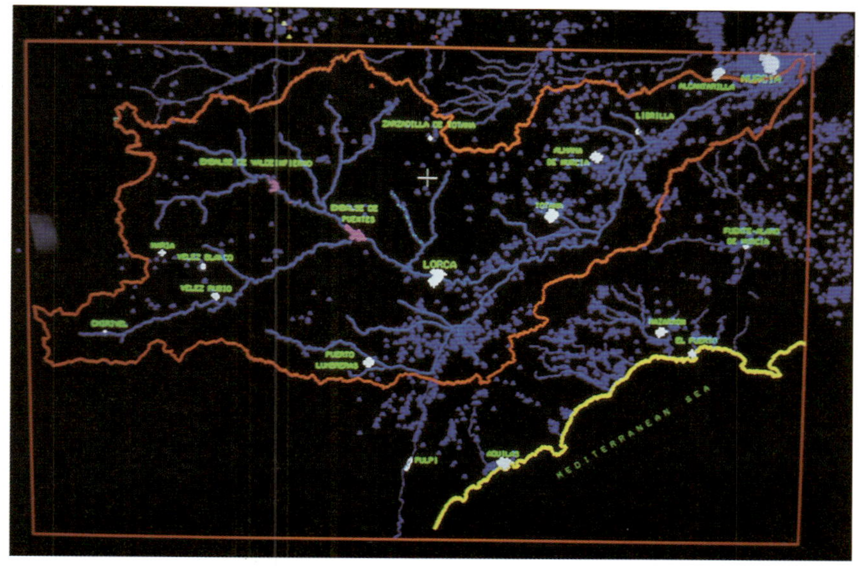

Figure 5.1.11 Hydrological extent of the Guadalentín Basin.

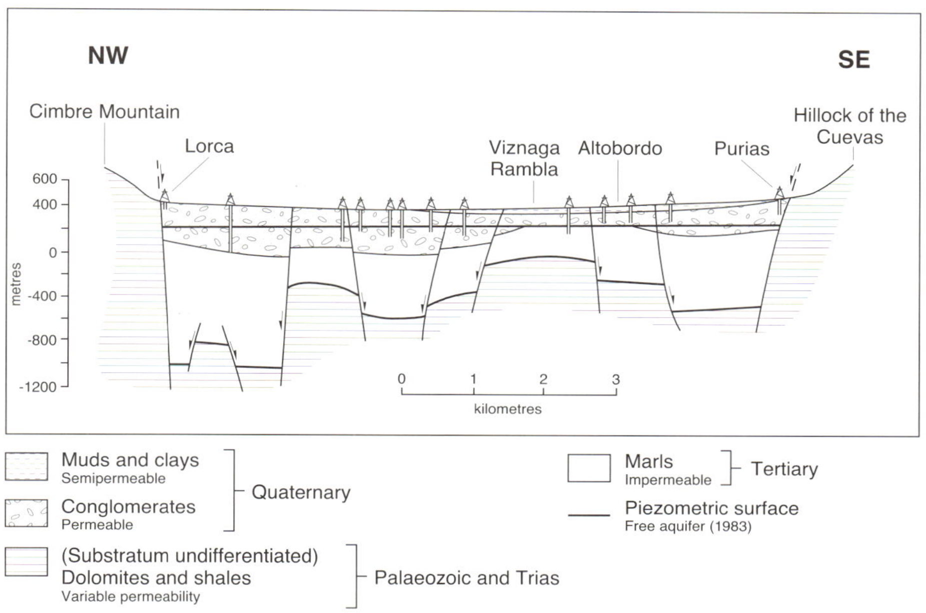

Figure 5.1.10 Hydrogeological section of the Guadalentín valley.

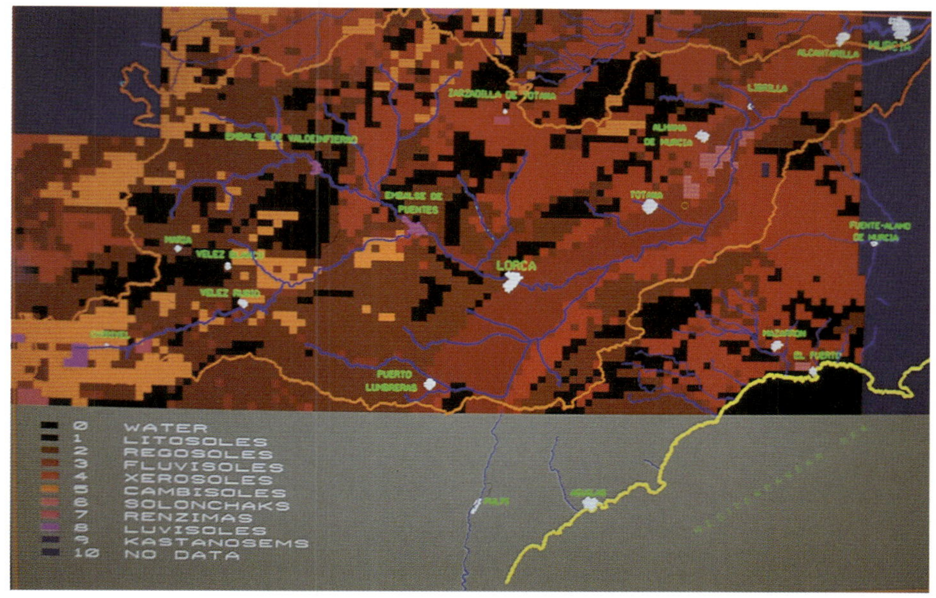

Figure 5.1.12 Main soil types in the Guadalentín Basin.

are 21 Hm^3/year, but withdrawals by pumping (103 Hm^3/year) yield a deficit of 82 Hm^3/year. The effects of such over-exploitation are a continuous decrease of the piezometric level at a maximum rate of 5 m/year, and contamination due to the release of gas and salts from the alluvium and the underlying rocks.

Sustained use of these aquifers at this rate would require water supplies from outside regions or the reinstatement of fresh water flows now dammed up by reservoirs in the catchment.

Soils

The soils of the Guadalentín Basin are closely related to lithology and topography. In areas with ephemeral rivers, fluvisols occupy 14.9% of the area. These are sometimes saline, and salinity reaches a maximum on the floor of the Guadalentín Valley.

In areas of Miocene marls and on Palaeozoic materials (slates, schists and phyllites) regosols have developed. These are lightly-consolidated soils over recently deposited material, that are very prone to erosion and occur most commonly in the Campo de Lorca, in Torrecilla and in the Almenara mountains, covering 34% of the basin. On limestone such as the España, Madroño, Carrascoy and Estancias mountains, xerosols occur which are usually shallow and very stony, frequently with a petrocalcic horizon, occupying a further 24% of the area. On north-facing slopes and at higher altitudes, cambisols can be found which are more evolved and characterised by an alteration horizon. These soils cover only 8% of the area, in the María, Estancias, Gigante, Almirez and Torrecilla mountains. At higher altitudes in all the mountains, mainly on south-facing slopes and areas denuded by erosion, are lithosols with profiles less than 10 cm deep. Locally other soils associated with particular topographic, hydrologic or microclimatic conditions occur, such as the rendzinas in the Sierra Espuña, ruvisols in the Sierra de Estancias, castanosems in Carrascoy and solonchaks in the Guadalentín Valley.

Soil aggregation

L. H. Cammeraat

In the Guadalentín Basin three research areas were selected to study the hierarchical linkages of processes of land degradation (Imeson *et al.*, 1994, 1995).

Research sites were selected to represent the main lithological units of the Guadalentín Basin and vegetation covers, i.e. areas with shales and slates (Rambla Chortal), marls (Cañada de Cazorla) and limestone, and related pediment surfaces (Sierra de Madroño). Hillslope transects were established covering areas with different land use or land-use history. For each transect the soil aggregate distribution, stability and microbial biomass were determined seasonally, as these parameters show temporal fluctuations (Cerdà *et al.*, 1994). Infiltration was also determined by rainfall experiments on large and small scales.

Soil aggregation, in particular, is considered a key parameter (Cammeraat and Imeson, 1995) that affects the infiltration and water-retention capacity of the soil. It is also easy to measure and is strongly related to both substratum and ecological parameters, like vegetation and soil faunal activity. Furthermore, the spatial distribution of vegetated patches in relation to bare areas, to a great extent determines the spatial distribution of soil aggregation and infiltration. This in turn determines where overland flow, and therefore erosion, occurs. The aggregate size distribution directly influences the pore space distribution of the soil. Larger soil pores will increase the magnitude and intensity of infiltration along preferential flow paths, instead of superficial wetting of the soil matrix for soils with smaller soil pores. The first situation is more favourable for plants as eventually more water will be available.

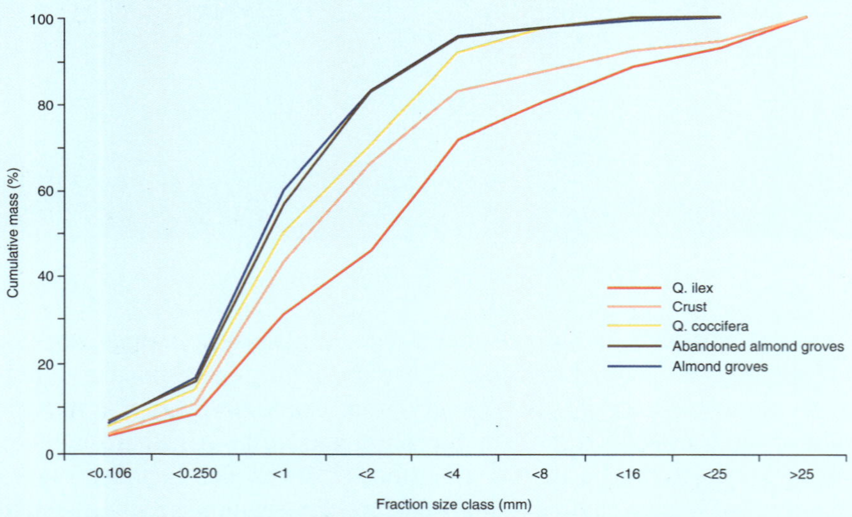

Figure 5.1.13 Aggregate size distribution under different soil cover types in the Cerro de Jarita area.

Differences in aggregate distribution have emerged for two locations, highlighting the difference among bare surfaces and several vegetation cover types.

The first area (Cerro de Jarrita site in the Rambla Chortal) is increasingly being planted with almonds. However, remnants of semi-natural vegetation are still present, either on south-facing slopes with very shallow soils, or on remote steep locations. Soil samples were collected in a semi-natural area and different types of vegetation cover on slates and greywackes. The semi-natural vegetation cover consists of an open holm oak woodland with patches of long-abandoned almond groves. It has been observed that (1) the mass of coarser aggregates is higher under semi-natural vegetation (crust versus holm oak), and (2) there is a shift to coarser soil aggregates due to recent tillage.

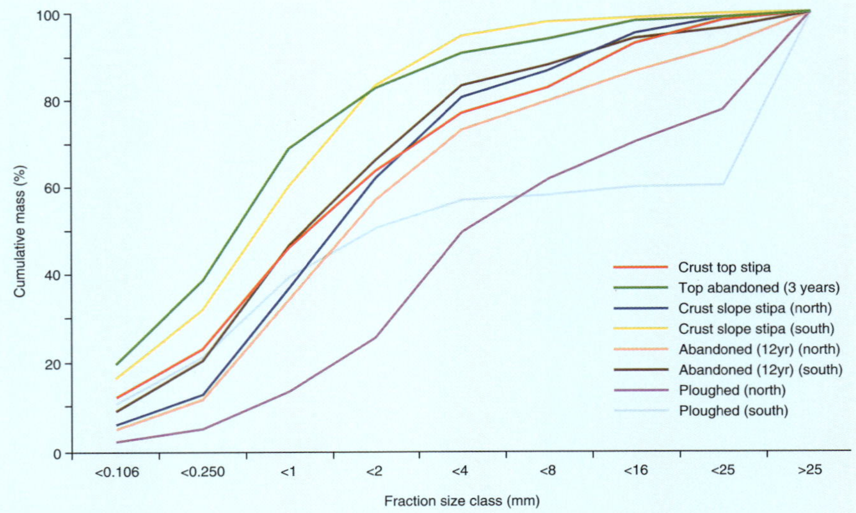

Figure 5.1.14 Aggregate size distribution from different soil cover types and aspects in the Cañada Cazorla area. Differentiation is made between southern and northern facing slopes, and flat surfaces.

In the second area (located in the vicinity of Cañada de Cazorla), characterised by marl lithology, the effect of both soil tillage and aspect was studied for different vegetation covers. Aspect is important as soil aggregation is influenced by differences in solar radiation, especially in the winter. It has been observed that more coarse aggregate sizes occur in ploughed and north-facing areas.

Vegetation

The Guadalentín Basin is located in the Mediterranean Botanical Region, which, despite variations due to altitude, aspect and geological substrata, is dominated by aridity as the most relevant factor in vegetation distribution.

The climax arboreal vegetation is evergreen oak woodland dominated by *Encina* (*Q. ilex* L. subsp. *rotundifolia* [Lam.] Schwartz ex T. Morais). It is typically limited to the Guadalentín Basin because of the aridity. These woodlands reach only limited height, and they typically have four layers (tree, shrub, grass and moss). They are dense in the most favourable sites but more and more open as they approach the bioclimatic limit of their natural distribution. This limit is generally about 350 mm average annual precipitation, so evergreen oak woodlands are found above 600–800 m a.s.l. *Encina* woodlands used to be dominant and after their degradation different retrogressive vegetation series with distinct floristic composition developed, depending on altitude, precipitation, location and substrate. However, the dynamics and stages of degradation of the various soil series are structurally similar.

Pine forests (*Pinus pinaster* Aiton, *P. nigra* Arnold, and mainly *P. halepensis* Mill.) are important in sites with shallow soils. Re-afforestation began in 1889 (Sierra de Espuña) with the intention of restoring and, in some areas, increasing

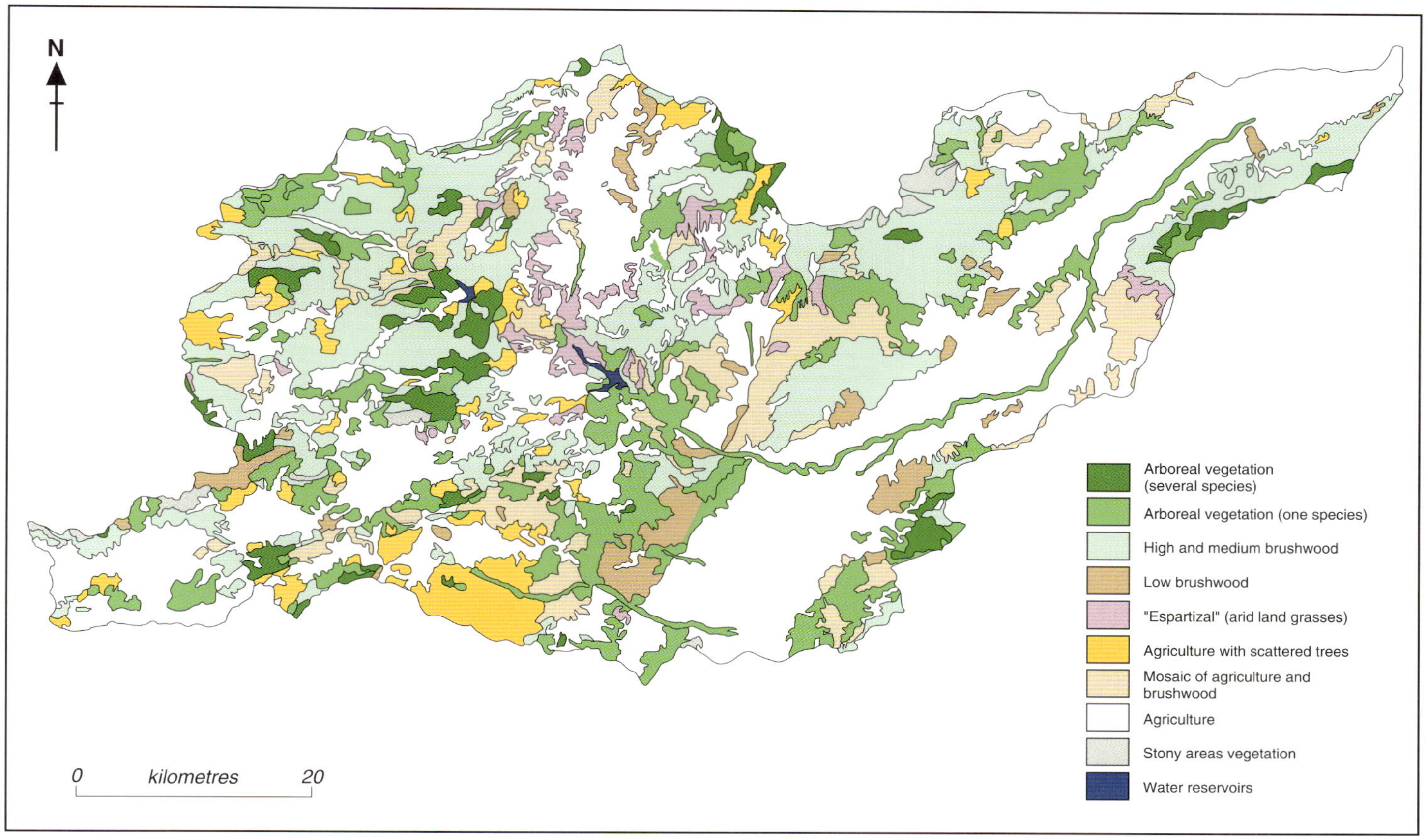

Figure 5.1.15 Actual vegetation and land use in the Guadalentín Basin. The map has been compiled by L. Rojo-Serrano, J. A. Martinez-Artero, F. García-Robredo and A. Martinez-Ruiz of the ICONA. Cartographic base: Mapa topografico SEG (1:200 000). (Source: Mapa Forestal de España, ICONA, 1990, sheets Baza (6–10) and Murcia (7–10).)

beyond the area of original pine forest. These forests are currently the dominant arboreal vegetation in the mountains above 800–1000 m a.s.l.

Below 600–800 m a.s.l., the potential vegetation would be a more or less dense, *maquis* type (coscojares) dominated by *Coscoja* (*Quercus coccifera* L.), *Espino negro* (*Rhamnus lycioides* L.), *Acebuche* (*Olea europaea* L. var. *sylvestrys* Brot.) and *Lentisco* (*Pistacia lentiscus* L.).

Throughout the whole Mediterranean Basin, ancient and intense human impact has strongly modified the original vegetation. In the Guadalentín too, the majority of the original vegetation disappeared a long time ago, particularly at the lower and middle altitudes where most of the land has been used for thousands of years. Natural vegetation is replaced by crops, grazing lands, and vast areas no longer used for agriculture, where the vegetation is recovering from abandonment and is in various stages of regeneration. Here progressive development of the vegetation has continuously been slowed down or arrested by new types of land-use activity.

The most extensive form of degraded semi-natural vegetation is the *Espartal,* a steppe with *Esparto* (*Stipa tenacissima* L.) and other similar grasses, because this species occurs in all the different stages of the vegetation degradation in the Guadalentín. In the uplands of the eastern part of the basin the *Retamar*, from *Retama común* (*Retama sphaerocarpa* [L.] Boiss.), along with *Aliaga* (*Genista scorpius* [L.] DC in Lam et DC), appears in the first stage of degradation of the original of *Encina* woodland. If degradation continues, *Esparto* appears. *Tomillar*, from *Tomillos* (*Thymus* spp.), is also common in the Guadalentín Basin. This is a garigue-type vegetation dominated by dwarf bushes; typical component species of this community are *Albaida* (*Anthyllis cytisoides* L.), *Tomillo* (*Thymus hyemalis* Lange), *Espilego* (*Lavandula dentata* L.) and, in the most degraded areas, *Boja* (*Artemisia barrelieri* Besser and *Artemisia campestris* L.), *Escobilla* (*Salsola genistoides* Juss. *ex* Poiret.), and *Bolaga* (*Thymelaea hirsuta* [L.] Endl.). *Tomillar* occupies the most degraded, arid and unstable localities and, in very nitrogen-poor soils, its species act as colonisers, after or even simultaneously with the pioneer therophytic phase.

Environmental processes and natural hazards

There are four major geomorphological regions:

1. The floor of the Guadalentín Valley, formed by the graben whose characteristics are outlined above;
2. Talus cones and debris flows, such as the alluvial fans of the Carrascoy and Espuña mountains and the extensive calcified pediments dissected by drainage network in the foothills of the Madroño, Pericay, Gigante and María mountains;
3. Interior basins, filled with soft materials (sandstones, marls and clays) and deeply eroded into badlands, e.g. near the Puentes reservoir or in the rambla of Algeciras;
4. Mountains, with upland processes and forms, especially on limestone terrains, such as karren, gorges and springs. In areas such as the María, Gigante and Espuña mountains, debris flow and periglacial deposits occur on very steep areas of north-facing slopes.

Soil erosion

Erosion is one of the most important processes of soil degradation in the Guadalentín Basin because of its agricultural and environmental implications. It is a chronic, dynamic and complex problem.

According to applications of the Universal Soil Loss Equation (USLE) by F. López-Bermúdez,

- only 15% of the territory has slight or no sheet erosion and a stable drainage system;
- significant sheet erosion, with rills and small gullies, and poor drainage occurs in 28% of the area;
- soil erosion with big gullies and estimated soil losses of between 50 and 100 t/ha/year affects 40% of the basin;
- high erosion with widespread well-developed big gullies, very active drainage and soil losses between 100 and 250 t/ha/year, occur in 15% of the area;
- very high erosion with significant mass movement and soil losses over 250 t/ha/year affects 2% of the area.

These figures must, of course, be treated with great caution. Nevertheless, they do give some idea of the relative severity of the erosion problem, which is supported by sediment transport rates and reservoir siltation.

Characteristically there is a sharp contrast between extensive badlands and cultivated areas with little or no erosion (cf. also Section 4.3). Moreover, there is growing evidence that some of the gullied areas have little erosion for very long periods of time and enormous erosion rates in very short, very infrequent, very intense rainfalls. It is therefore a question of the magnitude and frequency of erosional events.

However, serious erosion problems do affect marginal agricultural drylands with gradients over 12% where no soil conservation measures are implemented, and areas with sparse shrub vegetation, poor and shallow soils and unresisting lithologies with slope gradients over about 25%.

Figure 5.1.17 Valdeinfierno Dam full of sediments (March 1994). Of its initial capacity, 25 Hm^3, only 1 Hm^3 remains.

Salinisation

Salinisation in the Guadalentín Basin has increased over time because of the high salt content of water in areas that have been exploited for a long time. This is caused by the increase of free carbon dioxide (CO_2) liberated by falling water-tables. An increase in temperature and in the content of bicarbonate and sulphate ions also occurs.

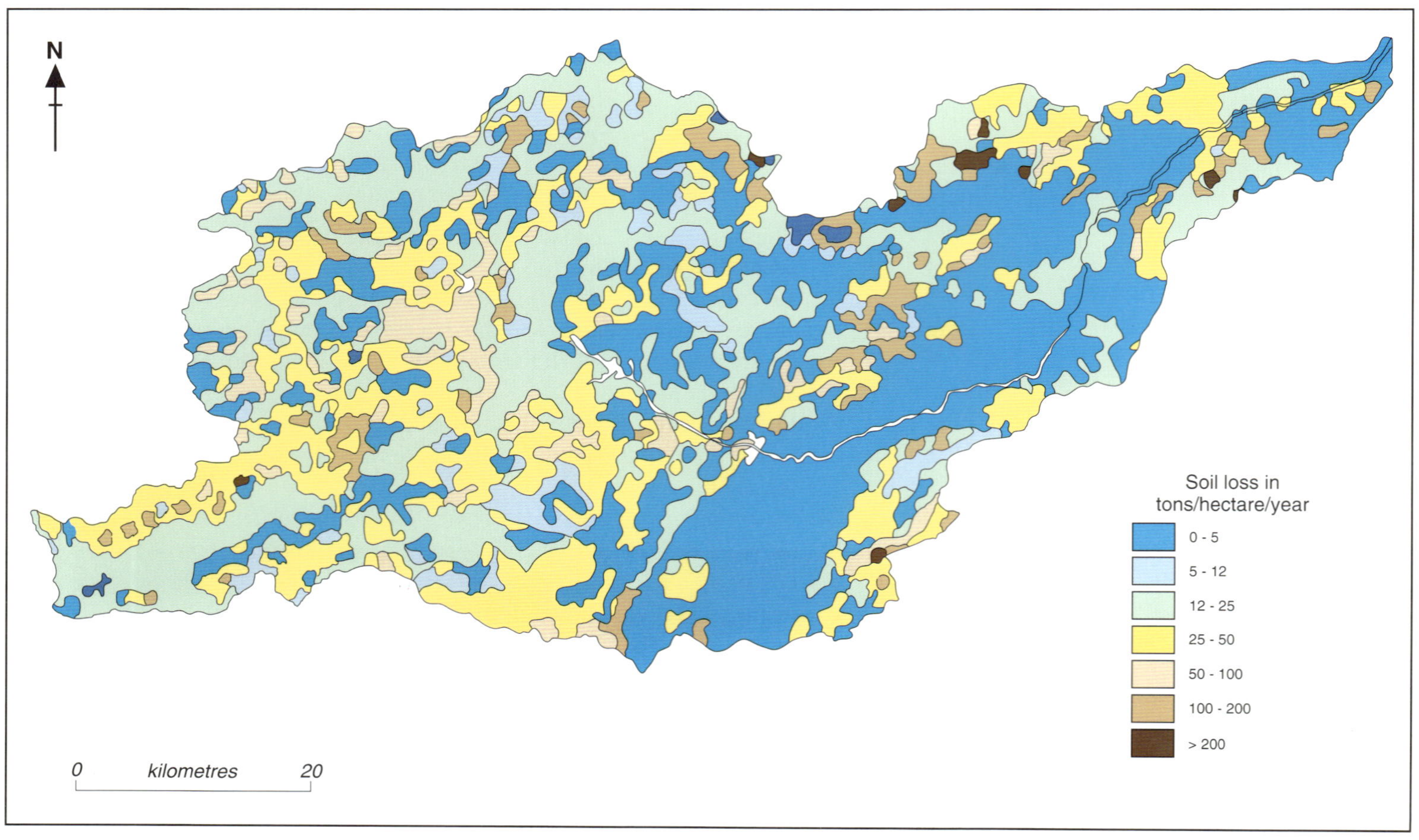

Figure 5.1.16 Map of rainfall erosion. The map has been compiled by L. Rojo-Serrano, J. A. Martinez-Artero, J. García-Robredo and A. Martinez-Ruiz of ICONA. Cartographic base: Mapa topografico SEG (1:200 000). Source: Mapa de Estados Erosivos, ICONA, 1988, Cuenca Hidrografica del Segura.

Figure 5.1.18 Soil salinisation in citrus crops due to irrigation with water of bad chemical quality.

Figure 5.1.19 Impacts of the flood of October 1973 in Puerto Lumbreras.

Contamination of agricultural soils is indicated by a dark ferric carbonate crust that obstructs root aeration. Moreover, the chemical reactions between the soil salts and the bicarbonates below a particular depth, generate compounds (e.g. sodium carbonate) that are highly noxious to plants, leading to a lower productivity.

Floods

The Guadalentín River is usually dry because of the very low rainfall and evapotranspiration in the basin. However, the torrential character of the rainfall, the sparse vegetation cover and the dendritic shape of the systems of tributaries in the headwaters, as well as the contributions of gullies and ramblas along the main channel, result in the simultaneous arrival of flood waves at the confluence of the Guadalentín and Segura rivers. This gives rise to extremely severe floods downstream. The 'Santa Teresa' flood in October 1879 and the flood of October 1973 are the most important on record. The latter produced peak flows of 2200 m^3/s in the Guadalentín River and 2000 m^3/s (of which 800 m^3/s was transported debris) in the Rambla Nogalte. Two reservoirs were constructed in the Guadalentín river basin to contain floods: the Valdeinfierno in 1806 and Puentes in 1884. The Pareton deviation dam to the Mediterranean Sea was also constructed. The new Puentes and Romeral dams in the Guadalentín River, and the Algeciras dam in the Algeciras gully are currently under construction.

Selection of representative sub-basins for detailed research

Due to the size of the Guadalentín Basin it is practically impossible to conduct detailed research across the entire catchment. Therefore a representative population of sample sub-basins has been identified from readily available information relevant to hydrogeology and erosion (e.g. a digital elevation model (DEM), a digitised geological map and climate records). The DEM was used to obtain a quantitative description of the topography of each grid cell in terms of slope angle, slope aspect, specific catchment area, slope length, and potential insolation. Using multivariate statistics on these terrain attributes, and the map of estimated mean annual precipitation, the 165 000 grid cells of the

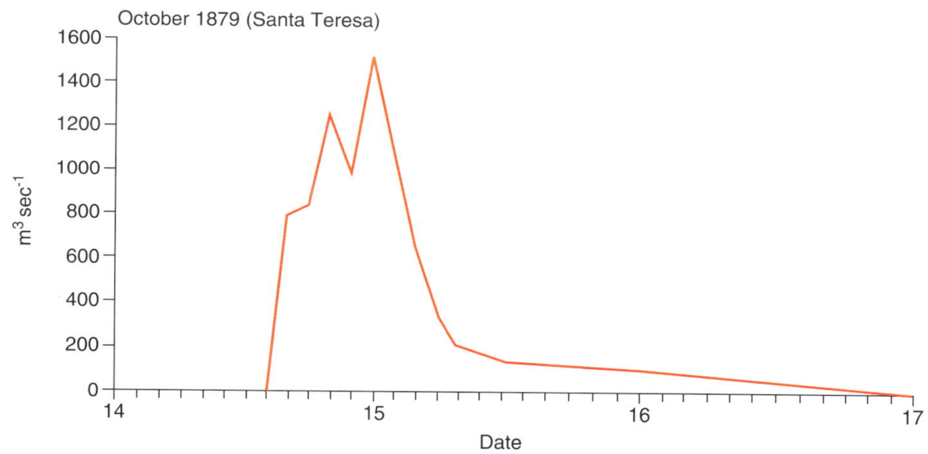

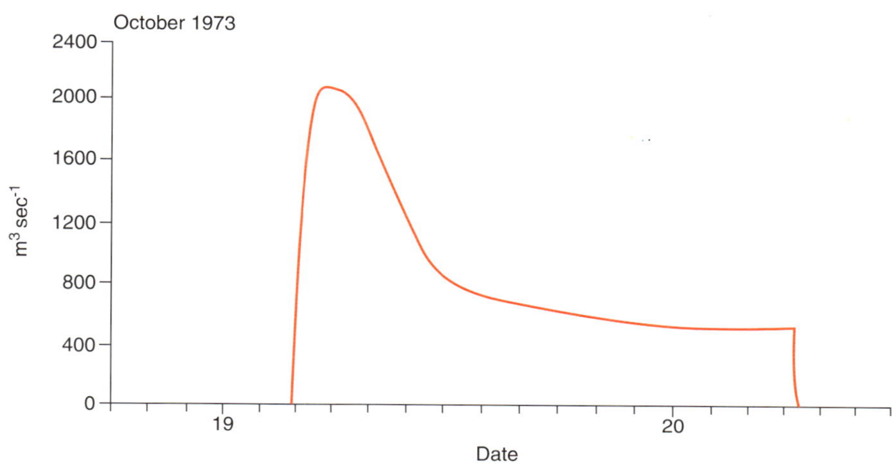

Figure 5.1.20 Flood hydrographs of the Guadalentín River at the Puentes reservoir for October 1869 (Santa Teresa) and October 1973.

Figure 5.1.21 A representative landscape view at the headwaters of the Guadalentín Basin.

Guadalentín were clustered into 10 groups, representing the main physiographic units in the basin.

A sub-basin map was also obtained from the DEM. First, a drainage network was derived from the map using a threshold of 5 km^2 to define the unit. Then a drainage algorithm was used to delineate the corresponding sub-basins and to compile a sub-basin map. This was then combined with a relief map and climatic data to provide the relative frequency of the broad landscape units within which sub-basins were clustered into seven main groups. These, in turn, were combined with a geological map, and one sub-basin (such as the Rambla del Chortal; see box) for each lithological group was selected for further research.

The impact of climate change on land abandonment in Rambla del Chortal

J. Obando

The impact of climatic change on land abandonment in Rambla del Chortal is being investigated using a simple model that assumes water to be the limiting factor in vegetation growth in semi-arid areas. The available soil moisture depends on the rainfall regime and the amount of vegetation. Vegetation growth is modelled as a function of evapotranspiration and water use efficiency. Evapotranspiration is calculated using the soil moisture deficit and the monthly potential evaporation values. Overland flow occurs when rainfall intensity exceeds the infiltration rate, or when the soil store is saturated. Erosion on the abandoned fields is calculated annually as a product of the annual overland flow and a reduction factor based on the vegetation cover.

Varying rainfall determines the success or otherwise of regeneration, the development of vegetation, and the rates of erosion on the abandoned land. The simulation of vegetation recovery for two different rainfall series follows a logistic curve and demonstrates the dependency of the rate of recovery on the rainfall. The long-term patterns of vegetation recovery simulated by the model reflect the amount of rainfall. For fields abandoned during dry periods (e.g. 1958 and 1961) the initial increase in vegetation cover is much less, and it takes much longer for a 'steady state' cover to be achieved compared with those abandoned in the wet periods (e.g. 1969 and 1971). 'Steady state' cover depends on the rainfall amount and history, and tends to fluctuate with the precipitation. The vegetation recovery curve obtained from measurements on abandoned land in Rambla del Chortal is consistent with these results.

Figure 5.1.22 Southern part of the Rambla del Chortal catchment.

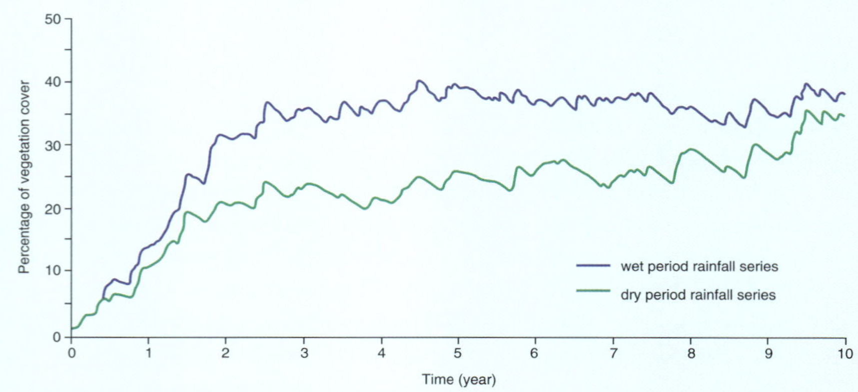

Figure 5.1.23 Simulation of vegetation recovery.

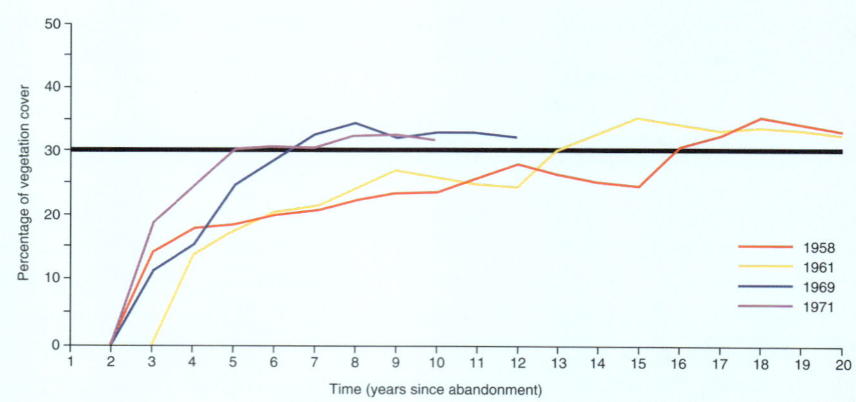

Figure 5.1.24 Pattern of vegetation recovery. Black line represents the 'steady state' cover for this rainfall series.

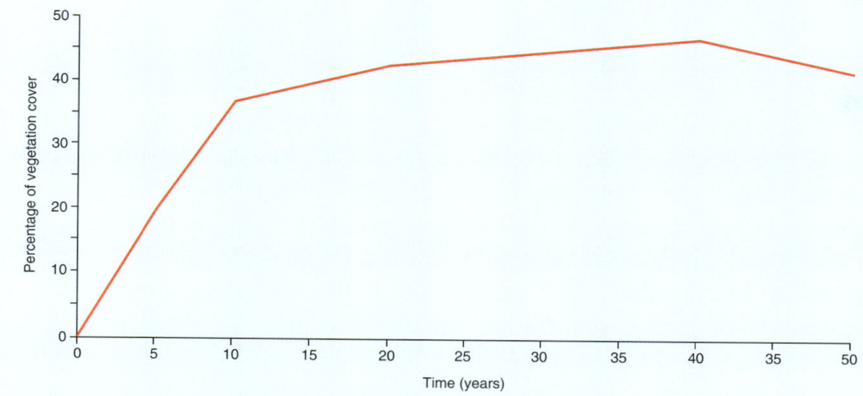

Figure 5.1.25 Vegetation cover after abandonment (field data).

Effective control of the erosion, and hence of the desertification problem, relies a great deal on the understanding of the interactions between the vegetation and the rainfall in these semi-arid areas. The total annual erosion depends on the overland flow and also varies with the amount of land abandoned in any particular year. By determining the areas of fields abandoned at different times and the different stages of recovery of these fields, the amount of sediment eroded can be estimated. Using the model a decrease is estimated in the sediment from the bare area (B) and in total sediment (T) over a period of 20 years of uniform abandonment for a catchment of 20 km^2. The sediment yield from the abandoned fields (A) is much lower after 20 years than would have been the case in a completely bare catchment. A simple model such as this one can be used for planning and managing the 'set-asides' in semi-arid areas within a changing climate without adversely affecting those already vulnerable areas.

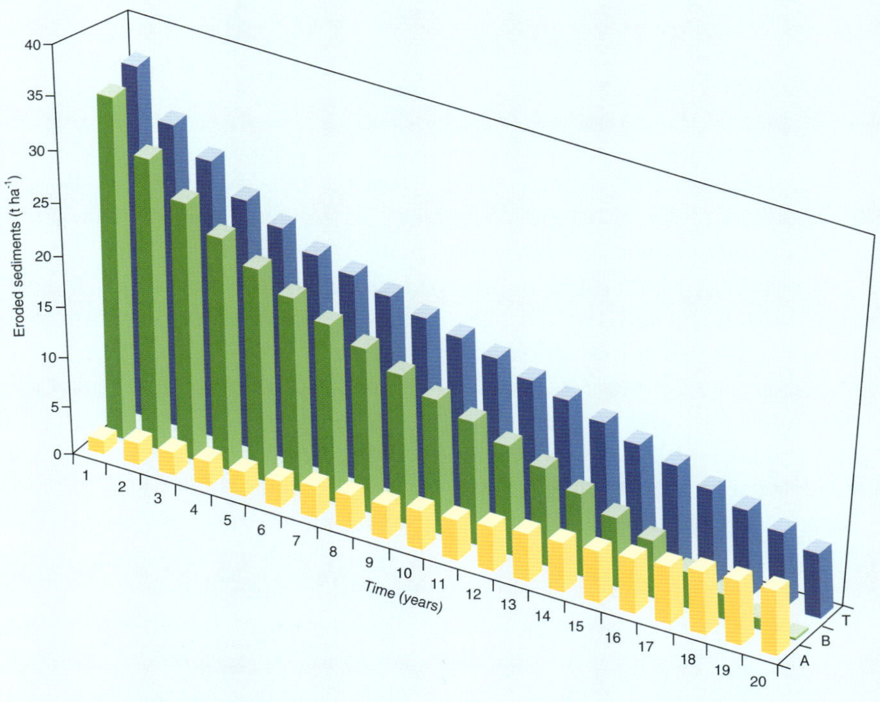

Figure 5.1.26 Model estimate of sediment eroded.

Parametrising the MEDRUSH model in the Guadalentín Basin

J. C. Hawkes

A basic assumption in the MEDRUSH modelling procedure is that field data collected from the Guadalentín can be used to derive predictions for the whole basin. In order to have confidence in these predictions, it is necessary to have a carefully designed sampling procedure. Thus, in the two field campaigns carried out in the Guadalentín, hillslopes were selected from the sub-basins identified by del Barrio and Böer (cf. previously in this Section) as being physiographically characteristic of the entire basin. Hillslopes were then selected to include a representative range of lithology and land use.

Since the model is very generalised, the aim of the field exercise was to capture the main environmental parameters not available from published data. Point measurements, such as stone cover, surface roughness and infiltration were taken, and the slope geometry of each site was determined, as a test of the theoretical basis of the model.

Results from the soil surface data suggest that predictions made by the model will dominantly reflect lithological and land-use history controls. However, local spatial variability means that there is likely to be a significant amount of uncertainty around the predicted values. This can be demonstrated by reference to the surface roughness data. For example, although surface roughness varies significantly with lithological class, there is still a large overlap between them. The importance of this uncertainty is dependent on the sensitivity of the model.

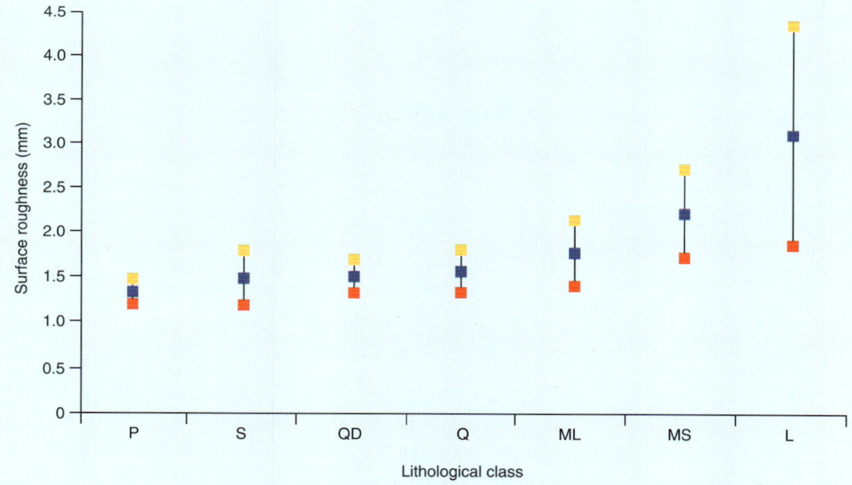

Figure 5.1.27 Variations of surface roughness with lithology, with within-class variance. Surface roughness is critical in the MEDRUSH model because it is an important control on sediment transport by overland flow and rill-wall collapse. In a changing erosional environment, it constantly adjusts to changing conditions (Kirkby *et al.*, 1994). P, Phyllites; S, sandstones; QD, Quaternary deposits; Q, quartzites; ML, marls and marly limestones; MS, mica schists; L, limestones.

The slope geometry (strip) measurements made in the field generally support the hillslope theory which underpins the MEDRUSH model. In theory, gradient increases along with the contributing area/unit width (a) in the convex part of the slope near the drainage divide, whereas the gradient (s) decreases relatively on the concave part of the slope near the

Figure 5.1.28 A characteristic strip on dolomite in the Guadalentín. The lines represent the changing area of the strip as the slope gradient (*s*) changes with distance from the top of the slope. The area (*a*) increases markedly in the long concavity leading down to the stream bed.

stream channel. This slope profile can be represented by a logged value of *a*/*s*, plotted against slope gradient (cf. Kirkby *et al.*, 1994). Log (*a*/*s*) continuously increases as the discharge load on the system increases. The gradient increases in the convex part of the slope near the drainage divide, and then decreases in the concave part near the stream channel.

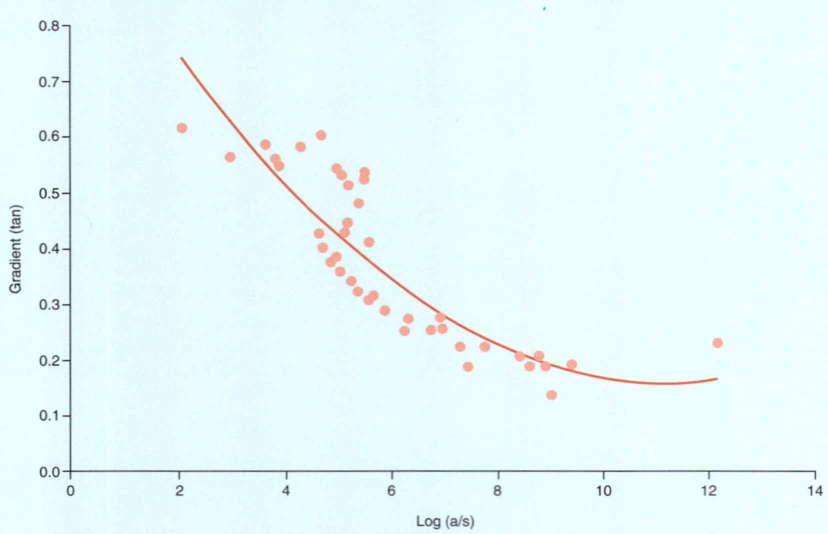

Figure 5.1.29 Slope profile in terms of log(*a*/*s*) for the hillslope shown in the photograph. The gradient decreases as log(*a*/*s*) increases, indicating that it is a long concave slope.

Population and land use

The Guadalentín Basin has been subjected to heavy population out-migration to other parts of Spain and to foreign countries. This movement has clearly influenced the evolution and present status of the population in the Guadalentín. The population showed a moderate but sustained increase from the beginning of the 18th century until the mid-19th century, when the growth accelerated probably because of the rise in mining. However, this increase was shortlived and by 1900 the population began to decrease again due to strong emigration. During the 1920s and 1930s, a further increase in mining, the development of industry and good agricultural prices reversed the trend. In subsequent decades, however, population has decreased markedly by emigration. Once more, from 1970 onwards population increased again and in 1990 population amounts to 113 000 inhabitants in the five municipalities of the Guadalentín Basin; Lorca being the largest, with 64 361 inhabitants.

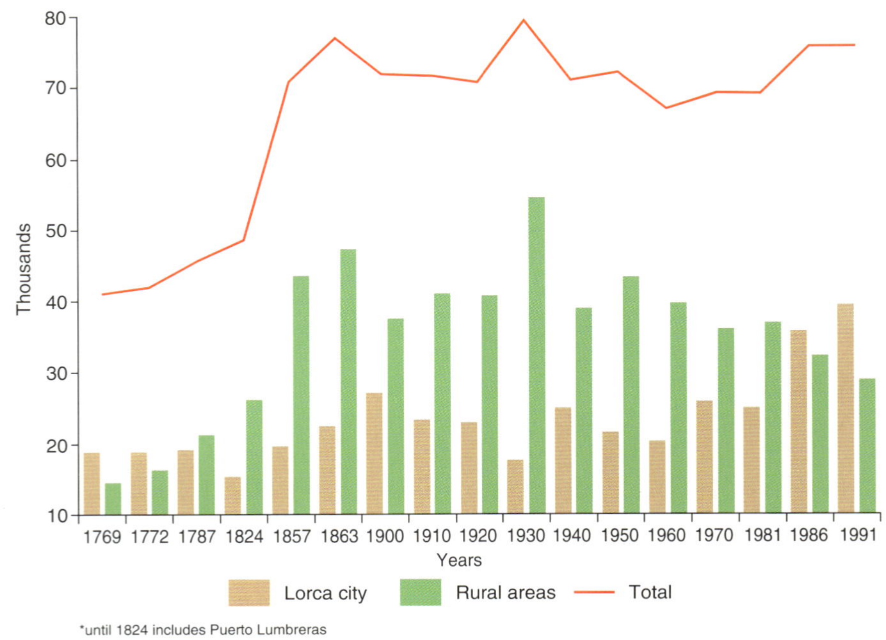

Figure 5.1.30 Population change in the municipality of Lorca over the period 1769–1991.

The expansion of irrigation, industry and the tertiary sector are the three basic elements of the development strategy in the Guadalentín Basin and the cause of population increase due to lower emigration rates.

Present climatic conditions in the basin barely favour tree crops. In such a harsh environment, any disturbances to the vegetation cover have important implications for soil conservation and protection.

At present, agricultural lands take up the major share of the total area (49%). Land where fallow and rain-fed agriculture of low income and uncertain profits is practised *(secano),* occupies 30% of the basin. Irrigated agriculture (*regadío*), with either permanent or arable crops, occupies 19.5%.

The area occupied by shrublands and open pine forests amounts to 31% of the total area, while other land uses (open spaces with virtually no agriculture, forests, grazing land and urban areas) occupy the remaining 20%.

The high moisture stress on the plants is caused by high evapotranspiration and insufficient rainfall except in occasional wet years, so irrigation is essential to assure a good harvest. Therefore, a large proportion of land has not enough vegetation cover as a result of a long regime of over-exploitation and disturbance of the natural vegetation by timber extraction, grazing and fire. This not only favours

Figure 5.1.31 Boquera. A system of channels to take flood flow to dry crops on terraces.

Table 5.1.1 Different land uses (in hectares) in some municipalities of the Guadalentín Basin in 1991.

Municipality	Non-irrigated land	Irrigated land	Forest/ land	Uncultivated shrubs	Other land	Total land
Aledo	1 907	432	1 601	200	875	5 018
Alhama	8 403	8 100	9 180	400	5 100	31 183
Librilla	702	3 158	1 223	0	585	5 668
Lorca	55 107	22 187	53 651	4 000	32 755	167 700
Puerto Lumbreras	4 377	5 270	2 373	0	2 332	14 352
Totana	4 487	10 133	9 909	600	3 638	28 767
Total	74 983	49 280	77 937	5 200	45 285	252 688

Source: Agricultural Regional Statistics (1991), Murcia.

Figure 5.1.32 Re-afforestation of *Pinus halepensis* Mill. with the terrace system on Muela Hill (Alhama de Murcia). Re-afforestation programmes have recovered and expanded the natural area of *P. halepensis*, which presently is the dominant vegetation in the mountains, over 800–1000 m a.s.l.

strong soil erosion, but also contributes to an increase in flood magnitude, which ultimately provokes large economic losses (including the filling of reservoirs) due to the sudden mobilisation of huge amounts of water and solid materials.

Until 1970, traditional and dominant crops (in terms of area) were cereals, olive and almond trees and vines; this was the traditional agriculture. During the last 25 years major land-use changes have occurred in the Guadalentín Basin as a consequence of intensification of agriculture and animal husbandry (pigs) activities, which also relate to the decrease of emigration.

The introduction of intensive crops, either in the field or in greenhouses (e.g. artichoke, cotton, pepper, tomato, flowers), fodder crops (alfalfa), tree crops (citrus and almond), besides having a great impact on land-use pattern, has also created serious water-management problems. Most of these crops, in fact, need to be irrigated. The increased demand for water supply has resulted in the over-exploitation of groundwaters on the one hand, and in the need for the importing of water from other river basins (such as the construction of the Tajo–Segura–Guadalentín channel), on the other. However, when long droughts affect the south of the Iberian Peninsula, the water disappears from the fluvial network of the Guadalentín, transfer from the Tajo is practically limited to domestic and industry supply, wells are exhausted, and as a consequence in those periods, a shift occurs, within the agricultural areas from irrigated to non-irrigated use.

In synthesis, the Guadalentín Basin is an extremely stressed system affected by serious environmental problems and a water deficit that renders unsustainable human activities oriented towards short-term socio-economic development.

Suggestions for a sustainable development

The satisfaction of human needs and aspirations is the major objective of development. Sustainable development is not a fixed state of harmony, but rather a process of change in which the use of resources, the direction of investments, the aims of technological development, and institutional change are made consistent with present as well as future needs. This process is not easy or straightforward and sustainable development must rest on political will.

The Guadalentín Basin is an example of Mediterranean environmental stress, reflecting physical and human pressures that are the basis of social conflict and unsustainable development. Water scarcity is the most pressing problem in the Mediterranean lands of southern Spain. It has been, and continues to be, the limiting factor to development, and one of the causes of land desertification.

The basin is a territorial unit and a system of dynamic and mutual interaction between nature and society. It is affected by very severe problems of soil and ecosystem degradation because of the physical and biological conditions and the systems of exploitation of soil, water and vegetation.

The complex causes of land degradation in the basin make them difficult to evaluate. However, aridity, drought recurrence, rain erosivity, soil erodibility, groundwater over-exploitation, soil and water salinisation, the impact of human activities such as deforestation, cultivation of marginal land, overgrazing, inadequate agricultural practices, the abandonment of lands, and soil conservation practices all play a part.

At the end of the millennium, adverse rainfall conditions and excessive human pressure on a vulnerable ecosystem and limited natural resources are more evident than ever. Thus, for example, the impact of drought, a normal and unavoidable event in the climatic regime of arid regions, is more severe than before because it affects a larger number of people and more demanding and complex economic activities, particularly intensive agriculture and industry.

Drought is a very important component of desertification as The International Convention to Combat Desertification in Countries Experiencing Serious Drought and/or Desertification (United Nations, 1994) makes clear. In fact, it was during the prolonged and historic droughts (1911–1917, 1925–1929, 1935–1941, 1944–1945, 1950–1956, 1981–1984 and 1993–1995)

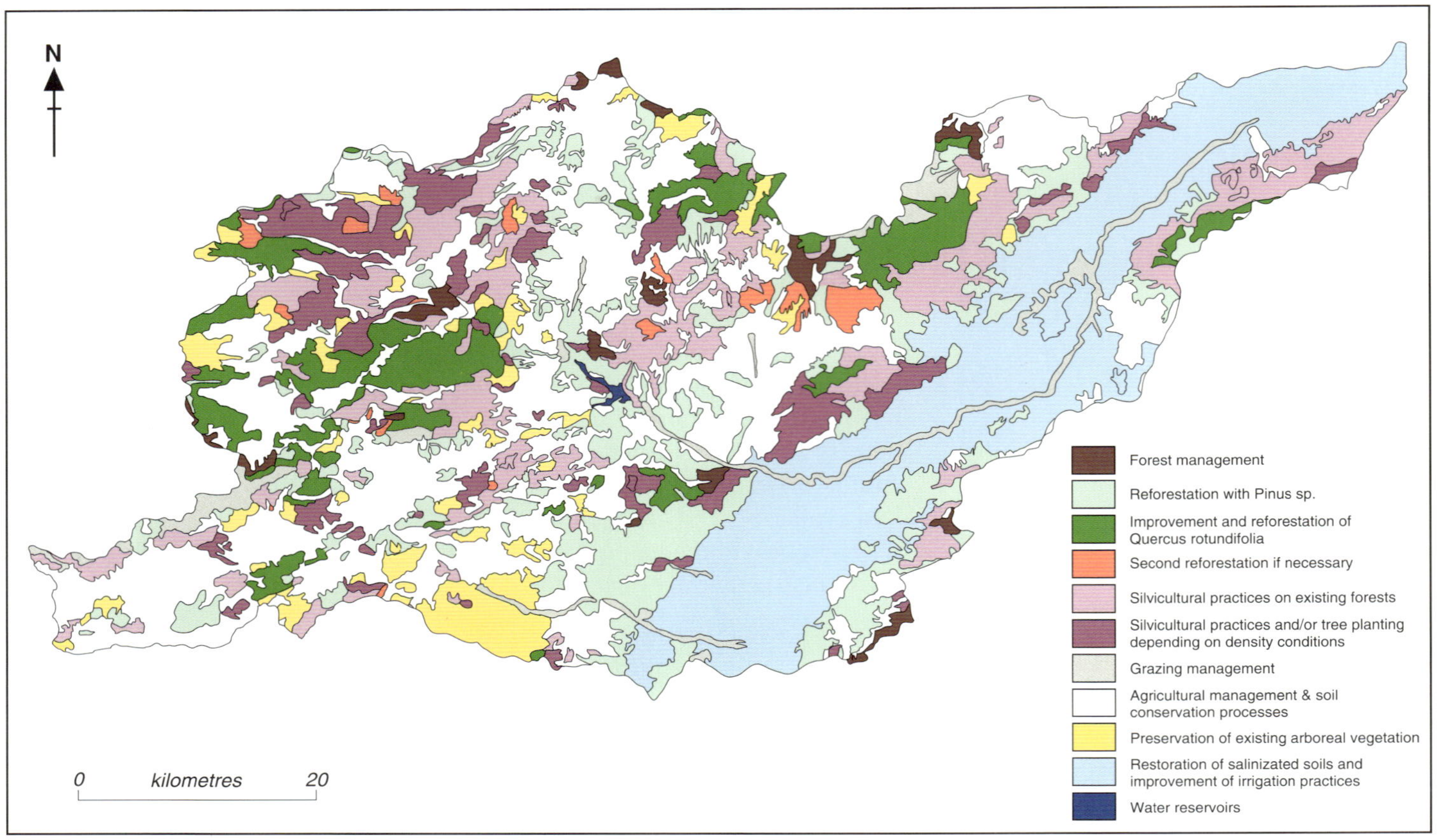

Figure 5.1.33 Map of soil and vegetation restoration and management. The map has been compiled by L. Rojo-Serrano, J. A. Martinez-Artero, F. García-Robredo and A. Martinez-Ruiz of ICONA. Cartographic base: Mapa topografico SEG (1:200 000). (Source: Mapa Forestal de España, ICONA, 1990, sheets Baza (6–10) and Murcia (7–10).)

Figure 5.1.34 Lorca city looking towards the south.

that desertification progressed by pulses. Nevertheless, droughts alone are not the core of degradation process. In arid environments, in the absence of human pressure, ecosystems have the necessary resilience to overcome these climatic crises. They have done so for centuries, or even millennia in some places, because they are adapted. It is the human beings that have created the conditions leading to desertification, and dry climates and drought recurrence are simply favourable circumstances. For all of that, reducing the vulnerability of society and natural systems to drought is an urgent task and a social need. To mitigate drought and combat desertification is to invest in the future and to contribute to sustainable development in the Guadalentín Basin.

La sequía y los incendios pueden convertir la Región en un desierto

MURCIA

Los regantes temen este año la «ruina total» por falta de agu

17.000 hectáreas de regadíos se perderán este año por la sequía

MURCIA

LA SEQUIA NO CESA

A faltá de agua, cascada de ideas

La Confederación triplica las licencias para abrir pozos en la cuenca

COMUNIDAD

RECURSOS HÍDRICOS

Los regantes demoran la petición de un trasvase y pronostican una «ruina total»

«No pediremos agua del Tajo cuando quieran los políticos, sino cuando lo creamos necesario»

Seis millones de españoles en la mitad sur padece restricciones en los abastecimientos de agua

EL DESIERTO LLEGA A LA REGION

Desesperada búsqueda de agua

40.000 cabezas de ganado pueden morir de hambre y sed

«Sin agua no vamos a ninguna parte»

La España reseca

La mitad sur del país pasará la Navidad con graves restricciones de agua debido a la falta de reservas

La polémica

Los efectos de la sequía

La falta de agua y los incendios empujan hacia la desertización

Expertos en gestión ambiental reunidos en Murcia auguran serios problemas para el medio natural si no se toman medidas urgentes

Figure 5.1.35 Contemporary trends and conflicts.

5.2 The Agri Basin, Basilicata, Italy

F. Basso, E. Bove, M. Del Prete and M. Pisante

The Agri Basin is located in the Region of Basilicata in southern Italy. Except for a few specific localities, it is economically and socially one of the less favoured areas of Europe and is suffering from a land degradation problem that arises from its seasonal extremes of climate and the nature of the geology, with its unstable marine clays and easily eroded sands and conglomerates. It is a climatically marginal region, parts of which have already experienced desertification and parts of which are threatened by land abandonment and the potential results of global warming. For this reason it was selected as a Target Area for investigation in MEDALUS II and will continue to be researched in MEDALUS III.

The Agri begins its 120 km course on the slopes of Lama Mountain (1200 m a.s.l.) in Potenza Province and stretches across Basilicata from north-west to south-east. Its course is interrupted by artificial reservoirs, the Pertusillo and Gannano, and the Marsico Nuovo, still under construction. It reaches the Ionian Sea in the Province of Matera. The catchment is 1700 km^2 comprising 17% of the region of Basilicata and a small part of neighbouring Campania. Twenty of its municipalities are in Potenza Province and nine in Matera Province, together making up about 20% of the area of Basilicata. The average population density is approximately 48 inhabitants/km^2.

The Agri is divided into the Upper, Middle and Lower Agri by both physical–environmental and socio-economic criteria. The Upper Valley, above the Pertusillo reservoir, has an average altitude above 600 m, an area of just under 60 km^2 and is characterised by a valley-floor plain. The plain is the former Reistovene lake bed and is surrounded by rugged mountains. It has a population density of 54 inhabitants/km^2.

The Middle Valley stretches from the Pertusillo dam down to the confluence of the Sauro and Agri rivers, in the municipality of Stigliano. This is an area of badlands called '*calanchi*', with a population of 31 inhabitants/km^2 at the 1991 census, and covers about 47% of the entire catchment.

The Lower Agri Valley, from the Sauro junction to the sea, is about 25% of the basin and includes the area of fertile Metaponto soils. It has a population density of 72 inhabitants/km^2.

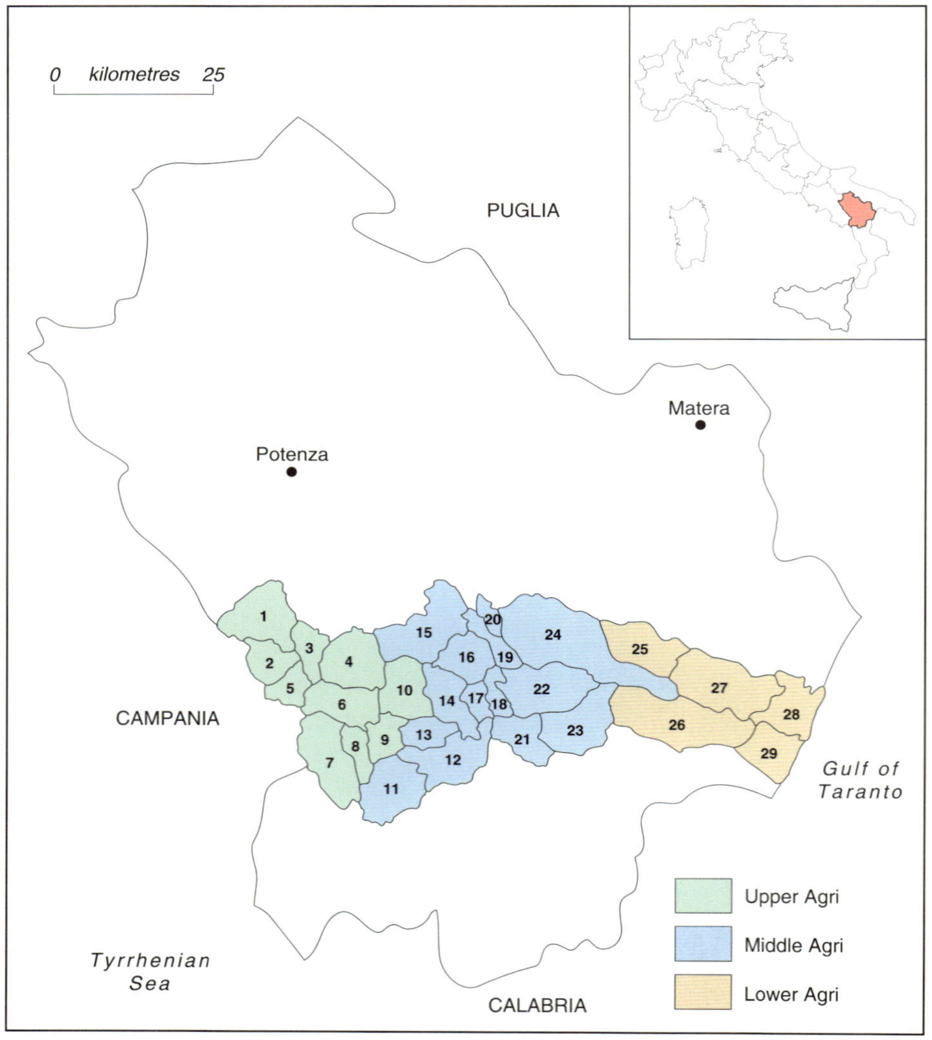

Figure 5.2.1 Location map of the study area. Redrawn from the map provided by E. Bove.

Relief

The basin is situated in the heart of the Basilicata Apennine Chain and the river rises in the springs of the Monte Maruggio and Monte Volturino and reaches the sea near Policoro. The relief of the Upper Valley is rough and indented as a result of the broken terrain of the faulted and folded limestone and dolomite. The Upper Valley has steep slopes, but the good vegetation cover restricts erosion. The Middle Valley is dominated by flysch and arenaceous lithologies, and badlands are common, though the relief is generally less dramatic than the Upper Agri. Landslides occur in the clay–marl lithologies, exacerbated by the past and present deforestation. Where Pliocene and Quaternary rocks are present, the morphology is more rounded and convex and contrasts strongly with the badland areas. In the Lower Agri, wide marine-cut platforms and terraces form the Miocene hills with their gentle morphologies interfacing with the more rugged landscapes of sands and conglomerates. Along the river Agri there is a well-marked suite of river terraces, which interdigitate with the main surfaces.

Geology

The Agri exhibits all of the main structural elements of the Apennine Chain and the Bradanic Foreland.

The Apennine Chain, present in the Upper Valley, is formed of a complex of nappes made up of numerous tectonic units derived from the fragmentation of larger palaeogeographic units. These broke up in the Miocene period to form the following main units:

1. the Liguride Complex, comprising pellites, calcareous clays and arenaceous conglomerates and flysches;
2. the Sicilide Complex, represented by the Nocara Flysch, variegated clays and the Tusa tuffs;
3. the Carbonate Complex, formed by limestones and dolomites deposited in a shallow sea platform environment;
4. the Lagonegro Complex of siltstones, loam rock, limestones, radiolarian schists and red flysch;

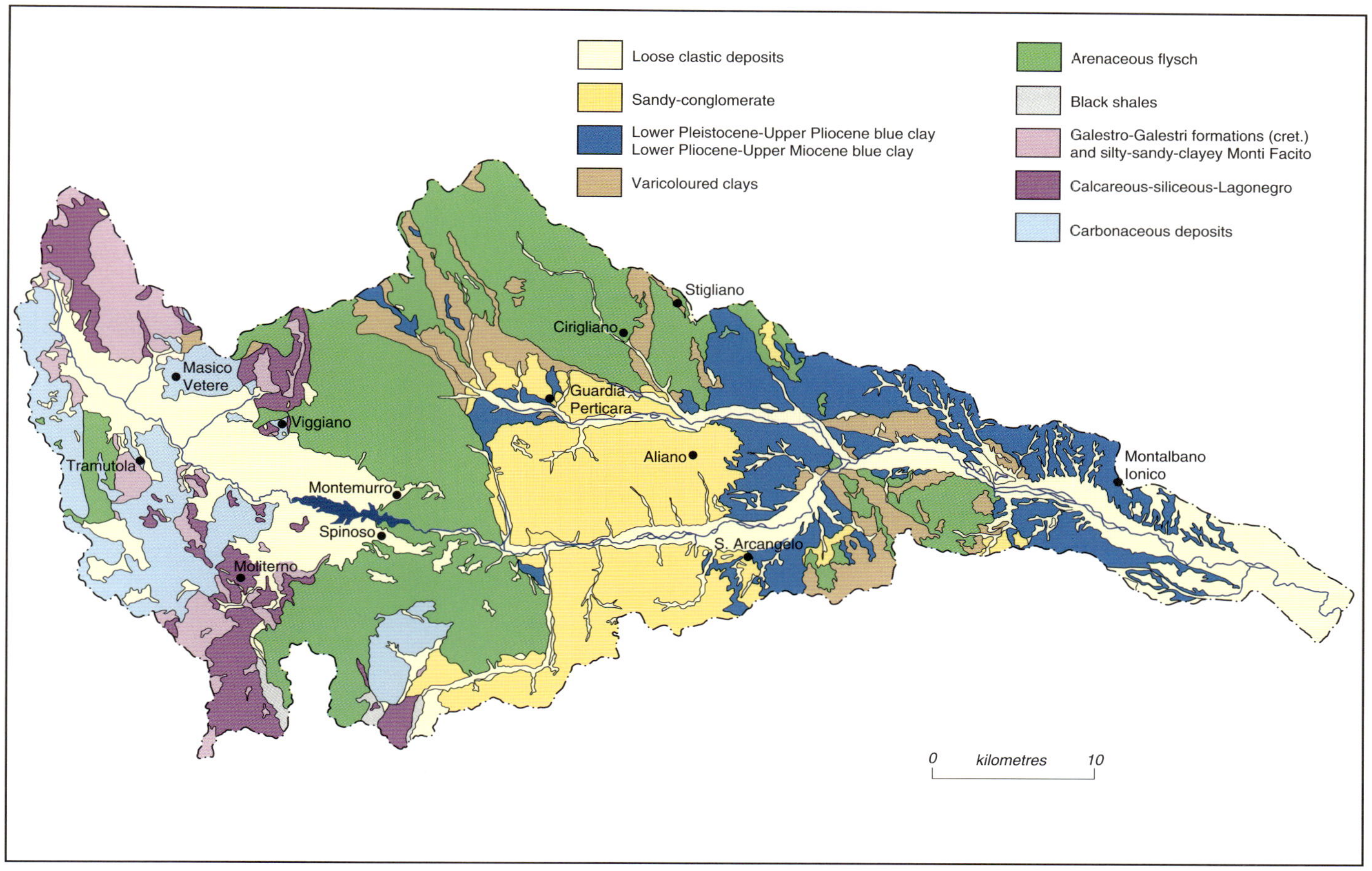

Figure 5.2.2 Lithological map of the study area. Redrawn from the map compiled by M. Del Prete.

5. the External Flysch Complex, almost entirely made of flysch and sandstones.

To the east of the Apennine Chain is the Bradanic Foreland with its Plio-Quaternary sequence, dominated by blue clays of marine origin that pass laterally into the Montemarano sands and Irsina conglomerates. Embedded in the Blue Clays is the allochthonous Metaponto Nappe. The Blue Clays coupled with the relatively recent tectonics and continuing earthquakes give rise to extensive mass failures. These are of several 'cycles' of formation and create a significant hazard in both built-up and rural areas in the region.

In the Bradanic (Middle) part of the Agri Valley *calanchi* are widespread and responsible for extensive loss of productive soil and the marginal economic nature of the region. The *calanchi* are developed in areas of deep incision of the Blue Clays by the tributaries to the Agri. In these hilly areas two distinctive morphological environments can be identified: (1) high-energy relief, formed

Figure 5.2.3 Large-scale arcuate failures and runout.

Figure 5.2.4 Shallow debris flow tracks formed in a single storm.

of the Blue Clays overlain by sandy-conglomeratic facies, with spectacular *calanchi* landscapes of nested gully systems separated by knife-edge ridges and (2) medium- to low-energy relief formed exclusively of the Blue Clays and progressively degraded forms towards the valley axis. Here rounded, isolated forms called *biancane* occur.

The combination of lithologies, tectonic activity and aspect determine the development and genesis of the *calanchi* forms. In particular, aspect exerts a strong control on the processes of erosion, piping, splash and mass failure to produce a strong asymmetry in slope form. South-facing slopes are steep, bare and intensively dissected whereas north-facing slopes are more gentle with *macchia* vegetation cover and low runoff and sediment yields.

Badlands

J. B. Thornes

Badlands are areas of intense linear erosion, which produce large amounts of runoff and sediment during intense storms. They are almost invariably free of vegetation and sometimes occupy quite large areas. Almost always they are associated with easily eroded rock or soil types, such as clays or marls (silts with a high calcium content). Among the MEDALUS sites and target areas, the Guadalentín Valley and the areas around the El Ardal and Rambla Honda sites, and the Agri are most afflicted. Small areas also occur in the Piomba and Cigno basins in the Abruzzo and around the Vale Formoso sites.

There has been much debate about the origin and evolution of these features on a world-wide scale (Bryan and Yair, 1983), particularly since they are often cited as prime evidence of desertification. They occur widely in the arid and semi-arid areas of the United States, in the grasslands of East Africa and in Turkey, Israel and the Mahgreb states. Occasionally they are also found in temperate environments where the vegetation has been removed from clayey or silty soils and rocks.

In some areas they do yield very large quantities of sediment which contributes to the rapid filling of reservoirs; for example, in the Guadalentín Basin, where they are undoubtedly very active today. In other areas, they appear to be changing very little. Thornes and Gilman (1983), for example, found that some of the badland gully systems in south-east Spain were formed before Bronze Age times and still contain the *in situ* remains of houses, burial chambers and even irrigation systems. Equally the recent development of aerial photography enables us to compare gully systems over 30–40 years. These comparisons generally dispel the notion that gully systems are advancing rapidly, even though this is occasionally the case. They tell us that we must be careful not to automatically associate badlands with high rates of erosion. The probable reality is that they expand rapidly in very intense but very rare rainfalls, maintain a fresh appearance by slope erosion in occasional less intense storms, and most of the time experience very little change.

In the Agri there are two major types: *calanchi* and *biancane*. The former are more like the characteristic gullies of other areas—sharp-crested, steep-floored, narrow channels with very steep sides and almost devoid of vegetation. They are often located on the sides of deep valleys. The *biancane* badlands, by contrast, comprise rounded forms separated by flattish floors but again are vegetation-free. Both forms are found on marine clays, e.g. the *biancane* around Aliano and the *calanchi* around Stigliano. Attempts to explain differences in the forms as a result of lithological differences in the materials have generally been unsuccessful (e.g. Alexander, 1983) and it seems likely that their forms owe a great deal to their position with respect to the main drainage lines and the extent to which these have been incised. The *biancane* then appear as residual features formed after a long period of evolution.

The development and growth of badland gullies has been studied in many parts of the world. Some develop from chance removal of the vegetation cover, e.g. as a result of human activity, such as leaks from irrigation ditches. Others seem to form from headward migration in expanding stream networks, growing rapidly in great but rare storms in amounts that are proportional to the drainage area above the gully headcuts. Yet others form and advance by mass failure of the clays or marls in which they are usually formed. Prediction of their growth can be obtained from

Figure 5.2.5 *Biancane*, near Aliano, Agri Valley.

Figure 5.2.6 *Calanchi*, near Atri, Pescara Valley.

computer models, but the timing of growth and its amount can only be predicted as well as we can predict the occurrence of large storms.

Badlands present a serious management problem because it is often impossible to use heavy machinery. There are only two possibilities. One is to build check dams in the beds of the channel. This is a costly and risky business because if the dams are not well designed they may be quickly undermined by rare but massive flows and seepage. Well-designed check dams quickly fill with sediments which increase infiltration losses into the bed and also encourage phreatophytic plants to grow. Once established they do an excellent job. The other attack is by ensuring there is no runoff into existing gully channels, by discouraging tillage and grazing in the headwaters and by building vegetated bands to limit runoff. It has to be admitted, however, that lands that are heavily dissected are often intrinsically unsuitable for agriculture and should be left to regenerate naturally. In that way costs are kept down and off-site effects mitigated.

Climate

Basilicata as a whole and the Agri Valley in particular can be characterised as having a cool temperate Mediterranean climate. However, there is a strong gradient from the coastline to the interior mountains. In the coastal areas the summer drought (June–September) is strong, with a rainfall of less than 100 mm and a mean monthly temperature in the warmest months of more than 23°C. Inland at higher altitudes (800–1000 m), the climate becomes cooler (20–23°C) though rainfall in summer remains about the same. Above 1600 m the rainfall in the critical summer months is above 150 mm.

Local variability can be summarised by the De Martonne (1926) aridity index ($P/(T + 100)$), by the Walter and Leith (1960) type diagrams for the highest and lowest meteorological stations in the basin, and by the Pavari phytoclimatic index regions for the Agri basin. The latter (Pavari, 1923) is based on the mean annual temperature, the mean temperature of the hottest and coldest months and the mean annual and summer rainfalls. The Agri Basin falls within the *Lauretum, Castanetum and Fagetum* classes of the system.

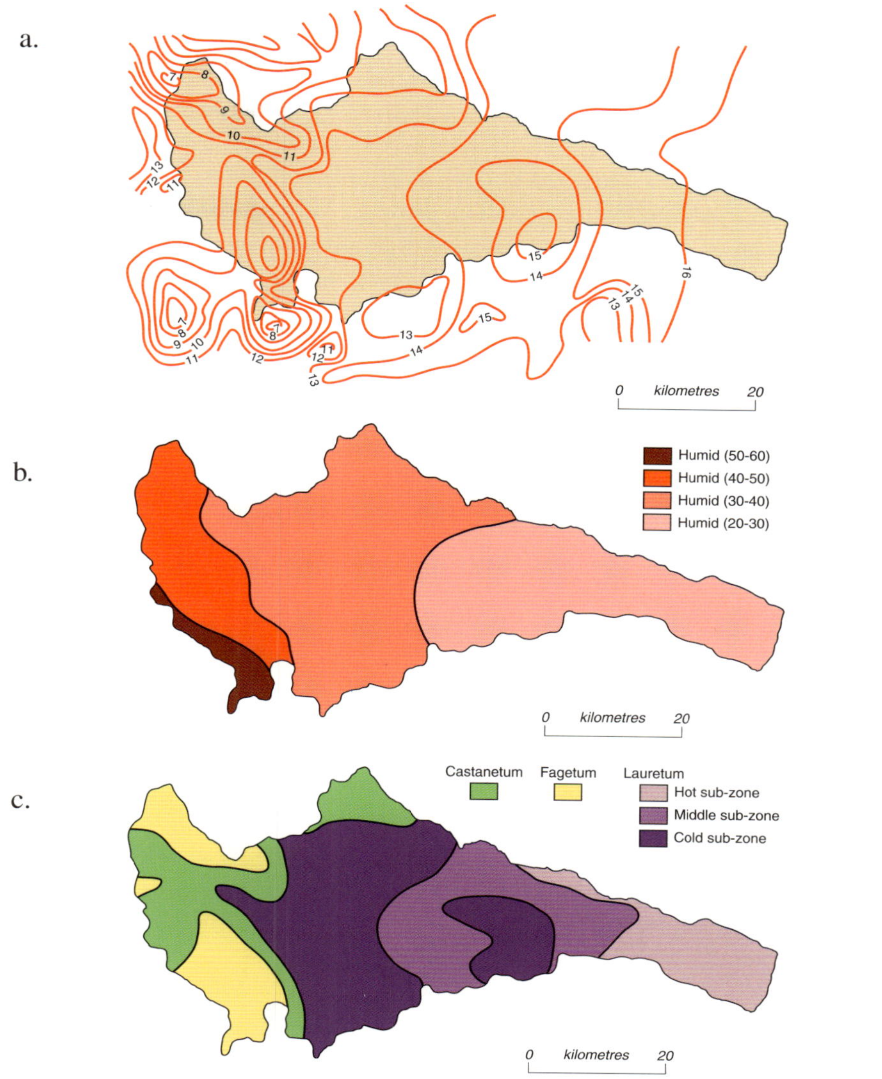

Figure 5.2.8 (a) Average annual isotherms for the Agri Basin. (Source: Ministero LL.PP. Servizio Idrografico, 1966.) (b) Spatial distribution of the De Martonne (1926) aridity index in the Agri Basin. (c) Spatial distribution of the phytoclimatic belts defined (Aita *et al.*, 1977) according to Pavari (1923) in the Agri Basin.

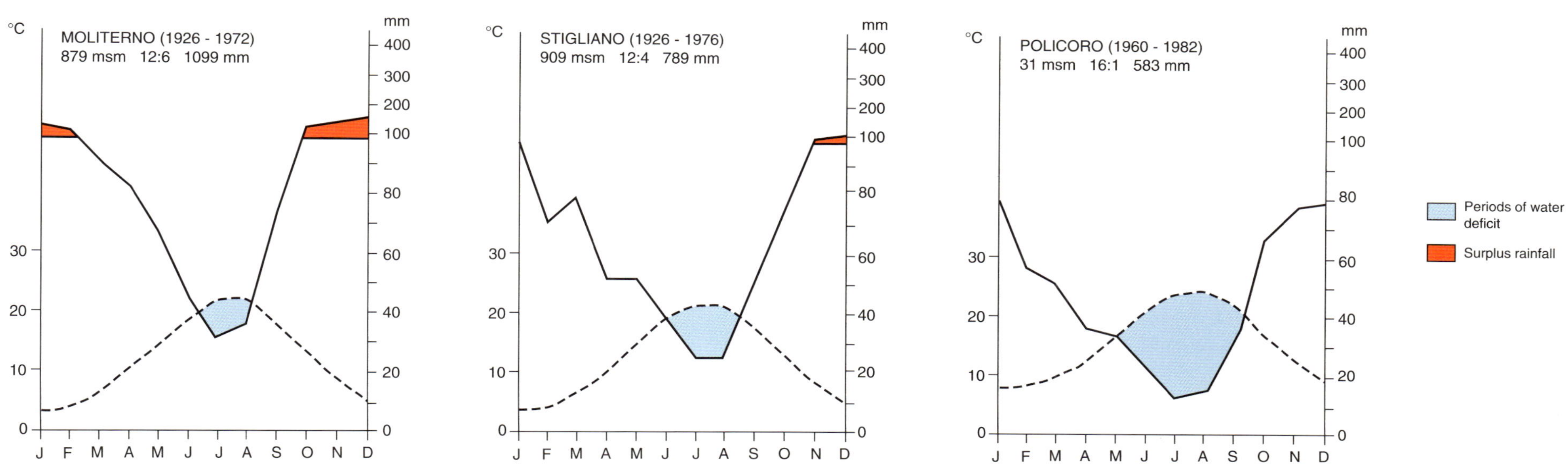

Figure 5.2.7 Climograms of mean monthly rainfall and temperature for the municipalities of Moliterno (upper Agri), Stigliano (middle Agri) and Policoro (lower Agri), calculated respectively over the periods 1926–1972, 1926–1976 and 1960–1980, following Walter and Leith (1960).

Temporal and spatial variations in the vegetation cover of the Sauro Basin

L. Postiglione, C. Ruggiero, M. Fagnano, G. Landi and M. Pisante

Three transects (1 km × 10 km) were established in the Sauro Basin to investigate herbaceous vegetation growth. The Sauro Basin is a sub-catchment of the Agri Basin, located in its middle part. The transects are 10 km apart from one another and are perpendicular to the river (W–E), crossing it at altitudes of 487, 370 and 270 m a.s.l., respectively. In each transect, six sample areas (1 ha) were selected to have the same herbaceous vegetation cover (winter cereal crops) on the south-facing slopes as well as on the north-facing slopes at various altitudes.

In the sample areas, above-ground biomass was sampled from two plots (1 m^2) and its fresh and dry weight measured twice a month during a 2-year period (May 1993–April 1995). From these data, leaf area index (LAI), crop growth rate (CGR) and net assimilation rate (NAR) were calculated according to Hunt (1978). Average monthly values for each transect and each position were subsequently used to calculate correlation between productivity and growth parameters.

Figure 5.2.9 Upper (a) and lower (b) parts of the Sauro Basin with contrasting above-ground biomass.

Dry biomass weight amounted to 373, 326, and 181 g/m^2 respectively, in the first, second and third transect. Biomass production was, on average, higher (400 ± 36 g/m^2) in the south-facing areas, warmer and receiving a larger amount of rainfall; it was intermediate (253 ± 49 g/m^2) in the north-facing areas, and it was lower (144 ± 57 g/m^2) in the bottom of the valleys.

The highest values of biomass were measured in May and June on all the transects and in both years, showing a fast decrease after July. The maximum value (1452 ± 36 g/m^2) was measured for the first transect in June 1994. In the other months biomass values were very low, showing a notable increase only as from March.

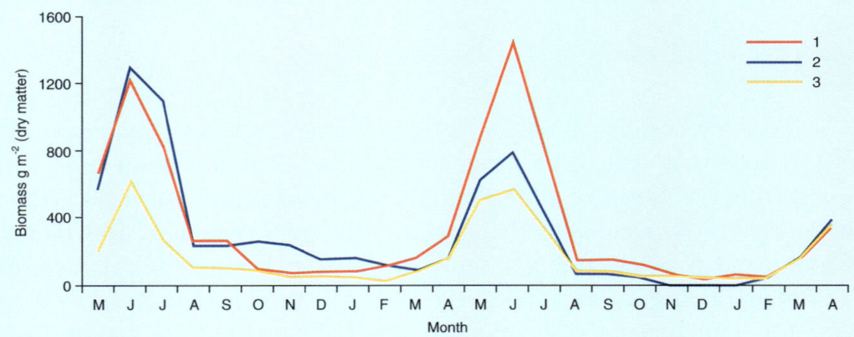

Figure 5.2.10 Seasonal variation of biomass for the three transects established in the Sauro sub-basin.

Herbaceous vegetation did not have green leaves from June to October in the first and second transect, and from June to January in the third transect. LAI is higher than 1 from March to May, with an increasing trend. Herbaceous vegetation is only present from winter to spring, with decreasing values from the first to the third transect.

Biomass production and growth patterns are related to the rainfall regime and soil moisture (Postiglione *et al.*, 1994). The rainy season starts in September and ends in May, with virtually no rainfall in the period June–August.

CGR was high during the spring, while decreasing to negative values in July, after the harvest of the wheat. Higher values were reached in the first transect (22 g/m^2 per day, April–May 1994) than in the second and third transects. NAR followed the same trend as CGR.

Biomass of the different sites in the three transects was, on average, directly correlated with plant height and LAI. Biomass and altitude show a curvilinear regression ($y = -651.004 + 2.942x - 0.002x^2$) with increasing values of biomass up to 760 m a.s.l. (400 g/m^2) and decreasing values from 760 to 900 m a.s.l. (334 g/m^2).

Vegetation and land use

The process of impoverishment of the forests of Basilicata in the last 150–200 years had less impact in the Agri, at least in the Upper and Middle Valley. Here conservation practices have produced mature forests dominated (Aita *et al.*, 1977) by Turkey oak (*Quercus cerris* L.), *farnetto* (*Quercus frainetto* Ten.) and chestnut (*Castanea sativa* Mill.). Other species are widely distributed, such as maples (e.g. *Acer campestre* L., *A. monspessulanum* L.), white hornbeam (*Carpinus betulus* L.), cherry (*Prunus avium* L.), Italian maple (*Acer opalus*, Mill.) and hazel (*Corylus avellana* L.). At higher elevations, such as the slopes of Monte Volturino, dense beechwoods (*Fagus sylvatica* L.) are found. Forming an undergrowth to these are hellebores, dentates, sweet woodruffs, wild orchid and two species of cyclamen, *Cyclamen repando* Sibth. *et* Sm. blooming in spring and *C. napoletano* Ten. blooming in autumn. Other beech forests occur at the foot of Monte Stella Cessuta in Moliterno, where 30 m high beech trees alternate with Italian maple, European holly (*Ilex aquifolium* L.) and the spindle tree (*Euonymous europaeus* L.).

The forests are often coppiced, as in the case of the chestnut, white hornbeam, maple and oaks found around the municipality of Grumento and Pertusillo lake. Here the trees are coppiced on a 14-year cycle.

The phyto-sociological associations have been characterised by Corbetta (1974) and these match up with the climatic parameters of the valley.

The Upper Valley is rich in water resources but these are mainly used to supply the neighbouring region of Apulia and to irrigate the coastal agricultural land rather than for local use. Since ancient times the abundance of water has favoured a system of scattered self-sufficient small family farms in alluvial zones near villages such as Marsico Nuovo, Paterno and Sarconi. This kind of agriculture is made possible by a dense network of irrigation canals managed by the farmers themselves without formal written agreements. This '*coltura mista*' is based on vegetables, herbs, vines and scattered fruit trees. In addition to this labour-intensive cultivation, the remaining area of production is largely occupied by cereal and fodder crops.

Figure 5.2.11 The Pertusillo Dam.

The expansion of irrigation, started in the 1950s on the plains, has brought about significant changes in the farming system. There has been an increase in livestock for milk production and a reorganisation of the dairy industry as well as the introduction of intensive horticulture and fruit growing. The hilly areas still remain under the traditional extensive agricultural system based on wheat cultivation and sheep husbandry. For example, Moliterno successfully markets *canestrato,* a soft cheese made of goat and sheep's milk (Bove *et al.*, 1993b). Vineyards, olive and chestnut groves and oak coppice also occupy significant amounts of land, while on the mountains coppiced beechwoods alternate with pastures, replacing former cultivation of rye and potatoes.

The agricultural landscape of the Middle Agri and the Sauro sub-basin is characterised by the presence of badlands. Semi-natural vegetation is represented by the *macchia mediterranea* (herbs and shrubs) which has, in many cases, replaced former forest on clay lands. Utilised agricultural area is dominated by durum wheat and pasture, which at higher elevations give way to animal husbandry (sheep). Olives are cultivated in hilly areas with excellent results in areas such as Aliano. Intensive agriculture in the Middle Agri is restricted to the riparian zones of the Agri and the Sauro and around Sant'Arcangelo. *Coltura mista* is also practised on small allotments and based on a wide range of horticultural and fruit crops.

Land use was affected for centuries by the presence of malaria that had infested the coastal zones, but it was eradicated in the 1950s, mainly as a result of land reclamation. The agricultural landscape, previously dominated by extensive land uses such as cereal crops and pasture, was radically altered by the introduction of irrigation. This made possible the intensive cultivation of horticultural, industrial and fruit crops and the spread of greenhouses. The products are mainly exported.

In the Upper Agri, contemporary trends are towards consolidation around the existing centres of population, and mountain areas are being re-afforested. In this area agriculture is likely to continue to dominate socio-economic development in conjunction with an emergent tourist industry. The latter will be encouraged by the formation of the Val d'Agri and Lagonegrese National Park. One strategy for the growth of agriculture is the expansion of horticulture (Bove *et al.*, 1993a) using farm machinery and practices that will not damage environmental and water resources. Regulation of animal husbandry should be used to enhance quality and to diversify production. Traditional agricultural activities in the area could be complemented by the cultivation of small fruits. Moreover, marginal and abandoned mountain lands could be restored by the organic cultivation of typical crops now in demand, such as high mountain potatoes and pulses. The use of resources in mountain areas, however, must rely on the balance between forestry and grazing.

Alternative strategies for the Middle Agri are more difficult to define because of the physical degradation in badland areas. Here the only solution, and it seems critical to undertake it as a matter of urgency, is to revegetate the slopes wherever possible. This process needs to be based on adjustments of soil-water regimes, better cultivation methods and more suitable herbaceous species for pastureland improvement.

In the Lower Agri, intensification of existing agriculture appears to be the best option in economic terms because of the favourable climate and the availability of water for irrigation. However, such an option needs to be tested for sustainability.

Agricultural changes and land degradation in the Agri Basin

P. Mairota

Socio-economic and cultural underdevelopment, unemployment and rural depopulation, persisted in Basilicata despite the political effort of the 1960–1980s (Cafiero, 1975; Leonardi *et al.*, 1987) towards the modernisation of agriculture, the key economic sector in the area, and were further exacerbated by the disastrous effects of the earthquake of 1980 and the severe droughts of both 1981 and 1982. In the 1980s the sources of the subsidies partially changed, and Community (e.g. Reg. CEE 797/85, Reg. CEE 1204/82) and national (e.g. DM 30/12/83, L 752/86) legislation was enforced. The regional government has thereby encouraged co-operation between young farmers, planting of fruit and citrus crops, rediscovery of formerly used arable crops, expansion of irrigation and mechanisation, and husbandry of sheep and goats aimed at typical dairy products.

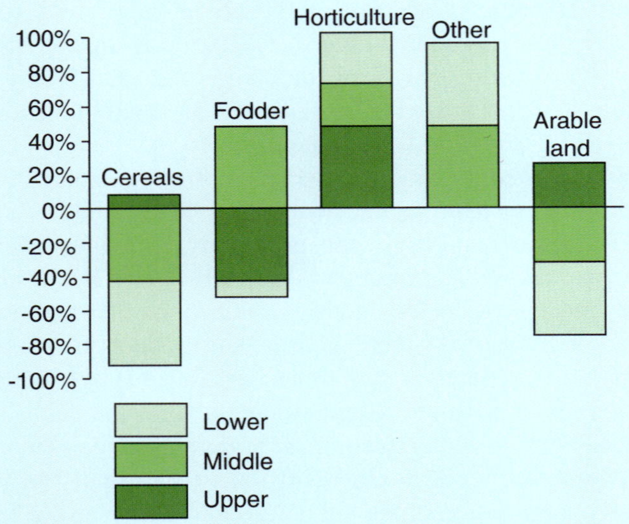

Figure 5.2.12 Intensive arable crop types have increased throughout the basin, particularly in areas with steep slopes or erosion-prone lithologies (most of the upper and middle Agri) where mechanised tillage occurs along maximum slope gradients. Arable land in Basilicata is the land use most prone to soil erosion (Cavazza, 1962), which is estimated (Boenzi and Giura Longo, 1994) to be of the order of 12 t/ha per year. In addition, horticultural crops require irrigation.

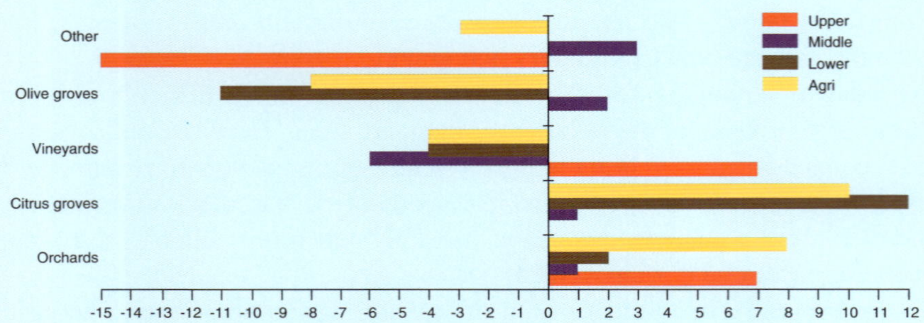

Figure 5.2.13 Fruit and citrus tree plantations can be compared hydrologically to afforestation, which, in the short term, reduces soil erosion and slope instability at the basin scale due to planting, management and harvesting operations. Unlike afforestation, fruit plantations do not provide, in the long term, slope stabilisation, reduced runoff and splash detachment. The canopy never closes in, formation of litter is not allowed, frequency of maintenance and harvesting operations is much higher, and ultimately the plantation has to be replaced at relatively short time intervals. Moreover, fruit plantations are demanding in terms of water, fertilisers and pesticides.

As a result of these policies, land-use changes have occurred in the Agri Basin in the last decade (Mairota, 1995a). Although possibly effective in terms of short-run socio-economic development, some of the resulting changes threaten land degradation directly or indirectly.

The trends towards intensification in agriculture and expansion of irrigation are very clear in the middle and the lower parts of the Agri Basin, whereas in the upper part of the basin agricultural activities complement other sources of income.

This is the typical pattern of agricultural intensification in Mediterranean countries (see Section 3.5), involving the transfer of profitable agricultural activities from the uplands to the plains, producing marginalisation of agricultural land at the basin and landscape scales. Agriculture intensification, including animal husbandry, also degrades the physical environment, mainly affecting the quantity and quality of water at the basin scale (Calder, 1993). In the Agri, the observed intensification in animal husbandry will increase the amount of farm wastes (e.g. manure, dip liquors of various kinds) because of the concentrations of animals in feedlots. Irrigation expansion does not seem feasible in the long run because future water demands for agriculture will not be satisfied by the existing reservoirs (Pertusillo and the Gannano), which also supply urban and industrial areas and neighbouring regions, and which are already silting up. New reservoir construction is unlikely, not least

Figure 5.2.14 A decrease in woodlands and an increase in grasslands, less hydrologically efficient, also enhances the risk of soil loss and slope instability, and certainly affects both plant and animal species richness.

because downstream aquifer recharge is decreased by the existing reservoirs, and coastal sediments are not replenished (cf. Boenzi and Giura Longo, 1994). Economic development alone has been pursued and achieved in the Agri Basin, as in many other Mediterranean rural areas, as this was needed to relieve the extreme under-development of these areas and raise living standards. However, such a development is far from being sustainable and greater attention needs to be paid from now on to both the inherent local characteristic and the spatial and functional policy outcomes at the basin and landscape scales. The ultimate result of unsustainable policies is land abandonment. For the Agri Basin and its people in particular this would mean an unaffordable step back to massive migration, abandonment of farm areas and possibly soil erosion.

Changes in population

E. Bove

The changes in the population of the Agri Basin that have occurred since the political unification of Italy (1861) are similar to those observed for the whole of Basilicata.

The moderate and slow increase in population that occurred between the two World Wars was preceded and followed by two more marked emigration phases.

Similar patterns are shown by the upper and the middle parts of the basin, where between 1861 and 1991 the resident population decreased by 21% and 13% respectively. The most impressive cases are those of the municipalities of Montemurro, in the Upper Agri, and Armento, in the Middle Agri, where the population has decreased by 50% and 70%, respectively, in 130 years.

In more recent times some areas have experienced increases in population (municipalities of Paterno, Sarconi and Marsicovetere, Upper Agri, and Roccanova, Stigliano and Sant'Arcangelo, Middle Agri) due to the presence of administrative offices, commercial activities and tourism, in addition to a special law passed in order to support communities hit by the earthquake of 1980.

However, geographical isolation still remains an important cause of out-migration of the youngest residents. This is evident from the records of enrolment in compulsory education. For example, in the municipality of San Chirico Raparo, in the Middle Agri, the already small number of students, 161 in 1982/83, dropped to 54 in 1993/94 (ISTAT, 1983, 1993).

Conversely, in the Lower Agri the population has increased fourfold from 1961 to 1991. Great improvements in the physical as well as in the socio-economic environments have created the conditions for thousands of people to move to the coastal plains. The municipality of Policoro has now approximately 15 000 inhabitants as opposed to 599 in 1861. Craco is the only municipality in the Lower Agri that has experienced a strong migration due the great instability of the land particularly prone to landslides and formation of badlands.

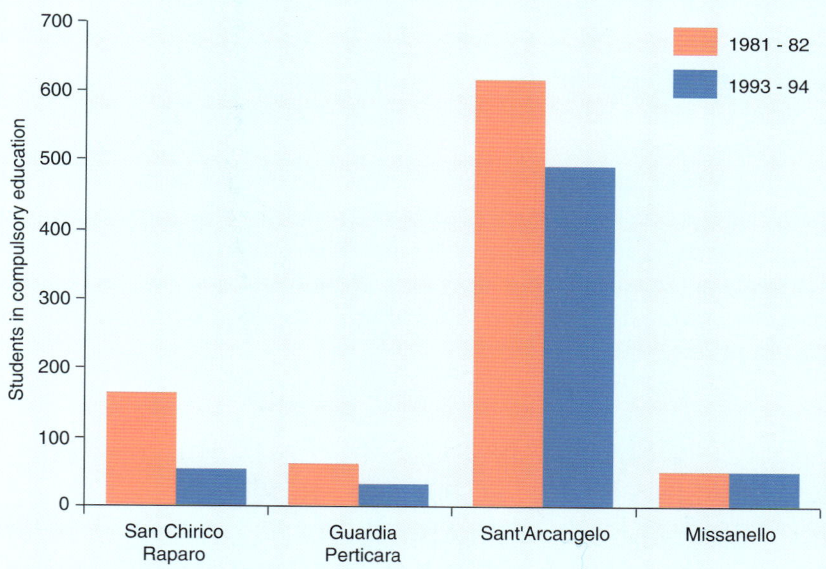

Figure 5.2.15 Evolution of the student population in compulsory education in some municipalities of the Agri Basin. (Source: ISTAT, 1983, 1993.)

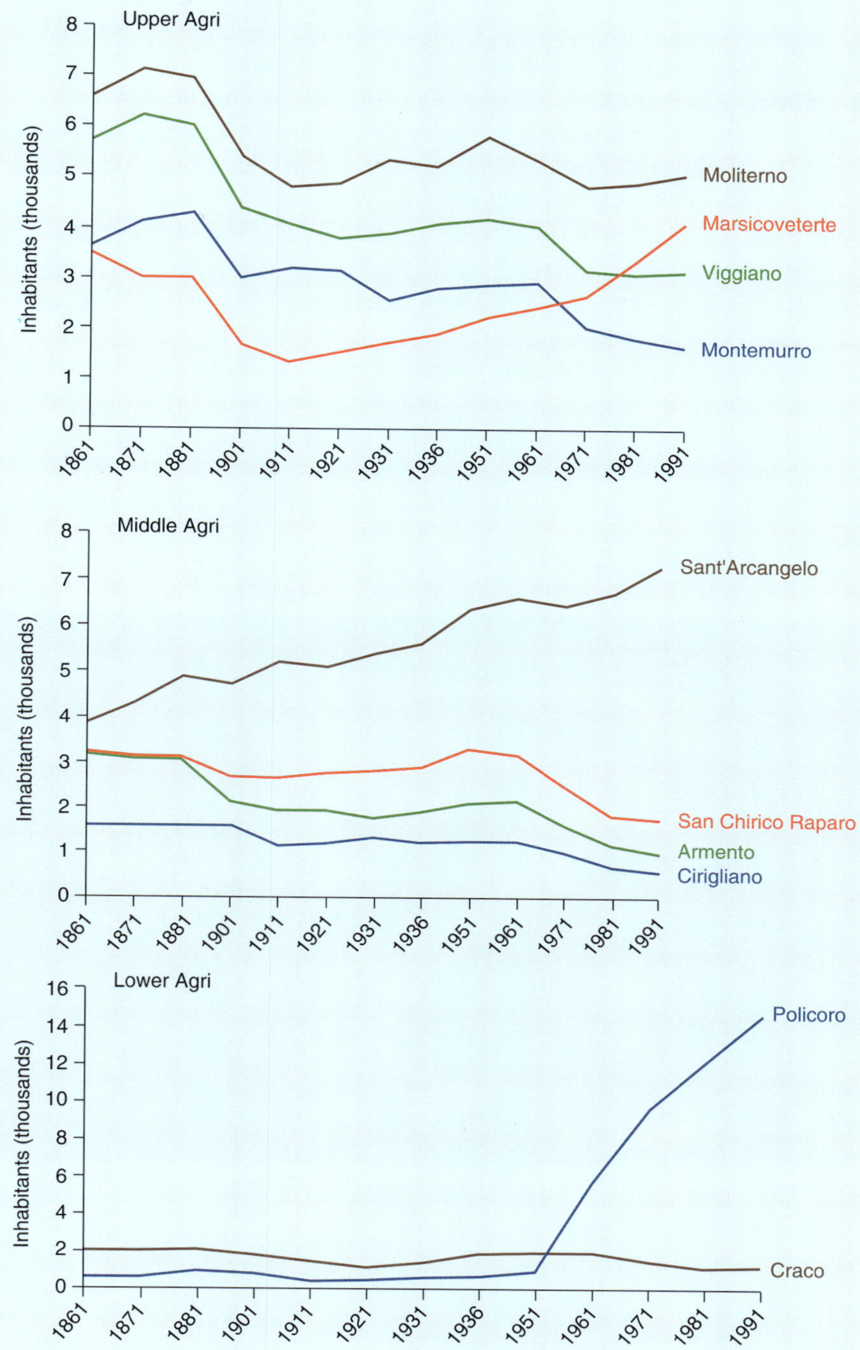

Figure 5.2.16 Population changes in the upper, middle and lower Agri over the period 1861–1991 (ISTAT, 1985, 1993).

5.3 The Abruzzo Region (Italy). Investigation of Desertification-threatened Areas

C. Conese

Introduction

The research project, carried out within the framework of the MEDALUS II Project, aims at evaluating and testing an integrated application of methodologies already developed for indentifying and alleviating desertification over wide areas. Particular attention has been devoted to the fact that in most areas affected by problems of environmental degradation few of the data needed to face the problem are usually available.

The Abruzzo area is not, at present, affected by obvious signs of desertification. It is potentially threatened by desertification processes for several reasons, including the neglect of rural areas, changes of land use, human pressure and so on.

Consequently, study areas in two valleys of the Abruzzo region were chosen as good test sites for long-term monitoring of the evolution of the environmental degradation.

This work deals with the study of the main desertification parameters in this area, such as land use, climatic conditions, soil erodibility and socio-economic aspects.

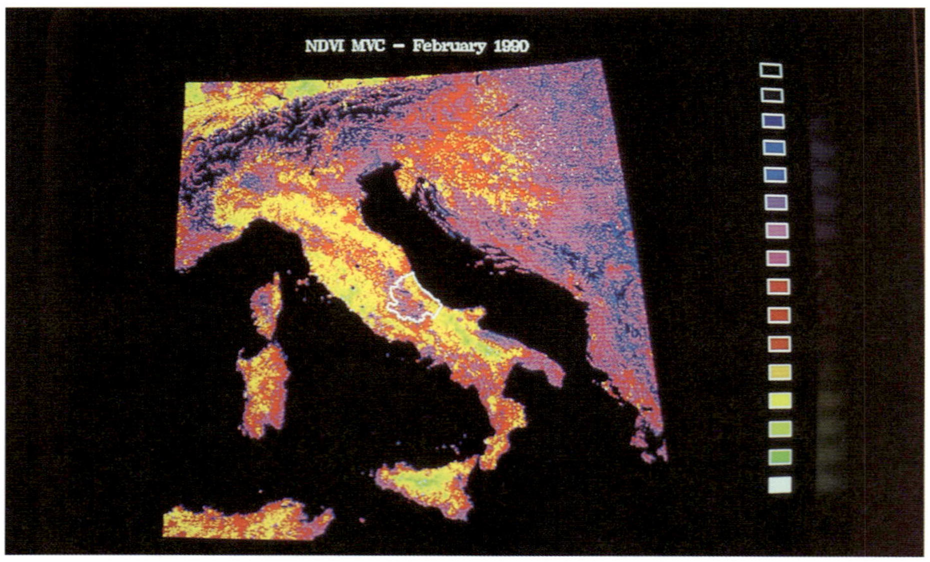

Figure 5.3.1 NDVI MCV image of Italy derived from NOAA 11 in 1990. The Abruzzo region is outlined. (Photograph: Telespazio.)

General overview of the research area

The study areas correspond to the basins of the Piomba and Cigno rivers, situated in the Abruzzo region, central Italy.

The Piomba is a first-order basin, while the Cigno is a tributary to the Pescara River. They flow from west to east, like most of the Adriatic rivers, and the basins have areas of approximately 100 and 60 km^2. The topography is quite varied, with elevation ranging from 10 to 740 m a.s.l. and from 80 to 1440 m a.s.l. respectively.

These basins are characterised, in the central-western part, by rugged terrain with steep slopes and narrow deep valleys, and there are a large number of failure features, which are usually small, with the exception of rare large landslides affecting whole hillsides. The central-eastern part of the landscape is smoother, with fewer steep slopes, often dissected by diffuse gully systems forming badlands.

The topography of the two basins was derived from 15 orthophotomaps at a scale of 1:10 000 (eight for the Piomba and seven for the Cigno Basin). These maps were of high quality and have contour lines every 10 m (produced by Regione Abruzzo, 1987–1989).

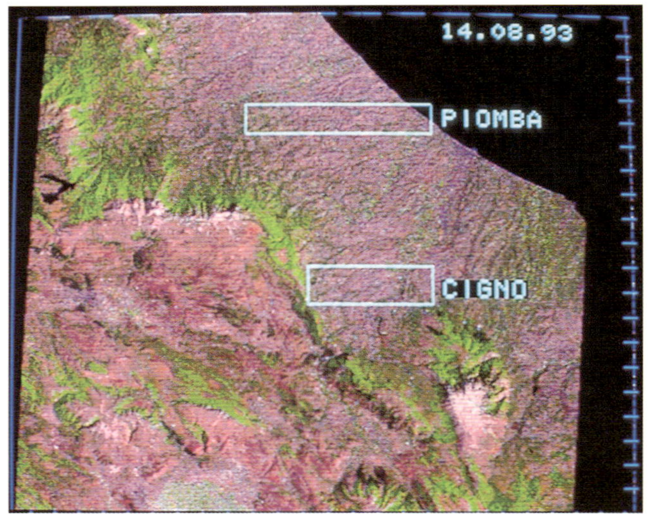

Figure 5.3.2 The geographical position of the Piomba and Cigno basins.

A digital terrain model (DTM) of the two basins has been constructed by digitising and processing the contour lines from the orthographic photomaps. Three raster files (elevation, slope and aspect) have been obtained.

Land uses in the basins are mainly agricultural land, pasture, and coniferous and deciduous forests. The main agricultural crops are maize, winter wheat, alfalfa, grapes and olives.

PRODUCTION OF LAND-USE INVENTORIES IN THE STUDY BASINS BY INTEGRATED PROCESSING OF REMOTELY SENSED AND ANCILLARY DATA

C. Conese, T. De Filippis, A. Rodolfi, F. Maselli, L. Petkov, G. Vernazza, S. B. Serpico and F. Roli

The success of any mitigation policy for environmental degradation depends strongly on the availability of detailed and updated information for the areas monitored, particularly on vegetation types and conditions. Remote-sensing techniques can be successfully employed to produce land-use and/or land-cover inventories thanks to the improved spectral and spatial resolution

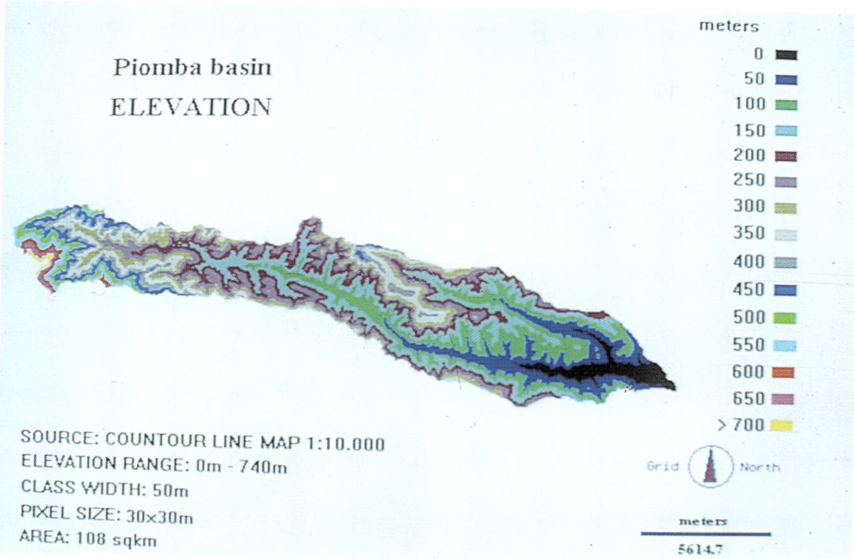

Figure 5.3.3 Digital terrain model of elevation for the Piomba basin.

of modern sensors. These techniques have numerous advantages compared with traditional monitoring methods such as synopticity, objectivity, relatively low cost and easy updating of the information produced (Swain and Davis, 1978). Nevertheless, the application of remote-sensing procedures to the study of complex areas in Europe faces several problems because of the irregularity of the terrain and the fragmentation of the cover types. Landsat 5 Thematic Mapper (TM) data, in particular, offers important opportunities for monitoring vegetated landscapes, but its full exploitation generally requires non-conventional processing methods capable of exploiting several information sources.

In the framework of the MEDALUS Project, an integrated investigation of the use of TM spectral and textural imagery and ancillary data for land-use inventory has been carried out on the two study basins in the Abruzzo region. Two TM scenes, taken in spring and summer 1993, constituted the multi-temporal spectral data set. Information layers for the morphology of the two basins were derived from the digital terrain model (DTM). Accurate ground references of 11 cover categories were collected in spring–summer 1993 and used for training and testing classification procedures. The first stage was devoted to the selection of appropriate spectral, textural and ancillary features for the automatic classifications. The spectral images were transformed by means of statistical procedures and vegetation indices. The texture images were derived from parameters of the Grey-Level Concurrence Matrix (Haralik *et al.*, 1973). The three raster files of a DTM with elevation, slope and aspect were considered as ancillary data layers (Hutchinson, 1982). The information content for class discrimination of all of these layers was then evaluated in order to select a restricted data set. On this integrated data set several advanced classification methods were tested, based on both parametric and non-parametric approaches. In the former, a Gaussian maximum likelihood classifier with modified prior probabilities was considered, as recently suggested by Maselli *et al.* (1992). In the latter, a K-nearest neighbour and some neural network classification procedures were tested (Bishof *et al.*, 1992). The best results were obtained from the integrated use of spectral, textural and ancillary data calibrated on the specific study set. Among the classifiers examined, the non-parametric procedures performed slightly better than the modified maximum likelihood classifier, probably due to the irregularity of the cover types in the two basins. The classification accuracy reached in the best cases were approximately 75–80%, which can be considered suitable for operational applications also. The land-use maps produced can be used both for planning land-management practices and as input for the evaluation of land-degradation phenomena. A specific application in the present context is the utilisation of vegetation cover as a parameter for the assessment of potential soil loss in the study basin.

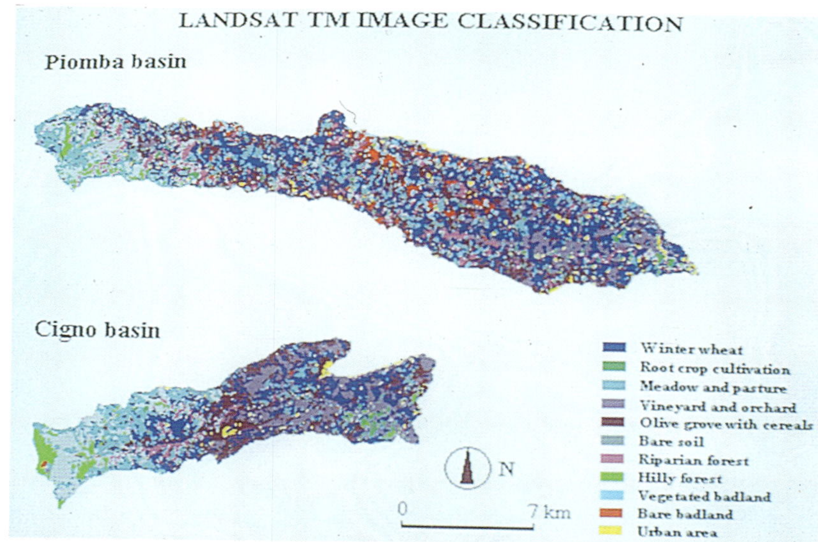

Figure 5.3.4 Contemporary land use map of the two basins, obtained by Landsat-TM image classification.

LAND COVER CLASSIFICATION

L. Bottai

Normalised Difference Vegetation Index (NDVI) time series derived from National Oceanic and Atmospheric Administration (NOAA) satellite data relevant to the years 1986–1989 have been used to map vegetation. Using remote-sensing information from all the available dates in a period, it is possible to define areas of homogeneous NDVI profiles using unsupervised classification procedures. This way landscapes with similar environmental features can be identified.

The database used contains 840 daily images of NDVI derived from the NOAA 9 and, since November 1988, NOAA 11 satellites. A simple Maximum Value Composite (MVC) method to analyse NDVI time series has been applied to the original images (Holben, 1986), thus considering the maximum value taken by the vegetation index in a time period as a good estimator of the real NDVI value. A 10-day sampling period, which is supposed not to neglect important dynamic vegetation features, has been used. Putting together the 10-day maximum value of each month, the monthly MVCs have been produced.

A non-hierarchical cluster analysis procedure, based on an interactive stabilisation algorithm, was applied to the same data set and a classification on the four first principal components has been created.

The clusters plotted on the map show the features of the Abruzzo region that have similar photosynthetic activity. This procedure is more efficient in the description of vegetation phenology (Benedetti *et al.*, 1994).

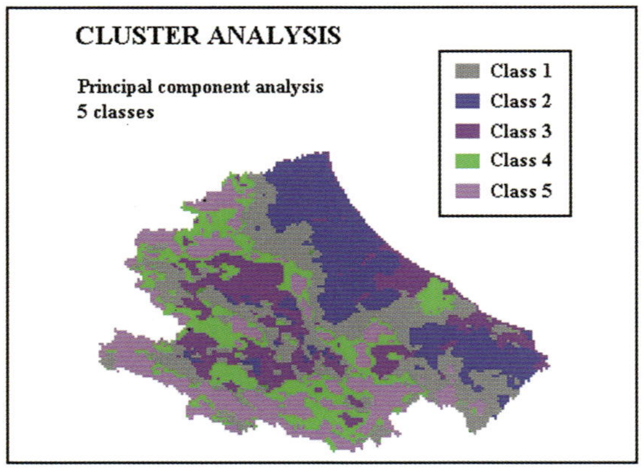

Figure 5.3.5 Thematic map (land use) obtained by an iterative cluster analysis procedure applied to the first four principal component images. Class 1: thermophyllous deciduous forest; Class 2: agricultural land with winter crops; Class 3: agricultural land with summer crops; Class 4: mesophyllous deciduous forest; Class 5: grassland of the cryo-Mediterranean plain.

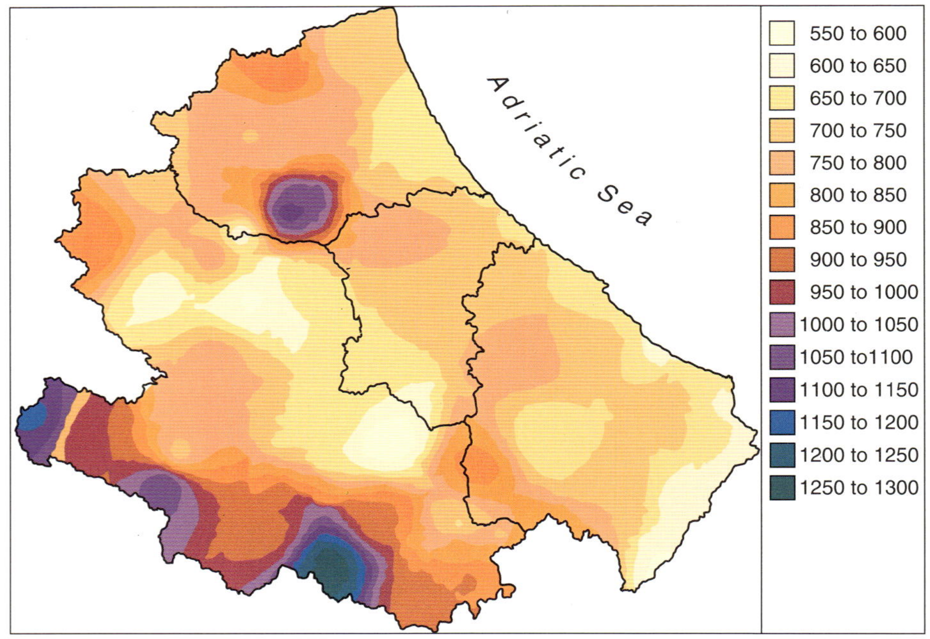

Figure 5.3.6 Yearly mean rainfall (in millimetres) in the Abruzzo region.

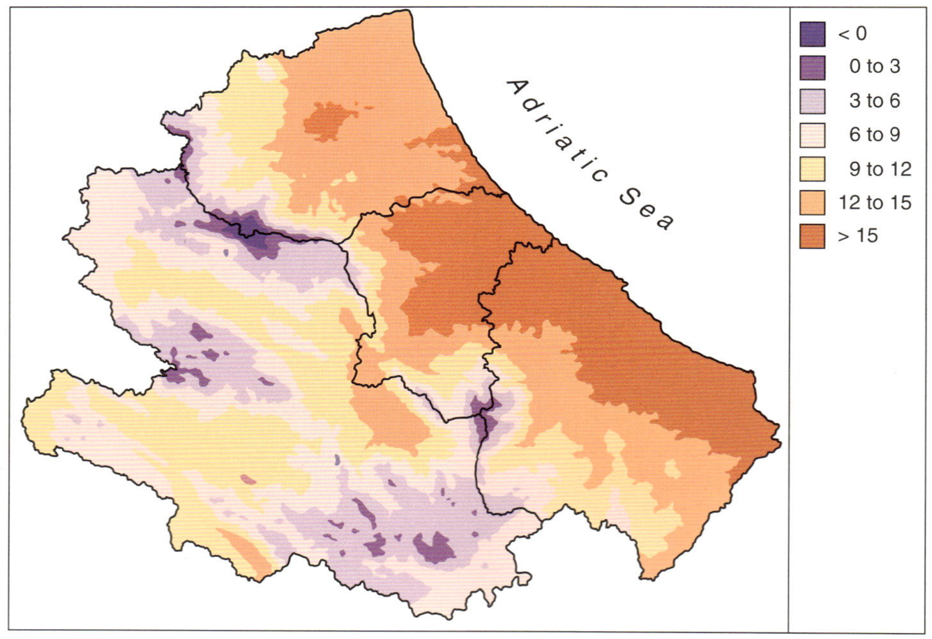

Figure 5.3.7 Yearly mean temperature (°C) in the Abruzzo region.

GENERAL CLIMATIC FEATURES OF THE ABRUZZO REGION

L. Mariani, V. Menichini and G. P. Minardi

The barrier effect of the Alpine–Apennine range and unstabilising and mitigating effect of the Mediterranean are the two principal geographical factors influencing the climate of the Italian Peninsula. Depressions that affect Italy are less deep and move less regularly than those over the Atlantic. A first subdivision is between depressions coming from the Atlantic and depressions derived from on-site cyclogenetic activity of the Mediterranean. Mediterranean cyclogenesis is generally produced by the movement of polar maritime air from the north-west in the Occidental Mediterranean area; polar and Arctic continental air from the north-east can also produce cyclogenesis. In the Mediterranean the principal cyclogenetic region is the Gulf of Genoa, producing 15–20% of the total cyclogenesis in the winter.

From a dynamic point of view, the part of the Italian Peninsula situated to the east of the Apennine watershed can by considered as an independent climatic region called the Adriatic climatic region. Over this region the main part of the rainy periods in autumn and winter are associated with (1) depressions coming from the Gulf of Genoa, crossing the Po Valley and reaching the area with the maximum frequency in autumn, the wettest season in the main part of the region; and (2) cold air that flows from the north-east (the Balkans region), which can produce very active depressions in the Adriatic and Balkans, with rainfall and snowfall events in the winter period.

The autumnal disturbances are enhanced by the destabilising action of the sea, which is usually very warm in this period.

In spring the weather is not greatly influenced by Atlantic depressions because of the dominance of anticyclones over the Balkans; however, the northern part or the region is sometimes influenced by lee depressions and the final part of spring is characterised by a relatively high frequency of thunderstorms.

In summer, depressions are rare and have trajectories that influence only the northern Adriatic. As a consequence there is usually fair anticyclonic weather with local instability associated with convective activity. The result is a rapid daily variation of cloudiness, with daytime cumulus activity over the hills and mountains and a high frequency of thunderstorms in August.

The air flow near the soil is influenced by the Apennines, which give protection from westerly winds, and by the Gargano Peninsula, which attenuates the southerly winds. Moreover, Pescara represents the southern limit of the region hit by the Bora, a very violent wind in the North Adriatic area caused by cold air bursting through the Carnick Gate.

The analysis of the thermal field shows that the Adriatic region is colder than the Tyrrhenian region at the same latitude and in particular the mean annual isotherm of 15°C is found entirely along the Tyrrhenian coast but appears in the Adriatic area only on the Abruzzo coast. This is produced by the barrier effect of the Apennines which protect the Tyrrhenian area from the cold air blowing from the north-east and by the small climatic effect of the Adriatic Sea, which is surrounded by the Italian Peninsula and the Balkans, and is shallow in the northern and central parts (with a maximum of 100–200 m) of the Abruzzo. The rainfall regime of the central Adriatic is characterised by a minimum in summer (July)

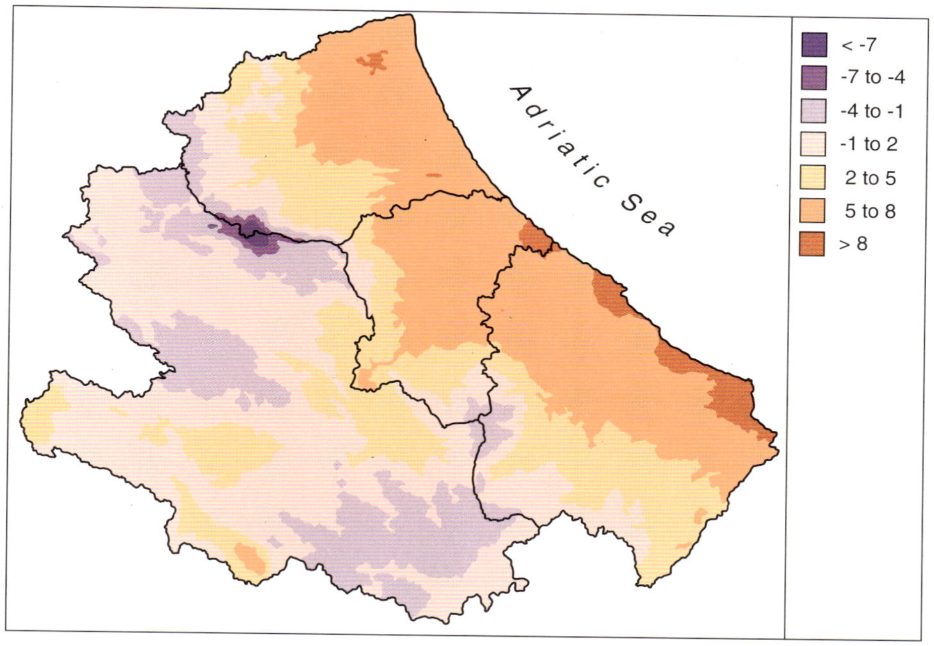

Figure 5.3.8 January mean temperature (°C) in the Abruzzo region.

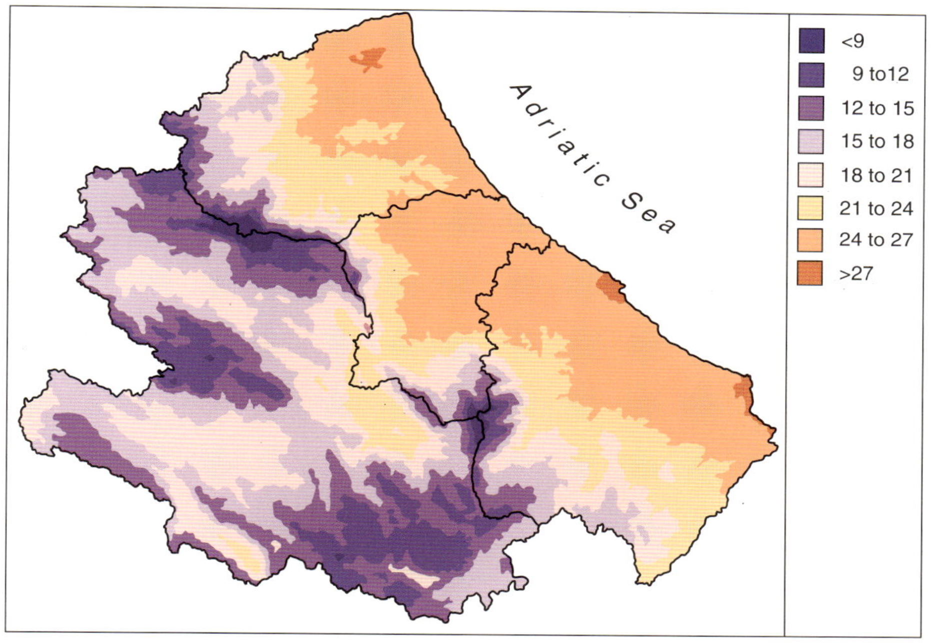

Figure 5.3.9 July mean temperature (°C) in the Abruzzo region.

and a maximum in autumn (November). A secondary maximum in spring (April–May) decreases in importance from the north to the south of the region.

The climate of the Abruzzo

The Abruzzo is enclosed between the Apennines watershed and the Adriatic Sea. From the coast to the interior the mean annual rainfall shows a general increase with altitude and a strong orographic effect. A rainfall maximum is observed over the east side of the mountains due to the lifting of humid air over the orographic barrier. Behind the range of mountains there is a rain-shadow area where the minimum precipitation values are attained (less than 600 mm). The maxima are observed in the eastern part of the region and are produced by the interaction of humid air flowing from the south-west and from the Apennines range.

Yearly mean temperature generally increases form the coast inland; the discontinuity in the Apennine range created by the Pescara valley allows the 12°C isotherm to reach the interior of the mountain area. The expansion of the 15°C isotherm over the south of the coastal part of the region can also be observed.

Extreme temperatures for most of the Abruzzo territory are in January and July. Some meteorological stations in the mountain area record the minimum and maximum temperature of the year in February and August respectively. In January the two experimental areas (the Cigno and Piomba basins) are on the 4°C isotherm. In July, only a few restricted areas along the coast have a mean temperature over 27°C.˙ The two experimental areas are within the 21°C isotherm.

APPLICATION OF MICROWAVE REMOTE SENSING FOR THE MONITORING OF DEGRADATION-AFFECTED SURFACES

L. Chiarantini, P. Coppo, S. Gagliani and G. Luzi

Degradation processes such as soil erosion and mass movement are closely related to surface changes of soil moisture content and surface roughness. However, such data are usually sparse, despite their high accuracy, and therefore they are not really suitable as input parameters for distributed physical models. For these reasons microwave remote sensing for soil moisture and surface roughness estimations seems to be more appropriate to the MEDALUS II objectives concerned with monitoring soil-degradation processes in agricultural areas.

The methodology for processing satellite Synthetic Aperture Radar (SAR) images described in Section 2.2 of the Atlas is being developed for a multitemporal analysis of European Remote Sensing Satellite (ERS1) SAR images of the Piomba and Cigno basins. The research aims to integrate remote sensing and field microwave probe data with territorial databases in order to improve spatial and temporal representativeness of data for models of soil erosion and mass movement predictions. Also, ultimately erosion predictions and recommendations for conservation should benefit from contributions of microwave technologies. This approach has been studied and developed at the Fondazione Scienza per l'Ambiente (FSA), Firenze, in the framework of the MEDALUS II Project (Chiarantini *et al.*, 1993, 1994).

In order to validate inversion algorithms developed for soil moisture and roughness estimation, field campaigns were carried out at the same time as ERS1 satellite overpasses. The surface dielectric constant was measured by means of a microwave portable probe operating at 1.42 GHz in order to estimate moisture content and to obtain soil dielectric constant data for SAR image interpretation. Real-time surveys of soil mosture were performed with the same probe after a laboratory calibration precedure with soil samples of different textures collected in a test area as described in Section 2.6 of this Atlas. Because of the influence of tillage practices, weathering and consolidation on soil erosion processes, surface roughness and soil bulk density data were also collected. Surface roughness parameters as an input to electromagnetic and soil erosion models were collected and successively processed as explained in Section 2.6. Finally soil bulk density data were computed by measuring volumes and weights of dried soil samples collected by means of a cylinder device.

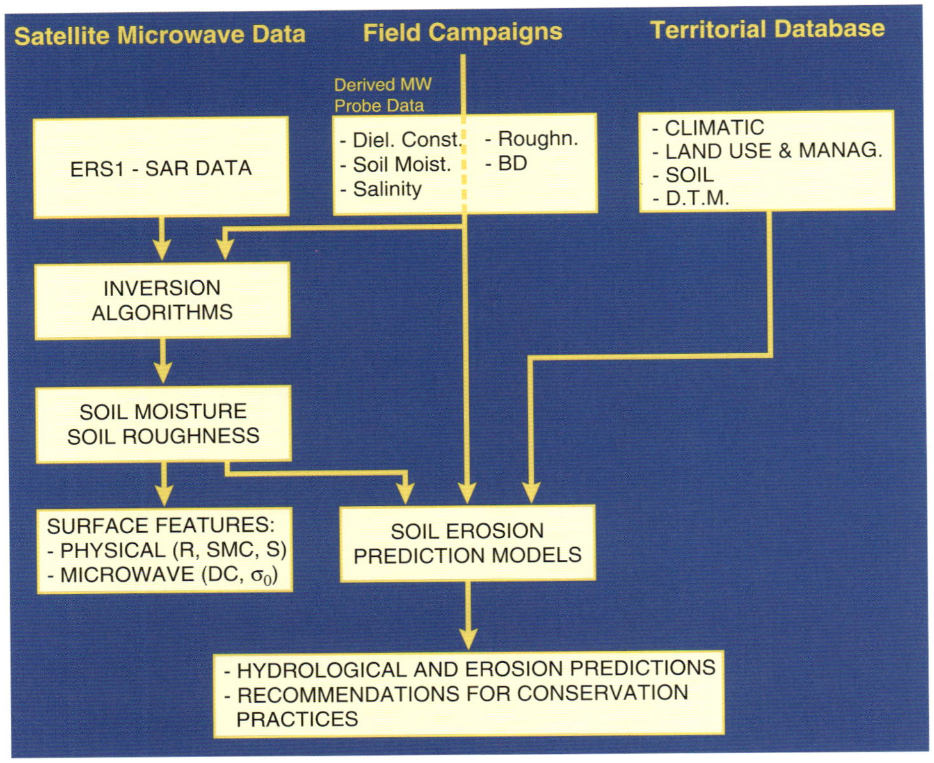

Figure 5.3.10 Proposed methodological approach for the application of microwave remote-sensing data.

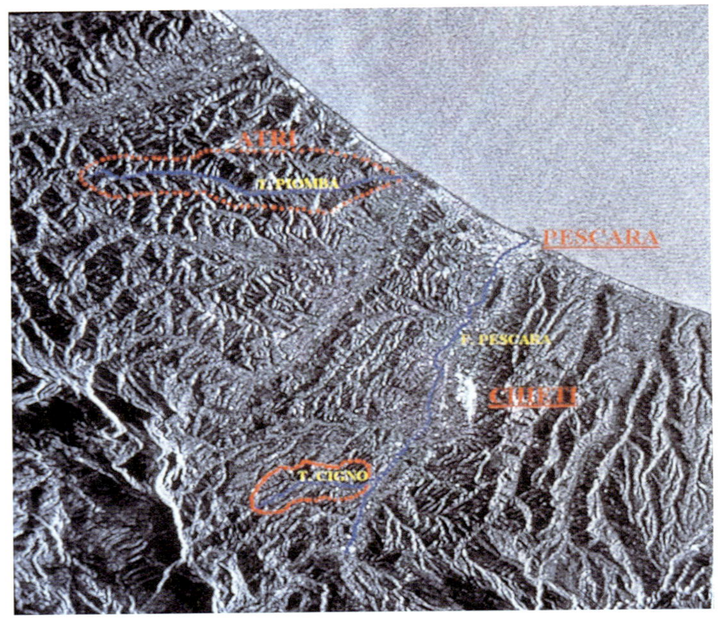

Figure 5.3.12 ERS1-SAR image of the Abruzzo research area.

The first map represents the mean monthly precipitation in October, and the second map represents the mean monthly soil moisture deficit for May, calculated over the 40-year period from 1951 to 1990. Daily soil moisture deficit was estimated at the nodes of a 5 × 5 km grid using a hydrological model. The grid-point values were then used to obtain the isoline map.

SOIL MOISTURE DEFICIT AT REGIONAL AND BASIN SCALES

A. Girolamo, F. Battistini, B. Cittarelli and A. Becker

The maps present some results obtained by applying a hydrological balance model over the Abruzzo region on the basis of the historical meteorological data stored in the National Agrometeorological Data Base (SIAN) of the Italian 'Ministero delle Risorse Agricole, Alimentari e Forestali', with a grant from UCEA (Ufficio Centrale di Ecologia Agraria). Additional meteorological data were provided by the Consorzio REDAT (LLPP weather stations).

Figure 5.3.11 Surface roughness measurement using a photograph of the profile.

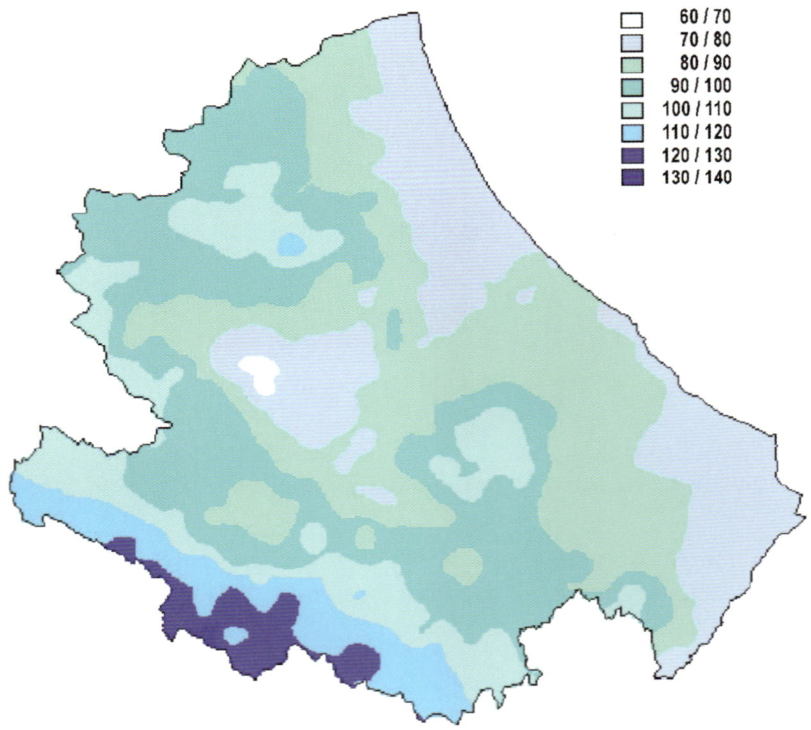

Finsiel
TERRITORIO E AMBIENTE

Figure 5.3.13 Mean monthly rainfall (in millimetres) for October in the Abruzzo. This is defined as the average total precipitation from 1 October to 31 October, calculated over the period 1951–1990. Daily total rainfall was estimated at the nodes of a 5 × 5 km grid (optimum interpolation). Grid point values were used to obtain the contour line map.

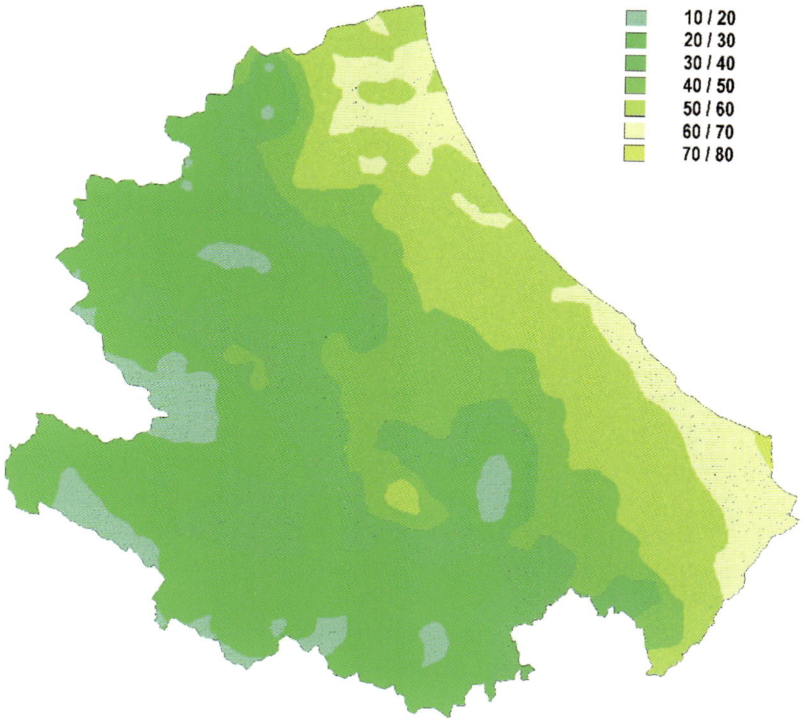

Finsiel
TERRITORIO E AMBIENTE

Figure 5.3.14 Mean monthly soil moisture deficit (%) for May in the Abruzzo region. This is defined as the average, from 1 May to 31 May, of the daily soil moisture deficit calculated over the period 1951–1990. Daily total rainfall was estimated at the nodes of a 5 × 5 km grid using a hydrological model. Grid point values were used to obtain the contour line map.

The two months were chosen in order to represent two different important situations: spring, when crops begin to use soil water resources, and autumn, when rainfall begins to replenish soil water resources.

The following methodology was applied to calculate soil moisture deficit:

- estimations of the meteorological parameters were made at the nodes of a 5 × 5 km grid (daily minimum and maximum temperature; total daily precipitation; day and night relative humidity; average day and night wind velocity; normalised daily sunshine hours);
- the hydrological model was applied using specialised meteorological parameters for the estimation of the potential evapotranspiration; actual evapotranspiration; soil moisture deficit; hydrologically effective rainfall.

The applied hydrological model uses a time-step of one day with three main operations:

- calculation of the potential evapotranspiration (PET) of the crop;
- evaluation of the surface resistance of the canopy and estimation of the actual evapotranspiration (AET) of the crop;
- estimation of the hydrological balance of the crop–soil system based on the actual evapotranspiration and on rainfall.

Both potential and actual evapotranspiration were calculated directly using the Penman–Monteith equation. Actual evapotranspiration was calculated, modifying bulk surface resistance according to the available water content. The modelling takes into account meteorological parameters and other factors such as: albedo, which varies depending on the canopy and the crop coverage through of the leaf area index (LAI); aerodynamic roughness length, depending on crop height; crop development and canopy LAI.

The term 'potential evapotranspiration', in this context, means the evapotranspiration of the specific crop with an unlimited reservoir of water. The term 'actual evapotranspiration' means the evapotranspiration of the same crop in the real conditions of water availability.

The hydrological balance was estimated through the following formula:

$$B = R_{\text{eff}} - AET$$

where B = water (in millimetres) that must be added, with its sign, to the soil reservoirs; R_{eff} = effective rainfall (in millimetres), estimated taking into account condensation and leaf interception; and AET = actual evapotranspiration.

This model was applied separately to the day and night estimated meteorological data, i.e. it was applied day by day to the regular grid of 5 × 5 km defined over the Abruzzo region.

THERMAL PROPERTIES OF SURFACES AND SOIL EROSION PROCESSES

A. Capraro

The knowledge of surface thermal properties is very useful in the study of desertification processes. Thermal properties of the surface provide information about the physical characteristics of the soils, e.g. composition, moisture content, degree of weathering, and thickness of regolith. The thermal balance map is related to the thermal inertia of the surface materials. This map is obtained from two airborne images that were obtained at minimum and maximum solar elevation. The images on which the map is based were collected and processed by the Alenia Remote Sensing Group, Rome. They were calibrated, geometrically corrected and subtracted one from the other to obtain a good estimation of the surfaces' thermal properties.

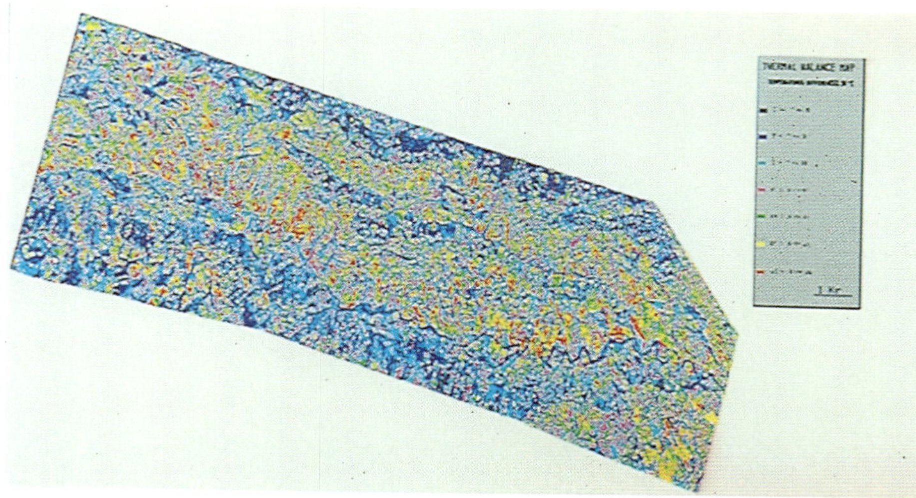

Figure 5.3.15 Thermal balance map of the Piomba river basin, Abruzzo, Italy.

A further map was obtained from near-infrared and middle-infrared airborne images, geometrically and radiometrically corrected. Infrared (IR) features provide important information on soil and vegetation conditions.

The map was obtained by means of spatial analysis, to derive textural

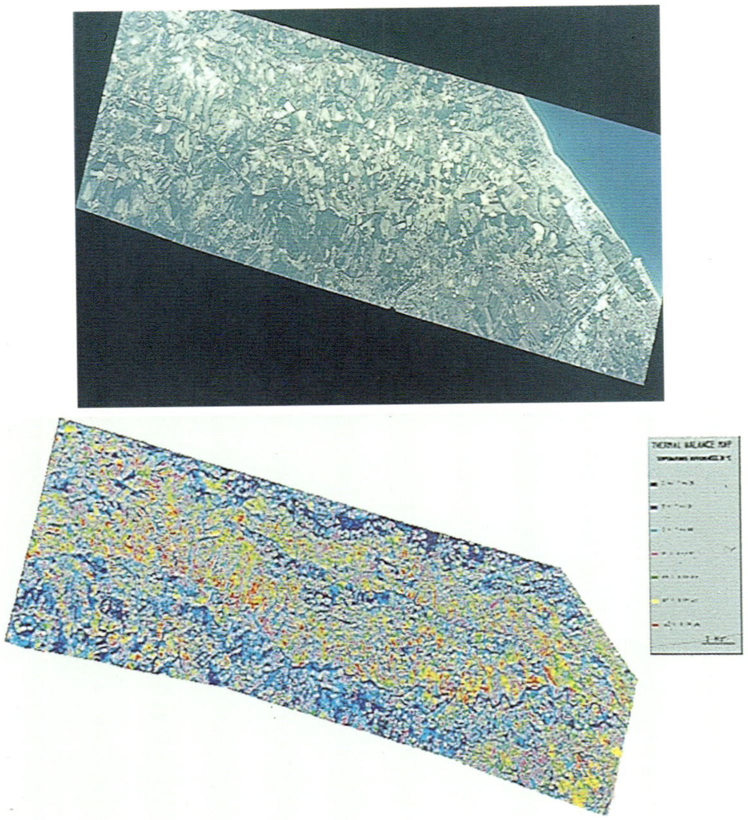

Figure 5.3.16 Near-infrared and middle-infrared images of Piomba Basin area, the Abruzzo, Italy. Reproduced with permission: Concessione dello Stato Maggiore Aeronautico (no. 096 20-4-1994).

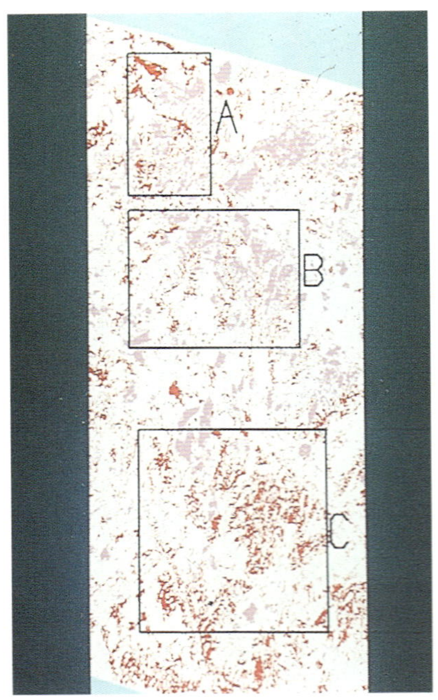

Figure 5.3.17 Detail of the 'soil erosion and degradation map' obtained from airborne multispectral data: Piomba Basin area, the Abruzzo, Italy. The effects of gully erosion are very visible. Images were collected and processed by Alenia, Remote Sensing Group, Rome, Italy.

information from the images. Multispectral images are calibrated by means of field measurements taken during the airborne campaign. Textural analysis is an effective tool for highlighting landscape erosional processes. These data are then superimposed on the data sorted from middle-infrared (MIR) and thermal-infrared (TIR) bands, to obtain a complete picture of the area.

This map describes the areas affected by strong erosion, mainly represented by gully eroded areas, and gives the most direct information for desertification studies.

CHARACTERISATION OF THE CIGNO AND PIOMBA CATCHMENTS BY AN INDEX COMBINING SOIL MOISTURE DEFICIT AND LENGTH OF THE ARID SEASON

M. A. Campanile and C. Aliventi

Soil moisture deficit was calculated as a function of the transpiratory demand for herbs in the dry season. The latter is defined as the period of at least five days when the average deficit is below a critical threshold value (typical for each soil textural class). Both the deficit and the length of the arid season are defined for the 75th percentile of the cumulative frequency distribution function for a 30-year time series.

An index was calculated to characterise the study areas with respect to soil moisture deficit and length of the arid season. The index is

$$I_a = (\mathrm{SMD_{min}}/\mathrm{SMD}_a) \times (\mathrm{LAS_{min}}/\mathrm{LAS}_a)$$

where I_a = index of aridity for each weather station; SMD = soil moisture deficit (as defined); LAS = length of the arid season (as defined); a = average and min = minimum. I_a can vary from 0 to 1.

A high value indicates that soil aridity is not a problem, whereas a low value indicates a serious problem.

The problem of soil moisture becomes less critical further from the coast as the land increases in elevation.

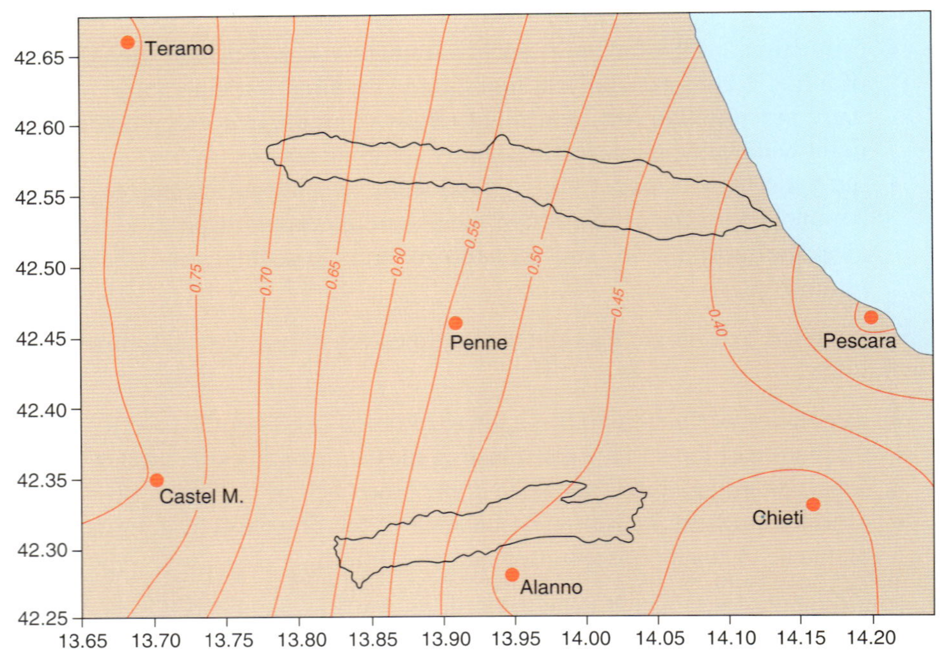

Figure 5.3.18 Spatial distribution of relative index of aridity values by means of linear Kriging.

LAND CAPABILITY CLASSIFICATION MODEL: APPLICATION TO CIGNO AND PIOMBA RIVER BASINS

M. Crespi, S. Lessi, A. Bonini Baraldi and F. Bovo

The model applied, called CAT ('Classificazione Agronomica del Territorio' (Land Agronomic Classification); Giardini *et al.*, 1987), was used to evaluate, at both regional and basin scales, the agricultural land productivity, on the basis of potential for agriculture, defined as the capability of land for cultivation of the main crops (herbaceous and trees) in a given climatic area.

To apply CAT, a thorough knowledge of the factors (such as soils) which influence productivity is essential.

The two tested areas, the Cigno and Piomba basins, have been divided into a mesh of 60 × 60 cells (pixels) and, on this basis, the model defines land agronomic value, identifying and quantifying possible limitations to current agricultural land uses. Limitations are ranked according to their importance.

The limitations considered are grouped within four components: soil, relief, hydrology and climate. The parameters considered include, for example, depth of the soil, soil texture, organic content, pH, exchangeable sodium percentage, salinity, water stagnation, risk of flooding, erosion and seasonal rainfall deficit.

The first step of the model application is data acquisition from different sources (maps, databases, direct investigations and bibliography) and the creation of a data set.

The next step is the classification of the land (such as suggested by Giardini *et al.*, 1987) which includes the following:

- assigning a score to each land element (pixel) according to the classes and subclasses to which it belongs;
- evaluation of interrelations between different limitations and definition of a new rank that corresponds to the land capability, based on weighting the constraints and summing these for each cell of the grid;
- if there are other restrictions, their values are compared with all the previous limitations and the sum adjusted in the same way.

Table 5.3.1 Land agronomic value.

Agronomic class	Constraint rank		Agronomic value
1	0–9.9	>90–100	Very good; wide range of crops; high productions
2	10–19.9	>80–90	Good; many possible crops; one or two weak constraints to agricultural use
3	20–29.9	>70–80	Fairly good; large number of possible crops, but for many others some constraints do not permit normal cultivation
4	30–39.9	>60–70	Satisfactory; some difficulties for normal agricultural use; only a few crops yield sufficient productions
5	40–49.9	>50–60	Marginal; reduced choice for possible crops
6	50–59.9	>40–50	Low; agricultural practices must be prudent; crops produce up to 50%
7	60–69.9	>30–40	Very low; agricultural uses are restricted and reduced to pasture or forest plantations
8	70–79.9	>20–30	Normal cultivation is unsuitable; practices must promote to environmental conservation
9	80–89.9	>10–20	No agricultural use possible; pasture and forest plantations
10	>90	>0–10	Difficulties for both pastures and forest plantations; actions for soil conservation to be taken

The process described is implemented in a geographical information system so it is possible to obtain graphical representations such as base and synthesis thematic maps and statistical data.

The Cigno Basin

The effectiveness of relief as a constraint is evident from the absence of the first two classes in the hilly Cigno Basin.

The percentage of the land classified in class 3 is also rather low (7.8%). This class comprises areas with a number of limitations, e.g. high silt and soil carbonate content, exchangeable sodium percentage between 2% and 6% and slope gradients of 4–10%. The occurrence of these and other limitations, such as the summer rainfall scarcity (200–300 mm), results in a score of 20–29.9. Class 3 lands are mainly located in the eastern-central part of the basin.

Class 4 occupies 13.3% of the area, in particular at low altitudes and also in the eastern part of the area. Generally pixels are assigned to this class as a result of the combination of some weak limitations, such as clay texture and gravel content.

Class 5 dominates mainly in the central-eastern part of the basin (occupying 25.8%) and particularly reflects high slope gradients.

Class 6 occupies 17.7% of the Cigno Basin, particularly the central-western locations which are either higher or in the vicinity of the confluence of the Cigno and Pescara rivers where stony soils or soils with high gravel content are present.

Only 1.9% of the area falls in class 7 in which the middle and upper river valley bottoms are included.

Class 8 (5.6%) is found in the north-eastern part of the basin, near the municipality of Nocciano. This area has shallow soil and very stony soils with high total carbonate contents and sub-alkaline reaction, in addition to relatively high slope gradients (4–10%).

The last two classes occupy 17.2% (class 9) and 10.6% (class 10) of the area. The former includes mainly river banks close to the river bed in the central-western part of the basin; the latter, the high parts of such river banks, characterised by rock outcrops and shallow soil, are found in the western part of the basin.

The Piomba Basin

The results of the classification are less well-defined as far as the estimation of the productivity of the agricultural land is concerned. The percentage of land classified as class 3 (1.2%) and class 4 (5.5%) is very small. Generally these areas are located at the bottom of the valleys and in the lower part of the basin. About a quarter of the land is in class 5 (28.0%). These hilly lands lie mainly in the eastern part of the basin and are characterised by slope gradients of 10–25%, a clay soil texture and a strong alkalinity.

Class 6 (14.5%) occurs particularly along the courses of the Piomba and Fossa del Gallo rivers, both characterised by stony soils with high gravel contents. However, given the texture and the shallowness of the soil, stagnation determines the constraints to productivity.

The production of the area in classes 7 and 8 (3.9% and 1% respectively) is also very low, while 32.3% of the area is in class 9. Steep slopes and a number of other limitations of different kinds render a great part of the area unsuitable for agriculture. Finally, 13.6% of the area falls within class 10: these lands are characterised by gully erosion, are generally very prone to erosion and are also characterised by steep slope gradients.

LAND CLASSIFICATION OF THE COASTAL ZONE OF THE ABRUZZO REGION BASED ON FARMING TYPE

P. Vignaroli

Another contribution to the knowledge of the background conditions of the area being studied is the identification and characterisation of the different agro-environmental typologies present in the research area. This has been done for 139 municipalities of the Abruzzo region by means of factor analysis, using 22 indicators for both components of the physical environment and the structure of agricultural sector. The socio-economic study has been restricted to the coastal area. This is defined as lying at an altitude between 0 and 600 m a.s.l. (ISTAT, 1982). The municipalities selected are classified as coastal hills and interior hills. In the coastal area soil surface erosion phenomena and suspended solid transport in the rivers are very pronounced. The mountainous part of the region is characterised by solid mass movement and it is not climatically marginal with respect to desertification.

Indicators for the socio-economic component have been derived from the data recorded in General Census of Agriculture and the General Census of Population (ISTAT, 1982, 1983). Nine types have been identified which are briefly described below.

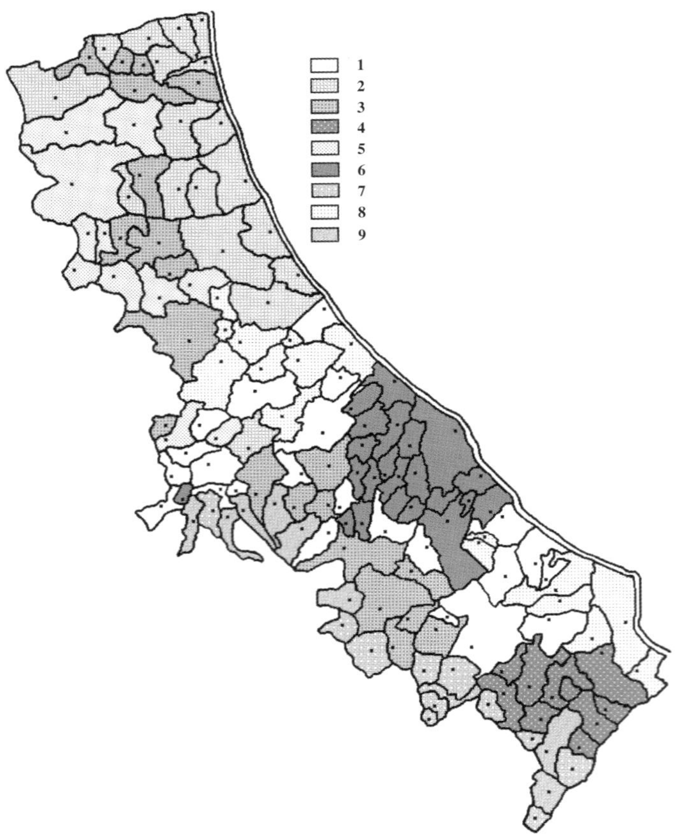

Figure 5.3.19 Land classification of the coastal zone of the Abruzzo region. 1, Productive family agriculture; 2, commercial, extensive and residential agriculture; 3, extensive family agriculture; 4, traditional family farming; 5, extensive unproductive agriculture; 6, intensive family agriculture; 7, traditional agriculture of the upland environment; 8, urban and suburban agriculture; 9, marginal agriculture of the high country areas.

1. *Productive family agriculture*. This comprises the 24 municipalities located in the coastal and internal hill country of the provinces of Pescara and Chieti. They are characterised by the presence of family-owned farms and non-intensive agriculture. The main activities are traditional ones, such as cereal cropping and animal husbandry, but intensive and high-profit crops (grapevine and orchard) also occur.
2. *Commercial, extensive and residential agriculture*. The 18 municipalities in this group are mainly situated in the coastal hill country and interior of the provinces of Teramo and Chieti and are characterised by extensive agriculture. The farms are medium- to large-sized (20–30 ha), and are largely managed by non-family labour. The low productivity of the agricultural sector is reflected by the residential character of the area.
3. *Extensive family agriculture*. This group includes 22 municipalities situated in the internal hill country of the province of Teramo. In contrast to the preceding group, the agricultural sector shows a greater degree of vitality and a closer relationship between ownership and farm operation.
4. *Traditional family farming*. This group comprises eight municipalities, of which seven are located in the southern interior of the province of Chieti, characterised by a crop distribution dominated by cereals and by low investment of capital and labour per unit area. It is a poor system which has still not completely disappeared and is typical of the interior areas of southern Italy.
5. *Extensive unproductive agriculture*. The 11 municipalities of this group, located in interior hill country of the provinces of Teramo and Pescara, are characterised by abandonment of animal husbandry and extensive crop distribution, accompanied by widespread inactivity and deconstruction of the farming enterprises.
6. *Intensive family agriculture*. This group consists of 20 municipalities located mostly in the coastal areas to the north of Chieti. The agricultural sector is very important with intensive crop production, and the majority of farms are small (less than 5 ha) or medium–small (5–10 ha) under family management.
7. *Traditional agriculture of the upland environment*. This type includes 12 municipalities situated in the strip of Apennine foothills that are within the three provinces under study. The physical features of the area represent a constraint to agriculture, which is extensive and largely based on cattle breeding.
8. *Urban and sub-urban agriculture*. The 15 municipalities of this group are to be found in close proximity to the city of Pescara and in the coastal hills of the province of Chieti. Agriculture is distinctly residential in character and production is not a primary concern.
9. *Marginal agriculture of the high country areas*. Seven municipalities of the sub-Apennine part of the provinces of Pescara and Chieti are of this type. The agriculture sector is similar to that of type 7, only here the conditions of marginality are due to the physical environment, resulting in the out-migration of the rural population, and consequently in the strong breakdown of the productive base.

6

RESULTS AND PROSPECTS

Results and Prospects

J. B. Thornes

Adjusting to and developing from the status quo

As we have shown in many sections of this Atlas, the Mediterranean terrestrial environments are complex in every respect. The geology and relief are complex and variable as a result of the relatively recent Alpine mountain folding. Resistant granites are faulted up against easily eroded marls; mountains and plainland are separated by only a few kilometres and deltas and lagoons pass quickly into rocky coastlines.

Superimposed on these geological and physiographic variations, and adding to the complexity of landscapes, is the Mediterranean climate with its hot dry summers and cool wetter winters. Diversity is generated by rainfall variations, topography and aspect. South-facing slopes in lowland areas in the extreme south may have 140 mm of effective rain; north-facing slopes in the northern Mediterranean mountains may have 1500 mm. Equally, the natural plant response must have been everywhere quite varied although it is almost certain that large areas that are now bush, grass or farmland were once forest. Humankind has largely changed the face of the Mediterranean so that it is nowhere natural, in the sense that the surface is nowhere unaffected by human activity. This human activity itself has added to the rich diversity of the environment that we see today.

Some would argue that, at least until perhaps a century ago, Mediterranean people were in harmony with their environment. This seems a rather far-fetched hypothesis. First, there is a recorded history of natural disasters, floods and droughts, earthquakes, fire and famine. Second, there was a high level of rural poverty, a low expectancy of life and a low level of productivity in ecosystems well adapted to periodic drought after centuries of change. Third, the Mediterranean economies were early adjusted to the exigencies of international trade, first in minerals but soon after in agricultural products, though, with a few marked exceptions, not to the industrial revolution. The comfortable view of a stable rural population nurturing and conserving the life-giving environment is largely a construction of romantics. The Mediterranean environment has been, and in rural areas still is, harsh and demanding, hazardous and a continuing challenge to the relationships between nature and humankind.

Nevertheless, as the pressures on health, welfare and trade begin to be brought under control, we have to recognise that the impacts of the last 50 years have been environmentally more profound than those of the previous 5000 years, and that improved social and economic conditions do bring with them environmental responsibility.

Building on MEDALUS

The work of MEDALUS and earlier research shows that there have been 40 years of economic growth and change that has pervaded all the Mediterranean regions of the European Community. This has brought significant improvements in health and welfare but at some cost in creating new environmental problems to add to the list of existing burdens. These include pollution and waste, salinisation of soils and accentuation of the flooding problems by urban occupancy of the flood-plains and poor water management. Above all, the increased reliance on water for agriculture, industry and domestic use has generated a rapidly growing crisis. MEDALUS simulations show that the water demand is likely to continue to increase steeply, especially in advanced island economies such as Crete. The problem will be to sustain the improvements in standard of living, built largely on the back of agriculture and tourism, without causing further environmental degradation and associated resource depletion.

The climatic studies of MEDALUS indicate an increase in temperature, a decreased rainfall and a lower soil moisture content over the next 60 years. Though the changes will be less marked than further north in Europe, and their effects somewhat obscured by the inherent climatic variability, their effects on environment may be more significant because the base condition is more hazardous. Plant growth is likely to be most reduced in the driest areas (by perhaps as much as 25–35%), with the largest effects on annual herbaceous grasses which are very important for erosion control. In wetter areas, above about 600 mm, the effects will be much less, perhaps a 6–10% reduction. The enhanced CO_2 effect is likely to have little impact on plant growth. Our simulations show that sustained drought even for only 8–10 years is sufficient to cause extensive die-back in semi-natural vegetation, and this is confirmed by both rainfall exclusion experiments and field monitoring of plant health.

All the MEDALUS work supports the widely held view that plant cover is the most effective mechanism for reducing runoff in storms and reducing soil erosion. Decreases in vegetation cover are therefore hazardous. However, our work also shows that there are significant increases in woodland cover, in *matorral* and in land abandonment throughout the target areas. These changes, if sustained or even encouraged, are likely to have much more dramatic impacts than climate change, and in a positive direction. If properly managed, on the basis of natural watershed units, they can reduce erosion to acceptable (geological) levels. Re-vegetation (not necessarily re-afforestation) must be understood to be for conservation as well as profit, with the associated public loss-bearing, i.e. the community has to learn to bear the cost. If successful, re-vegetation will also reduce runoff in storms and decrease the probability of flooding. It may not of itself generate more available soil moisture, since plants under such rainfall regimes will tend to use all available water for growth. However, the model simulations also suggest that there will be a progressive increase in soil organic matter content, improved soil structure and increased infiltration rates, all leading to long-term improvements in soil status. Recovery of vegetation to levels sufficient to reduce soil erosion dramatically might require as little as 8–10 years in ungrazed abandoned areas. Our field experiments show that, even in the dry years of the late 1980s and early 1990s, 30% cover of perennial vegetation can be achieved in a small time provided that the land is not grazed—and this without elaborate range land treatment. Restoration of soil organic matter to levels sufficient to improve aquifer recharge rates might take another 40–50 years.

The restoration of surface cover will not solve the water resource problem in the foreseeable future because too much water has already been mined and there is no sign of an end to the demand. Such restoration, carried out in an integrated fashion at the watershed scale, would however markedly reduce the flood risk hazard, increase phenotypic and genetic diversity, restore soil organic matter and soil productivity and notably decrease soil erosion. Meanwhile, on badly eroded lands better soil management and specific mitigation treatments, e.g. through stone mulching, could significantly reduce sediment yield and put abandoned lands on the road to recovery. Although it is too early to tell, there is some evidence that measured soil erosion rates at field plots are generally significantly lower than is commonly supposed. Ironically this may be because the field experiments have been conducted in a generally very dry run of years. Together with results of the MEDALUS hillslope model simulations, this might be taken as support of the general proposition that once below about 300 mm, decreasing rainfall decreases runoff and thence erosion. Of course, this depends on the intensity of the storms that remain. Reducing a large number of small storms has quite a different effect from reducing a small number of very large storms in the simulated erosion.

There are some caveats in this rather optimistic picture. MEDALUS work shows that recovery will require grazing exclusion to be effective, though this is not to say that all grazing is necessarily destructive. Secondly, more bush and tree cover is likely to significantly increase fire risk, and management will be needed to accommodate this problem. MEDALUS work shows that soon after fire runoff and sediment rates are high, but work elsewhere seems to indicate that this quickly recovers to pre-fire levels if recovery is allowed to occur and if there is sufficient rainfall. There is as yet an unquantified risk that further land abandonment will feedback positively to create rural wilderness over which fire control will be difficult if not impossible to manage. Thirdly, our work seems to show that the planting of exotics, and notably eucalyptus species, for restoration or for profit, is likely to be less successful than the use of native species. In practice, the low canopy cover beneath eucalyptus seems to encourage higher levels of erosion than under other covers. *Matorral*, with a biomass of more than 3 kg/m^2, seems to be capable of providing as good a protection from erosion as young fully grown pines.

Finally, and above all, the great diversity alluded to at the beginning of this Section implies that there is no fixed and unique recipe for addressing the Mediterranean types of desertification problems. Neither is there a unique 'technological fix'. Rather it is a matter of having a thorough understanding of the current state of the diverse environments, through mapping, remote sensing and regional level indicators with the aim of identifying Desertification Environmentally Sensitive Areas (DESAs). Then strategies for their development will have to be designed from the ground up, in conjunction with local and regional authorities. Mediterranean people have for too long been subject to ill-informed central control and a high level of public participation will be required for successful development. The methodology to achieve this level of management, in relation to centrally-funded mitigation policy, is being developed in MEDALUS III. Moreover, flexibility of response will be the keynote; Mediterranean environments and the social and economic forces controlling them are as variable in time as the Mediterranean environment is in space.

GLOSSARY

ablation rate: quantity of soil eroded from slopes and transported by running water. Generally expressed in t/ha/year.

absorption feature: a range of wavelengths (or frequencies) in the electromagnetic spectrum within which radiant energy is absorbed by a substance.

aggregate distribution: distribution of classes of soil aggregates (particles adhering to one another) according their size.

agro-ecosystem: an ecological system which results from agricultural cultivation.

agro-environmental typologies: land classes with similar environmental–physical characteristics (e.g. topography, land use).

albedo: proportion of incident solar radiation reflected by the clouds surrounding the Earth.

alfisols: soils that have an ochre skin around the pedological horizons, an argillic horizon, moderate to high base saturation, and in which water is held at < 15 bar-tension for at least three months each year when the soil is warm enough for plants to grow (see also soil taxonomy).

alluvial soils: soils formed on newly accumulated fluvial deposits.

anchor station: an observation site at which quantities are measured that are needed to calibrate measurements made from satellites and to validate the information inferred from these measurements.

ancillary data: in an environmental context, information, e.g. morphological, geological and climatological data, useful in support to remote-sensing data to obtain more accurate land-cover categories.

annual flow: volume of flow at the outlet of a catchment during one year. Generally expressed in millimetres depth as if spread over the entire basin.

Ap-horizon: the surface mineral horizon in cultivated soils.

apparent photosynthesis: The net result of photosynthetic CO_2 uptake and respiratory CO_2 losses ($molCO_2/m^2/\Delta t$).

aqueduct: artificial surface channel for conveying water.

aquifer: a geological formation of water-bearing rock with sufficient porosity and permeability to yield economic supplies of groundwater.

arable land: portion of farm area devoted to herbaceous annual crops: cereals (including fallow lands), horticultural, fodder crops and other arable crops (e.g. potato, sugar beet, and other industrial crops) (ISTAT, 1982).

atmospheric boundary layer: the air layer adjacent to the surface, the structure of which (wind, temperature, and moisture profiles as well as aerosol content) is strongly influenced by the surface; its extent may range from a few decametres to 3 km.

atmospheric circulation: atmospheric motion extending over a part or the whole of the Earth (official definition from WMO vocabulary).

badland: an area where gullies adjoin each other and cover all, or nearly all, of the surface.

biancane: a form of erosion with a typical dome-shaped configuration and a radial drainage network.

blown-sand dunes: a hill or ridge of blown-sand piled up by the wind.

calanchi (badlands): erosion form that occurs in blue clay without vegetation.

calcareous chernozem: meadow soil with finely dispersed calcium carbonate content.

canestrato: goat and sheep milk cheese.

catena: a repeated sequence of soil profiles that is geographically related to and associated with relief features. The term is derived from the Latin (= 'chain'), since the soil profiles were found to be linked together in the same fashion as a hanging chain when traced laterally down a slope from the interfluve to the valley floor (Whittow, 1984). A catena may also be of topodrainage or plant community sequences.

CD-ROM: Compact Disc–Read-Only Memory. Compact disc containing large sets of digital (computer) information.

change (in connection to variability and fluctuation): transition from one climatic condition to another, over a specified geographical region, due to variations and/or fluctuations of the atmospheric circulation at timescales from decades to centuries.

climate scenarios: internally consistent pictures of a plausible future climate, not predictions of future climate.

coltura mista: different crops (annual and perennial) cultivated on the same plot.

compaction: the development of a dense, compact surface soil layer (e.g. due to cultivation with heavy machinery, overgrazing), characterised by a much lower permeability so impeding the movement of water and air, and the growth of plant roots.

complex area: a spectrally complex and heterogeneous environment where several factors, e.g. mixed soil compositions, terrain irregularities and differences in substrates, make the different ground classes barely distinguishable on the base of the only spectral responses.

confined aquifer: groundwater reservoir overlain and underlain by impervious or almost impervious rock formations.

cortijo: farmhouse in southern Spain.

crusting: development of a surface layer on soils ranging in thickness from a few millimetres to a few centimetres, which is much more compact, hard and brittle when dry than the material immediately beneath it (see also compaction).

dielectric constant: for an isotropic medium, the dielectric constant is the ratio of the capacitance of a capacitor filled with a given dielectric to that of the same capacitor having only a vacuum as dielectric.

digital elevation model: an ordered collection of topographic elevation data for a particular area.

ECHIVAL: European international project on Climatic and Hydrological Interactions between Vegetation, Atmosphere and Land surfaces.

ecosystem (**ecological system**): an open system that includes all the organisms that function together in a given area (ranging in scale from very small to the whole globe) interacting with the physical environment so that a flow of energy leads to clearly defined biotic structures and cycling of materials between living and non-living parts (Odum, 1983).

edaphic: a term referring to the soil, particularly with respect to its influence on organisms, in which it is regarded as one of the major factors, together with the climatic factor, affecting plant growth (Whittow, 1984).

EFEDA: ECHIVAL Field Experiments in Desertification-threatened Areas.

emissivity: ratio of the radiation emitted by a surface to the radiation emitted by a perfect 'black body' at the same temperature; according to the referred region of the electromagnetic spectrum it is related to some geometrical and physical characteristics of the surface.

entisols: soils that have little or no evidence of development of pedogenic horizons (see also soil taxonomy).

erodibility: the ease with which natural materials break down when they are attacked by agents of denudation (whose capacity is called *erosivity*, especially for rainfall events of short duration–high intensity). Erodibility is dependent upon susceptibility to removal and transportation, and the term is often used in the context of the potential of a soil or sediment to gullying or rill formation (Whittow, 1984).

erotope: land unit for erosion assessment.

evapotranspiration: the combined loss of water from a given area and during a specified period of time, by evaporation from the soil surface and by transpiration from plants. In arid conditions most of precipitation returns to the atmosphere, and does not participate in the recharge. *Potential evapotranspiration* is the water lost by evapotranspiration from a growing crop without restriction on the available water.

fallow land: land that is temporarily without agricultural practices, namely ploughing. Land is just abandoned, the land cover being only the spontaneous herbaceous cover, on which grazing sometimes occurs. The objective is to allow the soil to recover some of its properties before the next cereal crop. On poor soils, such as in the Alentejo, fallow could last as long as 7–10 years.

field radiometry: the study of the interrelationships between the spectral characteristics of objects and their attributes in the field environment.

fluvents: brownish to reddish entisols that formed in recent water-deposited sediments, mainly on flood plains, fans, and deltas of rivers and small streams but not in back swamps where drainage is poor.

garden: a piece of ground, often walled, devoted to growing flowers, fruit and vegetables, usually irrigated, sometimes intensively.

gauging station: section along a stream where water flow and sediment discharge are measured.

general circulation models (GCMs): complex, three-dimensional computer-based models of the atmospheric circulation developed from numerical forecasting models and used to investigate future climate change.

geo-management system: system superimposed on existing software applications to retrieve and combine data, ordered via their geographical co-ordinates and maps.

geographic information system (GIS): system with digitised computer maps; these maps can be combined with other maps, or can be processed.

gota fría: very cold, low-pressure air mass at upper air elevations (500 mb) that produces a high atmospheric instability and torrential precipitation.

graben: a mass of rock let down between parallel faults.

gradient: transect characterised by at least one variable increasing or decreasing upwards or downwards.

grasslands: portion of farm area devoted to permanent meadows and pasture (permanent meadows are normally harvested and pastures are normally grazed) (ISTAT, 1982).

gravimetric method: a method of soil moisture estimation using the difference between the wet and dry weight of a soil sample.

greenhouse effect: greenhouse gases in the atmosphere, such as CO_2, are largely transparent to short-wave solar radiation, but absorb long-wave radiation from the Earth and so maintain the Earth at a temperature higher than it would be in their absence. This natural effect is enhanced by the release of greenhouse gases from human activities such as the burning of fossil fuel.

grey level co-occurrence matrix: matrix of the spatial grey tone dependence frequencies. It describes the average spatial relationship which the grey tones in an image have to one another.

gross domestic product (GDP): direct indicator of economic performance of a country or a region. For example, analysis of GDP regional trends have been used (Dunford, 1993) to investigate the recent regional economic dynamics in Europe.

groundwater: subsurface water that occurs beneath the water-table occupying the pores of the soils and geological formations that are fully saturated.

groundwater-table: surface within the zone of saturation of a phreatic (unconfined) aquifer along which the pressure is atmospheric.

Guadalentín: an arabic word that means 'river of mud'.

gullies: deep, water-worn channels, cutting through soil into weathered material and/or rock.

habitat: the place where an organism (plant or animal) or a community lives (survives and reproduces), i.e where its environmental requirements (ecological niche) are fully or partially met.

haploxeralfs: relatively thin, reddish to brownish xeralfs that have a clear or gradual upper boundary to an argillic horizon or have a loamy particle-size class throughout the argillic horizon.

hierarchical modular modelling: the process of constructing a model by selecting and linking a number of modules, each of which represents some component or relationship in the modelled system, and each of which can contain within it further modules, nested to any depth.

humic sand soils: sand soils with less than 1% humus content.

hydraulic conductivity: the capacity of a porous medium to let water flow through it. It is the combined property of a porous medium and the fluid moving through it in saturated flow, that determines the relationship, called Darcy's law, between the specific discharge and the head gradient causing it.

inceptisols: soils of humid regions that have altered horizons which have lost bases or iron and aluminum but retain some weatherable minerals (see also soil taxonomy).

inertia: the amount of stress a system can withstand before it is disturbed.

infrared region: portion of the electromagnetic spectrum just beyond the red end of the visible spectrum, such as radiation emitted by a hot body.

inhomogeneities: in the context of land-surface processes, the variability of soil moisture, roughness, soil type, slopes and vegetation that affect the exchange of heat and momentum between the surface and the atmosphere.

interception: fraction of rainfall lost to evapotranspiration, due to retention on the canopy. Interception is equal to rainfall minus throughfall.

key parameter: in an environmental context, a property that is an important indicator for geo-ecological processes.

LAI: leaf area index. The area of leaves above a given area of ground (usually one square metre).

land capability: the range of uses for a tract of land or the capacity of land to produce crops.

land evaluation: the process of assessment of land performance when used for specified purposes. The land evaluation concept involves the definition of profit-related land qualities, the end product of which is a decision on optimum land use, whether private or public.

land suitability: the fitness of a given type of land for a specified kind of use.

land suitability system: methodology for, directly or indirectly, evaluating land suitability for a specified purpose or practice.

land unit: an area of land, cartographically delimitable at the operational scale, reasonably homogenous in the geological, pedological, hydrological, vegetational and land-use characteristics.

LANDSAT: acronym which indicates a series of Earth resources scanning satellites; the data recorded by the sensors are widely used for land resources assessment.

latent heat flux: the heat of evaporation that is carried with the flow of moist air.

leaf water potential (Ψ): the energy status of the water contained in leaves.

lithology: description of the physical character of a rock.

loess: fine-grained, permeable, unstratified Pleistocene aeolian deposit.

logistic curve: this function refers to growth characterised by an initial rapid increase and ultimately limited by the availability of a resource. For instance, in semi-arid environments, vegetation growth is limited by the availability of rainfall and can only grow to some sort of capacity which is defined here as the steady-state cover.

macchia mediterranea: bushy vegetation made up of shrubs and low trees (see also matorral, maquis, Mediterranean scrubland).

macrofauna: animals living in and on the soil which are larger than 1 cm. This size class of animals is populated among others by larger arthropods (e.g. millipedes), worms and snails.

maquis: a vegetation type of the Mediterranean area, mainly composed of evergreen broadleaved shrubs, less than about 5 m high.

Markov chain: system that can exist in one of a number of states and in which the probability of transition from one state to another depends only on the current state. It can be used to model changes of vegetation types during successional processes.

matorral: type of shrubby vegetation cover as found in south-east Spain, comparable to the French *maquis* and *garrigue* and *macchia mediterranea*.

Mediterranean scrubland: Mediterranean scrublands have resulted from the interaction between natural factors and very ancient human disturbance. The main control on the ecosystem is the annual summer drought. This intense hydric stress imposes a set of adaptations (such as sclerophylous leaves) and characteristic structures of the plants.

mesofauna: animals living in and on the soil, with a size between 0.2 mm and 1 cm. Some of them play an important role in the first phase of breakdown of leaves and litter on the soil surface, like arthopods. Other animals belonging to this class are nematodes and mites.

mesoscale: the time and space scale of meteorological processes that are influenced by the regional topography and the diurnal course of insolation ranging from 10 to a few hundred kilometres and an hour to a few days.

Mezzogiorno: this is comprised of the the southern Italian administrative regions of Abruzzo, Molise, Campania, Puglia, Basilicata, Calabria, Sicilia and Sardegna as per the Italian D.P.R.1523/67. All these regions, except Sardegna, used to be part of the former Kingdom of Naples.

microwave backscattering coefficient: ratio of the power scattered, in a backward direction, by an equivalent isotropic scatterer, to the incident power for a unit area incident on the scatterer and the area illuminated by the incident beam, in the microwave spectral region of the electromagnetic spectrum.

microwave probe: portable probe operating in the microwave region at 1.42 GHz frequency, which measures the dielectric constant of the soil at the contact of the probe's tip.

microwave region: region of the electromagnetic spectrum corresponding to wavelengths from few tenths of a centimetre to some millimetres equivalent to frequencies from 1 to 100 GHz.

microwave remote sensing: see microwave region.

module: a computational unit that can be used in constructing a model.

montado: savanna in Portugal, typically with cork oaks (*Quercus suber*) and/or live oaks (*Quercus rotundifolia*) and some cultivated fields.

multispectral: generally used for acquisition of remote-sensing data in two or more spectral bands.

multitemporal data: data acquired over the same ground site in different times. They are useful to monitor in-progress phenomena.

natural: 'the word "natural" is here (as elsewhere) a verbal cloak for ignorance' (Hardin, 1963).

neural networks: computing architectures, inspired by biological nets, constituted by a massively parallel interconnection of simple computational elements.

neutron probe: equipment used to measure the soil moisture through the scattering and slowing down of fast neutrons from a radioactive source.

nutrient: any mineral element participating in the structure of the living plant matter.

ochrepts: mainly the light-coloured, brownish, more or less freely drained inceptisols of mid to high latitudes.

orthents: these are primarily entisols on recent erosional surfaces; the erosion may be geological or may have been induced by cultivation or other factors, but any former soil that existed has been completely removed or so truncated that the diagnostic horizons for all other orders are absent.

other land uses: portion of farm area devoted to built-up areas, roads, canals, rocks, ornamental gardens and 'non-UAA' within the farm area. The latter in particular is defined as the part(s) of the farm area currently not cultivated but susceptible to cultivation by the standards of the farm at the time of the census. Non-UAA therefore includes abandoned lands, but not fallow land (ISTAT, 1982).

palexeralfs: reddish xeralfs that have a thick argillic horizon; these soils are in relatively stable landscape positions on gentle slopes and their genesis began before the late Pleistocene.

parametrisation: a simplified mathematical expression to describe complex processes in models.

pastoral value: a natural pasture qualitative index determined along representative sampling transects. The pastoral value is obtained using the relation: $Pv = CR \times$ sum of $SCC \times IS$; where CR (covering rate) is the presence or absence of vegetation, the SCC is the Specific Contact Contribution ratio between the number of contacts for each species identified and the total number of contacts collected on the transect and, IS is the specific index for each species related to its palatability characteristics.

permanent crops: portion of farm area devoted to vineyards, olive groves, fruit orchards, citrus groves and 'other permanent crops' (i.e. nurseries for agrarian and forestry crops, chestnut groves for the production of fruits and poplar stands) (ISTAT, 1982).

phreatic aquifer: porous water-bearing formation in which the groundwater-table forms the upper boundary.

phrygana: undershrubs or 'dwarf shrubs', e.g. thyme, sage, the many species of broom, species of *Cistus* and *Phlomis*. Undershrubs are not potential trees; they are short-lived and reproduce by seed; most species are gregarious; most are unpalatable; after a fire most of them die, and are often replaced by different species of undershrubs.

pixel: abbreviation for picture element, which is the data sample in the output product (e.g. image) to which a radiance value is assigned.

plant community: an assemblage of individuals of one or more species that influence one another and modify their environment (after Mueller-Dombois and Ellenberg, 1974).

porosity: ratio of the volume of soil pores to the total volume of the soil (consisting of both soil particles and pores).

pressure field: spatial representation of the distribution of the atmospheric pressure, with its depression, anticyclones, etc. (Official definition from WMO vocabulary).

rambla: ephemeral water course (Spain).

recharge: infiltration of the rain, first into the soil, then deeper, towards aquifers.

reflectance: ratio of the radiant flux reflected by a surface to that incident upon it.

relational database: (large) set of data where data can be combined and retrieved following specific desired relationships, often with the help of queries.

relative water content: the percentage of water in plant tissues with respect to the maximum that these can retain.

remote sensing: the acquisition of information related in some way to the Earth's natural resources or environment by a recording device that is not in physical or intimate contact with the object under study.

resilience: the ability of a system to return to its original status after applying (environmental) stress.

rhodoxeralfs: more or less dark red xeralfs that form in limestone, basalt and other highly basic parent materials.

saladares: ecosystems typical of arid and semi-arid environments that can be named crypto-wetlands, as water is very rare over the soil surface but is the driving force responsible for their origin and the accumulation of salts in the soil which is an essential characteristic. They are easy to identify because of the presence of indicator vegetation communities where halophytes are the main component (Spain).

salinity: concentration of dissolved salts in water, expressed in grams per litre, when the organic matter has been oxidised, the carbonates converted to oxides, and the bromine and iodine replaced by chlorine.

salt-water intrusion: penetration of salt-water into freshwater aquifers under the influence of groundwater development.

savanna: grassland and/or shrubs with scattered trees which do not form a complete canopy.

scattering: physical phenomenon due to the presence of inhomogeneties and anisotropies which determine the re-irradiation of the energy impinging along an initial direction on a medium, the scatter, with different characteristics from the initial ones.

scenario: internally consistent picture of a plausible future state (e.g. a model simulation of a possible future state, a projection of future climates). A scenario is not a prediction of a future state.

sea breeze: wind system induced by the heating of land surfaces and the resulting pressure gradients.

sedimentary body: valley or trough filled with sediments.

semi-natural conditions: areas where human disturbance is slight, e.g. where there is extensive cattle grazing.

semi-natural ecosystem: a unit of the environmental system modified by the indirect impacts of human activities or resulting after the release of a direct human activity.

simulation model: simulation of (a part of) the physical environment with the help a computer program following rules which describe or fit actual processes.

sink: geographical location where water on sediments are stored or swallowed.

soil horizon Ap and CB: a layer of soil more or less parallel to the ground surface, but differing from adjacent genetically related layers in any of the following properties: soil colour, consistency, soil structure and soil texture—in addition to biological, chemical and mineralogical differences. Four horizons are recognised (A, B, C and D); only horizons A and B represent the true soil, while C is the subsoil of weathered parent material, and D is the relatively unweathered parent material or bedrock. The Ap is the ploughed horizon, in the A horizon; CB is the transitional zone between B and C (Whittow, 1984).

soil moisture regime: the presence or absence either of groundwater or of water held at a tension < 15 bars in the soil or in specific horizons by periods of the year.

soil taxonomy: a hierarchy of classes developed by soil scientists for comparing soils from different locations. The best known (and used here) was developed by The United States Department of Agriculture.

soil temperature regime: the temperature regime can be described by the mean annual soil temperature, the average seasonal fluctuations from that mean, and the mean warm or cold seasonal soil temperature gradient within the main root zone, which is the zone from a depth of 5 to 100 cm.

spectral image: image of the energy backscattered from the Earth in a given range (also called spectral band) of the electromagnetic spectrum.

spectral mixture modelling: technique that converts encoded radiance from all bands of any multispectral sensor to fractions of spectral end members which corresponds to material in the scene such as green vegetation, non-photosynthetic vegetation, soil, shade, etc.

spectral signature: spectral distribution of the flux incident at and reflected by a ground surface.

stability: in the context of plant communities/community regeneration is defined as resistance to irreversible change in species composition and vegetation structure.

stability (instability): stability in the context of soils is defined as the resistance to any sudden temporal or spatial physical and/or chemical change and hence with enduring characteristics. Unstable is the converse of this.

standard gross margin (SGM): unit of measurement of the economic size of the agricultural sector. It has been defined as 'the difference between the standardised monetary value of gross production and the standardised monetary value of certain special costs' (Hecke, 1992).

stemflow: fraction of rainfall which runs down tree trunks and plant stems.

structural hillslope: hillslope with a morphology controlled by the underlying rock structure.

subsistence agriculture: agriculture whose productivity is only enough to support the farmer's family as compared with commercial agriculture.

suction cups: porous device enabling the sampling of the water from an unsaturated soil.

supervised classification: digital information extraction technique in which the operator provides training site information that a computer uses to assign pixels to categories.

sustainable strategy: a plan of action that can be maintained in the long term.

SVAT model: a mathematical model that describes the transfer of energy and water (and sometimes carbon) from the soil surrounding the roots via stem and leaves into the atmospheric boundary layer.

tensiometer: porous cup filled with water buried in the soil to measure soil suction through capillary tension.

terrace: narrow surface plane or with shallow slope build-up following the contour lines, with the aim of increasing the water-holding capacity and accumulating fertile soil.

texture (soil): size distribution of the individual particles in the soil. Four main classes in particle sizes are generally distinguished: clay (< 2 μm); silt (2–63 μm); sand (63 μm–2 mm); gravel and boulders (> 2 mm). For instance, fine-textured soils contain many clay particles, coarse-textured soils contain many sand size particles or gravel.

texture image: image expressing how the grey level of an image (e.g. spectral image) is spatially distributed.

thermic temperature regime: the mean annual soil temperature is 15°C or higher but lower than 22°C, and the difference between mean summer and mean winter soil temperature is more than 5°C at the depth of 50 cm or at a lithic or paralithic contact, whichever is shallower (see also soil temperature regime).

throughfall: fraction of rainfall that falls through the tree canopy.

torric moisture regime: in the torric moisture regime, the moisture control section in most years is (1) dry in all parts more than half the time (cumulative) that the soil temperature at a depth of 50 cm is above 5°C; and (2) never moist in some or all parts for as long as 90 consecutive days when the soil temperature at a depth of 50 cm is above 8°C (see also soil moisture regime).

total dissolved solids (TDS): the TDS in groundwater is determined by weighing the solid residue obtained by evaporating a measured volume of filtered sample to dryness.

transect: sequence of locations following a geographical line.

unsupervised classification: digital information extraction technique in which the computer assigns pixels to categories with no instruction from the operator.

utilised agricultural area (UAA): portions of farm area devoted to crops (arable land and permanent crops) as well as to grasslands. Definitions of land-use categories vary slightly from country to country, however it appears from the available sources that as from 1981, an effort has been made by the EU countries towards the harmonisation of these definitions in order to render it possible to make comparisons between the countries. UAA as defined here corresponds to the 'agricultural area' of FAO statistical records, where under UAA only arable lands and permanent crops are included.

vegetation index: in this context, the difference between the signals measured in the near-infrared and red part of the reflected spectrum divided by the sum of both signals (there are various other possibilities to construct vegetation indices).

visible region: portion of the electromagnetic spectrum interval to which the human eye is sensitive, the spectral interval being from approximately 0.4 to 0.7 μm.

weathering: physical disintegration and chemical decomposition of earthy and rocky materials on exposure to atmospheric agents, producing an *in situ* residue called soil.

woodlands: portion of farm area devoted to tall forests, coppices and *maquis* (ISTAT, 1982).

xenobiotics: molecule whose presence in the environment results solely from human activities.

xeralfs: mostly reddish alfisols of regions that have a Mediterranean climate, characterised by xeric moisture regime.

xeric: of or pertaining to a habitat having a low or inadequate supply of moisture.

xeric moisture regime: in a xeric moisture regime the soil moisture control section is dry in all parts for 45 or more consecutive days within the four months that follow the summer solstice. It is moist in all parts for 45 or more consecutive days within the four months that follow the winter solstice (see soil moisture regime).

xerochrepts: soils that are poorly developed but more mature than the soil order entisols (e.g. xerorthents) with xeric moisture regime and variable organic matter content (the reddish or brownish ochrepts of Mediterranean climates).

xerofluvents: soils that have little or no evidence of pedogenic alteration, form on slopes of less than 25%, have xeric moisture regime and variable organic matter content. Fluvents which have a xeric moisture regime.

xerophytic: a plant species which has developed means of combating drought by changes in its anatomical and morphological structures. These include the reduction of the leaf surface; the substitution of spines or thorns for leaves; the development of hard, glossy leaves; the ability to produce waxy secretions; the growth of thick bark; the formation of thick bulbous stems and other water-storing devices; the growth of deep roots.

xerorthents: soils that have little or no evidence of development of pedogenic horizons, xeric moisture regime and often low organic matter content. Orthents of Mediterranean climates which have a xeric moisture regime.

REFERENCES

Abbott, M. B., Bathurst, J. C., Cunge, J. A., O'Connell, P. E. and Rasmussen, J., 1986a. An introduction to the European Hydrological System—Système Hydrologique Européen, 'SHE', 1: History and philosophy of a physically-based, distributed modelling system. *Journal of Hydrology* **87**(1/2): 45–59.

Abbott, M. B., Bathurst, J. C., Cunge, J. A., O'Connell, P. E. and Rasmussen, J., 1986b. An introduction to the European Hydrological System—Système Hydrologique Européen, 'SHE', 2: Structure of a physically-based, distributed modelling system. *Journal of Hydrology* **87**(1/2): 61–77.

Adrian, G. and Fiedler, F., 1995. Mesoscale modelling. In: Noilhan, J. (co-ordinator), Annual Progress Report, Contract EV5V-CT93-0269, EFEDA Group VI on surface processes and atmospheric modelling.

Aita, L., Corbetta, F. and Orsino, F., 1977. Osservazioni fitopatologiche sulla vegetazione forestale dell'Appennino lucano centro-settentrionale. I La cerreta. *Archivio biogeografico Italiano* **3**(4): 53.

Agricultural Regional Statistics, 1991. Conserjeria de Agricultura, Pesca y Alimentacion. Region de Murcia. Murcia.

Alberto, F., 1971. Considérations sur la pierrosité des sols bruns à croûte calcaire du bassin de l'Ebre. *Bull. Rech. Agron. Gembloux* **6**: 180–185.

Alexander, D., 1983. Difference between 'calanchi' and 'biancane' badlands in Italy. In: Bryan, R. and Yair, A. (eds), *Badland geomorphology and piping*. Geobooks, Norwich, pp. 71–85.

Ambroise, R. *et al*., 1989. *Paysages de Terrasses*. Edisud, Aix-en-Provence.

Aronne, G. and Wilcock, C. C., 1992. Reproductive characteristics and breeding system of shrubs of the Mediterranean region. *Functional Ecology* **8**: 69–76.

Aru, A., 1993. The problem of heavy metals pollution in mining areas of Sardinia. Preliminary findings. *Studi Sassaresi* **35**: 293–300.

Aru, A., Baldaccini, P. and Vacca, A., 1990. *Carta dei suoli della Sardegna*, scale 1:250 000. Regione Autonoma della Sardegna, Università di Cagliari, Cagliari, Italy.

Aru, A., Puddu, R., Tomasi, D., Usai, D., Vacca, A. and Pinna, G., 1993. Experimental plots at Is Olias (Rio S. Lucia catchment area). Some data and results of MEDALUS I and MEDALUS II (first year). Working Paper No. 6, MEDALUS Project, King's College London, London.

Aru, A., Baldaccini, P. and Barrocu, G., 1994a. Desertification in the Mediterranean: cases and remedial measures. *International Symposium and Workshop on Desertification in Developed Countries*, Tucson, AZ, 24–29 October 1994.

Aru, A., Brundu, G., Camarda, I., Kahanamoei, A., Madeddu B., Roxo, M., Puddu, R., Satta, V. and Vacca, A., 1994b. Assessment and management criteria of desertification in selected Mediterranean areas, Working Paper No. 31, MEDALUS Project, King's College London, London.

Aschmann, H., 1973. Man's impact on the several regions with mediterranean climates. In: di Castri, F. and Mooney, H. F. (eds), *Mediterranean type ecosystems.* Chapman and Hall, London, Springer-Verlag, Berlin, pp. 363–371.

Axelrod, D. I., 1973. History of the mediterranean ecosystem in California. In: di Castri, F. and Mooney, H. F. (eds), *Mediterranean type ecosystems.* Chapman and Hall, London, Springer-Verlag, Berlin, pp. 225–277.

Bachtler, J. and Michie, R., 1993. The restructuring of regional policy in the European Community. *Regional Studies* **27**(8): 719–725.

Bagnouls, F. and Gaussen, H., 1953. Saison sèche et indice xérothermique. *Doc. Carte Prod. Vegetation* **3**: 1–8.

Barrett, G. W. and Peles, J. D., 1994. Optimizing habitat fragmentation. An agro-landscape perspective. *Landscape and Urban Planning* **28**: 99–100.

Barrocu, G., Sciabica, M. G. and Paniconi, C., 1994. Three-dimensional model of salt water intrusion in the Capoterra coastal aquifer system (Sardinia). Proceedings of 13th Salt Water Intrusion Meeting, Department of territorial engineering, University of Cagliari, Villasmius-Cagliari.

Bartola, A., Sotte, F. and Sorci, S., 1989. Un'analisi regionale dell'agricoltura nella CEE. *Rivista di Economia Agraria* **4**: 3–20.

Bastiaanssen, W. G. M. and Roebeling, R. A., 1993. Analysis of land surface exchange processes in two agricultural regions in Spain using thematic mapper simulator data. In: Bolle, H. J., Feddes H. A. and Kalma K. A. (eds), *Exchange processes at the land surface for a range of space and time scales.* IAHS Publication No. 212, Wallingford, pp. 407–416.

Batista, J. A. *et al*. 1976. Alguns aspectos geológico-petrográficos da bordadura sul do complexo eruptivo de Beja, in *Comissão dos Serviços Geológicos de Portugal*, no. **60**, Lisboa, pp. 203–213.

Bazire, P. and Gadant, J., 1991. *La forêt en France*. La documentation française, Paris.

Beer, A. R., 1990. *Environmental planning for site development.* E & F N Spon, London.

Bellot, J. and Escarre, A., 1993. Canopy interception and rain water distribution in forest soil: influence of stems density in a catchment approach, rapport. Final Report DM2E Project.

Benedetti, R., Rossini, P. and Taddei, R., 1994. Vegetation classification in the middle mediterranean area by satellite data. *International Journal of Remote Sensing* **15**(3): 583–596.

Berry, J. and McGreal, S. (eds), 1995. *European cities: Urban planning and property markets.* E & F N Spon, London.

Bignami, M., Facciotto, P. M. and Mancini, A., 1985. *Cave e territorio.* L'Arciere, Torino.

Biraben, J. N. and Le Goff, J., 1969. La peste au Moyen Age. *Annales économies, sociétés, civilizations* **24**: 1484–1508.

Bishof, H., Schneider, W. and Pinz, A. J., 1992. Multispectral classification of Landsat images using neural networks. *IEEE Transaction of systems, man and cybernetics* **30**(3): 482–490.

Black, P. E., 1991. *Watershed hydrology.* Prentice Hall, New Jersey.

Blacksell, M. and Williams, A. M. (eds), 1994. *The European challenge: Geography and development in the European Community.* Oxford University Press.

Blamey, M. and Grey-Wilson, C., 1993. *Mediterranean wild flowers*. Harper Collins.

Blanchemanche, P., 1990. *Batisseurs de Paysages: terrassement, épierrement et petites hydrauliques agricoles en Europe XVIIe–XIXe siècles*. Maison des Sciences de l'Homme, Paris.

Blaney, H. F. and Criddle, W. D., 1950. Determining water requirements in irrigated areas from climatological and irrigation data. *USDA (ARS) Technical Bulletin* **1275**: 59.

Boenzi, F. and Giura Longo, R., 1994. *Basilicata*. Edipuglia, Bari.

Boix, C., Calvo, A., Soriano, M. D. and Tiemessen, I., 1994. Variabilidad espacio temporal de la agregación de suelos en laderas mediterráneas. In: Arnaez Vadillo, J., Garcia Ruiz, J. M. and Gomez Villar, A. (eds), *Geomorfologia en España. Tomo II*, Logrono, pp. 289–302.

Bolle, H. J. *et al*. 1993. EFEDA: European field experiment in a desertification-threatened area. *Ann. Geophysicae* **11:** 73–189.

Bottema, S., Entjes-Nieborg, G. and van Zeist, W. (eds), 1990. *Man's role in the shaping of the eastern Mediterranean landscape.* Balkema, Rotterdam.

Bove, E., De Nigris, T. and Giardiello, A., 1993a. I fagioli di Sarconi. *Rassegna dell'Economia Lucana* **2**: 3–20.

Bove, E., De Nigris, T., Giardiello, A., Mazzaccara, M. and Verneau, F. 1993b. Il canestrato di Moliterno. *Rassegna dell'Economia Lucana* **2**: 21–54.

Bradbury, D. E., 1981. The physical geography of the mediterranean lands. In: di Castri, F. and Goodall, D. W. (eds), *Mediterranean-type shrublands.* Elsevier Scientific Publishing, Amsterdam, pp. 53–62.

Brandt, C. J. and Thornes, J. B. (eds), 1996. *Mediterranean desertification and land use.* John Wiley, Chichester.

Braudel, F. (ed.), 1975 edn. *The Mediterranean and the Mediterranean world in the age of Philip II.* Fontana/Collins, London.

Braudel, F., 1977. *La Méditerranée. L'espace et l'histoire.* Senil, Paris.

Brinkman, R. and Smyth, A. J., 1973. *Land evaluation for rural purposes.* Publs. int. inst. ld reclam. improv. no. 17. Quoted in: McRae, S. G. and Burnham, C. P. (eds), *Land evaluation.* Monographs on soil survey, Clarendon Press, Oxford, p. 1.

Bruce-Chwatt, L. J. and de Zulueta, J., 1980. *The rise and fall of malaria in Europe.* Oxford University Press.

Brunet, M., 1990. Terrasses de cultures antiques; l'example de Délos, Cyclades. *Méditerranée* **71**: 5–11.

Bruschi, S., 1983, Lo studio d'impatto ambientale nelle procedure decisionali. *Agricoltura ed Ambiente* **19**: 119.

Bryan, R. and Yair, A. (eds), 1983. *Badland geomorphology and piping.* Geobooks, Norwich.

Bunte, K. and Poesen, J., 1993. Effects of horseshoe vortex erosion on sediment yield from soils covered by rock fragments. *Zeitschrift Geomorphologie Neue Folge* **37**(3): 327–335.

Butcher, G. C. and Thornes, J. B., 1979. Spatial variations in runoff processes in an ephemeral channel. *Zeitschrift Geomorphologie Neue Folge Supplement Band.* **29**: 83–92.

Butzer, K. W., Mateu, J. F., Butzer, E. and Kraus, P., 1985. Irrigation agrosystems in eastern Spain: Roman or Islamic origins? *Annals of the Association of American Geographers* **75**: 479–509.

Cafiero, S., 1975. *La pianificazione regionale in Basilicata.* Giuffrè, Milano.

Calder, I. R., 1993. Hydrologic effects of land use change. In: Maidment, D. R. (ed.), *Handbook of Hydrology.* McGraw-Hill, New York, pp. 13.1–13.49.

Calvo-Cases, A., Soriano Soto, M. D., Boix Fayos, C. and Tiemessen, I., 1994. Suelos y procesos geomórficos en un gradiente climático altitudinal. (Alicante). In: Arnaez Vadillo, J., Garcia-Ruiz, J. M. and Gomez-Villar, A. (eds), 1994. *Geomorfologia en España,* Logroño **2**: 29–40.

Cammeraat, L. H. (compiler), 1992. *Field manual version 3.1 MEDALUS project.* With contributions from Clark, S. C., Cammeraat, L. H., Imeson, A. C., Melía, J. and Collin, J. J., University of Bristol.

Cammeraat, L. H., 1996. The MEDALUS core field programme: an overview of sites and methodology. In: C. J., Brandt and J. B., Thornes (eds), *Mediterranean desertification and land use.* Wiley, Chichester.

Cammeraat, L. H. and Imeson, A. C., 1995. Deriving indicators of soil degradation from soil aggregation studies in south east Spain and southern France. *Geomorphology,* in press.

Cammeraat, L. H. and Prinsen, H. (eds) 1995. *The MEDALUS application on CD-ROM.* Da Vinci (Chaumont-Gistoux; Belgium) and FGBL (Amsterdam, The Netherlands).

Cantore, V., Iovino, F. and Pontecorvo, G., 1987. *Aspetti climatici e zone fitoclimatiche della Basilicata.* Pubblicazione no. 2, CNR Istituto di Ecologia ed Idrologia Forestale, Cosenza.

Caprara, R., 1991. Archeologia della occupazione del suolo. In: Maciocco, G. (a cura di), *Le dimensioni ambientali della pianificazione urbana.* Franco Angeli, Milano.

Cardoso, J. V. J. de Carvalho, 1965. *Os solos de Portugal—sua caracterização e génese, 1. A sul do Tejo.* Lisboa, Direcção Geral Serviços Agrícolas.

Casimiro, P. C., 1994. *Concelho de Mértola—Geobiografia das mudanças de uso do solo.* MSc Dissertation, Universidade Nova de Lisboa, Lisboa.

Castles, S. and Kosack, G., 1973. *Immigrant workers and class structures in Western Europe.* Oxford University Press.

Cavazza, S., 1962. Sulla erodibilità dei terreni di alcuni bacini calabro-lucani. Atti del XVIII Congresso della Società Geografica Italiana.

Cenami, E., Doneddu, G. and Maciocco, G., 1991. Morfologia della proprietà e formazione dell'insediamento. In: Maciocco, G. (ed.), *Le dimensioni ambientali della pianificazione urbana.* Franco Angeli, Milano.

Centre International de Hautes Etudes Agronomiques Méditerranéens–Institut Agronomique Méditerranéen de Montpellier (CIHEAM-IAMM), 1994. *MEDAGRI—Annuaire des économies agricoles et alimentaires des pays méditerranéens et arabes,* Montpellier.

Cerdà, A., García-Alvarez, A., Cammeraat, L. H. and Imeson, A. C., 1994. Agregacion del suelo en una catena afecta da por el abandono del cultivo en la cuenca del Guadalentín (Murcia). In: Arnaez-Vadillo, J., Garcia-Ruiz, J. M., Gomez-Villar, A. (eds), *Geomorfologia en España,* Logroño 1.

Cernesson, F., Folton, N. and Lavabre, J., 1993. *Etude climatologique sur la façade méditerranéenne française,* 1st workshop DM2E, Aix-en-Provence, p. 15.

Cherry, J. F., 1990. The first colonization of the Mediterranean islands: a review of recent research. *Journal of Mediterranean Archaeology* **3**: 145–221.

Chiarantini, L., Coppo, P., Fini, S., Luzi, G. and Gagliani, S., 1993. Microwave remote sensing and hydrological modelling as support to soil erosion forecasting. *Proceedings of the XIII EARSeL Symposium,* Dundee, Scotland, July 1993, pp. 267–274.

Chiarantini, L., Coppo, P., Luzi, G., Gagliani, S. and Benvenuti, M., 1994. ERS1/SAR image analysis for hydrological studies in the Abruzzo Region (Italy): experimental activities for soil moisture and roughness estimations. *Proceedings of the European Symposium on Satellite Remote Sensing,* 26–30 September 1994, Rome, Italy.

Chiarantini, L., Coppo, P., Gagliani, S. and Benvenuti, M., 1995. ERS1/SAR image analysis for hydrological studies in the Abruzzo region (Italy): experimental activities for soil moisture and roughness estimation. In: Mougin, E., Ranson, K. J. and Smith, J. A. (eds), *Multispectral and microwave sensing of forestry, hydrology, and natural resources.* Proc. SPIE 2314, pp. 494–501.

Clark, S. C., 1988. Winter annuals: rewarding subjects for research. In: Villar, L. (ed.), *Homenaje a Pedro Montserrat.* Monografias del Instituto Pirenaico de Ecologia, Jaca No. 4. Instituto de Estudios Altoaragoneses Instituto Pirenaico de Ecologia (CSIC), Jaca.

Clemente, F., 1964. *La pianificazione territoriale in Sardegna.* Gallizzi, Sassari.

CNRS, 1987. Déplacements des lignes de rivage en Méditerranée d'après les donnés de l'archéologie. Colloques internationaux du Centre National de la Recherche Scientifique, Paris.

Coccosis, H., 1991. Historical land use changes: Mediterranean regions of Europe. In: Brewer, F. M. *et al.* (eds), *Land use changes in Europe.* Kluwer, Dordrecht, pp. 441–461.

Colacino, M. and Conte, M., 1993. Greenhouse effect and pressure patterns in the Mediterranean basin. *Il Nuovo Cimento* **16C**: 67–76.

Colacino, M., Conte, M., *et al.* 1991. The precipitation in Italy during the period Autumn 1988–Spring 1989. *Il Nuovo Cimento* **14**: 605–613.

Combes, F., Hurand, A. and Meunier, M., 1994. La forêt de montagne; un remède aux crues, SHF. *23èmes journées de l'hydraulique* 'Crues et inondations', Nîmes, p. 6.

Commission of the European Communities (CEC), 1985. *Soil map of the European Communities,* 1:1 000 000, explanatory text. Directorate General Agriculture, Coordination of Agricultural Research. Commission of the European Communities, Luxemburg.

Commission of the European Communities (CEC), 1987. CB49-87-381-C.

Commission of the European Communities (CEC), 1993. *Annual situation in the Community (1992) Report,* Bruxelles.

Commission of the European Communities DG XI, EC, 1992. *CORINE, Soil erosion risk and important land resources in the southern regions of the European*

Community. Luxemburg. Office for Official Publications of the European Community.
Conte, M. and Sorani, R., 1991. Variazioni climatiche. Un modello per prefigurare il futuro. *Cooperazione* **16**: 51–54.
Conte, M. and Giuffrida, A., 1991. L'Oscillazione Mediterranea. Possibli effetti sulle precipitazioni in Italia. In: *Proceedings of the LXI Runione Società Italiana Progresso dell Scienze*, Catalia, 1991, pp. 61–88.
Coppin, M. J. and Richards, I. G., 1990. *Use of vegetation in civil engineering*. CIRIA-Butterworths, London.
Corbetta, F., 1974. Un panorama della vegetazione Lucana. *Giornale Botanico Italiano* **108**(5): 211–234.
Coutinho, A. P., 1913. *Flora de Portugal plantas vasculares*. Bertrand, Lisbon.
Dachary, M. C., 1976. Contribuition de l'erosion sélective des sols à la sédimentation. Barrage collinaire. Lisbos, CEG, INIC, *Finisterrae* **11**(22): 179–186.
Daget, P. and Poissonet, J., 1969. *Analyse phitologique des prairies; application agronomique*. CNRS, CEPE, Montpellier, DOC, pp. 48–67.
Danalatos, N. G, Kosmas, C. S., Moustakas, N. and Yassoglou, N., 1995. Rock fragments II. Their impact on soil physical properties and biomass production under Mediterranean conditions. *Soil Use and Management* **11**:121–126.
D'Angelo, M. and Podda, P., 1995. Fire hazard assessment in Sardinia as a model for semi-arid Mediterranean environment. In: Enne, G., Aru, A. and Pulina, G. (eds), *Land use and soil degradation. MEDALUS II in Sardinia*. Proceedings of the Conference held in Sassari, 27 May 1994, University of Sassari–University of Cagliari, pp. 251–263.
Davies, P. and Gibbons, B., 1993. *Field guide to wild flowers of southern Europe*. The Crowood Press.
Davis, J. A., 1979. The South, the Risorgimento and the origins of the southern problem. In: Davis, J. A. (ed.), *Gramsci and Italy's passive revolution*. Croom Helm, London, pp. 67–103.
Delano Smith, C., 1979. *Western Mediterranean Europe: a historical geography*. Academic Press, London.
De Martonne, E., 1926. Une nouvelle fonction climatologique: l'indice d'aridité. *La metéorologie* **19**.
De Stefano, F. and Lombardi, P., 1976. Evoluzione delle strutture agrarie nel Mezzogiorno. In: Banca Nazionale dell'Agricoltura, *Storia dell' Agricoltura Italiana*. Etas Libri, Milano, pp. 264–307.
Dietz, F. D. and Hoogervorst, N. J. P., 1991. Towards sustainable and efficient use of manure in agriculture—The Dutch case. *Environmental and Resource Economics* **1**: 313–332.
Di Gregorio, F. and Massoli Novelli, R., 1986. Le Cave. In: Massoli-Novelli, R. (ed.), *Ecologia in Sardegna. Natura e Ambiente*. Edizioni della Torre, Cagliari, pp. 39–42.
Di Gregorio, F. and Massoli Novelli, R., 1987. Impatto ambientale delle attivite minerarie in Sardegna. *Boll. Soc. Sarda Sc. Nat., Sassari* **26**: 17–42.
Di Martino, P., 1993. Deforestation and natural regeneration of woodland: the forest history of Molise, Italy, over the last two centuries. In: Watkins, C. (ed.), *Ecological effects of afforestation*. CAB International, pp. 69–92.
Dunford, M., 1993. Regional disparities in the European Community: evidence from the REGIO Databank. *Regional Studies* **27**(8): 727–743.
Duwig, C., 1994. *Elément de modélisation de l'interception des pluies par 4 espèces d'arbres de type méditerranéen*. Université Joseph Fourier de Grenoble 1, UFR de Mécanique et Cemagref d'Aix-en-Provence, division Ouvrages Hydrauliques et Equipements pour l'Irrigation. Mémoire de DEA.
Ehret, G., Kiemle, Chr. and Simmet, G., 1993. Aerosol and water vapor sensing of the planetary boundary layer using an airborne DIAL system. In: Bolle, H. J. and Streckenbach, B. (eds), *EFEDA Final Report EC*, Contract EPOC-CT 90-0030, Berlin.
Esau, K., 1965. *Plant anatomy*. 2nd edn, John Wiley, New York.
Espirito-Santo, D. M., Costa, J. C. and Rego, F., 1994. Development of plant communities after seven years of wildfire at Serra dos Candeeiros (Cantral Portugal). In: *Proceedings of the 2nd International Conference on forest fire research*, 1994.
Estadística Agrária Regional, 1991. *Comunidad autónoma de la región de Murcia*. Murcia, Spain.
European Commission (EC), 1994. *Europe 2000+ Cooperation for European territorial development*. Office for Official Publications of the European Communities, Luxemburg.
EUROSTAT (Statistical Office of the European Communities), 1989. *Regions: Nomenclature of territorial units for statistics*. Luxembourg.
EUROSTAT (Statistical Office of the European Communities), 1993, *Regions: Nomenclature of territorial units for statistics*. Luxembourg.
Fantechi, R. and Margaris, N. S., 1986. *Desertification in Europe*. EUR10395, Reidel, Dordrecht.
FAO-UNESCO, 1974. *Soil map of the world* 1:5 000 000. Vol. 1, Legend. UNESCO, Paris, France.
Food and Agriculture Organization of the United Nations (FAO). *FAO Yearbook*, 1971 then annually. FAO, Rome.
Food and Agriculture Organization of the United Nations, 1976. A framework for land evaluation. *FAO Soils Bulletin* **32**, FAO, Rome.
Food and Agriculture Organization of the United Nations, 1983. Guidelines: land evaluation for rainfed agriculture. *FAO Soils Bulletin* **52**. FAO, Rome.
Foundation of Economic and Industrial Research (FEIR), 1991. Sectoral research and intelligence units. Report on: olive oil industries. Athens, Greece. **72**: 68.
Francis, C. F. and Thornes, J. B., 1990. Runoff hydrographs from three Mediterranean vegetation cover types. In J. B. Thornes (ed.), *Vegetation and geomorphology*. Wiley, pp. 363–385.
Garcia-Haro, F. J., 1994. Inversión de un modelo lineal de reflectividad para la extracción de parámetros relacionados con la vegetación en paisajes semi-áridos. MSc thesis, Universitate de Valéncia.
Gates, W. L., Rowntree, P. R. and Zeng, Q. C., 1990. Validation of climate models. In: Houghton, J. T. *et al*. (eds), *Climate change—the IPCC scientific assessment*. Cambridge University Press, Cambridge, pp. 93–130.
Giardini, L. *et al*., 1987. Proposta di una metodologia per la classificazione del territorio Italiano in funzione della sua potenzialità produttiva agricola—C.N.R.—'Nuova indirizzi colturali'.
Gil Olcina, A., 1971. *El Campo de Lorca*. Estudio de Geografia Agraria. Universidad de Valencia, C.S.I.C., Valencia.
Gilman, A. and Thornes, J. B., 1985. *Land use and prehistory in south-east Spain*. The London Research Series in Geography: 8. Allen and Unwin, London.
GMS-DECIDE, 1994. *User manual*, v. 1.2. Da Vinci Consulting. Chaumont-Gistoux, Belgium.
Goodess, C. M. and Palutikof, J. P., 1993. The development of regional climate scenarios and the ecological impact of greenhouse gas warming. In: Woodward, F. I. (ed.), *Global climate change—the ecological consequences*. Academic Press, London, pp. 33–62.
Gottfried, R. S., 1986. *The black death: natural and human disaster in medieval Europe*. Macmillan, London.
Grenon, M. and Batisse, M., 1989. *Futures for the Mediterranean Basin: the Blue Plan*. Oxford University Press.
Greppi, M. and Niedda, M., 1995. Small catchments equipped to measure and to model rainfall, discharge and soil loss. *20th General Assembly of the European Geophysical Society*, Hamburg, 3–7 April 1995.
Gripaios, P. and Mangles, T., 1993. An analysis of European super regions. *Regional Studies* **27**(8): 745–750.
Grove, A. T. and Rackham, O., 1993. Threatened landscapes in the Mediterranean: examples from Crete. *Landscape and Urban Planning* **24**: 279–292.
Grove, J. M., 1988. *The Little Ice Age*. Routledge, London.
Gustavsson, N., 1980. *A review of methods for objective analysis*. European centre for medium range weather forecast, Reading (UK), Seminar on Data Assimilation Methods, pp. 19–82.
Haase, P., Pugnaire, F. I., Fernández, E. M., Puigdefábregas, J., Clark, S. C. and Incoll, L. D., 1995. An investigation of rooting depth in the semi-arid shrub *Retama sphae-*

rocarpa (L.) Boiss. by labelling of ground water with a chemical tracer. *Journal of Hydrology* **170**, 1–24.

Hadley Centre, 1992. *The Hadley Centre transient climate change experiment*. UK Meteorological Office.

Hallikainen, M. T., Ulaby, F. T., Dobson, M. C., El Rayes, M. A. and Lin Kun Wu., 1985. Microwave dielectric behaviour of wet soil—Part I: Empirical models and experimental observations. *IEEE Transactions Geoscience and Remote Sensing* **GE-23**: 25–33.

Haralik, R. M., Shanmugan, K. and Dinstein, I., 1973. Textural features for image classification. *IEEE Transactions of System, Man and Cybernetics* **3**(6): 610–621.

Hardin, G., 1963. The cybernetics of competition. A biologist's view of society. *Perspectives in Biology and Medicine* **7**: 58–84.

Hecke, E. van, 1992. *Farm Structure—1985. Survey: analysis of results—regional structure of agriculture production—cartographic analysis*. EUROSTAT, Luxembourg.

Hidding, M. J., 1993. In search of new concepts of sustainable development of rural areas in the Netherlands. *Landscape and Urban Planning*, **27**: 259–264.

Hoffmann, G., 1987. *Holozänstratigraphie und küstenlienvarlagerung an der Andalusischen Mittelmeerküste*. Berichte au dem Fachbereich Geowissenschaften der Universität Bremen.

Holben, B. N., 1986. Characteristics of maximum-value composite images from temporal AVHRR data. *International Journal of Remote Sensing* **7**(11): 1417–1434.

Houston, J. M., 1964. *The Western Mediterranean world: an introduction to its regional landscapes*. Longman, London.

Hughes, J. P., Lettenmaier, D. P. and Guttorp, P., 1993. A stochastic approach for assessing the effect of changes in synoptic circulation patterns on gauge precipitation. *Water Resources Research* **29**: 3303–3315.

Hulscher, W. S. and Hommes, E. W., 1992. Energy for sustainable rural development. *Energy Policy,* June 1992, pp. 527–532.

Hunt, R., 1978. *Plant growth analysis*. Studies in biology, 96. Arnold, London.

Huntley, B. and Birks, H. J. B., 1983. *An atlas of past and present pollen maps for Europe:0–13 000 years ago*. Cambridge.

Hutchinson, C. F., 1982. Techniques for combining Landsat and ancillary data for digital classification improvement. *Photogrammetric Engineering and Remote Sensing* **48**(1): 123–130.

IGME (Instituto Geologico y Minero de España), 1970. *Mapa geologico de España* 1:200 000, sheets 78 and 79, Madrid.

Imeson, A. C., Calvo, A., Lavee, H. and Perez-Trejo, F., 1994. *Modelling and exploring the impact of climate change on ecosystem degradation, hydrology and land use along a transect across the Mediterranean*. Proceedings EC Meeting, Copenhagen.

Imeson, A. C., Bergkamp, G. and Cammeraat, L. H., 1995. Scaling up from the plot to the watershed. A dynamic systems strategy. In: *Proc. IGBP—Global Change in Terrestrial Ecosystems Workshop*, Corvallis 1.

Imeson, A. C., Perez-Trejo, F. and Cammeraat, L. H., 1996. The response of landscape-units to desertification. In: Brandt, C. J. and Thornes, J. B. (eds), *Mediterranean desertification and land use*, John Wiley, Chichester.

INE, 1990. Censo Agraria de España.

Institut Agronomique Méditerranéen de Montpellier (CIHEAM-IAMM), 1994. *MEDAGRI—Annuaire des économies agricoles et alimentaires des pays méditerranéens et arabes*, Montpellier.

ISTAT, 1955. *9° Censimento generale della popolazione*. Tomo I, Facicolo 92, Rome.

ISTAT, 1982. *XIX Censimento Nazionale dell'Agricoltura Italiana*. Fascicoli provinciali Matera e Potenza.

ISTAT, 1983. *12° Censimento generale della popolazione*. Tomo I, Fascicolo 92, Rome.

ISTAT, 1985. *Popolazione residente e presente dei comuni*. Censimenti dal 1861 al 1981. Rome.

ISTAT, 1993. *13° Censimento generale della popolazione*. Tomo I, Fascicoli 76 (PZ) e 77 (MT), Rome.

ITGE, 1989. *Sistema Aquifero Numero 23*, Mancha Occidental, Direccion de Aguas Subterraneas.

Jaarsma, C. F. and Van Lier, H. N., 1993. Fossil energy and rural reconstruction. Potential for sustainable development. *Landscape and Urban Planning* **27**: 123–130.

Jurgens, C. R., 1993. Strategic planning for sustainable rural development. *Landscape and Urban Planning* **27**: 253–258.

Kavadas, D., 1956. *Illustrated botanical dictionary*. Athens.

Kazanis, D. and Arianoutsou, M., 1994. Vegetation structure in a post-fire successional gradient of *Pinus halepensis* forests. In: *Proceedings of the 2nd International Conference on forest fire research*, pp. 979–993.

Kershaw, K. A. and Looney, J. H. H., 1985. *Quantitative and dynamic ecology*, 3rd edn. Edward Arnold, London.

Kertész, Á., 1993. Application of GIS methods in soil erosion modelling. *Computers, Environment and Urban Systems* **17**: 233–238.

Kertész, Á., Huszár, T. and Lóczy, D., 1995. Land use changes in Lake Balaton catchment. In: Simmons, I. G. and Mannion, A. (eds), *The changing nature of the people–environment relationship: evidence from a variety of archives*. Charles University, Prague, pp. 69–78.

Kertész, Á., Márkus, B. and Richter, G. B., 1995. Assessment of soil erosion in a small watershed covered by loess. *Geojournal* **36**: 2–3, 285–288.

Kessler, J. J. and Laban, P., 1994. Planning strategies and funding modalities for land rehabilitation. *Land Degradation and Rehabilitation* **5**: 25–32.

King, R. (ed.), 1993. *Mass migrations in Europe: the legacy and the future*. Belhaven Press, London.

King, R., 1994. Migration and the single market for labour: an issue in regional development. In: Blacksell, M. and Williams, A. M. (eds), *The European challenge: Geography and development in the European Community*. Oxford University Press, pp. 218–41.

Kirkby, M. J., McMahon, M. L., Abrahart, R., Thornes, J. B., Shao, X. and Hawkes, J. C., 1994. The MEDRUSH model for a single flow strip. In: *First Annual Report MEDALUS II*.

Kirkby, M., Lockwood, J. G., Thornes, J., Woodward, I., Baird, A., McMahon, M., Shao, J., Mitchel, J. and Sheehy, J., 1996. The MEDALUS hillslope model. In: Brandt, C. J. and Thornes, J. B. (eds), *Mediterranean desertification and land use*. John Wiley, Chichester.

Klingebiel, A. A. and Montgomery, P. H., 1961. *Land capability classification*. Soil Conservation Service, USDA Handbook no. 210, Dept. Agr., Washington, DC.

Kosmas, C. S., Danalatos, N. G., Moustakas, N., Tsatiris, B. and Yassoglou, N., 1993a. The impacts of drought on the wheat biomass production along catenas in the semi-arid zone of Greece. *Soil Technology* **6**: 337–349.

Kosmas, C. S., Moustakas, N., Danalatos, N. G. and Yassoglou, N., 1993b. The effect of rock fragments on wheat biomass production under highly variable moisture conditions in Mediterranean environments. *Catena* **23**: 191–198.

Kosmas, C. S., Moustakas, N., Danalatos, N. G. and Yassoglou, N., 1996. The effect of land land use change on soil properties and erosion along a catena. In: Brandt, C. J. and Thornes, J. B. (eds) *Mediterranean desertification and land use*. John Wiley, Chichester.

Lai, M. R., Loddo, S., Puddu, R., Aru, A. and Baldaccini, P., 1995. Geological, soil and environmental survey in the Rio Santa Lucia catchment. In: Enne, G., Aru, A. and Pulina, G. (eds), *Land use and soil degradation*. MEDALUS II in Sardinia, Proceedings of the Conference held in Sassari, 27 May 1995, University of Sassari–University of Cagliari, pp. 99–112.

Lansac, A. R., Zaballos, J. P. and Martin, A., 1994. Seasonal water potential changes and proline accumulation in Mediterranean shrubland species. *Vegetatio*, pp. 113, 141, 154.

Lavabre, J., Boennec, J. M. and Cernesson, F., 1993a. Interception de la pluie par la canopée. Premiers résultats après 6 mois d'expérimentation, 6ème colloque international de climatologie, Université Aristote de Thessalonique. *Publications de l'Association Internationale de Climatologie* **6**: 614–622.

Lavabre, J., Martin, C., Bernard-Allée, Ph. and Folton, N., 1993b. Comportement hydrologique des sols d'un petit bassin versant méditerranéen après la destruction de l'écosystème forestier par un incendie, premières analyses 2 cycles hydrologiques après incendie. Rencontres forestiers, chercheurs en forêt méditerranéenne, La Grande Motte, INRA.

Lavabre, J., Sempere-Torres, D. and Cernesson, F., 1993c. Hydrological consequences of fire: changes on the hydrological response of a little Mediterranean basin a year after the fire. *Journal of Hydrology* **142**: 273–299.

Le Houerou, H. N., 1981. Impact of man and his animals in mediterranean vegetation. In: di Castri, F. and Goodall, D. W. (eds), *Mediterranean-type shrublands*. Elsevier Scientific Publishing, Amsterdam, pp. 479–521.

Le Lannou, M., 1941. *Pâtres et paysans de la Sardaigne*. Arrault, Tours.

Ledda, L., Porqueddu, C., Sulas, L. and Caredda, S., 1995. Sward characterization and grazing value of a hilly Mediterranean natural pasture. *Proceedings of the 8th meeting of the FAO working group on Mediterranean pastures and fodder crops*, Avignon, France, 143–146.

Lee, J., 1990. Land resources, land use and projected land availability for alternative uses in the EC. In: Brower, F. M., Thomas, A. J. and Chadwick, M. J. (eds), *Land use changes in Europe*. Kluwer Academic, Dordrecht.

Legg, C. J., 1995. Heathland vegetation dynamics: matters of scale. In: Thompson, D. B. A., Usher, M. B. and Hester, A. (eds), *Heaths and moorlands: cultural landscapes*. HMSO, London, Chapter 10, pp. 117–134.

Leonardi, R., Putnam, R. and Nanetti, R. Y., 1987. *Il caso Basilicata*. Il Mulino, Bologna.

Leontidou, L., 1990. *The Mediterranean city in transition: social change and urban development*. Cambridge University Press, Cambridge.

Leontidou, L., 1995. Repolarisation of the Mediterranean: Spanish and Greek cities in neo-liberal Europe. *European Planning Studies* **3**(2): 155–172.

Liebe, P., 1994. *Assessment of the state of water resources of Hungary*. Institute of Hydrology, Water Resources Research Center, Budapest.

Llamas, M. R., 1988. Conflicts between wetlands conservation and groundwater exploitation: two case histories in Spain. *Environmental Geology and Water Sciences* **11**(3): 241–251.

Llamas, M. R., Cruces, J., Casado, M., Barroso, J. L. and de la Hera, A., 1993. The EFEDA experimental area; hydro-geological overview. EFEDA, Final Report EC Contract EPOC-CT 90-0030. Bolle, H. J. and Streckenbach, B. (eds), Berlin.

Llorens, P. and Poch, R., 1994. *Monitoring of water losses to the atmosphere at the Cal Parisa basin*. 2ème workshop, DM2E.

Lóczy, D. and Szalai, L., 1993. GIS application for land capability survey in Hungary. *Geografski Sbornik*, Ljubljana, **23**: 53–64.

Lóczy, D. and Szalai, L., 1995. Assessment of loess as parent material for agroecological potential. *Geojournal*, **36**(2/3): 275–280.

López-Bermúdez, F., Garcia-Ruiz, J. M., Romero-Díaz, M. A., Ruiz-Flano, P., Martínez-Fernandez, J. and Lasanta, T., 1991. Medidas de flujos de agua y sedimentos en parcelas experimentales. *Cuaderno Técnicos de la Sociedad Española de Creouiofología*, 6.

López-Bermúdez, F., Romero-Díaz, M. A. and Martínez-Fernandez, J., 1991. Soil erosion in a semi-arid Mediterranean environment. El Ardal experimental field (Murcia, Spain). In: Sala, M., Rubio, J. L. and García-Ruiz, J. M. (eds), *Soil erosion studies in Spain*. Geoforma Ediciones, Logroño, pp. 137–152.

Lozarto-Giotart, J. P., 1990. *Méditerranée et tourisme*. Masson, Paris.

Luzi, G., Chiarantini, L., Coppo, P. and Gagliani, S., 1993. Soil moisture measurements by means of a portable dielectric probe. EARSeL International Workshop on agrometeorology. *Advances in Remote Sensing*, June, **2**(2).

Mairota, P., 1995a. Analysis of land use changes in the Agri basin (Basilicata, Italy): the impact of rural change on land degradation. In: *Proceedings of the Conference on Erosion and Land Degradation in the Mediterranean*. Portugal University of Aveiro, 14–18 June 1995, pp. 433–449.

Mairota, P., 1995b. EC policy structures and socio-economic aspects of desertification. First contribution: the MEDALUS II regions and common policies. Working Paper No. 63, MEDALUS Project, King's College London, London.

Margaris, N. S., 1981. Adaptive strategies in plants dominating Mediterranean-type ecosystems. In: Di Castri, F., Goodall, D. W. and Specht, R. L. (eds), *Mediterranean-type shrublands*. Elsevier, Amsterdam, pp. 309–315.

Margaris, N. S., 1987. Desertification in the Aegean islands. *Ekistics* **323**: 132–136.

Margaris, N. S., 1992. Primary sector and environment in the Aegean Islands, Greece. *Environmental Management* **16**(5): 569–574.

Margat, J., 1992. L'Eau dans le Bassin Méditerranéen. Situation et prospective. *Les Fascicules du Plan Bleu, Economica*, 6, Paris.

Martin, C., Beguin, E., Levat, M., and Quillard, J., 1995. *Soil mechanical erosion after a forest fire in the Rimbaud basin*. Rapport final du projet, DM2E.

Martínez-Díaz, J. L. and Hernandez-Enrile, J. L., 1992. Tectónica reciente y rasgos sismotectónicos en el sector Lorca-Totana de la falla de Alhama de Murcia. *Estudios Geológicos* **48**: 153–162.

Martínez-Fernandez, J., López-Bermúdez, F., Romero-Diaz, M. A., Martínez-Fernandez, J. and Sarria, F., 1992. El matorral semiárido del Sureste de España. Aportación metodológica para su evaluación. *Studia Oecologica* **8**: 97–105.

Martínez-Fernandez, J., Romero-Diaz, M. A., López-Bermúdez, F. and Martínez-Fernandez, J., 1994. Parámetros estructurales y funcionales de Rosmarinus officinalis en ecosistemas mediterraneáneos semiáridos. *Studia Oecologica* **10/11**: 289–296.

Martínez-Fernandez, J., López-Bermúdez, F., Martínez-Fernandez, J. and Romero-Diaz, M. A., 1995. Land use and soil-vegetation relationships in a Mediterranean ecosystem. *Catena* **98**, 16–21.

Masala, G., Pulina, G. and Enne, G., 1995. Computerized modelisation of sustainable stocking rate for extensive cattle production in Mediterranean areas: BOVISOFT. In: Enne, G., Aru, A. and Pulina, G. (eds), *Land use and soil degradation. MEDALUS in Sardinia*. University of Cagliari, Sassari, pp. 221–229.

Maselli, F., Conese, C., Petkov, L. and Resti, R., 1992. Inclusion of prior probabilities derived from a nonparametric process into the maximum likelihood classifier. *Photogrammetric Engineering & Remote Sensing* **58**(2): 201–207.

Matheron, G., 1970. La théorie des variables regionalisées et ses applications. Les Cahiers de Centre de Morphologie Mathématique de Fontainebleau, Fascicule 5.

Mazzoleni, S. and Spada, F., 1990. Deciduous broadleaved versus evergreen sclerophyllous forests. Disturbance and local shifting dominance in Mediterranean environments. In: Teller, A., Mathy, P. and Jeffers, J. N. R. (eds), *Responses of forest ecosystems to environmental changes*. Elsevier, London, pp. 839–840.

Mazzoleni, S., Esposito, A., Castaldo-Cobianchi, R. and Blasi, C., 1993. Bryophyte populations in Mediterranean environments. Relationships to vascular vegetation and disturbance patterns. *Phytocoenologia* **24**: 369–385.

McFarlane, N. A., Boer, G. J., Blanchet, J. P. and Lazare, M., 1992. The Canadian Climate Centre second-generation general circulation model and its equilibrium climate. *Journal of Climate* **5**: 1013–1044.

McNeill, J. R., 1992. *The mountains of the Mediterranean world; an environmental history*. Cambridge University Press.

McRae, S. G. and Burnham, C. P., 1981. *Land evaluation*. Monographs on soil survey. Clarendon Press, Oxford.

Mendoza, M., 1990. Global assessment of desertification: world atlas of thematic indicators of desertification: proposal document. In: *Desertification revisited, 1990*. UNEP DC/PAC, Nairobi.

Mika, J., 1988. Regional properties of global warming in the Carpathian basin. *Idôjárás* **92**(2/3): 178–189 (in Hungarian).

Mika, J., 1991. Predictable properties of a major global warming in Hungary. *Idôjárás* **95**(6): 265–278 (in Hungarian).

Mika, J., 1994. On the problem of downscaling climate changes projected by general circulation models. In: Brazdil, R. and Kolar, M. (eds), *Contemporary Climatology*. IGU Symposium Brno, Czech Rep., 15–20 August 1994, pp. 387–394.

Ministero LL PP–Servizio Idrografico, 1951 (cont. 1971). Annali Idrologici parte prima. Sezione Idrografica di Bari, Catanzaro, Napoli. Istituto Poligrafico dello Stato, Roma.

Ministero LL PP–Servizio Idrografico, 1957. Precipitazioni medie, mensili e annue e numero di giorni piovosi per il trentennio, 1921–1950. Sezione Idrografica di Bari. Pubblicazione n. 24 del Servizio Fascicolo 3. Istituto Poligrafico dello Stato, Roma.

Ministero LL PP–Servizio Idrografico, 1966. Distribuzione della temperatura dell'aria in Italia nel trentennio, 1926–1955. Pubblicazione n. 21 del Servizio Fascicolo 3, Italia Meridionale ed Insulare. Istituto Poligrafico dello Stato, Roma.

Minshull, G. N., 1990. *The new Europe into the 1990s*. Hodder and Stoughton, London.

Molnár, K., 1994a. 110-year temperature and precipitation trends for 16 and 17 meteorological stations in Hungary. *Földrajzi Értesítô* (in Hungarian).

Molnár, K., 1994b. Geographical aspects of climatic scenarios for Hungary. *Földrajzi Értesítô* (in Hungarian).

Moulin, A., 1991. *Peasantry and society in France since 1789*. Cambridge University Press.

Mueller-Dombois, D. and Ellenberg, H., 1974. *Aims and methods of vegetation ecology*. John Wiley, New York and Chichester.

Müller, M. J. 1982. *Selected climatic data for a global set of standard stations for vegetation science*. Dr. W. Junk Publishers, The Hague, The Netherlands.

National Statistics Service of Greece (NSSG), 1961–1995. *Agricultural statistics of Greece. Year 1961* et seq., Athens.

Navarro Hervás, F., 1991. *El sistema hidrográfico del Guadalentín*. Cuadernos Técnicos, 6. Consejería de Política Territorial, Obras Públicas y Medio Ambiente, Murcia.

Netherlands, Ministry of Agriculture, Nature Management and Fisheries, 1990. *Nature Policy Plan of the Netherlands*. The Hague.

Nicolau, J. M., Solé, A., Puigdefábregas, J. and Gutiérrez, L., 1995. Effects of soil and vegetation on runoff along a catena in semiarid Spain. *Geomorphology*.

Niedda, M. and Greppi, M., 1994. Time series simulation of daily rainfall, wind and cloud cover. *Proceedings of XXII CIGR World Congress and AgEng'94*, **1**: 419–426.

Niedda, M. and Greppi, M., 1995. Erosion risk in Sardinia: outline of the isoerosivity map. *Proceedings of the Conference on erosion and land degradation in the Mediterranean,* Averiro, Portugal, 14–18 June 1995.

Noilhan, J., 1995. *Annual Progress Report*, Contract EV5V-CT93-0269, EFEDA Group VI on surface processes and atmospheric modelling.

Noin, D. and Woods, R. (eds), 1993. *The changing population of Europe*. Blackwell, Oxford.

Odum, E.P., 1983. *Basic ecology*. Holt-Saunders International Editions, CBS College Publishing, Philadelphia.

OECD, 1965. *Geographic and occupational mobility of rural manpower*, Vol. 75, Paris.

Orshan, G., 1964. Seasonal dimorphism of desert and mediterranean chamaephytes and its significance as a factor in their water economy. In: Rutter, A. J. and Whitehead, F. H. (eds), *The water relations of plants*. Blackwell, Edinburgh, pp. 206–222.

Palutikof, J. P., Guo, X., Wigley, T. M. L. and Gregory, J. M., 1992. Regional changes in climate in the Mediterranean basin due to global greenhouse gas warming. *UNEP Mediterranean Action Plan Technical Report Series* **66**: 172.

Palutikof, J. P., Goodess, C. M. and Guo, X., 1994. Climate change, potential evapotranspiration and moisture availability in the Mediterranean basin. *International Journal of Climatology* **14**: 853–869.

Papadimitriou, F., 1994. Land use change and underdevelopement in southern Europe: the role of the EC policies. In: Mairota, P. and Geeson, N. (eds), *MEDALUS Working Paper Series* **17**: 63–72.

Papadimitriou, F., 1995. MELAS (Mediterranean Landscape Sustainability). A simulation system for sustainability assessment in Mediterranean landscapes. Copyright (1995) by Fivos Papadimitriou. (Software available from the author.)

Papadimitriou, F. and Mairota, P., 1994. Landscape diversity change and land management in Mediterranean EU countries. A preliminary assessment. *Proceedings of the International Conference on land and water resources management in the Mediterranean region*. CIHEAM-IAMB Valenzano, Bari, 4–8 September, Vol. 2, pp. 535–541.

Papadimitriou, F. and Mairota, P., 1995. Landscape diversity and landscape protection in Mediterranean Europe. What Protection Policies? *Proceedings of the Third International Congress on Energy, Environment and Technological Innovations Caracas (Venezuela)*, 5–11 November 1995, **3**: 155–157.

Papagiannakis, M., Gersi, J., Allaya, M., Allaya, M. G. and Mantlain, G., 1986. *Nutrition and agriculture in the Mediterranean*. Agricultural Bank of Greece Scientific Studies, Athens.

Patten, D. T., 1978. Productivity and production efficiency of an Upper Sonoran desert ephemeral community. *American Journal of Botany* **65**: 891–895.

Pavari, A., 1923. Primo contributo allo studio sperimentale del problema dei rimboschimenti su terreni argillosi della Basilicata. Le formazioni delle argille eoceniche. *Annali dell'Istituto Superiore Forestale Nazionale* **8**: 174–200.

Pena, A. *et al.*, 1985. *Fauna e flora de Mértola*. Mértola, CAM.

Perez-Trejo, F., 1992. *Desertification and land degradation in the European Mediterranean*. European Commission, EPOCH programme, EUR 14850EN, Brussels.

Perrings, C. and Pearce, D., 1994. Threshold effects and incentives for the conservation of biodiversity. *Environmental and Resource Economics* **4**: 13–28.

Pignatti, S., 1982. *Flora d'Italia*. Edagricole, Bologna.

Pirazzoli, P. A. and Pluet, J., 1991. *World atlas of Holocene sea-level changes*. Elsevier, Amsterdam.

Poesen, J. and Bunte, K., 1996. Effects of rock fragments on desertification processes in Mediterranean environments. In: Brandt, C. J. and Thornes, J. B. (eds), *Mediterranean desertification and land use*. John Wiley, Chichester, 247–270.

Poesen, J. and Lavee, H., 1994. Rock fragments in topsoils: significance and processes. *Catena* **23**: 1–28.

Poesen, J., Torri, D. and Bunte, K., 1994. Effects of rock fragments on soil erosion by water at different spatial scales: a review. *Catena* **23**: 141–166.

Polunin, O., 1969. *Flowers of Europe: a field guide*. Oxford University Press, London.

Polunin, O., 1977. *Guida agli alberi e arbusti d'Europa*. Zanichelli.

Polunin, O. and Huxley, A., 1965. *Flowers of the Mediterranean*. Chatto and Windus, London.

Polunin, O. and Smythies, B. E., 1973. *Flowers of south-west Europe: a field guide*. Oxford University Press, London.

Pons, A. and Quezel, P., 1985. The history of the flora and vegetation and past and present human disturbance in the mediterranean region. In: Gomez-Campo, C. (ed.), *Plant conservation in the Mediterranean area*. Dr. W. Junk Publishers, Dordrecht, pp. 25–43.

Postiglione, L., Ruggiero, C., Carone, F. and Fagnana, M., 1994. Agri basin vegetation survey. Second Annual Report, MEDALUS II, King's College London, pp. 55–62.

Pounds, N. J. G., 1990. *A historical geography of Europe*. Cambridge University Press.

Puddu, R. and Lai, M. R., 1992. Indagini pedologiche e geoambientali nel bacino del Rio S. Lucia (Sardegna meridionale). *Geologica Romana* **30**.

Puddu, R. and Lai, M. R., 1995. Quarrying activities in the Rio Santa Lucia catchment basin (southern Sardinia, Italy). In: *Proceedings of the Conference on Erosion and Land Degradation in the Mediterranean*. Portugal University of Aveiro, 14–18 June 1995, pp. 129–139.

Puddu, R., Tomasi, D. and Vacca, A., 1995. Erosion in areas afforested with Eucalyptus: a case study in southern Sardinia (Italy). In: *Proceedings of the Conference on Erosion and Land Degradation in the Mediterranean*. Portugal University of Aveiro, 14–18 June 1995, pp. 289–299.

Pugnaire, F. I., Haase, P. and Puigdefabregas, J., 1996. Facilitation between higher plant species in a semi-arid environment. *Ecology* **77**: 1420–1426.

Puigdefábregas, J., Aguilera, C., Brenner, A. J., Clark, S. C., Cueto, M., Delgado, L., Domingo, F., Gutiérrez, L., Incoll, L. D., Lázaro, R., Nicolau, J. M., Sánchez, G., Solé, A. and Vidal, S., 1995. The Rambla Honda field site. Interactions of soil and vegetation along a catena in semi-arid SE Spain. In: Brandt, C. J. and Thornes, J. B. (eds), *Mediterranean desertification and land use*. John Wiley, Chichester.

Pulina, G., Masala, G., Zanda, A. and Enne, G., 1994. Analysis of biomass and stocking rate in cattle and sheep production systems in Mediterranean Areas. Paper presented at the MEDALUS II Scientific Conference, Athens, Greece.

Pulina, G., Zanda, A. and Enne, G., 1995. The impact of animal husbandry on the degradation of the soil. In: Enne, G., Aru, A. and Pulina, G. (eds), *Land use and soil degradation. MEDALUS II in Sardinia*. Proceedings of the Conference held in Sassari, 27 May 1994, University of Sassari–University of Cagliari, pp. 231–238.

Quélennec, R. E. (ed.), 1990. *Littoral 1990.* Château-Gombert, Marseille.
Quezel, P., 1981. Floristic composition and phytosociological structure of sclerophyllous matorral around the Mediterranean. In: di Castri, F. and Goodall, D. W. (eds), *Mediterranean-type shrublands.* Elsevier Scientific Publishing, Amsterdam, pp. 107–121.
Quezel, P., 1985. Definition of the Mediterranean region and the origin of its flora. In: Gomez-Campo, C. (ed.), *Plant conservation in the Mediterranean area.* Dr. W. Junk Publishers, Dordrecht, pp. 9–24.
Rackham, O. and Moody, J., 1995. *The making of the Cretan landscape.* Manchester University Press (in press).
Rameau, J. C., Mansion, D. and Dumé, G., 1989. *Flore forestière française guide écologique illustré. 1 Plaines et collines.* Institute pour le développement forestier Ministère de l'agriculture de la forêt.
Raven, P. H., 1973. The evolution of mediterranean floras. In: di Castri, F. and Mooney, H. F. (eds), *Mediterranean type ecosystems.* Springer-Verlag, pp. 213–224.
Regione Autonoma della Sardegna, 1993. *Ente autonomo del flumendosa, 1988. Piano delle Acque,* Cagliari.
Ribeiro, A. *et al.*, 1981. *Introduction à la géologie génerale du Portugal.* SGP, Lisbon.
Richter, G., 1995. Assessment of soil erosion in a small watershed covered by loess. *GeoJournal* **36**(2/3): 285–288.
Rodriguez-Estrella, T., 1979. Geología e hidrogeología del sector Alcaraz-Lietor-Yeste (Prov.de Albacete). Síntesis geológica de la Zona Prebética. *Memorias del Instituto Geológico y Minero de España* **97**: 560.
Romero-Díaz, M. A., Barberá, G. G. and López-Bermúdez, F., 1995. Relationships between soil erosion, rainfall and vegetation cover in the semiarid environment of south east of Iberian Peninsula. In: *Proceedings of the Conference on Erosion and Land Degradation in the Mediterranean.* Portugal University of Aveiro, 14–18 June 1995, pp. 59–73.
Roxo, M., 1994. A Acção Antrópica no processo de degradação de solos—A serra de Serpa e Mértola. Doctoral thesis, Universidade Nova de Lisboa, Lisboa.
Royal Meteorological Office, 1981. The Meteorological Office Rainfall and Evaporation Calculation System MORECS. Hydrological Memorandum no. 45, July 1981.
Runnels, C. N. 1995. Environmental degradation in ancient Greece. *Scientific American* March: 72–75.
Russell, G., Jarvis, P. G. and Monteith, J. L., 1989. Absorption of radiation by canopies and stand growth. In: Russell, G., Marshall, B. and Jarvis, P. G. (eds), *Plant canopies: their form and function.* SEB Seminar Series no. 31, Cambridge University Press, Cambridge.
Sallares, R., 1991. *The ecology of the ancient Greek world.* Duckworth, London.
Sánchez, G. and Puigdefábregas, J., 1994. Interactions between plant growth and sediment movement in semi-arid slopes. *Geomorphology* **9**: 243–260.
Sanz, G. M. and Crespo Villalba, M. B., 1990. *Claves para la Flora Valencia.* Del Cenia al Segura, Valencia.
Shannon, C. E. and Weaver, W., 1949. *The mathematical theory of communication.* University of Illinois Press, Urbana.
Sharma, S. and Singh, S. P., 1994. Energy use and sustainable development. A case study in rural landscape of the central Himalaya. *Landscape and Urban Planning* **29**: 19–24.
Shubert, A., 1990. *A social history of modern Spain.* Unwin Hyman, London.
Shuttleworth, W. J. and Wallace, J. S., 1985. Evaporation from sparse crops—an energy combination theory. *Quarterly Journal of the Royal Meteorological Society* **111**: 839–855.
Simmers, I. *et al.*, 1992. Selected papers on aquifer overexploitation. *23rd International Congress of the International Association of Hydrogeologists.* Puerto de la Cruz, Tenerife (Spain), 15–19 April 1991, Vol. 3, Heinz Heise, Hannover.
Stefanovits, P. and Várallyay, Gy., 1992. State and management of soil erosion in Hungary. *Soil erosion and remediation workshop,* Proceedings—Central and Eastern European Agro-Environmental Program One, Budapest, 27 April–1 May 1992, RISSAC, Budapest, pp. 79–95.
Stein, J. and Su, P. Y., 1980. *The Random House dictionary.* Ballantine Books, New York.
Swain, P. H. and Davis, S. M. (eds), 1978. *Remote sensing: the quantitative approach.* McGraw-Hill, New York.
Szalay, J. (ed.), 1994. Seasonal report on the situation of the shallow ground water in lowland areas of Hungary. *Bulletins of VITUKI,* Budapest (in Hungarian).
Szentimrey, T., 1994a. Inhomogeneity estimations of temperature data series for Hungary. *Éghajlati és Agrometeorológiai Tanulmányok* **2**: 42 (in Hungarian).
Szentimrey, T., 1994b. Statistical problems connected with the homogenization of climatic time series. *European workshop on climate variations,* 15–18 May 1994, Majvik, Finland, pp. 330–339.
Texeira, C. and Gonçalves, F., 1980. *Introdução à geologia de Portugal.* JNICT, Lisboa.
Thomas, D. S. G. and Middleton, N. J., 1994. *Desertification: exploding the myth.* John Wiley, Chichester.
Thornes, J. B., 1974. Anatomy of a flood. *Geographical Magazine,* April 1974.
Thornes, J. B., 1976. *Semi-arid erosional systems.* London School of Economics. Papers in Geography, 7.
Thornes, J. B., 1977. Channel changes in ephemeral streams: observations and models. In: Gregory, K. J. (eds), *River Channel Changes.* John Wiley, Chichester.
Thornes, J. B., 1980. Structural instability and ephemeral channel behaviour. *Zeitschrift für Geomorphologie* supplement bd. **39**, 136–152.
Thornes, J. B., 1987. Environmental systems: patterns, processes and evolution. In: K. J. Gregory and J. Clark (eds), *Horizons in physical geography,* Macmillan, pp. 27–46.
Thornes, J. B., 1993. Channel process, evolution and history. In: Abrahams, A. D. and Parsons, A. J. (eds), *Geomorphology of desert environments.* Chapman and Hall, London, pp. 288–317.
Thornes, J. B., 1995. Global change and regional response: the case of the Old World Mediterranean. *Transactions Inst. British Geogr.,* **20**(3).
Thornes, J. B. and Brandt, J. A., 1993. Erosion–vegetation competition in a stochastic environment undergoing climatic change stochastic rainfall variations. In: Millington, A. C. and Pye, K. J. (eds), *Environmental change in the drylands: biogeographical and geomorphological responses.* John Wiley, Chichester, pp. 306–320.
Thornes, J. B. and Brunsden, D., 1977. *Geomorphology and time.* Methuen.
Thornes, J. B. and Gilman A., 1983. Actual and potential erosion around archaelogical sites in south-east Spain. In: de Ploey, J. (ed.) Rainfall simulation, runoff and soil erosion. *Catena Supplement* **4**: 91–113.
Thornes, J. B., Shao, J., Diaz, E., Roldan A., Diamond, S., McMahon, M. and Hawkes, J. C., 1996. Testing the MEDALUS model. *Catena* **120**.
Thornthwaite, C. W., 1948. An approach towards a rational classification of climate. *Geographical Review* 38: 55–94.
Thran, P. and Broekhuizen, S. (eds), 1965. *Agro-climatic atlas of Europe.* Centre for Agricultural Publications, Pudoc, Wageningen.
Thrower, N. J. W. and Bradbury, D. E., 1973. The physiography of the Mediterranean lands with special emphasis on California and Chile. In: di Castri, F. and Mooney, H. F. (eds), *Mediterranean type ecosystems.* Springer-Verlag, pp. 37–52.
Tomasi, D., Puddu, R. and Vacca, A., 1995. Erosione ed uso del suolo: indicazioni per la gestione e la pianificazione del territorio. Un esempio nella Sardegna sud-occidentale. *Proceedings of the Conference on Il Ruolo della Pedologia nella Pianificazione e Gestione del Territorio,* Cagliari, Italy.
Trump, D., 1980. *The prehistory of the Mediterranean.* Allen Lane, London.
Tsirtsis, E., 1988. Prospects for the olive plantations of the Aegean Islands. In: Olive groves of the Aegean. Eleourgiki, Athens, pp. 55–63 (in Greek).
Tutin, T. G., Heywood, V. H., Burges, N. A., Valentine, D. H., Walters, S. M. and Webb, D. A., 1964. *Flora Europaea,* Vol. I. Cambridge University Press, Cambridge.
Tutin, T. G., Heywood, V. H., Burges, N. A., Moore, D. M., Valentine, D. H., Walters, S. M. and Webb, D. A., 1968–80. *Flora Europaea,* Vols II–V, Cambridge University Press.
Tutin, T. G., Heywood, V. H., Burges, N. A., Moore, D. M., Valentine, D. H., Walters,

S. M. and Webb, D. A. (eds), 1976. *Flora europaea,* Vols I–V. Cambridge University Press, Cambridge.

Tzedakis, P. C., 1993. Long-term tree populations in northwest Greece through multiple Quaternary climatic cycles. *Nature* **364**: 437–440.

United Nations Environment Programme (UNEP), 1992. *World atlas of desertification*. Edward Arnold, Sevenoaks.

United Nations Environment Programme (UNEP), 1994. United Nations Convention to combat desertification in countries experiencing serious drought and/or desertification, particularly in Africa. Elaborated by the Intergovernmental Negotiating Committee (INCD), signed June 1994.

USDA Soil Survey Staff–Soil Conservation Service, 1975. *Soil taxonomy. A basic system for soil classification for making and interpreting soil survey.* Agriculture Handbook no. 436, Washington.

USDA Soil Survey Staff, 1992. *Keys to soil taxonomy*, 5th edn. SMSS technical monograph No. 19, Pocahontas Press, Blacksburg, Virginia.

Vacca, A., Puddu, R., Tomasi, D. and Usai, D., 1995. Erosion measurements in three areas of the S. Lucia catchment with different land uses. In: Enne, G., Aru, A. and Pulina, G. (eds), *Land use and soil degradation. MEDALUS in Sardinia.* Proceedings of the Conference held in Sassari, 27 May 1995, University of Sassari–University of Cagliari, pp. 115–127.

Van Andel, T. H. and Runnels, C. N., 1987. *Beyond the acropolis—a rural Greek past.* Stanford University Press, Stanford, California.

Van Mansvelt, J. D. and Mulder, J. A., 1993. European futures for sustainable development—a contribution to the dialogue. *Landscape and Urban Planning* **27**: 67–90.

Van Wesemael, B., Poesen, J. and de Figueiredo, T., 1995. Effects of rock fragments on physical degradation of cultivated soils by rainfall. *Soil and Tillage Research* **33** 229–250.

Vidal de la Blache, P., 1934. *Géographie Universelle VII: Méditerrannée*. Colin, Paris.

Voss, E. G., Burdet, H. M., Chaloner, W. G., Demoulliu, V., Hiepro, P., McNeill, J., Meirle, R. D., Nicolson, D. H., Rollins, R. C., Silva, P. C. and Greuter, W. (eds), 1983. *International code of botanical nomenclature.* Adopted by the Thirteenth International Botanics Congress, Sydney, August 1981. Dr. W. Junk Publishers, The Hague.

Walter, H. and Leith, H., 1960. *Klimadiagram Weltatlas*. Jena, Fischer.

Werner, R., 1993. Ecologically and economically efficient and sustainable use of agricultural landscapes. *Landscape and Urban Planning* **27**: 237–248.

Whittow, J., 1984. *The Penguin dictionary of physical geography*. Penguin, London.

Wigley, T. M. L., Jones, P. D. and Kelly, P. M., 1986. Empirical climate studies: warm world scenarios and the detection of climatic change induced by radiatively active gases. In: Bolin, B. *et al.* (eds), *The greenhouse effect, climatic change, and ecosystems*. SCOPE 29, John Wiley, Chichester, pp. 271–322.

Williams, A. M., 1984. Southern Europe transformed: political and economic change in Greece, Italy, Portugal and Spain. Harper and Row, London.

Williams, A. M. and Shaw, G. (eds), 1991. *Tourism and economic development: Western European experiences*, 2nd edn. Belhaven Press, London.

Windley, B. F., 1984. *The evolving continents*, 2nd edn. John Wiley, Chichester.

Wischmeier, W. H., 1960. Rainfall erosion index for a soil equation. *Proceedings of the American Soil Science Society*, **23.**

Wischmeier, W. H., 1962. Rainfall erosion potential. *Agricultural Engineering* **43**.

Wischmeier, W. H. and Smith, D. D., 1978. *Predicting rainfall erosion losses—a guide to conservation planning*. USDA Agricultural Handbook 537, US Government Printing Office, Washington, DC.

Wischmeier, W. H. *et al.*, 1958. Evaluation of factors in soil loss equation. *Agricultural Engineering* **39.**

World Commission on Environment and Development, 1987. *Our common future.* Oxford, Oxford University Press.

Yassoglou, N., 1993. Land and desertification. Course on desertification in an European context: physical and socio-economic aspects, Alicante, Spain, Commission of the European Communities, DG XII.

Younis, M. T., 1993. Caracterización y discriminación espectral, mediante radiometría de campo e imágenes Landsat-5 TM, de unidades litológicas en el área de Alcantarilla, Murcia. PhD Thesis, Universidad de Zaragoza.

MEDALUS PUBLICATIONS

Abrahart, R. J., Kirkby, M. J. and McMahon, M. L., 1994. MEDRUSH—a combined Geographical Information System and large scale distributed process model. Proceedings 2nd National Conference on GIS research, Leicester, pp. 67–76.

Alonso-Sarria, F., 1992. Estimación de la precipitación sobre cuencas hidrográficas durante episodios lluviosos de diversa duración. In: López Bermúdez, F., Conesa Garcia, C. and Romero Diaz, M. A. (eds), *Estudios de Geomorfología en España*, Sociedad Española de Geomorfología, Universidad de Murcia, Murcia, pp. 665–673.

Alonso-Sarria, F. and López Bermúdez, F., 1993. Rainfall time and space variability during short storms in South-East Spain. *Geoökodynamik*, Bensheim, Germany **15**(3): 261–278.

Aru, A., 1993. Problems of heavy metal pollution in mining areas in Sardinia. Preliminary findings. Symposium on biogeochemical cycle of metals in soils, Sassari, 1–2 April 1993. In: Studi Sassaresi, 1993, Sez.III, *Annali Facoltà di Agraria* **35**: 2.

Aru, A., 1994. Evaluation of soil properties for safe wastes incorporation. VII Interantional Congress. L'approccio integrato della moderna biologia: Uomo, territorio, ambiente, Vieste, Italy, 22–25 September 1994.

Aru, A., 1994. Soil contamination by heavy metals from mines in the Iglesiente region. Conference on land use and degradation—MEDALUS in Sardinia, Sassari, 25 May 1994.

Aru, A., 1995. The MEDALUS project in Sardinia: activity to date, results and future work. In: Enne, G., Aru, A. and Pulina, G. (eds), *Land use and soil degradation. MEDALUS in Sardinia*. Proceedings of the conference held in Sassari, Italy, 27 May 1994, pp. 13–18.

Aru, A., Baldaccini, P. and Barrocu, G., 1994. Desertification in the Mediterranean: causes and remedial measures. International symposium and workshop on desertification in developed countries, Tucson, AZ, 24–29 October 1994.

Aru, A., Madeddu, B. and Kahnamoei, A., 1995. Soil contamination by heavy metals from mines. In: Enne, G., Aru, A. and Pulina, G. (eds), *Land use and soil degradation. MEDALUS in Sardinia*. Proceedings of the conference held in Sassari, Italy, 27 May 1994, pp. 265–284.

Atherden, M. A. and Hall, J. A., 1994. Holocene pollen diagrams from Greece. *Historical Biology* **9**: 117–130.

Bacci, G., Meneguzzo, F., Musetti, N. and Niccolai, M., 1993. Comparison of precipitation patterns forecast by the General Circulation Model operating at Offenbach German Center with data collected by a raingauge network. Proceedings of the International Symposium on precipitation and evaporation, Bratislava, 1993.

Baird, A., Thornes, J. B. and Watts, G., 1993. Extending overland flow models to very rough slopes. In: Parsons, A. J. and Abrahams, A. (eds), *Overland flow*. University College Press, London, pp. 127–154.

Baldaccini, P., Previtali, F., Madrau, S., Deroma, M., Dessì Baldaccini, P., Previtali, F., Madrau, S., Deroma, M., Dessì, G. and Goussikpe, Y. G., 1995. Study of the Rio Astimini basin and problems relating to desertification: pedological outlines. In: Enne, G., Aru, A. and Pulina, G. (eds), *Land use and soil degradation. MEDALUS in Sardinia*. Proceedings of the conference held in Sassari, Italy, 27 May 1994, pp. 77–86.

Baldaccini, P., Previtali, F., Madrau, S., Dessì, G. and Deroma, M., 1995. Valutazione dell'attitudine al pascolo dei suoli della Sardegna. Il bacino del rio d'astimini fiume santo Sassari. L'informazione pedologica nella prospettiva di un servizio regionale del suolo, Milano, 26–27 October 1994.

Basso, F., Pisante, M. and Basso, B., 1994. Biological and productive behaviour of Jerusalem Artichoke (*Helianthus tuberosus* L.) Cv. in a hilly environment of southern Italy. Fourth International Conference on desert development, Mexico City, July 1993.

Basso, F. and Postiglione, L., 1994. Soil conservation agronomic aspects in hilly areas: layouts and tillage methods. *Rivista di Agronomia* **4**: 273–296.

Basso, F., Postiglione, L. and Ruggiero, C., 1994. Soil tillage sustainable agriculture: experimental results in the hilly area of Southern Italy—Agri Basin. Third MEDALUS II Meeting, Athens, May 1994.

Basso, B., Ritchie, J. T. and Martin, E. C., 1994. Effect of nitrogen fertilization on nitrate leaching. Proceedings of the ESA Workshop on modelling the fate of agrochemicals in the environment, Venice, March 1994.

Bastida, J., Younis, M. T., Lopez-Buendia, A. M. and Melia, J., 1992. Application of Landsat TM images for the study of the Hydrographic Network density and its relation to lithology in the area of Alcantarilla, SE Spain. *Annales Geophysicae* **10**(2): 311.

Bathurst, J. C., Kilsby, C. and White, S., 1996. Modelling the impacts of climate and land use change on basin hydrology and soil erosion in Mediterranean Europe. In: Brandt, C. J. and Thornes, J. B. (eds), *Mediterranean desertification and land use*. Wiley, Chichester.

Battista, P., Benincasa, F., Duce, P. and Materassi, A., 1995. Design and construction of a microprocessor impactometer. In: Enne, G., Aru, A. and Pulina, G. (eds), *Land use and soil degradation. MEDALUS in Sardinia*. Proceedings of the conference held in Sassari, Italy, 27 May 1994, pp. 169–180.

Battista, P., Benincasa, F. and Fasano, G., 1991. Progetto e realizzazione di un pluviointensimetro: misuratore dell'intensità della pioggia. *Rivista di Ingegneria Agraria* **1**: 24–29.

Battista, P., Benincasa, F., Fasano, G. and Materassi, A., 1992. SIVE: Sistema Integrato per la Valutazione dell'Erosione. *Atti Convegno Informatica e Agricoltura*, pp. 641–649.

Battista, P., Benincasa, F. and Materassi, A., 1994. Progetto e realizzazione di un impattometro a microprocessore. *Rivista di Ingegneria Agraria* **1**: 27–33.

Belmonte-Serrato, F. and Romero-Díaz, M. A., 1992. Evaluación de la capacidad de interceptación de la lluvia por la vegetación y su relación con la erosión de los suelos en el Sureste semiárido español. Primeros resultados. In: López-Bermúdez, F., Conesa-Garcia, C. and Romero-Díaz, M. A. (eds), *Estudios de Geomorfología en España.* Sociedad Española de Geomorfología, Universidad de Murcia, Murcia, pp. 33–43.

Belmonte-Serrato, F. and Romero-Díaz, M. A., 1993. Instrumentos y métodos para el estudio de la capacidad de interceptación de algunas especies vegetales mediterráneas, El Ardal (Murcia). Nuevos procesos territoriales. Asociación de Geógrafos Españoles, Universidad de Sevilla, Sevilla, pp. 181–186.

Belmonte-Serrato, F. and Romero-Díaz, A., 1994. Distribución de flujos de agua en el proceso de interceptación en cuatro especies vegetales mediterráneas y su relación con la cantidad de agua disponible en el suelo. In: Arnáez-Vadillo, J., García-Ruis, J. M. and Gómez-Villar, J. (eds), *Geomorfología en España.* Sociedad Española de Geomorfología, Logroño, pp. 201–210.

Berael, C., Poesen, J. and van Wesemael, B., 1995. Rill and gully erosion in relation to rock fragment cover and surface slope along a hillslope in southeastern Spain. In: Fantechi, R., Peter, D., Balabanis, P., and Rubio, J. (eds), 1995. *Desertification in a European Context: Physical and socio-economic aspects*. Proceedings of the European School of Climatology and Natural Hazards Course, Alicante, Spain, 1993. EUR 15415, Commission of the European Communities, Luxembourg: 371–376.

Berardo, D., Asara, P., Piras, R., Baffigo, P. and Ibba, I., 1995. Regional laws for the works of land improvement and forestation: their effect on the soil degradation processes in Sardinia. In: Enne, G., Aru, A. and Pulina, G. (eds), *Land use and soil degradation. MEDALUS in Sardinia*. Proceedings of the conference held in Sassari, Italy, 27 May 1994, pp. 41–50.

Bergkamp, G., Cammeraat, L. H. and Martinez Fernandez, J., 1996. Water movement and vegetation patterns on shrublands and abandoned fields in desertification threatened areas. *Earth Surface Processes and Landforms* **21**: 1073–1090.

Bochet, E., Poesen, J. and Rubio, J. L., 1995. Influence of three Mediterranean plant species on the spatial and temporal variation of water erosion (Valencia, Spain): experimental setup and preliminary results. In: Fantechi, R., Peter, D., Balabanis, P., and Rubio, J. (eds), 1995. *Desertification in a European Context: Physical and socio-economic aspects*. Proceedings of the European School of Climatology and Natural Hazards Course, Alicante, Spain, 1993, EUR 15415, Commission of the European Communities, Luxembourg: 377–388.

Böer, M. M., 1996. Characterising desertification response units on semi-arid hillslopes in the Rio Mula basin (SE Spain). Submitted to *Agriculture, Ecosystems and Environment*.

Böer, M., Cammeraat, L. H., Imeson, A. C., Janssens, F. R., Martinez Fernandez, J., Lopez Bermudez, F. and Mücher, H. M., 1996. Soil properties and desertification in the Rio Mula Basin (SE Spain). Submitted to *Catena.*

Böer, M., Del Barrio, G. and Puigdefábregas, J., 1994. A hierarchical approach to desertification hazard assessment in the semi-arid Mediterranean. *Annales Geophysicae* **12**(2): 413.

Botti, P., Vacca, A. and Aru, A., 1995. Mass balance and sediment transport and its significance in land evaluation. The example of the tributary basin of the Mulargia reservoir central southern Sardinia. In: Enne, G., Aru, A. and Pulina, G. (eds), *Land use and soil degradation. MEDALUS in Sardinia*. Proceedings of the conference held in Sassari, Italy, 27 May 1994, pp. 157–168.

Brenner, A. J., Cueto, M., Garcia, J., Gilabert, M. A., Incoll, L. D., Martinez, J., Porter, E., Pugnaire, F. I. and Younis, M. T., 1995. A comparison of direct and indirect methods for measuring leaf and surface area of individual bushes. *Plant, Cell and Environment*, **18**: 1332–1340.

Brenner, A. J., Domingo, F., Clark, S. C., Aguilera, C., Gutierrez, L. and Incoll, L. D., 1996. Water relations and gas-exchange of three perennial species along a soil catena in semi-arid southeast. Spain. II Carbon Assimilation. Submitted to *Functional Ecology.*

Brenner, A. J. and Incoll, L. D., 1994. Scaling up in a sparsely vegetated shrubland in Southern Spain. Society of Experimental Biology, National Conference, Swansea, Wales, 11–15 April. *Journal of Experimental Biology* **45**: 3.

Brenner, A. J. and Incoll, L. D., 1995. The effect of clumping and stomatal response on theevaporation from a sparsely vegetated shrubland. *Annales Geophysicae* Supplement II, **13**:C501.

Brey, V. and Cammeraat, L. H., 1993. The determination of micro-aggregation with the use of the Quantachrome Microscan II Particle Size Analyzer. Internal report, University of Amsterdam.

Brunone, B. and Romano, N., 1994. Experimental and numerical analysis of one-dimensional water flow into layered soils. Summary of the poster presented at XII CIGR World Congress on Agricultural Engineering, Milano, 29 August–1 September, part 1, pp. 66–67.

Bruzzone, L., Roli, F. and Serpico, S. B., 1995. An extension of the Jeffries-Matusita Distance to Multiclass Cases for Feature-Selection. *IEEE Trans. on Geoscience and Remote Sensing* **33**: 6.

Bruzzone, L., Roli, F. and Serpico, S. B., 1996. Issues in feature selection for remote-sensing image classification. *International Journal of Remote Sensing* (submitted).

Bruzzone, L., Roli, F. and Zali, M., 1994. Feature-selection for forest classification purposes. International Symposium on Remote sensing in forest studies and management, Brixen, Italy, 1–2 December 1994.

Bunte, K. and Poesen, J., 1993. Effects of horseshoe vortex erosion on sediment yield from soils covered by rock fragments. *Zeitschrift für Geomorphologie* **37**(3): 327–335.

Bunte, K. and Poesen, J., 1993. Effects of rock fragment covers on erosion and transport of noncohesive sediment by shallow overland flow. *Water Resources Research* **29**(5): 1415–1424.

Bunte, K. and Poesen, J., 1994. Effects of rock fragment size and cover on overland flow hydraulics, local turbulence and sediment yield on an erodible soil surface. *Earth Surface Processes and Landforms* **19**: 115–135.

Burton, A. and Bathurst, J., 1994. Modelling shallow landslide erosion and sediment yield at the basin scale. Proc. International Association of Hydraul. Res. International Workshop on floods and inundations related to large earth movements, University of Trento, Italy, October 1994.

Burton, A., Kilsby, C. G., White, S. M. and O'Connell, P. E., 1993. The use of the Neyman-Scott rainfall model and data from the Canadian Climate Centre GCM to produce rainfall time series for hydrological modelling. Presented at European Geophysical Society VIII General Assembly, Wiesbaden, 3–7 May.

Camarda, I. and Satta, V., 1993. Fire, overgrazing and desertification processes in a sample area in Sardinia Italy. European School of Climatology and Natural Hazards, Desertification in European Context. Physical and socio-economic aspects, Alicante, 6–14 October.

Camarda, I. and Satta, V., 1995. Degradation of vegetation and desertification processes in Olias southern Sardinia. In: Enne, G., Aru, A. and Pulina, G. (eds), *Land use and soil degradation. MEDALUS in Sardinia*. Proceedings of the conference held in Sassari, Italy, 27 May 1994, pp. 201–210.

Cammeraat, L. H., 1992. Preliminary data report petralona Field site (MEDALUS). Report produced at University of Amsterdam.

Caredda, S., Ledda, L., Porqueddu, C. and Sulas, L., 1995. The protection role against soil erosion of the natural and artificial swards in the Mediterranean environment. In: Enne, G., Aru, A. and Pulina, G. (eds), *Land use and soil degradation. MEDALUS in Sardinia*. Proceedings of the conference held in Sassari, Italy, 27 May 1994, pp. 189–200.

Caruso, G. and Postiglione, L., 1993. Effetti dell'irrigazione con acqua a diversa concentrazione salina sul suolo e su cultivar di pomodoro da industria (*Lycopersicon lycopersicum* L.). *Rivista di Agronomia* **27**(3): 211–219.

Casimiro, P., 1993. Concelho de Mértola—Geo-biografia das Mudanças de Uso do Solo. Masters Thesis, Nova Universidade de Lisboa.

Casimiro-Mendes, J., 1991. Clima e actividades humanas. Preocupaçoes do presente; acçoes para o futuro in 2° Congresso Técnico sobre Conservaçyo da Natureza nos países do sul da Europa, Olho, Portugal.

Casimiro-Mendes, J., 1991. Indice normalizado de anomalia da precipitaçao na regio da Bacia do Rio Cobres, INMG, Lisbon, Portugal.

Casimiro-Mendes, J., 1991. Variability of the precipitation in the Alentejo region. In IV Congresso do Alentejo, Sines, Portugal.

Casimiro-Mendes, J., 1992. A problemática das secas em Portugal. In Painel SECA92, Lisbon, Portugal.

Casimiro-Mendes, J., May 1994. Climatic variability of precipitation in Portugal; anomaly index; trends; variability along ten and thirty years period. In: Proceedings of the European Workshop on Climate Variations held in Kirkkonummmi, Finland.

Casimiro-Mendes, J. and Coelho, M. F., 1991. Standardized anomaly index of seasonal and annual values of mean air temperature in the Alentejo Region, INMG, Lisbon, Portugal.

Casimiro-Mendes, J. and Coelho, M. F., 1991. Standardized anomaly index of seasonal and annual rainfall amounts in the Alentejo Region, INMG, Lisbon, Portugal.

Casimiro-Mendes, J. and Coelho, M. F., 1993. Variabilidade climática em Portugal continental quantidade de precipitaçao; indice regional de anomalia; tendência; variabilidade por decénios e triénios, INMG, Lisbon, Portugal.

Casimiro-Mendes, J. and Guerreiro, M. R., 1991. Testes estatísticos de tendência das séries climatológicas da quantidade de precipitaço na regio da Bacia do Rio Cobres, INMG, Lisboa, Portugal. *Monografia* **32**.

Cau, L., 1995. Common land and land degradation in Sardinia. In: Enne, G., Aru, A. and Pulina, G. (eds), *Land use and soil degradation. MEDALUS in Sardinia*. Proceedings of the conference held in Sassari, Italy, 27 May 1994, pp. 291–300.

Ciollaro, G. and Romano, N., 1995. Spatial variability of the hydraulic properties of a volcanic soil. *Geoderma* **65**: 263–282.

Comagna, V. and Basile, A., 1994. Temporal stability of spatial patterns of soil water storage in a cultivated Vesuvian soil. *Geoderma* **62**: 299–310.

Comegna, V. and Vitale, C., 1993. Space–time analysis of water status in a volcanic Vesuvian soil. *Geoderma* **60**: 135–158.

Conese, C., 1995. Advanced technologies for the environmental systems monitoring. In: Enne, G., Aru, A. and Pulina, G. (eds), *Land use and soil degradation. MEDALUS in Sardinia.* Proceedings of the conference held in Sassari, Italy, 27 May 1994, pp. 29–40.

Conese, C., Maselli, F., Pelegretti, P. and Rodolfi, A., 1994. Studio integrato di immagini da satellite per il monitoraggio delle risorse agricole e naturali in un'area collinare dell'Abruzzo. Proceedings of the Annual Congress AICA, Palermo, 21–23 September 1994, Vol. 2, pp. 877–883.

Conte, M., 1991. Climate forecasting with dynamic models. *Mem. Soc. Ital. Geogr.* **46**: 125–144.

Conte, M., 1991. The Mediterranean oscillation index. In: The global climate system monitoring, 1991 WMO CSM R 84/86, pp. 28–30.

Conte, M., 1994. Variazioni climatiche nell'area mediterranea. Proceedings of the meeting in May 1994 at CIRPS (Centro Interuniversitario di Ricerca sui Paesi in via di sviluppo), Rome.

Conte, M. and Colacino, M., 1992. Andamento della temperatura in Italia dalla seconda met à del secolo scorso ad oggi. *Boll. Geofisico* **16**: 777–795.

Conte, M. and Colacino, M., 1994. Greenhouse effect and pressure patterns in the Mediterranean Basin. Regarding the observed patterns and the statistical climatic forecasts of the precipitations in the western central Mediterranean: a scenario for Sicily. Proceedings of the meeting at University 'La Sapienza', Rome, on 'Physics and dynamics in the analysis and prognosis of precipitation fields'.

Conte, M. and Giuffrida, A., 1991. L'oscillazione mediterranea: possibili effetti sulle precipitazioni in Italia. *Società Italiana per il Progresso Scienze* **61**: 61–88.

Conte, M. and Giuffrida, A., 1991. The long term evolution of the Italian climate. *Mem. Soc. Ital. Geogr.* **46**: 329–343.

Conte, M., Giuffrida, A. and Tedesco, S., 1991. L'Oscillazione Mediterranea. *Mem. Soc. Ital. Geogr.* **46**: 115–124.

Cook, C. M., Sgardelis, S. P., Pantis, J. D. and Lanaras, T., 1994. Concentrations of Pb, Zn, and Cu in relation to urban pollution. *Bulletin of Environmental Contamination Toxicol*. **53**: 204–210.

Corona, P. and Ferrara, A., 1994. Study on the spatial relationships in forested areas of central Italy. Third MEDALUS II Meeting, Athens, May 1994.

Corte-Real, J., Wang, X. and Zhang, X., 1995. Modes of variability in the northern hemisphere's large-scale mid-tropospheric circulation. *Theoretical and Applied Climatology* **50**: 133–146.

Corte-Real, J., Zhang, X. and Wang, X., 1996. Large-scale circulation regimes and surface climate anomolies over the Mediterranean. *International Journal of Climatology* **15** (in press).

Corte-Real, J., Wang, X. and Zhang, X., 1996. Downscaling GCM information to regional scale: a non-parametric multivariate regression approach. *Climate Dynamics* **11** (in press).

D'Urso, G., Mancini, M., Romano, N., Rosso, R. and Santini, A., 1993. Comparison of different techniques for soil water content measurements at basin scale. Proceedings of the AIGR National Congress, Maratea (Italy), 7–11 June, pp. 353–362 (in Italian, with English abstract).

D'Urso, G., Mancini, M. and Santini, A., 1994. Satellite S.A.R. data sensitivity to soil dielectric and moisture patterns. Proceedings of the XII CIGR World Congress on Agricultural Engineering, Milano, 29 August–1 September, Vol. 1, pp. 35–43.

Dalaka, A., Diamantopoulos, J. and Sgardelis, S. P., 1994. Growth and reproductive characteristics of some grasses in an evergreen sclerophyll formation of Northern Greece. Proceedings of the 16th Conference of the Greek Association of Biological Science, 3.49 (in Greek with an English summary).

Dalaka, A., Papatheodorou, E., Cook, C., Lanaras, T., Stamou, G. P., Diamantopoulos, J., Pantis, J. D. and Sgardelis, S., 1994. Effect of air temperature, relative humidity and PAR on stomatal conductance of *Quercus coccifera* leaves. Proceedings of the 16th Conference of the Greek Association of Biological Science, 3.43–3.45 (in Greek with an English summary).

Dalaka, A., Sgardelis, S. and Diamantopoulos, J., 1993. Reproduction and growth patterns of the grass species *Stipa bromoides* (L.). Dorf. Proc. of the

15th Conference of the Greek Association of Biological Science, pp. 147–149 (in Greek with an English summary).

Dalaka, A., Papatheodorou, E., Sgardelis, S., Pantis, J. D., Diamantopoulos, J., Stamou, G. P., Cook, C. and Lanaras, T., 1994. Temporal variations of apparent photosynthesis rate in *Quercus coccifera*. Proceedings of the 16th Conference of the Greek Association of Biological Science, 3.40–3.42 (in Greek with an English summary).

De Falco, E., Amato, M. and Basso, F., 1994. Nitrogen losses in surface and deep runoff in slope cropped with faba bean (*Vica faba minor* Moench.) as a function of different tillage methods. 13 ISTRO Confernce, Aalborg (DK), July 1994.

De Franchi, A. S., Amato, M., Pisante, M. and Graziano, F. S., 1994. Variations of physical characteristics (structure, soil moisture, porosity) on sloping soil compared with different tillage methods and cultures. *Rivista di Agronomia* **4**: 215–224.

Del Barrio, G. and Puigdefábregas, J., 1993. Definición de unidades de paisaje utilizando taxonomía numérica sobre un sistema de información geográfica: Valle del Guadaletín, Murcia. IV Jornadas de la Asociación Española de Ecología Terrestre. Alicante, 27–29 September 1993.

Delgado-Castilla, L., 1991. Estudio sedimentológico de los cuerpos sedimentarios Pleistocenos en la parte inferior de la Rambla Honda (Almería). VII Reunión nacional sobre el Cuaternario (AEQUA), Valencia, 1991.

Delgado-Castilla, L., 1993. La sedimentacíon cuaternaria en el borde norte de la cuenca de Tabernas (SE se España). In: El Cuaternario en Andalucía Oriental Vol II, AEQUA, regional de Andalucia, Seville.

Delgado-Castilla, L., 1996. Estudio sedimentológico de los cuerpos sedimentarios Pleistocenos en la Rambla Honda, al Norte de Tabernas, provincia de Almería (SE de España). *Cuaternario y Geomorfología* **7**: 91–100.

Delgado-Castilla, L., 1995. Los sedimentos continentales cuaternarios en el borde Norte de la cuenca de Tabernas (SE de España). *Monografías de AEQUA* **2**.

Delgado-Castilla, L., Pascual-Molina, A. and Ruiz-Bustos, A., 1993. Geology and micromammals of the SERRA—1 site (Tabernas basin, Betic Cordillera). *Estudios Geológicos* **49**: 361–366.

Delgado-Castilla, L., Torrentó, J. R. and Pardini, G., 1991. Estudio de una secuencia de paleosuelos en la vertiente sur de la Sierra de los Filabres. En 18th Reunion Nacional de Suelos, Universidad de la Sierra de la Laguna, Tenerife, 22–28 September 1991, pp. 261–273.

Dettori, S., 1995. A forest management model to prevent fires and erosion. In: Enne, G., Aru, A. and Pulina., G. (eds), *Land use and soil degradation. MEDALUS in Sardinia*. Proceedings of the conference held in Sassari, Italy, 27 May 1994, pp. 241–250.

Dettori, B., Dore, M., Goussikpe, Y. and Deroma, M. A., 1995. The hydrogeology and permeability of the Rio Astimini basin. In: Enne, G., Aru, A. and Pulina, G. (eds), *Land use and soil degradation. MEDALUS in Sardinia*. Proceedings of the conference held in Sassari, Italy, 27 May 1994, pp. 87–98.

Diamantopoulos, J., Pantis, J., Sgardelis, S., Pirintsos, S., Dalaka, A., Papatheodorou, E., Iatrou, G., Mardiris, Th. and Stamou, G. P., 1993. The desertification process in the North Mediterranean area. The EPOCH MEDALUS Project. Proceedings of the 15th Conference of the Greek Association Biol. Sci., pp. 390–392 (in Greek with an English summary).

Domingo, F., Brenner, A. J. and van Gardingen, P. R., 1993. Estudio de la transferencia de calor y vapor de agua foliar en plantas situadas en ambiente semi-arido. IV Journadas de la Asociatión Española de Ecología Terrestre, 27–29 September, Alicante, Spain, 13.

Domingo, F., Brenner, A. J. and van Gardingen, P. R., 1994. Variation of leaf boundary layer conductance in sparse canopies of *Retama sphaerocarpa* and *Stipa tenacissima*. *Journal of Experimental Botany* **45**: 3.

Domingo, F., Puigdefábregas, J., Clark, S. C., Incoll, L. D. and Lockwood, J., 1991. Plant physiological behaviour in relation to water in a semiarid environment in Southeast Spain. *IAHS* **204**: 335–343.

Domingo, F., Puigdefábregas, J., Clark, S. C., Incoll, L. D. and Lockwood, J., 1996. Comportamiento fisiológico vegetal en relación al agua en medioambiente semiárido en el Sudeste español. *Mediterránea Ser. Biol.* (in press).

Domingo, F., Puigdefábregas, J., Clark, S. C., Incoll, L. D. and Lockwood, J., 1996. Relaciones hídricas fisiológicas de tres especies perennes, *Retama sphaerocarpa, Anthyllis cytisoides y Stipa tenacissima* en ambiente semiárido del sudeste español. *Mediterránea, Ser. Biol.* (in press).

Domingo, F. and van Gardingen, P. R., 1994. The importance of leaf shape and canopy structure in determining water use of natural vegetation in semi-arid regions of Southern Spain. *The Bulletin of the British Ecological Society* **25**(4): 251–252.

Duce, P., Battista, P., Benincasa, F. and Materassi, A., 1995. Design and development of a microprocessor impactometer. Atti del convegno uso e degrado del suolo: il MEDALUS in Sardegna.

Enne, G., Aru, A. and Pulina, G. (eds), 1995. *Land use and soil degradation. MEDALUS in Sardinia*. Proceedings of the conference held in Sassari, Italy, 25 May 1994. 303pp.

Enne, G., Masala, G. and Pisano, G. 1995. An integrated approach to the study of phenomenon of degradation connected to the use of soil. In: Enne, G., Aru, A. and Pulina, G. (eds), 1995. *Land use and soil degradation. MEDALUS in Sardinia*. Proceedings of the conference held in Sassari, Italy, 25 May 1994, pp. 19–28.

Escos, J. and Puigdefábregas, J., 1994. Detección de estrés según tipo de suelo en desarrollo de *Anthyllis cytisoides* por medio de simetría de escala. *Stvdia Oecologica* **10/11**: 201–208.

Escos, J. and Puigdefábregas, J., 1994. Detección de estrés según tipo de suelo en desarrollo de Anthyllis cytisoides por medio de simetría de escala. 4 Jornadas de la Asociación Española de Ecología Terrestre, Alicante, 27–29 September 1993.

Espírito-Santo, F., 1992. Drought in the Alentejo Region. In 5 Congresso do Alentejo, Évora, Portugal.

Espírito-Santo, F., 1993. Vigilancia e caracteristicas das secas. Teoria do Caos e Previsao a Longo Prazo. In: Proceedings of the Simpósio sobre Catástrofes Naturais, LNEC, Lisbon, Portugal.

Espírito-Santo, F., 1994. Vigilância e Características das Secas. Variabilidade da precipitaçao no Alentejo. In: Proceedings of the Simpósio sobre Catástrofes Naturais, Governo Civil de Setúbal, Setúbal, Portugal.

Friligos, N., Karydis, M., Giourga, C. and Margaris, N. S., 1994. Modern agricultural practices and eutrophication in Pagassitikos Gulf, Greece. *Ecologia Mediterranea* **20**(1/2): 59–65.

Gallart, F., Calvo, A., López-Bermúdez, F. and Puigdefábregas, J. Analyzing badlands dynamics as dependent on climate and lithology along the Mediterranean facade of Spain. Research design. Conference on methods of hydrological basin comparison, Oxford, pp. 152–154.

García-Haro, F. J., Gilabert, M. A. and Melia, J., 1994. Linear mixing model inversion for estimating vegetation fraction by remotely sensed data. Poster presented at MEDALUS Almeria meeting, Aguadulce, 18–22 November 1994.

García-Haro, F. J., Gilabert, M. A. and Melia, J., 1995. Application of a linear mixing model for the estimation of ground vegetation proportion from Landsat-5 TM data in the Guadalentin Basin. 6 Simposium Nacional de reconocimiento de formas y analisis deimagens, Cordoba, April 1995.

Gentileschi, M. L., 1995. Population growth in metropolitan and coastal areas and land use in Sardinia. Human geography studies in the Commune of Capoterra. In: Enne, G., Aru, A. and Pulina, G. (eds), 1995). *Land use and soil degradation. MEDALUS in Sardinia.* Proceedings of the conference held in Sassari, Italy 25 May 1994, pp. 285–290.

Gilabert, M. A., Younis, M. T. and Meliá, J., 1991. Utilidad de la radiometría de campo en el estudio de la reflectividad de los componentes de un sistema vegetal. IV Reunión Científica de la Asociación Española de Teledetección 'Teledetección y Medio Ambiente', Seville, 13–15 November 1991, pp. 179–185.

Gilabert, M. A., Younis, M. T. and Meliá, J., 1993. Spectral properties of weathered and fresh surfaces of some lithological units by means of laboratory measurements. 18th General Assembly of the European Geophysical Society, Wiesbaden, 3–7 May 1993.

Gilabert, M. A., Segarra, D. and Meliá, J., 1994. Simulation of citrus orchard reflectance by means of a geometrical canopy model. *International Journal of Remote Sensing* **15**: 2559–2582.

Ginesu S., Previtali, F. and Sias, S., 1995. Geological and geomorphological outlines of the Rio Astimini-Fiume Santo basin northwestern Sardinia. In: Enne, G., Aru, A. and Pulina, G. (eds), *Land use and soil degradation. MEDALUS in Sardinia*. Proceedings of the conference held in Sassari, Italy, 25 May 1994, pp. 65–76.

Giourga, C. E. and Margaris, N. S., 1994. The role of the contemporary consuming patterns and cultivation practices in agriculture and environment of the Aegean islands. In: *The Greek agriculture in the European Union Market*, 2nd Hellenic Conference of Agroeconomy, pp. 577–594 (in Greek).

Gomes, I., May 1994. Use of NOAA satellites in support of agricultural activities. Proceedings of Aplicaçao da detecçao remota ao estudo dos recursos naturais, LNEC, Lisbon, Portugal.

Gomes, I., 1994. The use of METEOSAT imagery for cloud climatological studies in Portugal associated with typical synoptic situations. Proceedings of 10th METEOSAT Scientific User's Conference, Cascais, Portugal.

Goodess, C. M., 1994. Investigation of future changes in the occurrence of high-rainfall events in SE Spain. Abstract prepared for the 19th General Assembly of the European Geophysical Society. *Annales Geophysicae* **12**(II): 366.

Greppi, M., Niedda, M. and Pisano, G., 1995. Analysis of hydroerosive processes. In: Enne, G., Aru, A. and Pulina, G. (eds), *Land use and soil degradation. MEDALUS in Sardinia*. Proceedings of the conference held in Sassari, Italy, 27 May 1994, pp. 141–148.

Grove, A. T., Moody, J. and Rackham, O. (eds), 1992. Stability and change in the Cretan landscape. Petromarula I. Department of Geography, Cambridge.

Grove, J. M., 1994. Climate in the eastern and central Mediterranean, 1675–1715. In: Frenzel, B. (ed.), Climatic trends and anomalies in Europe, 1675–1715. *European Palaeoclimate and Man* **8**: 275–285.

Grove, J. M. and Grove, A. T., 1992. Little Ice Age climates in the eastern Mediterranean 1992. In: Frenzel, B. (ed.), European climate reconstructed from documentary data: methods and results. *Palaeoclimate Research* **7**: 45–50.

Grove, J. M., Grove, A. T. and Conterio, A., 1992. Little Ice Age climate in the easterm Mediterranean. In: Mikami, T. (ed.), Proceedings of the international symposium on the Little Ice Age climate. Department of Geography, Tokyo Metroprolitan University, pp. 221–222.

Grove, J. M. and Switsur, R., 1994. Glacial geological evidence for the Medieval warm period. *Climatic Change* **30**: 1–27.

Gutiérrez, L., Puigdefábregas, J. and Nicolau, J. M., 1993. Estructura, biomasa y producción de las comunidades vegetales presentes en una catena representativa de un ambiente semiárido de la provincia de Almería. IV Jornadas de la Asociación Española de Ecología Terrestre, Alicante, 27–29 September 1993.

Gutierrez, M. M. and Pulina, P., 1995. Policies for the control of supply and efficiency in the protection of the natural environment. In: Enne, G., Aru, A. and Pulina, G. (eds), *Land use and soil degradation. MEDALUS in Sardinia*. Proceedings of the conference held in Sassari, Italy, 25 May 1994, pp. 181–188.

Haase, P., 1994. Dynamics of cladode cohorts in the semi-arid evergreen shrub *Retama sphaerocarpa* (L.) Boiss. In: Tallis, J. H., Norman, H. J. and Benton, R. A. (eds), Proceedings of the VI International Congress of Ecology, Manchester.

Haase, P., 1996. Spatial pattern analysis based on Ripley's K-function and its use in ecology: An introduction and an evaluation of differnt methods of edge correction. *Journal of Vegetation Science* (in press).

Haase, P., Pugnaire, F. I. and Incoll, L. D., 1995. Seed production and dispersal in the semi-arid tussock grass *Stipa tenacissima* L. during masting. *Journal of Arid Environments* **31**: 55–65.

Haase, P., Pugnaire, F. I., Incoll, L. D. and Clark, S. C., 1994. Diurnal, seasonal, and age-related changes in the rate of stem photosynthesis in the semi-arid shrub *Retama sphaerocarpa* (L.) Boiss. *Bulletin of the British Ecological Society* **25**(4): 252.

Harrison, A. R., Hurcom, S. J. and Taberner, M., 1991. Examining spectral mixing modelling using *in situ* spectroradiometric measurements. Proceedings of Spatial Data 2000, Oxford, pp. 242–249.

Hatzopoulos, J. N., Giourga, C., Margaris, N. S. and Koukoulas, S., 1992. Land Cover Classification of Olive Trees in the Greek Islands Using Landsat-TM Images. ASPRS/ACSM/RT92, Washington, DC **4**: 492–498.

Hurcom, S. J., Harrison, A. R. and Taberner, M., 1993. Mixture modelling using AVIRIS and SIRIS data. Proceedings of the 25th International Symposium on remote sensing and global environmental change, Graz, Austria, **1**: 123–134.

Hurcom, S. J., Harrison, A. R. and Taberner, M., 1994. Factor analysis of semi-arid vegetation spectral response using AVIRIS and airborne video data. Proceedings of the first airborne remote sensing conference, Strasbourg, France, **2**: 424–436.

Iatrou, G. D., Argyropoulou, M. D. and Stamou, G. P., 1993. Colonization process of decomposing *Quercus coccifera* litter in an evergreen-sclerophyll formation. Proceedings of the 15th Conference of the Greek Association Biol. Sci, pp. 186–188 (in Greek with an English summary).

Kilsby, C. G. and Bathurst, J. C., 1992. Parameter scale dependence in erosion and sediment yield modelling with SHETRAN-UK. Presented at European Geophysical Society XVII General Assembly, Edinburgh, 6–10 April.

Kilsby, C. G. and Bathurst, J. C., 1993. Assessing the effects of changes in land use and climate on the hydrology and erosion rates of the Cobres basin, Portugal, with SHETRAN-UK. Presented at European Geophysical Society XVIII General Assembly, Wiesbaden, 3–7 May.

Kirkby, M. J., 1992. An erosion limited hillslope evolution model. *Catena Supplement* **23**: 157–187.

Kirkby, M. J., 1994. Thresholds, instability and frequency distributions. In: Kirkby, M. J. (ed.), *Process models and theoretical geomorphology.* John Wiley, Chichester.

Kirkby, M. J., Baird, A. J., Lockwood, J. G., McMahon, M. D., Mitchell, P. L., Shao, J., Sheehy, J. E., Thornes, J. B. and Woodward, F. I., 1996. The MEDALUS slope catena model: a physically based process model for hydrology, ecology and land degradation interactions in MEDALUS I. In: Brandt, C. J. and Thornes, J. B. (eds), *Mediterranean desertification and land use*. John Wiley, Chichester.

Kirkby, M. J. and Cox, N. J., 1995. A climatic index for soil erosion potential (CSEP), including seasonal factors. *Catena Supplement* (Jan de Ploey volume) **25**(1–4): 333–352.

Kosmas, C. and Danalatos, N. G., 1994. Climate change, desertification and the Mediterranean region. In: Loveland, P. and Rounsevell, M. (eds), *NATO Advanced research workshop on soil responses to climate change: Implications for natural and managed ecosystems*. Springer-Verlag, Berlin.

Kosmas, C. S., Danalatos, N. G., Moustakas, N., Tsatiris, B. and Yassoglou, N., 1993. The impacts of drought on the wheat biomass production along catenas in the semi-arid zone of Greece. *Soil Technology* **6**: 337–349.

Lafeuille, J. L., 1994. Development of an analytical procedure for studying the fate of pesticides in the soil. MSc Report, ESPCI, 10 rue Vauquelin, 75231, Paris.

Lanaras, T., Sgardelis, S. P. and Pantis, J. D., 1994. Chlorophyll fluorescence in the dandelion (*Taraxacum* spp.): a probe for screening urban pollution. *The Science of Total Environment* **149**: 61–68.

Lebreton, M., 1994. Characterisation of water transfer in the unsaturated zone of the Férauds (Vidauban) wine-growing plot. Study of sites Féraud 1 and Féraud 2. Report A00030, ANTEA, BP 6119, F 45061 Orléans Cédex 2.

López-Bermúdez, F., 1983. Le programme MEDALUS (Mediterranean Desertification and Land Use) de la Communauté Européenne. Sécheresse (Science et Changements Planétaires), John Libbey. AUPELF UREF, Paris. **3**(4): 189.

López-Bermúdez, F., 1989. Geomorfología y evolución de las redes de drenaje en zonas erosionadas. En zonas aridas en España, real academia de ciencias exactas, físicas y naturales, Madrid, pp. 253–271.

López-Bermúdez, F., 1990. Soil erosion by water on the desertification of a semi-arid mediterranean fluvial basin, the Segura basin. *Agricultural, Ecosystems and Environment* **33**(2): 129–145.

López-Bermúdez, F., 1990. El clima mediterráneo semiárido como factor de erosión. *Estudios Geográficos*, Tomo LI, Madrid, **199/200**: 489–506.

López-Bermúdez, F., 1992. El riesgo de desertificación en las tierras mediterráneas y el Programa MEDALUS. Universidad de Barcelona. *Notes de Geografía Física* **20/21**: 113–122.

López-Bermúdez, F., 1994. Degradación del suelo fatalidad climática o mala gestión humana. Hacia una gestión sosteneible del recurso en el contexto mediterráneo. *Papeles de Geografía* **20**: 47–61. Universidad de Murcia.

López-Bermúdez, F. and Albaladejo, J., 1990. Factores ambientales de la degradación del suelo en el área mediterránea. In: Albaladejo, J. and Sticking, M. A. (eds), Degradación y regeneración del Suelo en condiciones ambientales mediterráneas, Consejo Superior de Investigaciones Científicas, CEBAS, Murcia, pp. 13–45.

López-Bermúdez, F., Alonso-Sarria, F., Romero-Díaz, M. A., Conesa-Garcia, C., Martínez-Fernandez, J. and Martínez-Fernandez, J., 1992. Caracterización y diseño del Campo Experimental de Los Guillermos (Murcia) para el estudio de los procesos de erosión y desertificación en litologías blandas. In: López-Bermúdez, F., Conesa-Garcia, C. and Romero-Díaz, M. A. (eds), *Estudios de Geomorfología en España*. Sociedad Española de Geomorfología, Universidad de Murcia, Murcia, pp. 151–160.

López-Bermúdez, F., Alonso-Sarria, F., Romero-Díaz, M. A. and Martínez-Fernandez, J., 1993. Dinámica y evolución de badlands. Campo Experimental de Los Guillermos, Rambla Salada (Murcia). I Congreso de la Naturaleza de la Región de Murcia, ANSE, Cartagena (Panel).

López-Bermúdez, F. and Romero-Díaz, M. A. 1989. Piping erosion and badlands development in South-East Spain. Arid and semi-arid environments, geomorphological and pedological aspects. *Catena Supplement* **14**: 59–73.

López-Bermúdez, F. and Romero-Díaz, A., 1989. Relación entre escorrentías superficiales y características físicas y ambientales en pequeñas cuencas fluviales (Alto Guadalquivir). In: *Avenidas fluviales e inundaciones en la Cuenca Mediterránea*. Universidad de Alicante, Centre Europeen de Coordination de Recherches et Documentation de Viena, Alicante, pp. 171–183.

López-Bermúdez, F., Romero-Díaz, M. A. and Martínez-Fernandez, J., 1991. Soil erosion in a semi-arid Mediterranean environment. El ardal experimental field (Murcia, Spain). In: Sala, M., Rubio, J. L. and García-Ruiz, J. M. (eds), *Soil erosion studies in Spain*. Geoforma Ediciones, Logroño, pp. 137–152.

López-Bermúdez, F., Romero-Díaz, A. and Martínez-Fernandez, J., 1997. Vegetation and erosion under semi-arid Mediterranean climate: a case study from Murcia (Spain). *Geomorphology* (in press).

López-Bermúdez, F., Romero-Díaz, A., Martínez-Fernandez, José, Martínez-Fernandez, Julia., 1994. The response of soil and vegetation cover, Murcia Field Site. In: Brandt, C. J. and Thornes, J. B. (eds), *Mediterranean desertification and land use*. John Wiley, Chichester.

Madrau, S., Deroma, M. A., Dessì, G. and Goussikpe, Y., 1995. Soil properties and trafficability of the Rio Astimini-Fiume Santo experimental plots north-western Sardinia. In: Enne, G., Aru, A. and Pulina, G. (eds), *Land use and soil degradation. MEDALUS in Sardinia*. Proceedings of the conference held in Sassari, Italy, 25 May 1994, pp. 211–220.

Mardiris, Th. A., Pantis, J. D. and Stamou, G. P., 1994. The effect of climate on the structure and abundance of the evergreen sclerophyll formations (maquis) of continental Greece. Proceedings of the 16th Conference of the Greek Association of Biological Science, pp. 3.05–3.07 (in Greek with an English summary).

Mardiris, Th. A. and Stamou, G. P., 1993. Distribution of woody species abundance within the communities of maquis ecosystems in continental Greece. Proceedings of the 15th Conference of the Greek Association of Biological Science, pp. 180–182 (in Greek with an English summary).

Margaris, N. S., 1992. Regions de montagne en Grece. Interpretation du declin social et économique. In: *Montagnes et Forêts Méditerranéannes.* International Center for Alpine Reseaerch (ICALPE), Le Bourget du Lac, France, pp. 37–44.

Margaris, N. S., Giourga, C. and Koutsidou, E., 1996. Adaptive strategies in plants dominating phryganic (dry Mediterranean) ecosystems. *Ekistics* (in press).

Margaris, N. S., Giourga, C., Lazzaro, T. and Koutsidou, E., 1996. Rhodes Island, Greece: State of the natural resources. INSULA publication, Giardini, Naxos (in press).

Margaris, N. S., Giourga, C., Loumou, A. and Theodorakakis, M., 1996. Socioeconomic developments versus desertification in Greek mountains. In: Zerefos, Ch. (ed.), *Proceedings of climate changes in the Balkan Peninsula*. Reidel, Netherlands (in press).

Margaris, N. S., Koutsidou, E., Giourga, C. and Theodorakakis, M., 1996. Agriculture and the environment in the Aegean Islands, Greece. In: Quezel

P. (ed.), *Threatened flora and fauna common to Mediterranean countries*. APRE, Bouc-Bel-Air, France (in press).

Mariani, L., Menichini, V. and Minardi, G. P., 1996. Frequency of weather types and climatic risk of extreme rainfall and thermal events. *International Journal of Climatology* (in press).

Martínez-Fernandez, J., 1990. Estudio de las características hidrodinámicas de los suelos. El método del recipiente de tensión hídrica. *Actas de la I Reunión Nacional de Geomorfología*, Sociedad Española de Geomorfología, Zaragoza, pp. 699–708.

Martínez-Fernandez, J., López-Bermúdez, F., Romero-Díaz, M. A., Martínez-Fernandez, J. and Sarria, F., 1992. El matorral semiárido del Sureste de España. Aportación metodológica para su evaluación. *Studia Oecologica* **8**: 97–105.

Martínez-Fernandez, J., Martínez-Fernandez, J. and López-Bermúdez, F., 1994. Evolución de algunas propiedades edáficas y de la vegetación en campos abandonados en ambientesemiárido mediterráneo. In: Garcia-Ruiz, J. M. and Lasanta T. (eds), *Efectos geomorfológicos del abandono de tierras*. Sociedad Española de Geomorfología, Instituto Pirenaico de Ecología, CSIC, Zaragoza, pp. 107–119.

Martínez-Fernandez, J., Martínez-Fernandez, J., López-Bermúdez, F. and Belmonte-Serrato, F., 1994. Crecimiento y producción primaria en *Rosmarinus officinalis* en relación con algunos factores ambientales. ICONA, Madrid 1. *Ecologia* **8**: 177–183.

Martínez-Fernandez, J., Martínez-Fernandez, J., López-Bermúdez, F., Romero-Díaz, M. A. and Belmonte-Serrato, F., 1996. Evolution of vegetation and pedological characteristics in fields with different age of abandonment. In: Rubio, J. L. and Calvo, A. (eds), Desertification and land degradation in Mediterranean environments. *Geoforma Ediciones*, Logroño, pp. 279–290.

Martínez-Fernandez, J., Romero-Díaz, M. A., López-Bermúdez, F. and Martínez-Fernandez, J., 1993. Parámetros estructurales y funcionales de *Rosmarinus officinalis* en ecosistemas mediterráneos semiáridos. Ediciones Universidad de Salamanca. *Studia Oecologic* **10/11**: 289–296.

Martínez-Fernandez, J., Sanchez-Perez, J. M. and Cammeraat, L. H., 1994. La medida de la humedad del suelo mediante el método TDR (Time Domain Reflectometry). Estado del arte y ejemplos de aplicación. In: Morell Evangelista, I. (ed.), *Investigación en Zona Saturada*. Universitat Jaume I. Collecció Summa, Castellò, pp. 17–27.

Mata Reis, R., Godinho, S. and Diogo, A., 1991. Precipitaçao e escoamento superficial na Bacia do Rio Cobres. INMG, Lisbon, Portugal.

Meliá, J., Bastida, J., Gandia, S., Gilabert, M. A., Younis, M. T. and Lopez-Buendia, A., 1994. La teledetección en el estudio de procesos de desertificación, El Proyecto MEDALUS. In: 5a Reunión Científica de la Asociación Española de Teledetección 'Ordenación del territorio y Medio Marino', 10–12 November 1993, Las Palmas de Gran Canaria.

Meliá, J., Gilabert, M. A. and García-Haro, F. J., 1994. Modelling soil spectral contribution to NDVI measurements in a semi-arid landscape. Poster presented at the International Symposium and workshop on 'Desertification in developed countries: Why can't we control it ?', Tucson, Arizona.

Meliá, J., Gilabert, M. A. and Younis, M. T., 1993. Remote sensing contribution to the study of Alcantarilla (MEDALUS area). 3rd International Conference in Geomorphology, Hamilton, 23–29 August 1993.

Meneguzzo, F. and Bacci, G., 1994. Estimation of the orographical component of rainfall in a complex orography area by means of the vertical velocity field recovered from local data and a mesoscale wind field model. Proceedings of the meeting on atmospheric physics and dynamics in the analysis and prognosis of precipitation fields, Rome, 15–18 November 1994.

Meneguzzo, F., Niccolai, M. and Bacci, G., 1994. Climatology of rainfall in Tuscany: statistical-dynamical analysis of drought and statistical analysis of annual maximum daily rainfall. Proceedings of the meeting on atmospheric physics and dynamics in the analysis and prognosis of precipitation fields, Rome, 15–18 November 1994.

Moustakas, N., Kosmas, C., Danalatos, N. and Yassoglou, N., 1996. Rock fragments I. Their effect on runoff, erosion and soil properties under field conditions. *Soil Use and Management Journal* **11**: 115–120.

Nicolau, J. M., Solé, A., Puigdefábregas, J. and Gutiérrez, L., 1993. Effects of soil and vegetation on runoff and sediment yield along a catena in semi-arid Spain. III International Geomorphology Conference, Hamilton, Ontario, 23–28 August 1993.

Niedda, M., Greppi, M. and Pisano, G., 1994. Utilizzo dei modelli digitali del terreno per lo studio dell'erosione a scala di bacino e confronti sperimentali. Sviluppi recenti delle ricerche sull'erosione e suo controllo, Bari, 17–18 February 1994.

Niedda, M., Greppi, M. and Pisano, G., 1995. The application of USLE on a basin scale to study silting in an artificial reservoir. In: Enne, G., Aru, A. and Pulina, G. (eds), *Land use and soil degradation. MEDALUS in Sardinia*. Proceedings of the conference in Sassari, Italy, 27 May 1994, pp. 149–156.

Palutikof, J. P., 1993. Observed and modelled climatic data for the MEDALUS study areas available November 1993. MEDALUS Working Paper No. 5, p. 13.

Palutikof, J. P., Conte, M., Casimiro Mendes, J., Goodess, C. M. and Espirito Santo, F., 1996. Climate and climate change. In: Brandt J. and Thornes J.B. (eds), *Mediterranean desertification and land use*. John Wiley & Sons, Chichester.

Palutikof, J. P. and Goodess, C. M., 1993. Climate data management and provision. MEDALUS I Final Report to the European Communities European Programme on Climate and Natural Hazards (EPOC-CT90-00140(SMA)). Chapter 1.2, pp. 39–43.

Palutikof, J. P. and Goodess, C. M., 1993. Mediterranean temperature and precipitation trends and assessment of the field-site data; extreme precipitation events; and development of GCM-based scenarios. MEDALUS I Final Report to the European Communities European Programme on Climate and Natural Hazards (EPOC-CT90-00140(SMA)) Chapter 1.3, pp. 44–96.

Palutikof, J. P., Goodess, C. M. and Guo, X., 1994. Climate change, potential evapotranspiration and moisture availability in the Mediterranean basin. *International Journal of Climatology* **14**: 853–869.

Palutikof, J. P., Winkler, J. A., Goodess, C. M. and Andresen, J. A., 1995. The simulation of daily time series from GCM output. Part 1: comparison of model data with observations. In: Proceedings of the Sixth Symposium on Global Change Studies, American Meteorological Society, Dallas, January 1995.

Papatheodorou, E., Dalaka, A., Lanaras, T., Cook, C., Sgardelis, S., Stamou, G.P., Diamantopoulos, J. and Pantis, J. D., 1994. Effects of temperature and light intensity on chlorophyll a fluorescence in *Quercus coccifera*. Proceedings of the 16th Conference of the Greek Association of Biological Science, pp. 3.46–3.48 (in Greek with an English summary).

Papatheodorou, E., Dalaka, A., Pantis, J. D., Sgardelis, S., Stamou, G. P., Diamantopoulos, J., Lanaras, T. and Cook, C., 1994. Effect of PAR on the quantum efficiency of PSII in *Quercus coccifera*. Proceedings of the 16th Conference of the Greek Association of Biological Science, pp. 3.49–3.51 (in Greek with an English summary).

Papatheodorou, E., Pantis, J. and Stamou, G. P., 1993. Effect of environmental variable and grazing pressure on the growth pattern of *Quercus coccifera*. Proceedings of the 15th Conference of the Greek Association of Biological Science, pp. 138–140 (in Greek with an English summary).

Papatheodorou, E., Pantis, J. D. and Stamou, G. P., 1993. The effects of grazing on growth, spatial pattern and age structure of *Quercus coccifera*. *Acta Oecologica* **14(5)**: 589–602.

Pardini, G., Aringhieri, R., Josa, R., Gispert, M., Torrentó, J. and Solé, A., 1991. Structural stability evaluation in some soils under mediterranean climate. 18th Reunión Nacional de Suelos, Universidad de la Laguna, Tenerife, 22–28 September 1991, pp.43–49.

Perez-Trejo, F., Clark, N. and Kazana, V., 1992. A dynamic framework for exploring linkages between the environment, tourism and agriculture. Submitted to *AMBIO, Journal of the Swedish Academy of Sciences.*

Perissoratis, G. *et al.*, 1992. Implications of expected climatic changes on the island of Rhodes, UNEP (OCA) MED. WG 55/2.

Petanidou, Th., Ellis, W. N., Margaris, N. S. and Vokou, D., 1994. Constraints on flowering phenology in a phryganic (east Mediterranean) ecosystem. *American Journal of Botany.*

Pisante, M., De Falco, E. and Basso, B., 1994. Losses of mineral nitrogen in surface and deep runoff from a durum wheat crop (*Triticum durum* Desf.) on sloping soil with different tillage methods. 13 ISTRO Conference, Aalborg (DK), July 1994.

Poesen, J., 1993. Effects of rock fragments on soil degradational processes in Mediterranean environments. Procesos de desertificacion en condiciones ambivalentes Mediterraneas. Universidad Internacional, Pelayo, Valencia, Spain, 24–28 June.

Poesen, J., 1995. Soil erosion in Mediterranean environments. In: Fantechi, R., Peter, D., Balabanis, P., and Rubio, J. (eds), *Desertification in a European Context: Physical and socio-economic aspects*. Proceedings of the European School of Climatology and Natural Hazards Course, Alicante, Spain, 1993. EUR 15415, Commission of the European Communities, Luxembourg: 125–152.

Poesen, J. and Bunte, K., 1996. Effects of rock fragments on desertification processes in Mediterranean environments. In: Brandt, C. J. and Thornes, J. B. (eds), *Mediterranean desertification and land use*. John Wiley, Chichester.

Poesen, J. and Ingelmo-Sanchez, F., 1992. Runoff and sediment yield from topsoils with different porosity as affected by rock fragment cover and position. *Catena* **19**: 451–474.

Poesen, J. and Lavee, H., 1994. Rock fragments in top soils: significance and processes. *Catena* **23**: 1–28.

Poesen, J., Torri, D. and Bunte, K., 1994. Effects of rock fragments on soil erosion by water at different spatial scales: a review. *Catena* **23**: 141–166.

Poesen, J. and Van Wesemael, B., 1995. Effects of rock fragments on the structural collapse of tilled topsoils during rain. In Derbyshire, E., Dijkstra, T. and Smalley, I. J. (eds), *Genesis and properties of collapsible soils*. Kluwer Academic Publishers, Dordrecht, NATO Advanced Science Institute Series, Series C: Mathematical and Physical Sciences, **468**: 333–343.

Postiglione, L., 1993. Agriculture and environmental problems in the Mediterranean area (with particular reference to Italy). Proceedings International Conference on environmental and sanitary problems of land and coastal waters in the Mediterranean area, Naples, 8–10 October 1992. *Medit* **4**(2): 35–42.

Postiglione, L., Barbieri, G. and Tedeschi, A., 1994. Response of a clay loam soil to continuous irrigations with saline water. In: *Problems and management of soil salinization-alkalinization in Europe*. Corso di stampa su Proceeding Conference, Budapest, April 1994.

Postiglione, L., Barbieri, G. and Tedeschi, A., 1995. Long-term effects of irrigations with saline water on some characteristics of a clay loam soil. *Rivista di Agronomia* (in press).

Postiglione, L. and Basso F., 1993. Erosione superficiale e desertificazione dei suoli. Risultati di 15 anni di ricerche sperimentali nella collina interna del Mezzogiorno. In: *La salvaguardia dell'ambiente*. Napoli, April 1993, pp. 32–33.

Postiglione, L., Basso, F., Amato, M. and Carone, F., 1990. Effect of soil tillage methods on soil losses, on soil characteristics and on crop production in a hilly area of Southern Italy. *Agr. Med.* **120**: 148–158.

Puddu, R. and Lai, M. R., 1993. Soil, geological and environmental survey in the Rio Santa Lucia catchment basin. European School of Climatology and Natural Hazards, Desertification in European Context. Physical and socio-economic aspects, Alicante, 6–14 October.

Puddu, R. and Lai, M. R., 1995. Quarrying activities in the Rio Santa Lucia catchment basin Southern Sardinia. In: Enne, G., Aru, A. and Pulina, G. (eds), *Land use and soil degradation. MEDALUS in Sardinia*. Proceedings of the conference held in Sassari, Italy, 25 May 1994, pp. 129–140.

Pugnaire, F. I., 1994. Adaptation to environmental stress of the perennial tussock grass *Stipa tenacissima.* In: Tallis, J. H., Norman, H. J. and Benton, R. A. (eds), Proceedings of the VI International Congress of Ecology, Manchester.

Pugnaire, F. I., Haase, P. and Incoll, L. D., 1994. Effect of availability of water on the seasonal pattern of gas exchange in the perennial grass *Stipa tenacissima* in semi-arid south-east Spain. *Annales Geophysicae* **12**: C416.

Pugnaire, F. I., Haase, P. and Incoll, L. D., 1994. Effect of availability of water on the seasonal pattern of gas exchange in the perennial grass *Stipa tenacissima*. *Physiologia Plantarum* **91**(3): A18.

Pugnaire, F. I., Haase, P., Incoll, L. D. and Clark, S. C., 1994. Succession in the understorey of *Retama sphaerocarpa* (L.) Boiss. in a semi-arid environment. *Bulletin of the British Ecological Society* **25**(4): 252.

Pugnaire, F. I., Haase, P., Incoll, L. D. and Clark, S. C., 1996. Adaptation of the tussock grass *Stipa tenacissima* to watering in a semi-arid environment. *Functional Ecology*, **10**: 265–274.

Puigdefábregas, J., 1992. Mitos y perspectivas sobre la desertificación. Asociación Española de Ecología Terrestre, Madrid. *Ecosistemas* **3**: 18–22.

Puigdefábregas, J., 1992. Natural regeneration potential of vegetation in Mediterranean systems. Conference on restoration of vegetation cover in Mediterranean ecosystems. Instituto de Cultura Juan Gil-Albert and University of Alicante, Alicante.

Puigdefábregas, J., 1994. Desertificación: Una perspectiva sobre la cuenca mediterránea. CSIC. Madrid. *Fronteras de la Ciencia y la Tecnología* **4**: 15–19.

Puigdefábregas, J., 1994. Interactions between vegetation dynamics and slope hydrology and their significance in dryland degradation. *Annales Geophysicae* **12**(2): C417.

Puigdefábregas, J., 1996. Desertification: stress beyond resilience, exploring a unifying process structure. *Ambio* **24**(5): 311–313.

Puigdefábregas, J., Aguilera, C., Brenner, A., Clark, S., Cueto, M., Delgado, L., Domingo, F., Gutiérrez, L., Incoll, L., Lázaro, R., Nicolau, R. M., Sánchez, G., Solé, A., Torrentó, J. R. and Vidal, S., 1993. Mediterranean desertification: studies at the MEDALUS field site in Rambla. III International Geomorphology Conference, Hamilton, Ontario, 23–28 August 1993.

Puigdefábregas, J., Sánchez, J. G., Moro, S. C., Solé, A. and Gutierrez, L., 1994. Interactions between vegetation patchiness and slope hydrology in Mediterranean semi-arid climate. Santiago de Chile. *Noticiero de Biología* **2**(3): 46.

Puigdefábregas, J., Solé, A., Lázaro, R. and Nicolau, J. M., 1992. Factores que controlan la escorrentía en una zona semiárida sobre micaesquistos. In: López, Conesa y Romero (eds), Estudios de Geomorfología en España - Actas de la II Reunión Nacional de Geomorfología, tomo I. Sociedad Española de Geomorfología, Universidad de Murcia, Murcia, pp. 117–127.

Pulina, M. A., 1995. General climatic outlines of the Rio Astimini–Fiume Santo basin. In: Enne, G., Aru, A. and Pulina, G. (eds), *Land use and soil degradation. MEDALUS in Sardinia*. Proceedings of the conference held in Sassari, Italy, 25 May 1994, pp. 51–64.

Pulina, G., Masala, G., Zanda, A. and Enne, G., 1994. Analysis of biomass balance and stocking rate in cattle and sheep production systems in Mediterranean areas. *Medit.* December.

Queralt, I., Domingo, F. and Solé-Benet, A., 1993. The influence of local and distant sources in the mineral content of bulk deposition over an altitudinal gradient in the Filabres range (SE Spain). *Journal of Geophysical Research* - D (Atmospheres) **98**(D9): 16 761–16 768.

Rojo-Serrano, L., 1993. Afforestation for soil and water conservation in Spain. Seminar on desertification and land use in the Mediterranean basin, Almeria, 28–30 July 1993.

Rojo-Serrano, L., 1993. Criteria for the determination of protective objectives in the afforestation. Species selection and plantation density. 3rd Trimester 1993. *Montes* **33** (in Spanish).

Rojo-Serrano, L., 1993. Forest hydrologic restoration of degraded Mediterranean basins. School of Climatology and Natural Hazards: Desertification in Europe. Alicante, October 1993.

Rojo-Serrano, L., 1994. Experimental plots and watersheds. Definition of research objectives and processes. Seminar in erosion and desertification of agricultural and forest soils., CSIC, Madrid, 13–16 June 1994 (in Spanish).

Rojo-Serrano, L., 1994. The Spanish national plan to combat desertification. 2 Congreso Latinoamericano de Manejo de Cuencas Hidrogr ficas, Merida, Venezuela, 6–11 November 1994 (in Spanish).

Rojo-Serrano, L. and Serrada, R., 1992. Environmental impact assessment of the afforestations. ICONA. Madrid. In: *Forest Hydrology and Soil Protection* (in Spanish).

Roli, F. and Serpico, S. B., 1995. Land cover classification in remote-sensing images using structured neural networks. *EARSeL Journal of Advances in Remote-Sensing*, **4**: 3.

Romano, N., 1993. Use of an inverse method and geostatistics to estimate soil hydraulic conductivity for spatial variability analysis. *Geoderma* **60**: 169–186.

Romero-Díaz, M. A., López-Bermúdez, F., Thornes, J. B., Francis, C. and Fisher, G. C., 1988. Variability of overland flow erosion rates in a semi-arid Mediterranean environment under mattorral cover. Murcia, Spain. *Catena Supplement* **13**: 1–11.

Romero-Díaz, M. A., Martínez-Fernandez, J., López-Bermúdez, F., Martínez-Fernandez, J. and Alonso-Sarria, F., 1991. Algunas aproximaciones al estudio integrado del sistema atmósfera-planta-suelo en ambientes mediterráneos contrastados. XI Jornadas de Fitosociología, Universidad de Murcia, La Manga (Murcia).

Roxo, M. J., 1992. Erosion on agricultural soils - inner Alentejo. Presented at the International Symposium Farm lands erosion in temperate plains environment and hills, Centre de Biogeogaphie-Ecologie, ENS de Fontenay - Saint Cloud, 23–29 May 1992.

Roxo, M. J. 1994. A Acção Antrópica no Processo de Degradação de Solos—A Serra de Serpa e Mértola. Doctoral Thesis, Nova Universidade de Lisbon.

Roxo, M. J. and Casimiro, P., 1993. A monocultura cerealífera no Baixo Alentejo Interior e o processo de desertificação. Seminario sobre la desertificación y la utilización del suelo en la cuenca Mediterranea, Almería, Spain, 28–30 June 1993.

Roxo, M. J., Kilsby, C., Bathurst, J. C. and Casimiro, P., 1993. Assessing the effects of changes in land-use and climate on the hydrology and erosion rates of the Cobres basin, Portugal, with SHETRAN-UK. 18th General Assembly of the European Geophysical Society, Wiesbaden, Alemanha.

Ruggiero, C., De Pascale, S. and Barbieri, G., 1994. Effetto dell'irigazione con acque a diverso contenuto salino sullo stato idrico, sull'accrescimento e sulla produzione della melanzana (*Solanum melongena* L.). *Rivista di Agronomia* **28**(3).

Sánchez-Toribio, M. I., López-Bermúdez, F. and Del Amor, F., 1990. Estudio geoestadístico de la demanda evaporativa en la Región de Murcia. Asociación Española de Riegos y Drenejes, Mérida. *8th Jornadas Técnicas sobre Riegos*, pp. 93–102.

Sánchez-Toribio, M. I., López-Bermúdez, F. and Del Amor, F., 1990. Relación entre Eto gramíneas y Epan en las condiciones de clima mediterráneo de la Región de Murcia. Asociación Española de Riegos y Drenajes, Meridia. *8th Jornadas Tecnicas sobre Riegos*, pp. 121–127.

Sánchez, G. and Puigdefábregas, J., 1991. Discountinuous plant covers in semi-arid environments and their relation to runoff and sediment movement. *Annales Geophysicae* **9**: C582.

Sánchez, G. and Puigdefábregas, J., 1992. Interactive relations between plant growth and sediment movement in semi-arid slopes. Field symposium on steepland geomorphology. International Geographical Union—Commission on Measurement, Theory, and Application (IGU), 31 July–8 August 1992, Berkeley.

Sánchez, G. and Puigdefábregas, J., 1993. Relacciones entre el desarrollo de las macollas de esparto (*Stipa tenacissima* L.) y la arquitectura y el transporte de sedimentos en el Sudeste semiárido de la provincia de Almería. 4th Jornadas de la Asociación Española de Ecología Terrestre, Alicante, 27–29 September 1993.

Sánchez, G. and Puigdefábregas, J., 1994. Interactions of plant growth and sediment movement on slopes in a semi-arid environment. *Geomorphology* **9**: 243–260.

Sánchez, G., Lázaro, G., Cueto, R. M. and Puigdefábregas, J., 1991. Estructura de las comunidades terofíticas en medio semiárido. 3rd Jornadas de la Asociacion Española de Ecologia Terrestre, León, Spain, 16–20 September 1991.

Santini, A., 1992. Modelling water dynamics in the soil-plant-atmosphere system for irrigation problems. *Excerpta* **6**: 133–166.

Santini, A. and Romano, N., 1994. A simulation model for irrigation water management. In: Anaya-Garduno, M., Pascual-Moncayo, M. A. and Zarate-Zarate, A. (eds), *Proceedings of the IV International Congress on desert development sustainable for our common future*, México City, 25–30 July, **1**: 332–339.

Santini, A. and Romano, N., 1995. Laboratory and field soil hydraulic characterization by parameter estimation approach. *Excerpta* **8**: 179–211.

Santini, A., Romano, N., Ciollaro, G. and Comegna, V., 1995. Effect of tillage practices on soil hydraulic properties determined using different laboratory methods. *Soil Science*, **160**(5): 340–351.

Serpico, S. B., Pellegretti, P. and Bruzzone, L., 1994. Feature-selection for remote-sensing data classification, EUROPTO "Image and Signal Processing" Conference, Rome, Italy, October 1994.

Serpico, S. B. and Roli, F., 1995. Classification of multisensor remote-sensing images by structured neural networks. *IEEE Trans. on Geoscience and Remote Sensing*, **33** (3).

Serrano, L., Pardos, J. A., Pugnaire, F. I. and Domingo, F., 1994. Absorption of radiation, photosynthesis and biomass production in plants. In: Pessarakli, M. (ed.), *Handbook of plant and crop physiology*. Marcel Dekker, New York, pp. 243–256.

Sgardelis, S. P. and Margaris, N. S., 1993. Effects of fire on soil microarthropods of a phryganic ecosystem. Gustav Fischer Verlag Jena. *Pedobiologia* **37**: 83–94.

Solé, A., 1991. Runoff and infiltration in terraced slopes in semiarid Almeria (Spain). 16th General Assembly European Geophysical Society, Wiesbaden, Germany, 22–26 April 1991.

Solé, A., Adams, A., Domingo, F., Torrentó, J. R. and Puigdefábregas, J., 1991. Runoff and infiltration in terraced slopes in semiarid Almería (SE Spain). *Annales Geophysicae* **9**: C585.

Solé, A., Lázaro, R. and Puigdefábregas, J. (eds), 1994. *Field Guide for the Fourth Group Meeting of MEDALUS II*, Almería, 18–22 November.

Thornes, J. B. and Mairota, P., 1995. MEDALUS: Achievements and prospects. In: Enne, G., Aru, A. and Pulina, G. (eds), *Land use and soil degradation. MEDALUS in Sardinia*. Proceedings of the conference held in Sassari, Italy, 25 May 1994, pp. 1–12.

Thornes, J. B., Francis, C., López-Bermúdez, F. and Romero-Díaz, M. A., 1990. Reticular overland flow with coarse particles and vegetation roughness under Mediterranean conditions. In: Rubio, J. L. and Rickson, R. J. (eds), Strategies to combat desertification in Mediterranean Europe. Commission of the European Communities, Directorate General Agriculture, EUR 111 75 EN/ES, Brussels, pp. 228–243.

Thornes, J. B., López-Bermúdez, F., Romero-Díaz, M. A., Brandt, J., Watts, G. B. and Martínez-Fernandez, J., 1990. Modelo digital para el estudio de la relación vegetación-erosión y su aplicación a problemas ambientales, 10th Jornadas de Fitosociología. *Cartografía Vegetal*, Granada, pp. 19–21.

Thornes, J. B., Shao, J. X., Diaz, E., Roldan, A., Hawkes, C. and McMahon, M., 1995. Testing the MEDALUS hillslope model. *Catena*.

Torrentó, J. R., Sánchez, G. and Puigdefábrergas, J., 1991. Soil erosion and aggregate stability induced by fires in *Stipa tenacissima* communities. Conference of soil erosion and degradation as a consequence of forest fires (European Society for Soil Conservation). Barcelona-Valencia, 3–7 September 1991.

Vacca, A., Puddu, R., Tommasi, D. and Usai, D., 1995. Erosion measurements in three areas of Santa Lucia catchment with different land uses. In: Enne, G., Aru, A. and Pulina, G. (eds), *Land use and soil degradation. MEDALUS in Sardinia*. Proceedings of the conference held in Sassari, Italy, 27 May 1994, pp. 115–128.

Van Wesemael, B., Poesen, J. and de Figueiredo, T., 1995. Effects of rock fragments on physical degradation of cultivated soils by rainfall. *Soil and Tillage Research* **33**.

Van Wesemael, B., Poesen, J., de Figueiredo, T. and Govers, G., 1995. Surface roughness evolution of soils containing rock fragments. *Earth Surface Processes and Landforms*, **20**.

Vernazza, G., Serpico, S. B. and Pellegretti, P., 1994. Classificazione mediante reti neurali strutturate di aree agro-forestali sottoposte a fenomeni di degrado ambientale. AEI'94, 95ª Riunione Annuale AEI, Bologna, 2–6 October 1994.

Vidal, S., 1994. A device for simultaneous measurement of soil moisture and electrical conductivity. (Application nº at Sociedad Española de Patentes y Marcas: 9401681.)

White, S. M., 1993. Assessing the effects of changes in land use and climate on the hydrology and erosion rates of the Mula catchment, Spain, with SHETRAN-UK. Presented at European Geophysical Society XVIII General Assembly, Wiesbaden, 3–7 May.

White, S. M., Kilsby, C. G., Pisante, M., Roxo, M. and López-Bermúdez, F., 1994. Physically based modelling of runoff and erosion plots in the Mediterranean environment. Hydrological and ecological aspects of dryland degradation. European Geophysical Society, 19th General Assembly, Grenoble.

Winkler, J. A., Palutikof, J. P., Andresen, J. A. and Goodess, C. M., 1995. The simulation of daily time series from GCM output. Part 2: Development of local scenarios for maximum and minimum temperature using statistical transfer functions. In: *Proceedings of the Sixth Symposium on Global Change Studies*, American Meteorological Society, Dallas, January 1995.

Woodward, F. I., 1994. Predicting plant and vegetation responses to climatic change. Global change and terrestrial ecosystems (IGBP) science conference, Woods Hole, MA, May1994.

Woodward, F. I., 1995. Developing the potential for describing the terrestrial biosphere's response to a changing climate. In: Walker, B. and Steffen, W. (eds), *Global change and terrestrial ecosystems*. Cambridge University Press.

Woodward, F. I. and Kelly, C. K., 1995. Plant functional types: towards a definition by environmental constraints. In: Smith, T. M., Shugart, H. H. and Woodward, F. I. (eds), *Towards the development of a functional classification of plants*. Cambridge University Press.

Yassoglou, N., Kosmas, C. and Danalatos, N., 1994. Global change and land degradation. International Symposium on Eastern Europe and Global Change, Chalkidiki, Greece, 3–10 October 1994.

Yassoglou, N., Kosmas, C. and Moustakas, N., 1995. The red soils, their origin, use and management in Greece. *Catena* (in press).

Younis, M. T., Gilabert, M. A., Lopez-Buendia, A. M. and Meliá, J., 1991. Características espectrales de superficies frescas y lateradas de differentes tipos de rocas en una región arida. Proceeding of the 4th Reunión Científica de la Asociación Española de teledetección, Sevilla, 13–15 November 1991.

Younis, M. T., Lopez-Buendia, A. M. and Bastida, J., 1992. Caracterización radiomatrica de unidades litológicas de la Cuenca de Mula (Murcia, SE España): Su aplicación a la evaluacifón de imagenes Landsat TM. 3rd Congreso Geológico de España y VIII Congreso Latinoamericano de Geología, Salamanca, 23–26 June 1992.

Younis, M. T., Lopez-Buendia, A. M. and Bastida, J., 1992. Integration of seismic data and Landsat TM image interpretation for the Newtectonic in the area of Murcia, SE Spain. *Annales Geophysicae* **10**(1): 159.

Younis, M. T. and Meliá, J., 1992. Interpretation of Landsat TM images for mapping erosion susceptibility in the area of Murcia, SE Spain. *Geobserver* **92**: 353–356.

Younis, M. T., Meliá, J. and Bastida, J., 1991. Application of Landsat TM images interpretation for the study of the desertification factors in the area of Alcantarilla, Murcia, SE Spain. Presented in the 16th General Assembly of the European Geophysical Society, Wiesbaden, 22–26 April 1991.

Younis, M. T., Meliá, J., Bastida, J. and Gandia, S., 1991. Evaluación de imagenes Landsat TM para la discriminación litológica en el area de Alcantarilla (Murcia). 4th Reunión Científica de la Asociación Española de teledetección, Sevilla, 13–15 November 1991, pp. 160–167.

Younis, M. T., Meliá, J., Lopez-Buendia, A. and Bastida, J., 1991. Caracterización minerlalogica de las unidades litológicas differencidas mediante imagenes Landsat TM en el area de Alcantarilla (Murcia). *Boletín de la Sociedad Española de Mineralogía* **14**: 87–88.

Younis, M. T., Meliá, J., Lopez-Buendia, M. A. and Bastida, J., 1994. Aplicación de imagenes Landsat-5 TM al reconocimiento de la formación torosa del complejo alpujarride en la sierra de las estancias (Almería). 16th Reunion de Geologia y Minera, Galicia, 23–26 November.

Minutes of meetings

First Group Meeting of MEDALUS (I), 1–5 November 1990, Murcia, Spain (35 pp).

Minutes of the Second Group Meeting of MEDALUS (I), 15–19 May 1991, Cagliari, Sardinia (23 pp).

Minutes of the Third Group Meeting of MEDALUS (I), 30 September–3 October 1991, Leuven, Belgium (26 pp).

Minutes of the Fourth Group Meeting of MEDALUS (I), 1–5 April 1992, Thessaloniki, Greece (48 pp).

Minutes of the Plant Parameter Meeting, Almeria, 12–13 May 1992 (5 pp).

Minutes of aspects of MEDALUS modelling *ad hoc* Meeting No. 1, 4 June 1992, Newcastle upon Tyne, UK (5 pp).

Minutes of Aspects of MEDALUS modelling, *ad hoc* Meeting No. 2, 8 July 1992, Newcastle, UK (6 pp).

Minutes of aspects of MEDALUS Modelling *ad hoc* Meeting No. 3, 26 September 1992, Bristol, UK (3 pp).

Minutes of the Fifth Group Meeting of MEDALUS (I), 26–29 September 1992, Bristol, UK (34 pp).

Minutes of the First Start Up Meetings of MEDALUS II, Remote Sensing Meeting, Valencia, 4 February 1993; Project 1 Meeting, Lorca, 5 February 1993; Plenary Session, Lorca, 6–8 February 1993 (62 pp).

Minutes of the MEDALUS II, Project 4: Guadalentín Meeting, 1 March 1993, Murcia (1 p).

Minutes of the MEDALUS II Project 1 Meeting, Amsterdam, 18–19 March 1993 (12 pp).

Addition to the Minutes of the MEDALUS II Project 1 Meeting, Amsterdam, 18–19 March 1993 (Report of the Vegetation Sub-Group) (6 pp).

Minutes of the Second Start Up Meeting of MEDALUS II, Project 4, Basilicata, Italy, 29–31 March 1993; Abruzzo, 1–3 April 1993 (30 pp).

Minutes of the MEDALUS II, Proyecto 4: Guadalentín Meeting, Murcia, Spain, 15 April 1993 (in Spanish) (2 pp).

Minutes of the MEDALUS II, Project 3 Meeting, Cagliari, 2–6 May 1993 (1 p).

Short report on the MEDALUS Data-base Meeting, Paris, 14 May 1993 (3 pp).

Minutes of the First Group Meeting of MEDALUS II, Lisbon, Portugal, 21–25 May 1993 (53 pp).

Minutes of the Second Group Meeting of MEDALUS II, Montpellier, 19–23 November 1993 (66 pp).

Minutes of the Agri Basin Group Meeting, Basilicata, 11 February 1994 (21 pp).

Timetable, Agenda and Abstracts of the Third Group Meeting of MEDALUS II, Athens, 18–24 May 1994.

Minutes of the Third Group Meeting of MEDALUS II, Athens, 18–24 May 1994 (53 pp).

Minutes of the Fourth Group Meeting of MEDALUS II, Almeria, 18–22 November 1994 (11 pp).

Minutes of the Fifth Group Meeting of MEDALUS II, Maastricht, The Netherlands, 27–30 April 1995.

Minutes of the Start Up Meeting of MEDALUS III, Lisbon, Portugal, 3–6 November 1995 (53 pp).

Minutes of the First Plenary Meeting of MEDALUS III, Newcastle, UK, 19–22 April 1996 (76 pp).

Minutes of the Second Plenary Meeting of MEDALUS III, Lorca, Spain, 17–21 October 1996 (66 pp).

Reports

First Interim Report for MEDALUS I, covering the period 1 January–15 May 1991 (59 pp).

Second Interim Report for MEDALUS I, covering the period 16 May–30 September 1991 (109 pp).

MEDALUS I First Annual Report, covering the period 1 January–31 December 1991 (179 pp).

Third Interim Report for MEDALUS I, covering the period 1 October 1991–1 April 1992 (99 pp).

Fourth Interim Report for MEDALUS I, covering the period 1 April 1992–26 September 1992 (181 pp).

Mediterranean Desertification and Land Use, MEDALUS I Final Report, June 1993 (607 pp).

First Interim Report for MEDALUS II, covering the period 1 January 1993–21 May 1993 (152 pp).

Second Interim Report for MEDALUS II, covering the period 21 May–19 November 1993 (262 pp).

First Annual Report for MEDALUS II, covering the period 1 January–31 December 1993 (4 volumes: Project 1, 104 pp; Project 2, 166 pp; Project 3, 70 pp; Project 4, 203 pp).

Third Interim Report for MEDALUS II, covering the period 19 November 1993–20 May 1994 (236 pp).

Fourth Interim Report for MEDALUS II, covering the period 20 May–18 November 1994 (252 pp).

Second Annual Report for MEDALUS II, covering the period 1 January–31 December 1994 (4 volumes: Project 1, 113 pp; Project 2, 183 pp; Project 3, 89 pp; Project 4, 249 pp).

Final Report for MEDALUS II covering the period 1 January 1991–30 September 1995 (4 volumes: Project 1, 282 pp; Project 2, 364 pp; Project 3, 223 pp; Project 4, 715 pp).

First Interim Report for MEDALUS III, covering the period 1 January 1996–19 April 1996, 265 pp.

Second Interim Report for MEDALUS III, covering the period 20 April 1996–17 October 1996 (2 volumes: Projects 1 and 2, 244 pp; Projects 3 and 4, 398 pp).

Start up documents

MEDALUS I Start Up Document (prepared for the First Meeting, Murcia, 1–5 November 1990) (57 pp).

MEDALUS II Start Up Document (prepared for the First Start Up Meeting of MEDALUS II, Lorca, Spain, 5–8 February 1993) (77 pp).

MEDALUS II Start Up Document—Supplement containing Document 8: Presentation of the Abruzzo Research area (15 pp).

MEDALUS III Start Up Document (prepared for the First Meeting of MEDALUS III, Lisbon, 3–6 November 1995).

Directories and manuals

MEDALUS Field Manual, Version 3.1, February 1992 (115 pp).

Who's Who in MEDALUS I: A directory of all participating scientists and institutions (2nd edn), June 1991 (47 pp).

Who's Who in MEDALUS II: A directory of all participating scientists and institutions, May 1993 (104 pp).

Who's Who in MEDALUS III: A directory of all participating scientists and institutions, May 1966 (83 pp).

Newsletters

MEDALUS Newsletter Number 1, August 1991.
MEDALUS Newsletter Number 2, November 1991.
MEDALUS Newsletter Number 3, February 1992.
MEDALUS Newsletter Number 4, June 1992.
MEDALUS Newsletter Number 5, September 1992.
MEDALUS Newsletter Number 6, January 1993.
MEDALUS Newsletter Number 7, August 1993.
MEDALUS Newsletter Number 8, April 1994.

Other documents

Mediterranean Desertification and Land Use (full colour brochure describing the work of MEDALUS I), September 1992 (8 pp).

Mediterranean Desertification and Land Use: a synopsis of the MEDALUS I Project results (full colour booklet abstracted from the summary of the MEDALUS I Final Report), August 1993 (28 pp).

Mediterranean Desertification and Land Use: a synopsis of the MEDALUS I and II Project results (full colour booklet describing the work of MEDALUS I and II), June 1966 (30 pp).

Working papers

1. WORKING PAPER NUMBER 1, James C. Bathurst, *Report on visit to the Jornada Del Muerto Long Term Ecological Research (LTER) site, Las Cruces, New Mexico, USA, 1 to 5 February 1993, September 1993*, 7 pp.
2. WORKING PAPER NUMBER 2, Joao Corte-Real, Xuebin Zhang and Xialan Wang, *Large-scale circulation regimes and surface climate anomalies over MEDALUS area*, November 1993, 34 pp.
3. WORKING PAPER NUMBER 3, Angelo Aru, *The problem of heavy metals pollution in mining areas of Sardinia. Preliminary findings*, November 1993, 10 pp.
4. WORKING PAPER NUMBER 4, Piermaria Corona and Agostino Ferrara, *Study of the spatial relationships in forested areas of Central Italy*, November 1993, 12 pp.
5. WORKING PAPER NUMBER 5, J. P. Palutikof, *Observed and modelled climatic data for the MEDALUS study areas available November 1993*, November 1993, 13 pp.
6. WORKING PAPER NUMBER 6, A. Aru, R. Puddu, D. Tomasi, D. Usai, A. Vacca and G. Pinna, *Experimental plots at Is Olias (Rio S. Lucia catchment area). Some data and results of MEDALUS I and MEDALUS II (first year)*, November 1993, 70 pp.
7. WORKING PAPER NUMBER 7, Luigi Mariani, *Climatological study of the experimental area of the Pescara valley with particular reference to Cigno and Piomba basins (extended version of the Second MEDALUS II Interim Report)*, November 1993, 34 pp.
8. WORKING PAPER NUMBER 8, Michele Conte, Roberto Sorani, *Some severe drought events in the central Mediterranean (extended version of the Second MEDALUS II Interim Report)*, November 1993, 68 pp.
9. WORKING PAPER NUMBER 9, Claudio Conese, Tiziana de Filippis, Fabio Maselli, Maurizio Romani, Alessandra Rodolfi, Maria Antonietta Falchi, *Cigno stream, hydrologic preliminary study (Appendix to the Second MEDALUS II Interim Report)*, November 1993, 24 pp.
10. WORKING PAPER NUMBER 10, G. Bacci, F. Meneguzzo, N. Musetti, M. Niccolai, *Comparison of precipitation patterns forecast by the general circulation model operating at Offenbach German Center with data collected by a raingauges network (Appendix A to the Second MEDALUS II Interim Report)*, November 1993, 7 pp.
11. WORKING PAPER NUMBER 11, G. Frustaci, G. Bacci, F. Meneguzzo, A. Pellegrini, G. Clerici, S. Giarola, *Using a diagnostic numerical model in Antarctic mesoscale analysis: some case studies (Appendix B to the Second MEDALUS II Interim Report)*, November 1993, 5 pp.
12. WORKING PAPER NUMBER 12, A. T. Grove, *Erosion in the historic past; climate or man-made*, March 1994, 25 pp.
13. WORKING PAPER NUMBER 13, C. Conese *et al.*, *List of data available to date from the Abruzzo Group, for exchange with all MEDALUS members*, March 1994, 20 pp.
14. WORKING PAPER NUMBER 14, L. D. Incoll *et al.*, *Plant ecophysiology group, extended Second Interim Report on work with desertification response units*, November 1993, 9 pp.
15. WORKING PAPER NUMBER 15, L. D. Incoll *et al.*, *Plant ecophysiology group, extended Third Interim Report on work with desertification response units*, May 1994, 9 pp.
16. WORKING PAPER NUMBER 16, J. Casimiro Mendes and F. Espirito Santo, *Climatic variability of precipitation in the western Iberian peninsula*, May 1994, 9 pp.
17. WORKING PAPER NUMBER 17, P. Mairota and N. Geeson (eds), *Proceedings of the Workshop on: EC Policy Strategies for land use in the Mediterranean countries*, May 1994, 84 pp.
18. WORKING PAPER NUMBER 18, E. Koutsidou, C. Giourga, A. Loumou and N. S. Margaris, *Vegetation recovery in grazing exclosures in Greece*, May 1994, 11 pp.
19. WORKING PAPER NUMBER 19, L. Mariani, V. Menichini and G. P. Minardi, *Frequency of weather types and climatic risk of extreme rainfall and thermal events*, May 1994, 18 pp.
20. WORKING PAPER NUMBER 20, G. Vernazza, *Neural-network techniques for the classification of remote-sensing images*, May 1994, 45 pp.
21. WORKING PAPER NUMBER 21, P. Pellegretti, G. Vernazza, C. Conese, F. Maselli, A. Rodolfi and L. Petkov, *Classification of agro-forest areas in central Italy by the integrated use of multi-spectral, textural and ancillary data*, May 1994, 17 pp.
22. WORKING PAPER NUMBER 22, F. Meneguzzo, L. Petkov and M. Niccolai, *Statistical-meteorological applications to soil conservation and*

land use policies: spatial objective analyses of wind fields and rainfall-drought time series analyses. May 1994, 20 pp.

23. WORKING PAPER NUMBER 23, G. Pulina, G. Masala, A. Zandra and G. Enne. *Analysis of biomass balance and stocking rate in cattle and sheep production systems in Mediterranean areas*, May 1994, 11 pp.
24. WORKING PAPER NUMBER 24, J. Corte-Real. X. Wang and X. Zhang, *Modes of variability of the northern hemisphere's mid-tropospheric large-scale circulation*, May 1994, 31 pp.
25. WORKING PAPER NUMBER 25, J. Corte-Real, X. Wang and X. Zhang, *Large-scale circulation regimes and surface climatic anomalies over the Mediterranean*, May 1994, 28 pp.
26. WORKING PAPER NUMBER 26, M. Conte, S. Sorani and S. Velletri, *Heat waves in the central Mediterranean*, May 1994, 27 pp.
27. WORKING PAPER NUMBER 27, P. Battista, F. Benincasa, P. Duce and A. Materassi. *Design and construction of a microprocessor impactometer*, May 1994, 12 pp.
28. WORKING PAPER NUMBER 28, C. S. Kosmas, N. G. Danalatos, N. C. Moustakas and N. Yassoglou, *Effect of parent material and land use on soil erosion rates under Mediterranean conditions*, May 1994, 13 pp.
29. WORKING PAPER NUMBER 29, N. C. Moustakas, C. S. Kosmas, N. G. Danalatos and N. Yassoglou, *The effect of rock fragments on soil properties and erosion under field conditions*, May 1994, 14 pp.
30. WORKING PAPER NUMBER 30, N. G. Danalatos, C. S. Kosmas, N. C. Moustakas and N. Yassoglou, *The impact of rock fragments on soil physical properties and biomass production under Mediterranean conditions*, May 1994, 13 pp.
31. WORKING PAPER NUMBER 31, A. Aru, G. Brundu, I. Camarda, A. Kahnamoei, B. Maddeddu, M. Roxo, R. Puddu, V. Satta and A. Vacca, *Assessment and management criteria of desertification in selected Mediterranean areas*, May 1994, 20 pp.
32. WORKING PAPER NUMBER 32, C. D. Tavares and I. M. F. Gomes, *Vegetation monitoring and land use mapping with NOAA-AVHRR*, May 1994, 11 pp.
33. WORKING PAPER NUMBER 33, C. M. Goodess and J. P. Palutikof, *Regional scenarios of climate change: changes in the precipitation regime of the Guadalentín Basin*, May 1994, 10 pp.
34. WORKING PAPER NUMBER 34, M. Niccolai, L. Petkov and P. Vignaroli, *Socio-economic aspects of the coastal zone of the Abruzzo region—MEDALUS II*, May 1994, 25 pp.
35. WORKING PAPER 35, J. Diamantopoulos, E. M. Papatheodorou, J. D. Pantis and G. P. Stamou, *Assessment of the effect of grazing on nutrient concentration in the different parts of the Quercus coccifera shrublands*, September 1994, 8 pp.
36. WORKING PAPER 36, A. Dalaka, E. M. Papatheodorou, T. Lanaras, J. D. Pantis, C. M. Cook, G. P. Stamou, J. Diamantopoulos and S. P. Sgardelis, *Photosysnthesis in a* Quercus coccifera *stand on Mt. Hortiatis, Northern Greece*, September 1994, 6 pp.
37. WORKING PAPER 37, E. M. Papatheodorou, A. Dalaka, J. Diamantopoulos, C. M. Cook, S. P. Sgardelis, G. P. Stamou, J. D. Pantis and T. Lanaras, *The effect of grazing on some photosynthetic parameters of* Quercus coccifera, September 1994, 6 pp.
38. WORKING PAPER 38, G. P. Stamou, Th. A. Mardiris, J. D. Pantis and J. Diamantopoulos, *Effects of common management practices on Greek maquis ecosystems*, September 1994, 13 pp.
39. WORKING PAPER 39, M. D. Argyropoulo and S. P. Paraskevopoulos, *Desertification in Thessaly (Central Greece): a socio-economic perspective*, September 1994, 6 pp.
40. WORKING PAPER 40, J. N. Quinton, G. M. Edwards and R. P. C. Morgan, *Influence of some vegetation characteristics on runoff generation and sediment production in semi-arid S.E. Spain*, September 1994, 10 pp.
41. WORKING PAPER 41, S. J. Hurcom, A. R. Harrison and M.Taberner, *Mapping semi-arid vegetation at the regional scale using satellite data*, September 1994, 29 pp.
42. WORKING PAPER 42, A. Burton and J. Bathurst, *Modelling shallow landslide erosion and sediment yield at the basin scale,* September 1994, 15 pp.
43. WORKING PAPER 43, S. J. Hurcom, A. R. Harrison and M.Taberner, *Assessment of biophysical vegetation properties through spectral decomposition techniques*, October 1994, 27 pp.
44. WORKING PAPER 44, L. D. Incoll, S. C. Clark, F. I. Pugnaire, P. Haase, O. D. Salinas, E. M. Fernández and R. Bevan, *Plant Ecophysiology Group, report to the fourth MEDALUS II group meeting in Almeria*, November 1994, 12 pp.
45. WORKING PAPER 45, F. Basso, L. Postiglione and C. Ruggiero, *Soil tillage methods for a sustainable agriculture. Experimental results in the hilly area of southern Italy: Agri basin*, February 1995, 37 pp.
46. WORKING PAPER 46, L. Mariani, V. Menichini and G. P. Minardi, *The climate of the Abruzzo area in the 1993–1994 period*, February 1995, 71 pp.
47. WORKING PAPER 47, S. Ferroni and G. P. Minardi, *Automated weather types classification by means of neural networks*, February 1995, 14 pp.
48. WORKING PAPER 48, R. Sorani, M. Conte and S. Velletri, *Blocking systems in the Euro-atlantic region and extreme events in the Mediterranean area*, February 1995, 55 pp.
49. WORKING PAPER 49, E. Koutsidou, Ph. Xatzltheodoridis and N. S. Margaris, *Shepherds and overgrazing by their livestock: fires and desertification on Chios island, Greece: a cost-effective economic analysis*, February 1995, 19 pp.
50. WORKING PAPER 50, Ce.S.I.A., Firenze, Italia. *The soil in the Piomba and Cigno river basin*, February 1995, 29 pp.
51. WORKING PAPER 51, M. Crespi, S. Lessi, C. Aliventi and M. A. Campanile, *A soil water balance simulation model*, February 1995, 37 pp.
52. WORKING PAPER 52, J. Martinez-Fernandez, J. Martinez-Fernandez, F. Lopez-Bermudez, M. Romero-Diaz and F. Belmonte-Serrato. *Evolution of vegetation and pedological characteristics in fields with different age of abandonment*, February 1995, 21 pp.
53. WORKING PAPER 53, X. Zhang, J. Corte-Real and X. Wang, *Low-frequency oscillations in the northern hemisphere*, February 1995, 16 pp.
54. WORKING PAPER 54, J. Corte-Real, X. Zhang and X. Wang, *Downscaling GCM information to regional scale: a non-parametric multivariate regression approach*, February 1995, 17 pp.
55. WORKING PAPER 55, R. M. Trigo and C. C. Dacamara, *Long-term variability of circulation weather types affecting Portugal*, February 1995, 16 pp.
56. WORKING PAPER 56, D. Lóczy, L. Szalai and J. Szalay, *A sand region in Hungary particularly sensitive to environmental change*, February 1995, 13 pp.
57. WORKING PAPER 57, J. Szalay and D. Lóczy, *Some trends in groundwater level changes on the Danube-Tisza interfluvial region, Hungary*, February 1995, 14 pp.
58. WORKING PAPER 58, I. Matyasovszky, *Temperature and precipitation trends in the Hungarian great plain during the present century*, February 1995, 19 pp.

59. WORKING PAPER 59, L. Chiarantini, P. Coppo and S. Gagliani, *Soil surface roughness measurements and analysis*, July 1995, 4 pp.
60. WORKING PAPER 60, L. Chiarantini, G. Luzi and S. Gagliani, *Measurement of the soil dielectric constant and moisture content by means of a portable microwave probe*, July 1995, 3 pp.
61. WORKING PAPER 61, L. Chiarantini, P. Coppo, S. Gagliani and G. Luzi, *Contributions of microwave remote sensing for soil changes monitoring,* July 1995, 4 pp.
62. WORKING PAPER 62, P. Mairota, *Legislation abolishing feudalism and its consequences on landownership structure in southern Italy*, July 1995, 12 pp.
63. WORKING PAPER 63, P. Mairota, *EC policy structures and socio-economic aspects of desertification. First contribution, the MEDALUS II regions and common policies*, July 1995.
64. WORKING PAPER 64, A. Santini, N. Romano, G. Ciollero and V. Comegna, *Effect of tillage practices on soil hydraulic properties as determined by different laboratory methods*, July 1995, 26 pp.
65. WORKING PAPER 65, A. Kertesz, *Aridification in a region adjacent to the Mediterranean: objectives and outline of a scientific programme*, July 1995, 12 pp.
66. WORKING PAPER 66, D. Loczy and L. Szalai, *Climatic change and land capability in a sand region of Hungary*, July 1995, 8 pp.
67. WORKING PAPER 67, K. Molnar, *Temperature and precipitation trends for Hungary*, July 1995, 42 pp.
68. WORKING PAPER 68, S. Papp, *Changes in the soil–vegetation relationship at Fulophaza, Kiskunsag National Park, Hungary*, July 1995, 10 pp.
69. WORKING PAPER 69, A. Kertesz, *Soil erosion in Lake Balaton catchment*, July 1995, 13 pp.

PLANT SPECIES GLOSSARY

Latin	English	French	Portuguese	Greek	Italian	Spanish
Acer opalus Mill. [also mentioned: *A. campestris*, L. and *A. monspessulanum* L.]	Italian maple	Érable à feuilles d'obier	Bardo	Σφένδαμνος	Acero opalo	Asar
Ampelodesmos mauritanica (Poiret) T. Durand *et* Schinz					Saracchio	
Anthyllis cytisoides L.	Anthyllis cytisoides	Anthyllide		Αλογοθύμαρο		Albaida
Arbutus andrachne L.	Arbutus	Arbousier	Ervodo	Αγριοκουμαριά, Ανδράκα	Albatro	
Arbutus unedo L.	Strawberry tree	Arbousier	Medronheiro	Κουμαριά	Corbezzolo	Madroño
Artemisia barrelieri Besser	Artemisia				Artemisia	Bojantina
Artemisia campestris L.	Artemisia			Αρτεμισία των αγρών	Artemisia	Boja
Brachypodium retusum (Pers.) Beauv.	Mediterranean false broom				Palèo delle garighe	Cervero
Brachypodium sylvaticum (Hudson) Beauv		Brachypode de bois			Palèo silvestre	
Capparis spinosa L.	Caper	Câprier	Alcaparra	Κάπαρη	Cappero	Alcaparra
Carpinus betulus L.	White hornbeam				Carpino bianco	Carpe
Castanea sativa Mill.	Chestnut	Châtaignier	Castanheiro	Καστανιά	Castagno	Castaño
Cistus ladanifer L.	Gum cistus	Ciste	Esteva	Λαδανιά	Cisto	Jara
Cistus salvifolius L.	Cistus		Sargaço		Cisto	Jara
Corylus avellana L.	Hazel	Noisetier, Coudrier	Aveleira	Φουντουκιά Λεπτοκαρυά	Nocciolo	Avellano
Crocus sativus L.	Saffron	Crocus, Safran	Açafrào	Κρόκος	Zafferano	Azafrán
Cyclamen neapolitanum Ten.	Cyclamen			Πατατιά, Σκυλάκι Καφεμπρικούδι, Ορεσκοι	Ciclamino	Ciclamen
Cyclamen repandum Sibth. *et* Sm.	Cyclamen			Τουρκάκι Κотταρίδα, Αξινίδα	Ciclamino	Ciclamen
Erica arborea L.	Tree heather	Bruyère arborescente	Urze branca, Torga	Ρείκι, Ρείγκλα, Τσάρο, Ντρίζα,	Erica da scope	Brezo, Urce
Eucalyptus L'Hér. ssp.	Eucalyptus	Eucalyptus	Eucalipto	Ευκάλυπτος	Eucalitto	Eucalipto
Euonymus europaeus L.	Spindle tree	Fusain d'Europe		Πασχαλιά	Berretta da prete	Bonetero
Euphorbia dendroides L.	Tree spurge	Euphorbe		Φλόμος	Euforbia	Euforbia
Fagus heterophylla L.	Beech	Hêtre			Faggio	Haya
Fagus sylvatica L.	Common beech	Hêtre		Οξυά	Faggio	Haya

Latin	English	French	Portuguese	Greek	Italian	Spanish
Festuca drymeja Mert. *et* Koch in Röhling	Fescue	Fétuqueo			Setaiola	
Fraxinus ornus L.	Manna Ash	Frêne-à-fleurs	Frexio		Orniello	Fresno
Genista scorpius (L.) DC in Lam. *et* DC	Greenweed, Needle furze, Petty whin	Genêt scorpion		Αιάνα	Ginestrella	Aulaga
Hyparrhenia hirta L. Stapf. in Oliver	Hyparrhenia hirta				Barboncino mediterraneo	Cerrillo
Ilex aquifolium L.	European holly	Houx	Azevinho, Pica-folha, Visqueiro	Λιόπρινο, αρκουδοπούρναρο, λαύρος	Agrifoglio	Acebo
Juniperus oxycedrus L.	Juniper	Genévrier oxycèdre	Zimbro, Sabina da-praia	Κέδρο, Αρκεύθο	Ginepro	Enebro, Sabina
Lathyrus cicera L.	Red vetchling	Gesse	Araca Chicharos-miúdos	Λαθούρι, αφάκα, ρεβύθι	Fiorgalletto	Cicércula Galbaua
Laurus nobilis L.	Laurel, Sweet bay	Laurier	Loureiro	Δάφνη	Alloro	Laurel
Lavandula dentata L.	Lavender	Lavande	Lavanda	Λεβάντα	Lavanda	Espliego
Medicago polymorpha L.	Medick	Luzerne	Luzerna	Μηδική	Medica	Mielga, Carreton
Myrtus communis L.	Myrtle	Myrte	Murta	Μυρτιά	Mirto	Mirto,
Nerium oleander L.	Oleander	Laurier rose	Loendro, Sevadilha	Πικροδάφνη, ροδοδάφνη	Oleandro	Adelfa, Baladre
Olea europaea L. (var. *sylvestris* Brot.)	Wild olive	Oliver	Oliveira Zambujeiro	Αγριελιά, Ελιά	Olivastro	Acebuche
Ostrya carpinifolia Scop.	Hop-hornbeam	Charmehoublon		Γαύρος μελιόγαυρος, οστρύα	Carpino orientale	Carpino nero vegaruiente
Phillyrea latifolia L.	Phillyrea latifolia		Aderno Cadorno	Φιλλύκι	Fillirea	Labiernago
Phlomis fructicosa L.	Jerusalem sage			Ασφάκα, αφάκα		Salvione giallo
Pinus brutia Ten.	Calabrian pine	Pin ?	Pinheiro	Πεύκο, πεύκη τραχεία	Pino bruzio	Pino
Pinus halepensis Mill.	Aleppo pine	Pin ?	Pinheiro alepo	Πεύκο, πεύκη, χαλέπιος	Pino d'Aleppo	Pino carrasco
Pinus nigra Arnold	Black pine	Pin noir	Pinheiro	Αγριόπευκο	Pino nero	Pino negral
Pinus pinaster Aiton	Maritime pine	Pin maritime	Pinheiro bravo		Pino marittimo	Pino rodeno
Pinus sylvestris L.	Scots pine	Pin sylvestre	Pinheiro silvestre	Πεύκο, πεύκη, δασική	Pino silvestre	Pino albar
Pistacia lentiscus L.	Mastic tree	Lentisque, Arbre-du-mastic	Aroeira, Lentisco	Σχίνος	Lentisco	Lentisco
Prunus avium L.	Wild cherry	Merisier	Cerdeira		Ciliegio selvatico	Cerezo de monte
Pyrus communis L.	Pear	Poirier commun	Pereira Maçã craveira	Αγριαπιδιά, απιδιά	Pero	Peral

Latin	English	French	Portuguese	Greek	Italian	Spanish
Quercus cerris L.	Turkey oak	Chêne Chevelu		`Αγρια βαλανιδιά, ρουπακι, τσέρο	Cerro	
Quercus coccifera L.	Kermes oak	Chêne kermès	Carrasqueiro, Carrasco	Πουρνάρι, πρίνος	Quercia spinosa	Coscoja
Quercus faginea Lam.	Lusitanian oak	Chêne zeen	Carvalho-portugues		Quercia faginea	Quejigo
Quercus frainetto Ten.					Farnetto	
Quercus ilex L. subsp. *rotundifolia* (Lam.) Schwartz *ex* T. Morais	Holm oak	Chênes vertes, Yeuse	Azinheira, Azinho	Αριά	Leccio	Encina Carrasca
Quercus pubescens Willd.		Chêne pubescent			Roverella	Roble
Quercus suber L.	Cork oak	Chênes liège, Arbousiers	Sobreiro, Sobro		Sughera	Alcornoque
Retama sphaerocarpa (L.) Boiss.						Retama común
Rhamnus lycioides L.		Neprun		Ράμνος		Espigno negro
Rosmarinus officinalis L.	Rosemary	Romarin	Alecrim	Δεντρολίβανο	Rosmarino	Romero
Salsola genistoides Juss. *ex* Poiret.	Saltwort	Soude	Barrilha espinhosa, Soda		Soda, Erba-kali	Boja negra
Stipa tenacissima L.	Esparto grass	Plumet	Esparto		Stipa	Esparto
Thymelaea hirsuta (L.) Endl.		Passerine	Trovisco	Θερόκαλο	Spazzaforno	Bojalaga
Thymus hyemalis Lange	Thyme	Thym	Tomilho	Θυμάρι	Timo	Tomillo rojo
Vicia sativa L.	Common vetch	Poisette, Vesce	Ervilhaca	Βίκος	Veccia buona	Alverja

INDEX OF CONTRIBUTORS

SUBJECT INDEX